HANDBOOK OF FINANCIAL DATA AND RISK INFORMATION

Volume II: Software and Data

Risk has always been central to finance, and managing risk depends critically on information. In recent decades, derivative markets for parceling and shifting risks have burgeoned, while computational advances have increased the pace of trading. As evidenced by recent events, the need has never been greater for skills, systems, and methodologies to manage risk information in financial markets. Authored by leading figures in risk management and analysis, this handbook serves as a unique and comprehensive reference for the technical, operational, regulatory, and political issues in collecting, measuring, and managing financial data. It will appeal to a wide range of audiences, from financial industry practitioners and regulators responsible for implementing risk management systems, to system integrators and software firms helping to improve such systems.

Volume I examines the business and regulatory context that makes risk information so important. A vast set of quantitative techniques, internal risk measurement and governance processes, and supervisory reporting rules have grown up over time, all with important implications for modeling and managing risk information. Without an understanding of the broader forces at work, it is all too easy to get lost in the details.

Volume II sets out a detailed structural and operational framework for managing a financial risk data repository. As experience accumulates on managing modern risk systems, the knowledge base of practical lessons grows. Understanding these issues and leading practices may mean the difference between failed and successful implementations of risk systems.

MARGARITA S. BROSE, a former Senior Counsel in the Division of Enforcement of the US Securities and Exchange Commission, has almost two decades of experience in the financial markets. After leaving the SEC and earning an MBA from The Wharton School, she spent ten years consulting to financial services companies, with a focus on risk and compliance, with IBM Business Consulting Services (and its predecessor, PricewaterhouseCoopers Consulting). In that role, her clients included leading global investment banks and money center banks, as well as GSEs. Brose has since served in Director positions at Fannie Mae and Barclays investment bank, in their Finance and Risk Management groups, respectively.

MARK D. FLOOD did his undergraduate work at Indiana University in Bloomington, where he majored in finance (B.S., 1982), and German and economics (B.A., 1983).

In 1990, he received his Ph.D. in finance from the Graduate School of Business at the University of North Carolina at Chapel Hill. He has worked as Visiting Scholar and Economist in the Research Department of the Federal Reserve Bank of St. Louis, an Assistant Professor of finance at Concordia University in Montreal, a Visiting Assistant Professor of Finance at the University of North Carolina at Charlotte, a Senior Financial Economist in the Division of Risk Management at the Office of Thrift Supervision, a Senior Financial Economist with the Federal Housing Finance Agency, and most recently as a Research Principal with the US Office of Financial Research in Washington, DC. His research interests include financial markets and institutions, systemic financial risk, financial data management, securities market microstructure, and bank market structure and regulatory policy. His research has appeared in a number of publications, including the *Review of Financial Studies*, the *Annual Review of Financial Economics*, the *Journal of International Money and Finance*, *Quantitative Finance*, and the St. Louis Fed's *Review*.

DILIP KRISHNA is a Director with the Governance, Risk and Regulatory Consulting practice with a focus on risk architecture and information. He is involved in several regulatory initiatives such as stress testing and capital management implementations. He has 17 years of experience across a range of wholesale and retail banking products in banks across North America. Previously, Krishna was a partner with Teradata Corporation, leading its North-East Banking and Capital Markets consulting team. He worked with many large banks in the North-East to implement some of their largest data warehouse and data management programs. Krishna was also Director of Teradata's Risk Management Center of Expertise in the Americas, responsible for working with many large financial institutions across the USA and Canada implementing Basel II and risk management solutions. In this role he was also active with Teradata's Government Services team, helping to articulate the role of analytics in regulation and transparency, including being called upon to offer expert testimony to the US Congress. He has served as chief architect for Basel II at a major Canadian bank, and has implemented front-offices trading solutions in Canadian capital markets operations. Krishna is widely published in the area of risk information and risk architecture.

BILL NICHOLS has been at the intersection of technology and finance for 25 years. He co-founded and served as CEO of a research firm acquired by Thomson Financial in 1995, and spent the following seven years with Thomson. Subsequently, he spent three years in venture capital-related organizations, followed by consulting stints for major banks, asset management firms, and securities exchanges. He has participated as an expert in multiple ISO standards groups and is the former Convenor of the ISIN Standard for financial instruments. With Mark Flood, he co-chaired the Data Committee of the Committee to Establish the National Institute of Finance. Currently a Senior Advisor for Information Architecture and Innovation at the Office of Financial Research, he was previously Data Architect at Bridgewater Associates in their Data Research group.

HANDBOOK OF FINANCIAL DATA AND RISK INFORMATION

VOLUME II

Software and Data

Edited by

MARGARITA S. BROSE

MARK D. FLOOD

DILIP KRISHNA

and

BILL NICHOLS

CAMBRIDGE
UNIVERSITY PRESS

CAMBRIDGE
UNIVERSITY PRESS

University Printing House, Cambridge CB2 8BS, United Kingdom

Cambridge University Press is part of the University of Cambridge.

It furthers the University's mission by disseminating knowledge in the pursuit of education, learning, and research at the highest international levels of excellence.

www.cambridge.org
Information on this title: www.cambridge.org/9781107012028

First published 2014

Printed in Spain by Grafos SA, Arte sobre papel

A catalog record for this publication is available from the British Library

ISBN 978-1-107-01202-8 Hardback

Contents

Contributors

Marc Alvarez is a Senior Director with Interactive Data in New York responsible for Reference Data service infrastructure and enterprise data architecture. He is the author of *Market Data Explained: A Practical Guide to Global Capital Markets Information* (Elsevier, 2007) and an active industry commentator focusing on capital markets data content and management.

Nicola Antoniou is a Bank Expert at the European Banking Authority, where she specializes in policy and regulation for European cross-border banking supervision, in particular on the supervisory review and evaluation of bank risks. Nicola has previously worked at the UK Financial Services Authority focusing on international strategy and policy coordination.

Jim Blair is a Senior Consultant with Teradata, Inc. He is recognized globally for his experience in data warehouse development, implementation, and performance. He is dedicated to driving business value from analytical processing. Jim is an international speaker on data warehousing and a standing member of TDWI and the Society of Information Management.

Jefferson Braswell is the Founding Partner of Tahoe Blue Ltd. and has been active in the development of financial technology for over 30 years. He has delivered comprehensive risk management solutions to many of the world's largest financial institutions. Recently, he has served as a lead coordinator of the G20 LEI Private Sector Preparatory Group.

Margarita S. Brose has 20 years of experience in financial services. She began her career as an attorney at the US Securities and Exchange Commission. After earning an MBA, she specialized in risk and compliance consulting with PwC and IBM. She has degrees from Barnard College, GW Law School and The Wharton School.

Dianne Buonincontri is a data manager with the Federal Reserve Bank of New York. She is responsible for a team of business and data analysts performing functional and data analysis and creating data quality policies, standards, and processes. Dianne holds an executive MBA from Baruch College and has held previous roles in data management.

John Carroll is a well-known industry expert and experienced financial services operations and data manager currently working as a Managing Consultant at element22, a boutique advisory firm that specializes in data management within the financial services industry. John spent 22 years at Merrill Lynch, where he was responsible for Data Solutions and Integrations.

Thomas Dang is a software/database developer working for the University of British Columbia. He holds a Masters in computer science, with a subspecialization in human computer interaction and has been actively involved with research on visual analytics. Thomas's other active interests include data integration, high-density information displays, and interactive video.

Brian Fisher is Associate Professor of Interactive Arts and Technology and Cognitive Science at Simon Fraser University and Associate Director of the Media and Graphics Interdisciplinary Centre at the University of British Columbia. His research explores the cognitive science of interaction with visual information systems (visual analytics).

Mark D. Flood is a graduate of Indiana University in Bloomington (B.S. in finance, and B.A. in German and economics). He earned his Ph.D. in finance from the University of North Carolina at Chapel Hill. He has taught finance and business at universities in the USA and Canada, and has been active in policy development and research publication as an economist in the risk management and research departments for a number of regulatory bodies.

Janine Forsythe is a Vice President at Goldman Sachs leading the technology teams that support the firm's global insurance businesses. Janine has over 25 years in the financial services industry in line management roles and consulting services focused on a broad spectrum of clients and business functions including insurance, capital markets, and commercial banking.

Martijn Groot is an expert in financial data management and has worked in this space in product management roles from the investment banking, financial technology, and services perspectives. His specialties include instrument, entity and analytics data services. Martijn is the author of *Managing Financial Information in the Trade Lifecycle* (Elsevier, 2008).

John Patrick Hunt is Acting Professor of Law at UC Davis School of Law (King Hall). His scholarly interests include credit rating agencies and securitization.

Alan King is a Research Staff Member at the IBM Thomas J. Watson Research Center in Yorktown Heights, New York, in the Industry Solutions Research Department with a focus on the Financial Services Sector. Alan's research focus is on analytics middleware technology in risk management processes and applications of stochastic programming.

Laura Kodres, an Assistant Director in the IMF's Monetary and Capital Markets Department, oversees the analytical chapters of its Global Financial Stability Report, which frequently focuses on systemic risk monitoring and mitigation techniques. Her research centers on financial contagion and interactions of macroeconomic and financial policy issues.

Alejandro Komai is a Ph.D. candidate at the University of California, Irvine. His research interests include economic history and monetary theory.

Dilip Krishna is a Director with Deloitte & Touche LLP, focusing on risk and regulatory analytics, data and architecture in financial services firms. He has previously held roles implementing large data warehouse and data management programs for initiatives such as Basel II.

Jarl Kure has worked as a supervisor for over 20 years; he is currently with the European insurance authority, EIOPA, where his responsibilities include project lead on the Solvency II Preparatory Guidelines. Previously, Jarl coordinated the EIOPA Financial Stability area, including EIOPA Risk Dashboard, stress testing and relations with the European Systemic Risk Board.

Simon Kwan is Vice President and Head of Financial Research at the Federal Reserve Bank of San Francisco. He conducts research and policy analysis in the areas of financial institutions and financial markets. Simon has published research articles in leading finance journals and book chapters.

Victoria L. Lemieux is an Assistant Professor at the University of British Columbia's iSchool with a research focus on financial informatics. She founded and directs the Centre for the Investigation of Financial Electronic Records and is also Acting Director of the University of British Columbia's Media and Graphics Interdisciplinary Centre.

Irina S. Leonova is a member of the Financial Stability Board Secretariat, where she currently specializes in market structure questions related to financial data infrastructures, LEI, and OTC and exchange traded derivatives among others. Previously she held positions at the US Department of the Treasury, CFTC, and Bureau of Economic Analysis.

John C. Liechty is a Professor at the Smeal College of Business at Penn State University. He is an expert in derivative pricing and asset allocation, computational statistics and high performance computing, and marketing research. He has a Ph.D. from the Statistical Laboratory at Cambridge University.

Robert Mark is the Founding Partner of Black Diamond Risk, serves on several boards, has led Treasury/Trading activities, and was a CRO at Tier 1 banks. He was the Founding Executive Director of the MFE Program at UCLA, has co-authored two books on risk management, and holds a Ph.D. in applied mathematics.

Michael J. McCabe is a Director with Deloitte Financial Advisory Services LLP, specializing in the application of analytic and forensic technology in the financial services. He has held previous roles consulting on regulatory and compliance issues, business processes and controls, data collection and retention, as well as transaction monitoring in financial firms.

Karla McKenna is Chair of the International Organization for Standardization Technical Committee 68 (ISO TC68), Financial Services, a position she has held since 2006. When not handling ISO assignments, Karla manages Market Practice and Standards for Citibank's Global Transaction Services, Securities, and Fund Services businesses. She sits on numerous other standards committees worldwide.

Bill Nichols is Senior Advisor at the US Office of Financial Research. Prior to joining the OFR, he was at Bridgewater Associates. He has worked at the intersection of finance and technology for the past 25 years and been active in the global financial standards community for ten years.

Jim Northey (B.S. University of Indianapolis, M.S. Illinois Institute of Technology) is a co-founder and senior partner in The LaSalle Technology Group, LLC. He is co-chair of the FIX Americas Regional Committee, representing FIX within ASC X.9. Jim is a member of the US Treasury OFR Advisory Committee.

Per Nymand-Andersen is an adviser to Senior Management at the European Central Bank. He specializes in European banking, financial markets, statistics, securities settlement systems, management and communication. Per is a key editor of the Statistics Paper Series and lectures at the Goethe University. He is a former international consultant and has an MBA in Economics and Management Science from Copenhagen Business School, Denmark.

John C. Pattison is a financial consultant specializing in regulatory and risk governance matters. He is a former banker and has taught at the Ivey School of Business, University of Western Ontario. His publications are mainly in the field of international finance and economics including the regulation of financial institutions.

Gary Richardson is a professor at the University of California, Irvine. He has recently been named the Federal Reserve System Historian. He is also a research associate at the National Bureau of Economic Research. His research interests include economic history, macroeconomics, the Great Depression, the Industrial Revolution, property rights, and economic development.

Nicholas Robson is an IT consultant with 18 years of experience across Canada, the USA, UK and Italy. He has a depth of experience in Enterprise Data Management, including database design, administration, architecture, and project management of EDM implementations. Nick holds an M.S., Computer Science, from the University of Hertfordshire, UK.

Clifford V. Rossi is an Executive-in-Residence and Tyser Teaching Fellow at the Robert H. Smith School of Business, University of Maryland. Prior to entering academia, he had nearly 25 years of experience in banking and government. He has also held senior risk management positions at Freddie Mac and Fannie Mae. He received his Ph.D. from Cornell University.

David M. Rowe is founder and president of David M. Rowe Risk Advisory, a risk management consulting firm. David has spent almost 40 years in the risk management technology, banking, and economic forecasting industries. He has written the monthly Risk Analysis column in Risk magazine since late 1999.

Jonathan Sparks is an attorney at King & Spalding in Atlanta, Georgia. He recently graduated from the George Washington University Law School in Washington, DC, where he specialized in corporate, regulatory, and securities law and policy.

Richard Stanton is Professor of Finance and Real Estate and holds the Kingsford Capital Management Chair in Business at the Haas School of Business, University of California at Berkeley. He is best known for his research on mortgage markets, term-structure modeling, mutual funds, and employee stock options.

Deborah Stockdale is an Engagement Partner at Teradata focusing on Enterprise Data Management, Data Governance, and Data Architecture. She has held previous roles as Chief Data Architect designing information architectures for satellite constellations, designing and implementing large scale data warehouses for Fortune 100 firms, and teaching data strategy and data management courses.

Charles Taylor is Deputy Comptroller of the Currency for Capital and Regulatory Policy. Previously, he was Director of the Financial Reform Project at Pew Charitable Trusts and Executive Director of the Group of Thirty.

Nancy Wallace is a Real Estate and Finance Professor, Haas School of Business, University of California Berkeley and the Lisle and Roslyn Payne Chair in Real Estate and Capital Markets. She co-chairs the Fisher Center for Real Estate and Urban Economics and co-faculty directs the Berkeley Center for Law, Business and the Economy.

Jesse T. Weintraub is a broker-dealer and regulatory enforcement attorney at Bracewell & Guiliani in New York City, specializing in the representation of various financial institutions in matters arising under federal securities laws and self-regulatory organization rules. Jesse is a graduate of the George Washington University Law School.

Preface

On August 9, 2007, a large European bank announced unexpectedly that it could not obtain valuations on subprime securities underlying three of its investment funds. Accumulated anxiety about the quality of the collateral in repurchase agreements (repos) turned into a full-blown panic attack. Lenders withdrew from this market in large numbers, sending dealers on an urgent hunt for new funding. The world was witnessing a full-scale run on the shadow banking system. At that time, the shadow banking system was not well understood by practitioners outside its arcane confines. Yet this episode demonstrated its ability to quickly spread contagion not only across the financial system, but also to the real economy, culminating in the events of the fall of 2008, where extreme measures had to be taken by the US and European governments to stabilize the economy. All told, the failures represented trillions of dollars in assets and hundreds of billions of dollars in losses, with central banks, finance ministries, and the US Treasury piling resources into the breach.

This was all a prelude to the full catastrophe.

In September 2008, a second intense wave of failures and near-failures claimed a number of victims, mostly in the USA, and many supposedly "too big to fail." The episode sent financial markets into a free fall that lasted for six months. The ensuing failures, interventions, conversions, and mergers have permanently and fundamentally altered the financial industry landscape. Much time, effort, and discussion has been devoted to analyzing these events; yet even today, years later, there are several conflicting theories about what exactly happened and who was responsible. It is facile to lay the blame wholly on first-time homebuyers in subprime and Alt-A mortgage markets, as these borrowers represented only one half of a transaction at the front end of a much longer securitization chain. How did loanable funds make their way into these deals? Observers of various persuasions have laid responsibility at the feet of such varied parties as the US and Chinese governments, Government Sponsored Enterprises, investment banks, and credit ratings agencies. In hindsight, one of the surprising things was how unprepared the world economic community was for the scale and speed of these events.

Anyone who has sat through a college investments course knows that information plays a central role in finance. Markets can be seen as an information-processing system. The Efficient Markets Hypothesis asserts that markets should reliably react and adjust to available information. They may make mistakes for a time, but they should not be wildly or consistently biased. Yet what the events leading up to 2007 showed us was that the markets did not, in fact, properly react to information that was available, and continued to build up stresses. How could sophisticated institutional investors have been so wrong for so long about the values of mortgage-backed securities?

Starting in February 2009 – as the economic situation continued to deteriorate despite extraordinary measures – a small group of academics, supervisors, and financial industry participants began a series of conference calls to discuss the "what happened" question. The initial core group of Professors John Liechty and Arthur Small of Penn State University and Mark Reesor of the University of Western Ontario, Joe Langsam at Morgan Stanley, and Allan Mendelowitz and Mark Flood at the Federal Housing Finance Agency, began soliciting further expertise among their professional networks. Many of the authors and editors of the Handbook you are reading participated in those discussions. Like a garage band, we settled on a name before we learned how to play our instruments: *The Committee to Establish a National Institute of Finance* (CE-NIF, `www.ce-nif.org`). We agreed from the start that a basic problem – although surely not the only challenge – was the supervisors' lack of wherewithal: they simply did not have data of sufficient quantity and quality, nor the analytic capacity required for the scale, pace, and sophistication of modern financial markets. We had the hubris to propose a new federal agency to rectify this shortcoming. The group grew quickly through professional connections. There was never a formal charter, outside funding, or well-defined authority within the CE-NIF. Instead, the incentive structure ensured that the most dedicated participants would take the lead: significant effort was required, with zero remuneration guaranteed and little hope of success or glory.

The idea for a Handbook emerged from the conversations of the Data Subcommittee of the CE-NIF, which assembled a wealth of technical and professional expertise and wisdom. This "data committee" included chief data and technology officers, senior risk managers, and economists, researchers, and engineers from government, industry, and academia. Although the original impetus for the CE-NIF and its data committee was a systemic financial crisis, the discussion quickly moved beyond systemic risk to the broader risk and data management issues in finance. Where economic theory speaks in abstractions about public information sets and conditional distributions, the data committee discussions focused on more mundane topics and concrete realities such as end-user computing, metadata management, messaging formats, data validation, ISO standards, and semantic modeling.

In sharing ideas and swapping war stories, two themes emerged. The first was the extent of disarray in financial data, not just for supervisors but at financial firms themselves. While it is common in policy and management circles to proclaim that it is the "What" that matters, not the "How," it became abundantly clear that the converse was often true: that the How had in fact influenced the What in a profound way.

The second theme was that we all had a lot to learn from one another. Too many facts and lessons were passed along as oral traditions within subspecialties, or re-learned the hard way by individuals confronting a problem that others had seen before. It was observed repeatedly during these sessions that, "this stuff just isn't written down anywhere."

This Handbook is our attempt to write some of those facts and lessons down. We have in mind the financial economist who needs to know about the main industry messaging standards, the database administrator who wants to learn what "volatility clustering" is and why it is important, the risk manager working with the mortgage supply chain for the first time, and the systems engineer trying to understand bank regulatory reporting. In short, we envisioned this Handbook as a practical companion and reference sitting on desks of the wide range of people who need to come together to solve the problems of information infrastructures in large financial institutions and regulatory organizations

The ambition to collect this much information in such a wide-ranging and fast-moving area in one place creates an inevitable tension between comprehensiveness and timeliness. We hope that the resulting compromises and omissions are at least tolerable, and that the Handbook will achieve its desired status as a dog-eared companion for the reader. We encourage the reader to make liberal use of references and sources provided to extend the lessons provided herein.

The Handbook is presented in two volumes. This satisfies the needs of practicality; two 600-page books are undoubtedly less daunting and more tractable than one 1,200-page book, but there is logic to this as well. Any attempt to address information infrastructure in a large organization (be it commercial or regulatory) must stand on a foundation of requirements. Volume I provides this basis in three parts. Part I provides the context and rationale for the need for financial data, describing the various data needs of risk and regulatory processes. Part II describes the supply-chain of data in the financial markets and the real economy, as well as the intersection and transmission of data across their common boundaries. Part III covers a broad range of topics relating to the use of data by regulators.

Volume II addresses topics that a financial institution or regulator would find necessary to implement change in their infrastructure. Part IV addresses a critical component of implementation by describing how data emanates and are modified within an organization. Part V addresses various topics relating to the rapid changes in technology for data processing that makes these problems much more tractable. Part VI covers topics relating to the management of data remediation programs and environments.

The Handbook brings together an eclectic range of experts from a wide range of fields, as befits this expansive topic. Early on in its development we felt it would better serve our readers to let the voices of these experts shine through unadulterated by attempts to standardize the presentation across the Handbook. Accordingly, each chapter employs its own presentation style, with the attempts at normalization restricted to the superficial "look-and-feel" of the chapter.

Bringing this Handbook together has been at once a challenging and rewarding task. We hope that you find as much value in reading and referencing the book as we did in putting it together. Nothing would please us more than if this Handbook were to render itself obsolete over time by playing a small part in fixing the present unwieldy information infrastructures in the financial industry.

Margarita Brose
Mark Flood
Dilip Krishna
Bill Nichols

VOLUME II

SOFTWARE AND DATA

PART IV

DATA OPERATIONS IN FINANCIAL INSTITUTIONS

EDITORS' INTRODUCTION TO PART IV

Dilip Krishna

The first volume of the Handbook described the need and context for data infrastructure changes in financial institutions (FIs) arising from regulatory pressures and increasing sophistication in risk management. The volume also described the major types of data found in FIs. This volume turns to the issue of how to implement an infrastructure to satisfy the risk management needs raised previously. One obvious aspect of implementation – technology – is addressed in Part V. Part VI follows by describing processes to assist in effectively implementing and running an environment to manage an analytics on financial data.

Most (but not all) data in an FI are created at the point of transaction. These data are then used in different parts of the organization. In addition, there are processes to derive data from underlying transactional data – this is common in risk measurement activities for example. Much of the challenge of implementation arises with how data are created or modified in an FI. If data are created with a high degree of fidelity at the point of origination, then the challenge of data integration and usage is much simplified. Unfortunately, however, many major FIs generate transactional data of sometimes questionable quality. In addition, operations in these FIs involve creating aggregate reports by drawing from source systems in ways that are often not consistent, resulting in concerns about the fidelity of the resulting reports and analytics generated from this information. For these reasons, an important ingredient in addressing the data quality problem lies in a good understanding of the business processes responsible for generating and modifying data.

Part IV aims to give the reader an understanding of the business processes in the financial industry. Business processes within the financial industry can be complex given the many different subindustries involved. In addition, the plethora of financial products available results in rather different considerations for processing. The section accordingly begins with Chapter 15, by John Carroll and Jonathan Sparks, describing the various players in the financial industry such as firms active in the capital markets and retail banking. Additionally, ancillary institutions are described, including service providers like exchanges and custodians, as well as the gamut of regulatory organizations that oversee the financial industry in the USA. Chapter 16, by Janine Forsythe, describes generic processes in the financial industry and how these generate or use data. This is done by creating a process model of a generic FI and mapping the components of this model to specific types of businesses within the financial industry. Chapter 17, by Dianne Buonincontri and Nicholas Robson, focuses on the data requirements of these business processes and presents a framework for understanding how this data is managed within a financial organization. Detailed treatment of how five important kinds of financial data are produced is accompanied by a discussion of common problems and solutions seen in FIs.

Since capital markets products are a distinct and complex topic, Chapters 16 and 17 focus on these products alone. To bring a retail financial products perspective, Chapter 18, by John Hunt, Richard Stanton and Nancy Wallace, rounds out the section with a detailed discussion on data in the mortgage industry. The chapter describes whole-loan mortgage data, and shows how mortgages transfer through the supply-chain to help yield new products that generate their own data.

In summary, this section provides the foundation for improvement in generating risk information from financial data by describing the processes that are used to generate and modify that data. In combination with the following sections on technology and implementation procedures, this volume should give the interested reader a solid foundation of the disciplines required to address many important challenges in managing financial data.

15

Financial market participants

John Carroll and Jonathan Sparks

15.1 Interactions of institutions in the financial industry

There are many unique financial institutions in the USA, with each fulfilling a specific role in financial transactions. Information providers, financial intermediaries, buy-side and sell-side firms, commercial banks, Government Sponsored Entities (GSEs), and regulators, all interact with one another to form the complex web that is the financial industry today (see Figure 15.1). Broker-dealers, Investment Banks, Commercial Banks and Asset Managers transact their own and their clients' business in the financial services markets.[1] Custodians, Agents, Exchanges and Depositories[2] perform diverse functions throughout the financial industry.

Maintaining a fluid transactional system requires institutions to communicate with one another using the same or similar data types. Both investors and regulators need to understand the data provided to them in transactions. The best way to accomplish this is through the use of industry-wide standards for data capture and reporting. Unfortunately, this does not exist in most cases; data collection methods are diverse, and processing the sheer amount of data available is difficult – if not impossible – for most individual investors. For example, classes of derivatives are often mislabeled as other instrument types. This throws off the data stream, and makes for bad data collection.

Data need to be collected and processed in a timely manner so as to be available and "fit for purpose" for participants in the financial marketplace. To the extent that

[1] Agency trades occur when a firm buys or sells a security on behalf of one of its clients. In this situation, the firm does not own the actual security itself, but rather, is working as an agent for its client. Normally, the firm charges its client a fee for this service. Conversely, a principal trade occurs when the firm buys or sells its own securities for its own profit. For more information, see Johnsen (2011).

[2] Each of these functions is further defined in Chapter 16.

Handbook of Financial Data and Risk Information, ed. Margarita S. Brose, Mark D. Flood, Dilip Krishna and Bill Nichols. Published by Cambridge University Press.

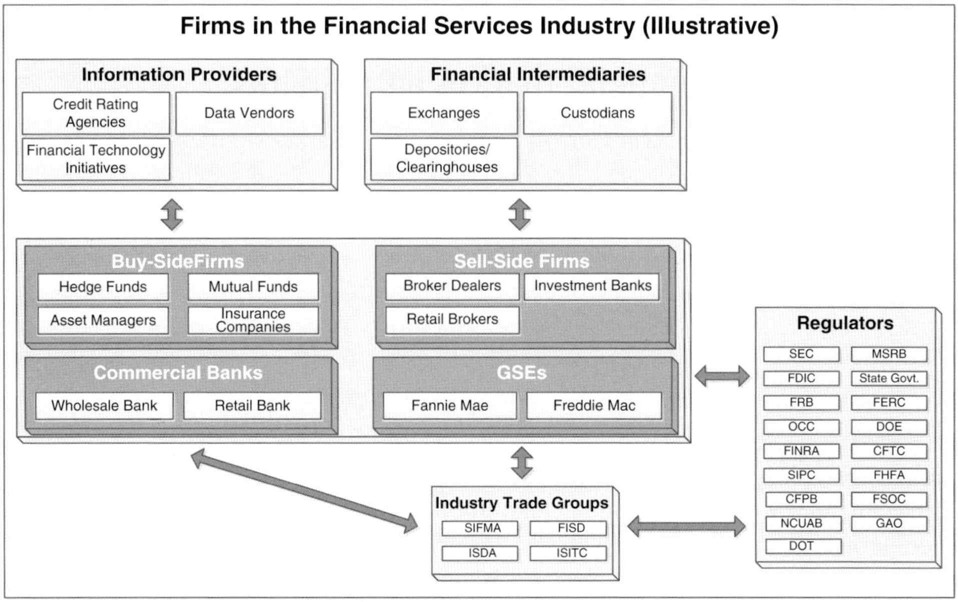

Figure 15.1 Interactions between firms in the financial industry.

participants do not or cannot create or update the data themselves,[3] the data must be acquired from Data Vendors who all have their own unique system of data collection and reporting. While several industry participants have attempted to create comprehensive data stores over the years, a lot of work remains to achieve the goal of establishing an enterprise-wide standardized data repository for all data domains that is fully integrated in the financial industry's data process systems. The work required for an institution to use this data within its organization across businesses, product lines, and global regions is extensive, complex and expensive. Those that have had a degree of success in establishing the data stores have found integrating the data, if only internally, to be a nearly impossible task. Further, once success is achieved internally it must be continually maintained and enhanced as businesses and regulatory laws change. Even then, although a single organization would benefit from such capability, it would not solve all the financial industry's data problems. Using one single standardized data collection method and collection entity can better inform the investing public and reduce redundant information costs associated with data research.

This chapter will describe the various financial market participants and their unique roles and relationships to one another. Additionally, the logistics and data needed to support financial transactions will be discussed.

[3] Updating and processing data includes evaluating instrument values, entities and their hierarchies, applying prices, defining corporate actions terms and conditions, providing positions and transactional information, and determining finance and risk measurements.

15.2 Sell-side financial institutions

There is an important distinction between "sell-side" and "buy-side" institutions. Sell-side refers to all financial institutions or individuals that make their money from transaction fees and commissions they charge their clients for trading securities. For example, if a portfolio manager for a mutual fund purchases 1,000 shares through a broker-dealer, the broker-dealer makes a commission on the purchase. In this example, the broker-dealer is considered "sell-side" because he makes his money from fees on the transactions themselves. Conversely, the portfolio manager is considered "buy-side" because his money comes from buying stocks.

15.2.1 Broker-dealers

Broker-dealer is the term for any individual or firm that buys and sells securities for the firm or others. Broker-dealers are on the "sell-side" of the industry. While working as a broker, broker-dealers execute orders on their client's behalf; clients can be individuals or entities. When acting as dealers, broker-dealers engage in trades for clients or their firm's own account. Securities purchased for the firm's own account may be sold to clients or other firms or kept as property for the firm itself. Although many broker-dealers are independent firms offering only broker-dealer services (some well-known examples include Raymond James and Ameriprise), others are business units or subsidiaries of investment banks, commercial banks, or investment companies. These firms generate their revenue from (1) fees and commissions that clients pay for security and commodity transactions undertaken on their behalf, and (2) interest charged on debit balances in margin accounts.[4]

During the recent financial crisis, a few larger broker-dealers applied for and received a bank charter which allowed them to borrow at the Federal Reserve Discount Window[5] when funds were not readily available to them elsewhere.

15.2.2 Retail brokers

Retail brokers primarily serve smaller "retail" investors rather than large companies or "institutional" investors. A good example of a retail broker is ETrade,

[4] Margin accounts are accounts where a broker-dealer temporarily lends to a client so that the client can purchase securities, usually on a securities exchange.

[5] The Federal Reserve Discount Window refers to the ability of certain banking institutions to borrow, usually short term, loans from the Federal Reserve at discounted rates in response to liquidity problems. The discount rate offered is controlled by the Federal Reserve. See Federal Reserve Discount Window website, available at `www.frbdiscountwindow.org/discountwindowbook.cfm?hdrID=14&dtlID=43` (last visited March 23, 2012); see also Tanselle (2011).

which allows customers to trade and sell their own stocks on major stock exchanges, without necessarily having large amounts of capital to invest with. Other popular retail brokerage companies include Charles Schwab and Merill Lynch.

15.3 Buy-side financial institutions

15.3.1 Asset and investment management

Asset and investment management refers to the professional management of various assets[6] and securities, such as shares and bonds. The goal of asset and investment management is to help investors meet their specific investment goals. These investors may be institutions (pension funds, insurance companies, corporations, educational establishments, charities, etc.), private investors who work directly through investment contracts, or more commonly, investors that use collective investment schemes like mutual funds.[7]

Defining some terms associated with asset and investment management may be helpful. "Asset management" refers to the management of collective investments for private investors – this is considered "buy-side," referred to in Section 15.2. More generically, the term "fund management" refers to any form of institutional investments, including asset management. Investment managers that specialize in either discretionary or advisory management for private investors usually call their services "wealth management," or "portfolio management."[8] All of these terms are included under the umbrella term "private banking."

"Investment management services" specifically refers to financial statement analysis, the selection of assets and stocks to purchase, and constantly monitoring investments. The large global investment management industry supervises trillions of dollars, yuan, euro, pounds, and yen. Many of the world's biggest companies are, at least partly, investment managers.[9] Aggregated, these companies employ millions of employees that create billions of dollars in revenue. "Fund managers," also

[6] The term "asset" can refer to many different things, and is often incorrectly used in reference to liabilities. The basic definition of an asset is property or an interest in something that produces money at a profit, rather than a loss. The term generally refers to real estate, but can also refer to an owned business or collector's item that is gaining value. While the technical definition of an asset requires that the property appreciate in value, at a profit, the term is also used to refer to property that is a liability. For example, in 2007, many people purchased their own homes, which was at that time an "asset" because it appreciated in value. The real estate then depreciated in value, costing the purchaser money, and therefore becoming liability, but is still referred to as an "asset." `www.investopedia.com/terms/a/asset.asp#axzz1qvmqoZUG`; see also *Black's Law Dictionary*, "asset" ("An item that is owned and has value").

[7] `http://dictionary.sensagent.com/asset+management/en-en/` (last visited July 5, 2012).

[8] `http://english.turkcebilgi.com/Asset+Management` (last visited July 5, 2012).

[9] Examples include Blackrock Inc., and State Street Global Advisors, with \$3.35 trillion and \$1.9 trillion respectively.

referred to as "investment advisers," indicate both firms that provide investment management services and individuals that control fund management decisions such as portfolio managers for mutual funds or trustees for pension funds.[10]

15.3.2 Agents

An agent is commissioned by a corporation or mutual fund to maintain shareholder records. This includes maintaining the client's purchases, sales, and account balances, and managing changes resulting from corporate action events – dividend payments to stockholders, or principal and interest payments to bondholders – on behalf of the stock or bond issuer. Often, the agent is the bank itself, although separate institutions are sometimes created for this specific purpose.

15.4 Commercial banks

Commercial banks are large entities that include smaller subcompanies or banks. Large banks like Wells Fargo or JP Morgan Chase have wholesale bank, investment bank, and retail bank divisions, all of which are included under the commercial bank's company name. This section will speak to the three different types of banks that are normally included under the umbrella term "commercial bank."

Previously, commercial banks were not allowed to do both retail and investment banking at the same time. From the 1933 Glass–Steagall Act until the passage of the Gramm–Leach–Bliley Act in 1999, the USA maintained a distinct separation between investment and retail banking, such that each type of bank had a unique function; overlap was not allowed. This created a unique system in the USA; other industrialized countries, including the remaining G8 countries, did not have this separation.

The theory behind the separation was to avoid public disclosure of sensitive information. Larger banks were usually split into public and private functions via a "Chinese wall." This "wall" separated the private from the public functions, preventing information from crossing over or being shared. Investment banks utilized private inside information that could not be publicly disclosed to retail banking customers. Conversely, the retail banks, including the offices of the stock analysts, only dealt with public information. The Glass–Steagall Act mandated such a separation, but the mandate was overturned in 1999 with the Gramm–Leach–Bliley Act.[11]

[10] `http://en.wikipedia.org/wiki/Investment_management` (last visited July 5, 2012).
[11] For more information, see Congressional Research Service (2010).

15.4.1 Retail banks

Retail banks accept deposits, make business loans, and offer related services to the general public. They provide various types of accounts, including checking, saving, and term deposits.[12] While retail banks do offer some investment services to individuals, they are primarily concerned with receiving deposits and lending to small businesses rather than making large investments in the securities markets. Retail banks are often engaged in the following activities.

1. Processing payments, transfers, and deposits through telegraphic transfers,[13] EFTPOS,[14] internet banking,[15] or other systems.
2. Issuing regular bank checks and bank drafts, such as cashier's checks, that are guaranteed by the bank.
3. Accepting money deposits or term deposits.
4. Lending money by overdraft, instalment loans, or other means.
5. Keeping documents and other items secure in safe deposit boxes.
6. Offering a variety of financial services – operating as "financial supermarkets."[16]

15.4.2 Wholesale banks

Wholesale banks work almost exclusively with large financial institutions and companies, as well as other banks, investment firms, securities dealers, pension funds,[17] and lenders. These banks help corporations, individuals, and governments to raise capital by underwriting[18] and acting as their client's agents when issuing

[12] Term deposits are, essentially, very short-term loans, made by the consumer to the bank itself. These usually last 1–12 months, at which time the lender agrees not to withdraw the money until the specified term matures, www.investopedia.com/terms/t/termdeposit.asp#axzz1pwceGy4w (last visited July 5, 2012).

[13] Telegraphic transfer or "telex transfer," sometimes abbreviated as TT or T/T, transfers money oversees by sending an electronic cable message from one bank to another, http://en.wikipedia.org/wiki/Telegraphic_transfer (last visited July 5, 2012).

[14] EFTPOS stands for Electronic Funds Transfer at Point Of Sale. It specifically refers to debit card transactions made during the sale itself. For example, paying for a coffee with a debit card from your local commercial bank is considered an EFTPOS, http://en.wikipedia.org/wiki/EFTPOS, (last visited July 5, 2012).

[15] Internet banking refers to making transfers or engaging in financial transactions online, through the use of secure websites that are run and monitored by commercial banks. If, for example, you have a checking and credit card account at a commercial bank, and want to make a payment to the credit card online, this transfer can be done remotely via online banking.

[16] Financial supermarkets are institutions or companies that offer a wider range of financial services than the normal commercial bank. Typically, financial supermarkets provide banking, stock brokerage, insurance, and even real estate brokerage. The basic rationale behind the financial supermarket concept is to generate more fees per client and create more business by fostering customer loyalty, www.investopedia.com/terms/f/financialsupermarket.asp#axzz1pwceGy4w (last visited July 5, 2012).

[17] http://en.wikipedia.org/wiki/Wholesale_banking (last visited July 5, 2012).

[18] Most underwriting deals use a "firm offer." In a firm offer, the underwriting investment bank will buy a certain amount of shares from a company making a public offering – selling its stocks in public stock exchanges like the New York Stock Exchange. With firm offers, the investment bank doing the underwriting will purchase shares from the company at a discount, usually 8% below the expected market price. In this way, when the public offering starts, the offering company has already sold most or all of its shares to the underwriter, leaving the

securities.[19] Additionally, wholesale banks may assist companies with mergers, and provide auxiliary services like trading derivatives, market making,[20] foreign exchange, fixed income instruments, commodities, and equity securities,[21] They also distribute or broker trades in securities or unit trusts,[22] and make markets in merchant banking,[23] private equity financing,[24] interest rates, and credit-related securities.

Currently, wholesale bank advisors that provide investment banking services in the USA have to be licensed broker-dealers; they are subject to regulation by the Securities and Exchange Commission (SEC) and the Financial Industry Regulatory Authority (FINRA), described in Section 15.8.2.

15.5 Government sponsored enterprises[25]

Government sponsored enterprises (GSE) are financial services corporations created by the US Congress aimed at enhancing the flow of credit to targeted sectors of the economy. Congress created the first GSE in 1916 with the creation of the Farm Credit System. Sallie Mae was chartered in 1972 to provide educational credit,

underwriter with the task of selling the shares to the investing public. At this point, any increase in the stock price becomes profit for the underwriting investment bank, see Federal Regulation (2004) ("Although the statute requires that an offer of credit be honored in order to qualify as a 'firm offer of credit,' . . . this element is not dispositive. To determine whether the offer of credit comports with the statutory definition, a court must consider the entire offer and the effect of all the material conditions that comprise the credit product in question. If, after examining the entire context, the court determines that the 'offer' was a guise for solicitation rather than a legitimate credit product, the communication cannot be considered a firm offer of credit.")

19 www.enotes.com/topic/Investment_banking (last visited July 6, 2012).

20 The act of "market making" promises investors that a certain amount of stock will be available to purchase at a price that is slightly higher than the market value of the stock, while simultaneously purchasing shares at a price set slightly lower than the market value of the stock with the purpose of keeping the security easily transferable or liquid. By promising to buy and sell the stock despite transfers in the open market, market makers are able to manipulate or force the transferability of the stock. For example, if Company X has a liquidity problem – for example, their shares are not being bought or sold – then a market maker could step in to increase the stock's transferability. Let us say that the fair market value of the stock is $20. In order to increase trading on the stock, the market maker promises to sell shares at $21 and buy shares at $19. These promises can induce investors to make trades; see Kaplan (2001).

21 See http://en.wikipedia.org/wiki/Investment_bank (last visited July 6, 2012).

22 Unit trusts are open-ended investments where the overall value of the asset is directly related to the total number of units issued, multiplied by each unit price, minus the transaction fee charged and other associated costs. Each fund has a specified investment objective that determines the management aims and limitations, http://en.wikipedia.org/wiki/Unit_trust (last visited July 6, 2012).

23 A merchant bank is a financial institution that provides capital to companies in the form of ownership in the form of shares rather than more traditional cash loans. Additionally, the banks often provide advice on corporate matters to the firms that they lend to, http://en.wikipedia.org/wiki/Merchant_bank (last visited July 6, 2012).

24 Private equity is an asset class consisting of equity securities in operating companies that are not publicly traded on a stock exchange. These investments are generally made by private equity firms, venture capital firms, or angel investors. While these different types of investors each have their own sets of goals, preferences and investment strategies, all provide similar working capital to the target company to foster expansion, product development, or restructuring of the company's operations, management, and ownership, http://en.wikipedia.org/wiki/Private_equity; see also http://money-investmentbank.blogspot.com/ (both last visited July 6, 2012).

25 See http://en.wikipedia.org/wiki/Government-sponsored_enterprise (last visited July 6, 2012).

although it relinquished its government sponsorship and become a fully private institution via legislation in 1995. The most significant of the sectors currently is in housing finance. Congress created the Federal Home Loan Banks in 1932, followed by the Federal National Mortgage Association (Fannie Mae) in 1938 and Federal Home Loan Mortgage Corporation (Freddie Mac) in 1970. These privately owned but publicly chartered corporations create a secondary market for mortgage loans through guarantees, bonding and securitization, allowing mortgage lenders to increase loan volume and availability of housing-related credit to the US consumer. Their activities have not been without controversy, however, and their future has been in doubt since their being put into conservatorship by the US Government in 2008.

15.6 Information providers: credit rating agencies

A credit rating agency is an organization that rates companies and other organizations. Specifically, these agencies are firms that provide opinions on the creditworthiness of companies and their financial obligations – such as preferred stocks, bonds and commercial papers – issued by the company. Broadly speaking, credit ratings make a distinction between the rated entity's investment grade and non-investment grade debt. Credit agencies may, for example, assign "triple A" credit ratings as a top "investment grade" rating for things like corporate bonds while giving a "double B" rating for "non-investment grade" or corporate bonds that are more risky.

Credit rating agencies registered with the SEC are known as Nationally Recognized Statistical Rating Organizations, or NRSROs. Presently, there are ten firms registered as NRSROs[26]:

1. A.M. Best Company, Inc.
2. DBRS Ltd.
3. Egan-Jones Rating Company
4. Fitch, Inc.
5. Japan Credit Rating Agency, Ltd.
6. Kroll Bond Rating Agency, Inc. (formerly LACE Financial Corp.)
7. Moody's Investors Service, Inc.
8. Rating and Investment Information, Inc.
9. Realpoint LLC
10. Standard and Poor's Ratings Services.

[26] Repeal of credit ratings agency exemption from regulation FD, `www.gibsondunn.com/publications/pages/RepealOfCreditRatingsAgencyExemptionFromRegFD.aspx` (last visited July 6, 2012).

Under the Credit Rating Agency Reform Act of 2006, an NRSRO may be registered to evaluate up to five different types of institutions for credit worthiness: (1) insurance companies, (2) broker-dealers, (3) issuers of asset-backed securities, (4) corporate issuers, and (5) issuers of government securities, municipal securities, and securities issued by foreign governments.[27]

NRSRO ratings are used by regulatory agencies, mutual funds, and insurance regulators to determine the risk involved in acquiring certain securities. For example, mutual funds, which are not insured by the FDIC, often promise customers that the mutual fund will only buy securities that meet certain NRSRO threshold ratings. Similarly, insurance regulators use the ratings to establish the risk involved in insuring securities transactions.[28]

15.7 Financial intermediaries

15.7.1 Stock exchanges

Stock exchanges are any organizations, associations, or groups that provide or maintain a marketplace where securities, options, futures, or commodities can be traded. Stock exchanges assess membership and transaction fees, and provide market data and listing feeds for a fee. All US stock exchanges are subject to US securities laws, referred to below in Section 15.8, and most have their own additional rules for selling stocks on their exchange. For example, the New York Stock Exchange has the NYSE Rules – its own set of guidelines that companies must comply with in order to sell shares on their stock exchange.

15.7.2 Depositories

A depository is a bank or company that holds funds or securities deposited by others to ensure that securities transactions are safe. Depository Trust and Clearing Corporation (DTCC) is the primary US depository. This entity works as a type of clearinghouse[29] and specializes in providing a number of services such as post-trade

[27] US financial reform: credit ratings agencies, www.duanemorris.com/alerts/credit_rating_agencies_3768.html (last visited July 6, 2012).

[28] For further reading, see Brownlow (2011). http://en.wikipedia.org/wiki/Nationally_recognized_statistical_rating_organization, (last visited July 6, 2012).

[29] A clearinghouse is a financial institution that provides clearing and settlement services for financial and commodities, derivatives, and securities transactions. A clearinghouse stands between two clearing firms – clearing participants – and its purpose is to reduce the risk of a clearing firm failing to honor its trade settlement obligations. A clearinghouse reduces settlement risks the risk of default between investors trading in securities by netting offsetting transactions between multiple counterparties, by trading parties, requiring collateral deposits, by providing independent valuation of trades and collateral, by monitoring the credit worthiness of the clearing firms, and in many cases, by providing a guarantee fund that can be used to cover losses that exceed a defaulting clearing firm's collateral on deposit. For further reading, see Bernanke (2011) and http://en.wikipedia.org/wiki/Clearing_house_(finance) (last visited July 6, 2012).

settlement, clearance, and data services. DTCC is cooperatively owned by its members and governed by a board chosen from among its owners.

15.7.3 Data vendors

A financial data vendor provides data to financial firms, traders, and investors. These data are collected from many sources, including stock exchange feeds, broker-dealers, and regulatory filings to the SEC. Additionally, the financial data include both pre-trade data, such as the bid/ask information necessary to accurately price a financial instrument, and post-trade data, such as the last trade price for a security on a given trading day. Each data vendor has its own unique way of processing information about entities, companies, instruments, and securities. Pricing data are sold separately from other related data, such as corporate actions and events, valuation information, fundamental data including company performance, and reference data for specific entities and instruments or securities. Aggregated, all the data from different types of instruments and securities create an extremely large pool of information that may be overwhelmingly broad for the average investor. Data vendors process these large data pools into smaller and more useful datasets, so that investors can use the information to make better, more informed, investment decisions.

A key function of data vendors is to cleanse data and ensure high levels of quality. Some firms like Bloomberg and Reuters are largely focused on providing data, while others collect data as part of their on-going businesses and act as data vendors for specific types of products. The latter group ranges from credit ratings agencies like Standard and Poors to mortgage processing firms like LPS and stock exchanges like the New York Stock Exchange.

15.7.4 Custodians

A custodian is a bank, agent, trust company, or other organization that holds and safeguards assets for an annual fee. State Street Bank & Trust, and Bank of New York Mellon are good examples of well-known custodians in the industry today. A custodian will collect and remit dividend and interest payments and provide the notification and entitlement processing[30] of corporate actions and events. Custodians also provide performance reports, periodic statements of accounts, and tax reporting statements.

[30] Entitlement processing refers to obtaining and verifying the rights one has with regard to his or her property. For example, if someone owns a piece of land and would like to place a business on it, entitlement processing would tell the person whether he or she had the right, in terms of compliance with zoning laws or municipal laws, to develop the land for that particular use.

15.8 Regulators

Broker-dealers, investment banks, asset investment managers, stock exchanges, and credit rating agencies doing business in the USA are all regulated by various regulatory agencies, as well as federal and state banking laws. Most of this regulatory authority is derived from the Securities and Exchange Act of 1934; however, some authority was issued by the Dodd–Frank Act in 2010.[31] This regulatory authority is mainly enforced by the Securities and Exchange Commission (SEC), but is also delegated to other agencies, such as the Financial Industry Regulatory Authority (FINRA). Many states also regulate broker-dealers under separate state securities laws, otherwise known as "blue sky laws." Commercial banks are regulated by either the Comptroller of the Currency, Federal Reserve Bank (FRB) or State Banking Commissions, and the Federal Deposit Insurance Corporation (FDIC). Custodian banks, agent banks, and depositories are generally regulated in a similar manner to commercial banks. State insurance commissions generally oversee the insurance industry. Lastly, because they only provide information, data vendors have traditionally been unregulated. The Dodd–Frank Act, however, has begun to add certain restrictions on rating agencies. For example, the Act requires that rating agencies report their rating methodologies, and maintain consistent terminologies and symbols throughout their reports.[32]

As a result of the 2007–2009 financial crisis, additional regulations are being enforced in the financial services industry through regulatory agencies. The Dodd–Frank Act requires a 2–5 year rulemaking process where roughly 250 new regulations must be researched and written by at least 12 regulatory agencies. The goal of this process is to ensure that final regulations are balanced and consistent with the intent of the initial legislation, in order to avoid any unintended consequences. Presently, the main regulatory agencies and institutions include the following.

15.8.1 Important financial industry regulatory institutions

US Securities and Exchange Commission (SEC)

The SEC seeks to maintain an efficient securities market by providing equal access to information in an attempt to make the market more transparent for the general investing public. The SEC does so by: (1) issuing rules and regulations regarding all publicly traded companies and the sale of securities, (2) preventing and litigating securities fraud cases, (3) preventing and litigating instances of insider trading, and

[31] The Dodd–Frank Wall Street Reform and Consumer Protection Act is a federal statute in the USA that was signed into law in July 2010. Passed as a response to the mid-2000s financial crisis, the Act brought the most significant changes to financial regulation in the USA since the regulatory reform that followed the Great Depression

[32] `www.sec.gov/spotlight/dodd-frank/creditratingagencies.shtml` (last visited July 6, 2012).

(4) ensuring that investors have equal access to standardized data concerning publicly traded securities. Mass communication and technology have made entering the securities market dramatically easier. SEC enforcement and rulemaking becomes all the more crucial as greater percentages of the general public invest in the securities market.

Federal Deposit Insurance Corporation (FDIC)

Created in the 1930s, the FDIC protects citizens when banks fail by: (1) insuring every deposit made in American banks from what was previously $100,000, to its current rate of $250,000, (2) stepping in when banks – both large and small – fail, and (3) examining banks and institutions for compliance with consumer protection laws. While the FDIC insures all deposits up to $250,000, it does not insure securities such as stocks or mutual funds. FDIC does not receive any federal funding; all of its funds come from insurance premiums with American banks.

Explaining the FDIC's role may be helped by another example. If Person X had money in a local bank that had to close its doors during the 2008 mortgage crisis, the FDIC would have stepped in to help by selling the local bank's deposits and loans to other banks. In this way, X, and all other customers of the closed bank, would seamlessly become customers of the bank assuming the debts, thus ensuring that all of X's money deposited in the local failed-bank remained safe and secure.[33]

Federal Reserve System[34]

The main objectives of the Federal Reserve System (known informally as the Fed) are (1) to keep employment rates high, (2) to keep prices stable, and (3) to moderate long-term interest rates. The Fed was created in 1915 by statute, and regulates US monetary policies and banking institutions. The Federal Reserve System's structure is composed of the presidentially appointed Board of Governors (also known as the Federal Reserve Board or FRB), the Federal Open Market Committee (FOMC), 12 regional Federal Reserve Banks[35] located in major cities throughout the nation, numerous privately owned US member banks and various advisory councils.

Although the Board is subject to congressional oversight, it functions independently of the government. Members of the Board must be approved by the Senate, but the Board's monetary policies and regulatory decisions do not have to be approved by the executive or legislative branch. The FRB can control the securities

[33] `www.fdic.gov/about/learn/symbol/` (last visited July 6, 2012).
[34] A detailed description can be found at `http://en.wikipedia.org/wiki/Federal_Reserve_System#Member_banks` (last visited July 6, 2012).
[35] The 12 Federal Reserve Banks include are located in Boston, New York, Philadelphia, Cleveland, Richmond, Atlanta, Chicago, St. Louis, Minneapolis, Kansas City, Dallas, and San Francisco.

markets and depository banks directly by monitoring the economy and changing lending rates for US Treasury bonds. By performing these functions, the FRB is charged with keeping the US economy stable.[36]

Financial Stability Oversight Council (FSOC)

The FSOC is a council comprising regulators that monitor systemic risk in the financial market and advise the President and other regulatory agencies on fiscal policy. The Chairman of the FSOC is also the Secretary of the Treasury, thus FSOC largely interacts with the Treasury Department. Other members of the Council include the Chairman of the Federal Reserve and Comptroller of the Currency. The main FSOC goals are to promote efficiency and greater transparency in the financial industry. To achieve this, the FSOC was given, among other powers, the authority to disapprove mergers of large financial institutions.

US Department of the Treasury[37]

The Department of the Treasury has broad powers to manage most of the US Federal Government's money. It regulates the overall US financial systems by controlling the production of currency, disbursements of payment to American citizens, and revenue collection. The Department also works with other foreign governments to facilitate economic growth and predict and prevent world economic crises. Some of its specific duties include advising the government on financial and economic decisions, as well as managing (1) the collection of federal taxes, (2) federal government debt, (3) currency and coinage, (4) government accounts and public debt, (5) national banks, and (6) the enforcement of federal tax laws. The US Treasury plays an important role in regulation via offices within the department like the OCC (see below) and the newly created OFR.

Office of the Comptroller of the Currency (OCC)

The OCC charters, regulates, and supervises all national banks and federal savings associations with the goal of increasing competition in the financial markets and reducing regulatory burden. The OCC has the power to (1) examine national banks, (2) approve or disapprove applications for new charters, capital branches, capital, or changes in corporate banking structure, (3) take supervisory actions against national banks and federal thrifts that fail to comply with laws and regulations, and (4) issue rules, regulations, and legal interpretations governing investments and lending.[38]

36 www.federalreserve.gov/pf/pdf/pf_1.pdf#page=4 (last visited July 6, 2012).
37 www.treasury.gov/about/role-of-treasury/Pages/default.aspx (last visited July 6, 2012).
38 www.occ.treas.gov/about/what-we-do/mission/index-about.html (last visited July 6, 2012).

Office of Financial Research (OFR)

Created by the Dodd–Frank Act of 2010, the OFR was created to work with the FSOC and the Treasury Department to improve the quality of financial data available to both the general public and policymakers with a robust and sophisticated data analysis of the financial system. In response to redundant systems of financial data collection and analysis from financial institutions, the OFR promotes and distributes standardized and comprehensive data concerning the financial markets in an effort to reduce the transaction costs of redundant data collection and data aggregation.[39]

Commodity Futures Trading Commission (CFTC)

Established in 1974, the CFTC is an independent federal agency that regulates commodity futures, derivatives, and option markets. Its powers and goals were substantially expanded and revised by the Dodd–Frank Act in 2010. Now the CFTC assures the economic utility of the futures markets by: (1) encouraging their competitiveness and efficiency, (2) protecting market participants against fraud, manipulation, and abusive trading practices, and (3) ensuring the financial integrity of the clearing – transaction – process.[40] Through effective oversight, the CFTC enables the futures markets to serve the important function of providing a means for price discovery and offsetting price risk.[41]

Federal Housing Finance Agency (FHFA)

The FHFA oversees the housing market, specifically Fannie Mae, Freddie Mac, and federal home loan banks. It does so by: (1) prosecuting banks that misrepresent mortgage values to Fannie Mae or Freddie Mac, (2) issuing regulations for the housing market, and (3) educating the public on housing mortgage loans.

Government Accountability Office (GAO)

The GAO, headed by the Comptroller General, monitors and regulates how federal tax dollars are spent. It advises the President and Congress on fiscal policies and decisions related to Congressional expenditures. The GAO also issues decisions on government contract disputes, including bid protest rulings and reports on Congressional agency rules.[42]

Bureau of Consumer Financial Protection (CFPB)

The goal of the CFPB is to help consumers. It is a relatively new government office, created by the Dodd–Frank Act in 2010. Its first director was only recently

[39] `www.treasury.gov/initiatives/wsr/ofr/Pages/default.aspx` (last visited July 6, 2012).
[40] See note on the clearing process, footnote 29.
[41] `www.cftc.gov/About/MissionResponsibilities/index.htm` (last visited July 6, 2012).
[42] `www.gao.gov/about/index.html` (last visited July 6, 2012).

appointed in January 2012. Thus far, CFPB duties include: (1) taking consumer complaints, (2) educating consumers about their rights in financial markets, (3) monitoring both consumer behaviors and common risks to consumers, and (4) enforcing laws against deception or abuse in the financial markets.[43]

The CFPB is positioned to help with common consumer issues in the banking industry. As an example of a practice that the CFPB may focus on, consider Person X who opens a checking account at a major bank. Unfortunately, X forgets to consider the cost of a small purchase in relation to his balance, and overdrafts the account by $5.00, creating an automatic $35.00 charge for overdraft coverage on the account. The $35.00 overdraft fee is charged, even though X has other accounts with the bank that could have easily paid the $5.00 overdraft had X known to "link" the accounts. The CFPB is designed to educate citizens about common issues like this one and enforce laws against banking practices that are deceptively designed to "catch" unwary citizens.[44]

National Credit Union Administration Board (NCUAB)

The NCUAB is an independent federal agency that was originally created in 1934 as part of the New Deal to make credit available to more Americans. Currently, the NCUAB supervises and charters federal credit unions by operating and managing the National Credit Union Share Insurance Fund (NCUSIF). This fund insures all account holder deposits in federal credit unions – as well as most state-chartered credit unions – by taking over failing credit unions and placing them in conservatorship.

15.8.2 Self-regulatory organizations

Financial Industry Regulatory Authority (FINRA)

Similar to the SEC, FINRA's goal is to promote equal access to information and prevent the fraudulent sale of securities. The main differences between the two entities are that: (1) the SEC is an executive agency and FINRA is an SRO (Self-Regulatory Organization) that is ultimately held accountable to the SEC, and (2) the SEC mainly protects the individual investor while FINRA is more concerned with regulating and licensing securities brokerage firms and dealers. Like the SEC, FINRA enforces anti-fraud securities laws and regulations while promoting equal access to standardized information for all investors. Specifically, FINRA registers and educates all brokers, examines securities firms, and writes and enforces rules for

[43] www.consumerfinance.gov/the-bureau/ (last visited July 6, 2012).
[44] www.consumerfinance.gov/ (last visited July 6, 2012).

all securities representatives. They also monitor trading in the US stock markets and administer their own securities-related dispute resolution forum.[45]

15.8.3 Agencies and state governments that regulate municipalities and public programs

Municipal Securities Rulemaking Board (MSRB)

As its name suggests, the MSRB monitors and regulates municipal securities by: (1) issuing rules with the force of federal law, and (2) monitoring all securities related to city and municipal governments. The MSRB regularly meets with the SEC and FINRA to discuss enforcement of the rules it issues.[46]

State governments

Each state government has its own unique set of securities laws to prevent fraud and regulate broker-dealers within the intrastate securities market. Traditionally called "blue sky laws," these laws complemented, and were largely based on, federal securities laws. However, in 1998, Congress pre-empted a large portion of state securities laws when it passed the Securities Litigation Uniform Standards Act. While the Act took away some state powers to regulate the sale of securities within its borders, it standardized securities laws into one cohesive body of laws. This standardization makes for a more fluid national securities market by establishing one set of laws that the investing public must comply with, rather than having to research and comply with each state's unique securities law variation. Still, the Act did not pre-empt all state regulatory powers over the securities industry. Today, state governments continue to regulate broker-dealers by requiring licensing in their state and compliance with their regulations. The states also provide investors with certain state-specific claims for securities fraud.[47]

Federal Energy Resources Commission (FERC)

The FERC regulates the sale of energy, monitors energy markets, and supervises regulated companies by monitoring their accounting and financial reports. Its reach is broad, encompassing both energy corporation mergers and acquisitions by energy companies.

45 `www.finra.org/AboutFINRA/` (last visited July 6, 2012).
46 `www.msrb.org/About-MSRB.aspx` (last visited July 6, 2012).
47 `http://en.wikipedia.org/wiki/Blue_sky_law` (last visited July 6, 2012). For further information, see Rogers.

15.8.4 Industry trade groups

Security Industry and Financial Markets Association (SIFMA)[48]

SIFMA represents numerous segments of the financial industry. It has become a voice of the industry that, according to its website, is "uniquely suited to provide trusted and expert information to policymakers, regulators, media, industry participants and the general public" about the financial industry. In 2006, the Security Industry Association (SIA) and the Bond Market Association (TBMA) merged to form SIFMA, a move that merged the two agency interests. Specifically, SIFMA is focused on several key issues related to the Dodd–Frank Act that its members have identified as critical, one of which is Systemic Risk Regulation, defined in Section 15.2.1.

The Software and Information Industry Association's Financial Information Services Division (FISD)[49]

The FISD is an international non-profit industry forum created to find and solve business or technical issues that harm the administration, distribution or use of financial data. On its website, FISD maintains that market data distribution, efficient trade execution, and regulatory compliance require a highly predictable and consistent service – each of which depends on the close cooperation of numerous financial organizations and systems. The FISD aims to help with these issues by providing investors with sound, up-to-date, and intuitive financial information.

International Swaps and Derivatives Association (ISDA)

Created in 1985, the ISDA oversees the US OTC derivatives market, with the goals of promoting market efficiency and security. The Organization monitors the OTC derivatives market and advises Congress on related policy issues.[50]

International Securities Association for Institutional Trade Communication (ISITC)

The ISITC is a group of leading institutions and individuals in the financial industry that aim to reduce inefficiencies in trading, lower investment risk, and add value to general investors in the global securities markets. Beginning in 1991, the ISITC was created to reduce errors and inefficiencies in trading mortgage securities by updating the aging Telex systems and replacing them with an advanced standard electronic

48 `www.sifma.org/about/history/` (last visited July 6, 2012).
49 `http://archive.fisd.net/FISD-YIR-08.pdf` (last visited July 6, 2012).
50 `www2.isda.org/about-isda/` (last visited July 6, 2012).

message format. Today, the ISITC's goal is to promote efficiency and lower risk in the broader financial industry.[51]

15.9 Current data-related challenges and technology initiatives

One of the major challenges facing the industry is how to develop and integrate a level of data standardization that facilitates straight-through processing[52] for all financial transactions. The lack of standardization continues to inhibit communications in the clearance and settlement of transactions between financial institutions. It also inhibits risk calculations within firms and across the industry. It is hoped that the Office of Financial Research (OFR)[53] will push the industry towards solving these problems and increase the transparency in financial transactions with the result of increasing investor and consumer confidence in financial institutions overall. Additionally, regulatory financial reporting and shareholder communications should be in standardized electronic formats so the information is easier to transmit electronically. Some of the larger problems with the current system include the following.

1. Multiple identification schemes are still used to identify and collect data on instruments.
2. A standard Legal Entity Identifier (LEI)[54] has been agreed upon but not yet implemented.
3. No standard taxonomy exists for Asset Product Categorization.
4. Standard methodologies continue to change for the evaluation of all instruments, positions, and transactions that are "hard to price."
5. No standard terms and definitions exist for reference data attributes stored in the master files of financial institutions and data vendors that pass among supply chain partners.[55]

[51] `www.isitc.org/history.cfm` (last visited July 6, 2012).

[52] Straight-through processing (STP) enables the entire trade process for capital markets and payment transactions to be conducted electronically without the need for re-keying or manual intervention, which is subject to legal and regulatory restrictions. `http://en.wikipedia.org/wiki/Straight-through_processing` (last visited July 6, 2012).

[53] The OFR is further discussed in Section 15.8.1.

[54] A Legal Entity Identifier (LEI) is a unique ID associated with a single corporate entity. The lack of a standard identification system for financial counterparties makes it difficult for financial firms and regulators to develop a consistent and integrated view of risks in specific firms and the financial industry generally. As of 2012 no common entity ID convention exists in the market. However, a range of regulatory initiatives is driving the creation of a universal LEI standard for financial markets. The importance of the LEI has been recognized in statements by the Financial Stability Board, IOSCO, and G20 finance ministers and leaders. In the USA, Dodd–Frank mandated initiatives to create standard LEIs. The main effort is being driven by the Office of Financial Research (OFR), with the Commodities and Futures Trade Commission (CFTC) also issuing LEI standards around derivatives. Regulators in Hong Kong, Canada, the Australian Securities and Investments Commission, as well as the International Organization of Securities Commissions, have all recognized the importance of developing a unique LEI. See `www.sifma.org/issues/operations-and-technology/legal-entity-identifier/overview/` (last visited July 6, 2012).

[55] Supply chain partners are networks of businesses that communicate and organize to produce end products for specific consumers.

6. No common "logical data model" exists across data domains or types.
7. No codified "best practices" exist to measure the effectiveness of enterprise data repositories.

Some progress has been made over the years by cooperation within the industry in the form of various industry groups, entities or agencies in the financial industry that join together to promote the awareness of certain issues or to promote changes in the financial industry. These trade groups and associations all work together with the Accredited Standards Committee X9 (ASC X9)[56] whose goal is to establish, manage, and foster higher standards for the financial data services industry in order to streamline and standardize the delivery of financial services and products. As it has become the main organization responsible developing standards for the financial industry, ASC X9 provides an important function in the creation and adoption of new banking, brokerage, and insurance industry technologies. Procedurally, ASC X9 proposes standards for the international committee's consideration for adoption as international standards, also called "ISO standards."[57] ASC X9 offers membership to all US companies and financial services industry organization.[58] Examples of Financial Industry Initiatives that work with ASC X9 to develop standards for the financial industry include the following.

15.9.1 eXtensible Business Reporting Language (XBRL)

The XBRL is an organization that promotes the use of a standard, XML-based computer language (also known as XBRL) for digitizing business reports according to accounting rules in various countries and banking regulations and performance benchmarking.[59] They argue that the use of a single computing language would help investors learn about companies more easily and, by doing so, increase market efficiency.[60]

15.9.2 FIXML

Similar to XBRL, FIXML is an initiative that promotes a "Financial Information eXchange" (FIX) format. Specifically, FIXML promotes a series of messaging specifications to electronically communication trade-related messages. According to its website, this was "developed through the collaboration of banks, broker-dealers, exchanges, industry utilities and associations, institutional investors and

[56] ASC X9, Inc. is an American National Standards Institute (ANSI) accredited standards developing organization. ASC X9 is the US Technical Advisory Group (TAG) to the International Technical Committee on Financial Services ISO/TC 68 under the International Organization for Standardization (ISO) of Geneva, Switzerland, `http://en.wikipedia.org/wiki/ANSI_ASC_X9` (last visited July 6, 2012).

[57] `http://en.wikipedia.org/wiki/ANSI_ASC_X9` (last visited July 6, 2012).

[58] `http://en.wikipedia.org/wiki/ANSI_ASC_X9` (last visited July 6, 2012).

[59] `www.xbrl.org/AboutXBRL` (last visited July 6, 2012).

[60] As valuations of securities become more precise and accurate, the market is considered more "efficient."

information technology providers."[61] Broadly speaking, FIX is a platform used for securities trades by many large financial institutions; most major stock exchanges and investment banks use the FIX format that FIXML promotes.[62] Each new institution that uses FIX makes the platform more valuable; the more it is used, the more investors and institutions understand it and use it themselves.

15.9.3 Enterprise Data Management (EDM) Council

The EDM Council is a non-profit trade association designed by leading financial industry participants. Its main purpose is to address the issues and challenges associated with managing data content as a business and operational priority. This helps ensure that the data used for business processing and regulatory oversight are trusted to be fit-for-purpose, accessible, easy to integrate into downstream data systems, and comparable across multiple data source-types.[63] EDM sponsors various other initiatives including the Data Management Maturity Model, Supply Chain Management method, and Data Quality Rulebook.

Although progress has been made, as evidenced by the development of FIX trade messages, XBRL for corporate financial reporting, and so on, much remains to be done. The establishment and integration of the required standards should be viewed by industry participants as an opportunity to provide greater transparency to their customers and regulators, in the hopes of establishing greater trust. While this may not be an opportunity to acquire competitive advantages, it would certainly create cost reductions throughout the industry.

15.10 Conclusion

This chapter has primarily covered the financial industry structure in the USA. The financial industry is diverse and complex. While the chapter has described the plethora of independent functions that are necessary for the smooth functioning of the financial industry, many firms are involved in several of these businesses. For example, a number of the large commercial banks have large investment banking arms. Several financial exchanges are prominent as data vendors. The regulatory landscape in the financial industry world-wide is changing rapidly, driven by new regulation as well as mergers and acquisitions. This is especially true in the USA where the larger firms are usually regulated by a number of regulatory agencies.

61 `http://fix-events.com/about.html` (last visited July 6, 2012).
62 `http://fixprotocol.org/what-is-fix.shtml` (last visited July 6, 2012).
63 `www.edmcouncil.org/about.aspx` (last visited July 6, 2012).

Understanding the complexity of interconnections between firms in the financial industry as well as their interactions with regulators is an important precursor to gaining an understanding of the data-flows between the various actors in the financial industry.

15.11 Appendix: Key terms and regulations from the Dodd–Frank Act

The Dodd–Frank Act of 2010 has had, and continues to have, a significant and wide-ranging regulatory impact on the financial industry in the USA. This section describes concepts that emanate from the regulation that are increasingly becoming implemented in financial institutions (as of 2012).

Systemic risk regulation

Systemic risk generally refers to the interdependency of institutions in global financial markets and the domino effect that can result from correlated risks[64] if any financial institution fails.

Volcker Rule

The Volcker Rule, as formulated in the Dodd–Frank Wall Street Reform and Consumer Protection Act of 2010 (Dodd–Frank), prohibits banks from engaging in proprietary trading[65] and restricts commercial banks and their affiliates from investing in hedge funds and private equity. In the Dodd–Frank Act, Congress identified certain banks, banking activities, and non-bank institutions as systemically risky. These include, among others, hedging, market making, risk management, and securitization.[66] The Volcker Rule also capped bank ownership in hedge funds and private equity funds at 3%.[67] The cap is meant to dissuade large banking firms from relying on the types of high-risk hedge funds that led to the 2008 housing market collapse. Institutions have seven years from the passage of Dodd–Frank, which became law in 2010, to become compliant with these regulations. There is a great deal of concern in the market about the final form that agency regulations interpreting the Act will take, and the unknown effects the regulations may have.

[64] Correlated risks arise from differences in the actual and projected value of two or more assets.

[65] Proprietary trading occurs when a firm trades its own stocks, bonds, currencies, commodities, derivatives, or other instruments, with the firm's own money rather than its client money. By doing so, the bank keeps profit for itself. Generally, proprietary trading is considered riskier than other forms of investment. For further reading, see Merkley and Levin (2011).

[66] www.sifma.org/issues/regulatory-reform/volcker-rule/overview/ (last visited July 6, 2012).

[67] www.sifma.org/issues/regulatory-reform/volcker-rule/overview/ (last visited July 6, 2012).

Recovery and resolution plans (RRPs or "living wills")[68]

Section 165 of the Dodd–Frank Wall Street Reform and Consumer Protection Act requires all systemically important financial institutions to create "living wills" to facilitate "rapid and orderly resolution, in the event of material financial distress or failure." The term living will is typically used to describe a plan that spells out how an individual would like to be treated medically should he or she become incapacitated, and was first applied to financial institutions in the UK in 2009. It quickly gained favor as a concept among financial regulators around the world.

As defined by Dodd–Frank, systemically important financial institutions include all bank holding companies with more than $50 billion in assets, plus other banks and non-banks whose distress or failure, regulators believe, could have an adverse effect on the US financial system. While Dodd–Frank focuses on resolving distressed companies, recent discussions by regulators appear to be leaning toward the creation of two plans as part of the living will – a recovery plan (for going concerns) and a resolution plan (for "gone" concerns).

OTC derivatives

Over-the-counter derivatives are financial contracts that are used to manage risk by transferring it from someone who wishes to reduce his or her exposure to someone willing to take on that exposure. Legislation was introduced in 2009 to increase oversight of the high risk derivatives market as part of the government's response to the role that credit default swaps[69] played in American International Group (AIG)'s financial troubles. That legislation later became Title VII of the Dodd–Frank Act. Under Title VII, swaps must be cleared by at least one central counterparty (CCP).[70] To comply with this requirement, the CCP will hold large amounts of capital and will be closely monitored. This, in turn, reduces the risk that one CCP default will trigger a chain of defaults in swaps markets.[71] Additionally, swaps must be traded on either stock exchanges or Swap Execution Facility (SEF)[72] platforms.

[68] See Deloitte Financial Reform Insights.

[69] Default credit default swaps are contracts in which a buyer makes a payment to a seller in return for the seller's promise that he will compensate the buyer if a specified credit 'trigger' event occurs, see Solman (2008).

[70] As modern trading in securities becomes faster than the actual transfer of property traded, investors can unknowingly become insolvent while making trades, placing them in default. The central counterparty is an organization that distributes this risk of default evenly among its members, in order to allow fast and efficient trading, see `http://en.wikipedia.org/wiki/Central_Counterparty_Clearing` (last visited July 6, 2012).

[71] An interest rate swap is a popular and highly liquid financial derivative instrument in which two parties agree to exchange interest rate cash flows, based on a specified notional amount from a fixed rate to a floating rate, or vice-versa. Interest rate swaps are commonly used for both hedging and speculating (Hazen, 2009).

[72] SEFs are trading platforms which were created by the Dodd–Frank Act. Under the Act, swaps must be traded on a Designated Contract Market or SEF. Swaps, which are regulated by the CFTC, are private agreements between any two individuals to exchange cash flows at specific intervals during the contract, see `www.marketswiki.com/mwiki/Swaps` (last visited July 6, 2012).

Fiduciary standard

The term "fiduciary relationship" refers to "the highest standard of customer care available under law."[73] Fiduciary duties include a duty of loyalty and a duty of care.[74] Generally speaking, these legally significant duties require a fiduciary – the person with the duty – to act in the client's best interest and fully disclose material facts and conflicts of interest. However, the current system, established in the 1940s, allows states to develop their own, often conflicting, definitions and requirements for fiduciary duties. Depending on which state companies are formed in, corporations can potentially be bound by different state's legal definitions of fiduciary duties, and therefore what duties are required. The inconsistent definitions and expectations in fiduciary duty laws can confuse investors.[75]

Fiduciary duties under the Dodd–Frank Act

Dodd–Frank engaged the Securities and Exchange Commission (SEC) in studying the need for creating a uniform federal fiduciary standard of care – or fiduciary duty – for brokers and investment advisers that provide personalized investment advice. The Act authorized the SEC to establish a new standard if it believed that one was necessary. Despite this authority, the SEC has yet to issue any new standards.

Currently, most publicly traded securities[76] are organized under Delaware state law because the majority of publicly traded companies are incorporated in Delaware.[77] In this way, Delaware law's definition of fiduciary duties has become the industry standard. There is a good argument to be made that the free market has pressured publicly traded companies to organize in Delaware, making their stocks more attractive to investors by conforming to the known, and now standard, corporate laws that Delaware has established.

In 2010 the Department Of Labor (DOL) dramatically revised its regulation that defined fiduciaries in the Employee Retirement Income Security Act (ERISA) and the Internal Revenue Code (IRC), a revision which would have conflicted with the duties described in Dodd–Frank. The proposed definition substantially changed

[73] www.securitiesexpert.org/wp-content/uploads/2011/08/3-d-Aronson-Fiduciary-Standard-Issues-SIFMA.pdf (last visited July 6, 2012).

[74] www.securitiesexpert.org/wp-content/uploads/2011/08/3-d-Aronson-Fiduciary-Standard-Issues-SIFMA.pdf (last visited July 6, 2012).

[75] See the State Governments section IV(b)(2) for more discussion of these laws.

[76] Securities, also called instruments, may be stocks, bonds, funds, options, futures, and currencies.

[77] Traditionally, each state had different laws for companies incorporating in their state. As major companies became publicly traded, investors began to view certain state laws as better for their business – or the corporation they invested in – than other state laws. Over time, Delaware became the state of choice to incorporate in because its laws were considered more favorable to businesses than other states. Today, more than 50% of publicly traded companies are incorporated in Delaware, and are therefore subject to Delaware state laws, www.corp.delaware.gov/aboutagency.shtml (last visited July 6, 2012).

who would be legally considered a "fiduciary."[78] It was also poised to change state law on the subject, by promulgating a set definition of fiduciary duties created by government agencies rather than the states, where they were traditionally created. In response to major backlash from investors with IRA accounts, in September 2011, DOL announced that it would review its rule on the definition of a fiduciary, and propose a new rule, which is expected in 2012.

Comparing Dodd–Frank to Basel III

Basel III refers to regulations made by the Bank of International Settlements, a council of regulators in Switzerland. The counsel's ideas focus on maintaining the long-term success of banks by keeping larger amounts of money in the banks themselves, rather than loaning the money out; under Basel III, banks will not be able to lend as much money as previous regulations allowed. This limitation is offset by the fact that the banks would be more prepared for large unexpected economic swings in the world economy like the recent financial crisis. Basel III has been widely adopted by European banks, and it is believed that US regulatory agencies will follow suit to create a more cohesive and synergistic world market.

The Dodd–Frank Act, however, seems to use different regulatory rules than the Basel III accords.[79] Specifically, Basel III is more focused on capital conservation, believing banks are better structured for safety.[80] The liquidity coverage ratios Basel III allows are based on external credit ratings. Conversely, The Dodd–Frank Act requires that federal regulations remove references to credit ratings altogether.[81] Whether or not the Dodd–Frank Act is implemented in a way that is consistent with Basel III policies is an issue that is yet to be decided.

15.12 Appendix: Securitization and housing finance

Securitization

Securitization means converting assets with predictable cash-flows – like real estate or businesses – into securities so that they can be traded in public financial markets. Put differently, securitization lets financial institutions bundle and convert normally

[78] Specifically, the proposed definition "affected whether retail brokers, prime brokers, institutional trading desks, swap dealers, and others who work with pension and 401(k) plans" would be deemed fiduciaries, `www.sifma.org/issues/private-client/fiduciary-standard/overview/` (last visited July 6, 2012).

[79] See `www.diazreus.com/news-articles-dodd-frank-wall-street-reform.html` (last visited July 6, 2012).

[80] See `www.diazreus.com/news-articles-dodd-frank-wall-street-reform.html` (last visited July 6, 2012).

[81] See `www.diazreus.com/news-articles-dodd-frank-wall-street-reform.html` (last visited July 6, 2012).

illiquid[82] assets on their balance sheets, like mortgage loans or credit card receivables, into more easily transferable – liquid – securities.[83]

The securitization market has become an enormous source for consumer financing. It helps consumers get credit to finance their houses, cars, or college education. It allows companies to create jobs by providing added sources of funding for lenders that extend credit to businesses and consumers.[84] Moreover, securitization "facilitates the flow of capital from investors located around the world into U.S. credit markets, making credit more widely available to consumers" at a lower cost.[85]

Housing finance

When real estate prices started rapidly declining in 2006–2008, many borrowers defaulted on their mortgage loans. As most of these loans were the underlying collateral for securitizations, the private-label securitized products market saw a decrease in liquidity and, often, a significant price drop. Agency Mortgage Backed Securities (MBS) markets – claims on the cash flows from mortgage loans – were also disrupted, but generally maintained liquidity. Consequently, the financing made available to consumers and businesses that was dependent on these markets became strained; lending costs increased rapidly, a condition known as "a credit crunch."[86] For example, if All-Star bank's cash comes from securitized mortgage loans, the housing market crash of 2008 would cause All-Star's securitized mortgage loans to go down in value in proportion to the amount of mortgage loans in default. Effectively, the securitized mortgage-loan securities sold become less valuable as the assets they are based upon drop in value. Thus, in response to the mortgage crisis, All-Star bank has less cash from the sale of its securitized mortgage-backed securities and therefore has substantially less lending power. Broadening this example to include all major banks involved in mortgage-backed securities explains why consumers were unable to borrow money as easily as before. Banks with mortgage-backed securities had less money to lend out because mortgages they loaned were in default.

On February 11, 2011, President Obama's administration released a report to Congress that analyzed the causes of our current economic situation and identified areas, particularly in housing finance. The report discussed the consequences of winding down Freddie Mac and Fannie Mae, and reducing the government's overall role in housing finance. The report also assured consumers that loans from

[82] An asset's "liquidity" or "illiquidity" refers to the ability to resell the asset on the open market. If an asset is illiquid, it is difficult or impossible to sell quickly, see Listokin (2011).
[83] See `www.sifma.org/events/2009/269/event.aspx?id=10614` (last visited July 6, 2012).
[84] `www.sifma.org/events/2009/269/event.aspx?id=10614` (last visited July 6, 2012).
[85] `www.sifma.org/events/2009/269/event.aspx?id=10614` (last visited July 6, 2012).
[86] `www.sifma.org/events/2009/269/event.aspx?id=10614` (last visited July 6, 2012).

Government-Sponsored Enterprises (GSE) – businesses owned by governments – would be honored.[87]

Bibliography

Bernanke, Ben S., 2011, Remarks of FRB Chairman *Clearinghouses, Financial Stability, and Financial Reform*, at 2011 Financial Markets Conference.

Brownlow, Benjamin H., 2011, Rating agency reform: preserving the registered market for asset-backed securities, *N.C. Banking Inst.*, **15**, 111.

Congressional Research Service, 2010, Permissible Securities Activities of Commercial Banks Under the Glass–Steagall Act (GSA) and the Gramm–Leach–Bliley Act (GLBA).

Deloitte Financial Reform, Financial Reform Insights: A blueprint for "living will" requirements, Insights Issue #2, at www.deloitte.com/assets/Dcom-UnitedStates/Local%20Assets/Documents/FSI/us_fsi_bking_Living%20wills_081110.pdf (last visited July 6, 2012)

Federal Regulation, 2004, Real Estate & Mortgage Lending § 9:25 (Fourth edition); *Cole v. U.S. Capital*, 389 F.3d 719, 727 (7th Cir. 2004)

Hazen, Thomas Lee, 2009, *Law of Securities Regulation*, section 1.7, Interest Rate Swap, Sixth edition, West, http://en.wikipedia.org/wiki/Interest_rate_swap (last visited July 6, 2012)

Johnsen, D. Bruce, 2011, Using bond trades to pay for third party research, *Virginia Law and Business Review*, **5**, 481.

Kaplan, Arthur M., 2001, Antitrust as a public-private partnership: a case study of the Nasdaq litigation, *Case Western Reserve Law Review*, **52**, 111, 112.

Listokin, Yair, 2011, Taxation and liquidity, *Yale Law Journal*.

Merkley, Jeff and Carl Levin, 2011, The Dodd-Frank Act Restrictions on Proprietary Trading and Conflicts of Interest: New Tools to Address Evolving Threats, *Harvard Journal on Legislation*.

Rogers, B., Blue sky laws, *American Law Reports*, **87**, 42.

Solman, Paul, 2008, Risky Credit Default Swaps Linked to Financial Troubles, PBS Newshour.

Tanselle, Ryan W., 2011, Some secrets do not hurt everyone: the case for additional discount window reform in the United States' Federal Reserve, *Indiana International and Comparative Law Review*, **147**, 169–170.

[87] See www.sifma.org/issues/capital-markets/securitization/housing-finance-and-securitization/overview/ (last visited July 6, 2012).

16

Functional model for financial data and risk information

Janine Forsythe

16.1 Background

This chapter defines a generic, functional business model for a typical financial institution and the way any financial institution creates, utilizes and maintains data related to risk. The model is focused on the types of data required within financial organizations to define and manage risk and on the business processes that create and consume these data. Underlying the model is the premise that financial data and risk information are created, consumed and matured by nearly every function in a financial institution.

Throughout this chapter we will be utilizing the model organization for thinking about the forms of data utilized by financial institutions. The model is meant to be applicable to any type of financial institution including banking, investment management, investment banking, insurance and securities organizations. To illustrate how business processes create and consume data, we will construct a simple model organization consisting of a Financial Holding Company with four consolidating subsidiaries. Each subsidiary is engaged in defined lines of business conducted in limited transaction types. We will assign a few key employees to each subsidiary and will examine the processing architecture and how data flow through the architecture as each transaction is processed. In addition, we will examine, at the highest level, each subsidiary's regulatory and risk environment to illustrate the use of data in the infrastructure layer to monitor and report on business activities. Finally we will conclude with some illustrative ways that the model can be utilized.

16.2 Overview of the functions and organization of a financial institution

As we examine our sample financial institution it is important to have some background on the functions and organization of any financial institution. Further in this chapter we

Handbook of Financial Data and Risk Information, ed. Margarita S. Brose, Mark D. Flood, Dilip Krishna and Bill Nichols. Published by Cambridge University Press.

will use an analogy to the manufacturing process and will compare a financial institution to an automobile assembly line. Let us use that analogy here as well.

Any financial institution conducts basic business processes or functions which fall into an organizational structure that is a common vernacular in the industry: front office, middle office and back office. This arbitrary breakdown is very focused on the core business transactions of the institution – supply and trading of financial products and services. In addition to the front, middle, back office breakdown are the functions required to manage the business. Although these are often set up as services shared across the organization, they may certainly be replicated in the component operating companies. These functions, although not a focus of our modeling discussion, include marketing, legal, human services, real estate and facilities management, public relations, investor relations, executive office functions and corporate accounting.

16.2.1 The front office: functions and processes

In any financial institution the front office is the group of processes and people engaged in facing the market. This will include the sales function and the trade and order execution functions. Staff providing these functions will include relationship managers, who are the primary staff assigned to and facing clients and counterparties. In a banking business this will include wealth managers, corporate bankers and loan officers. In an investment business it will include sales-people, traders, investment bankers and research analysts (whose product is research opinions). In an insurance business, this function will include agents and brokers. Included as well is a range of staff who play supporting roles in the sales and decision processes of the institution. To align to our automobile model, the front office consists of product design, engineering, sales and distribution and research and development.

16.2.2 The middle office: functions and processes

The middle office in a financial institution consists of the functions and staff who support the transaction between the front office, above, and the back office, below. Although much closer in function to the front office, a key point of differentiation is that the middle office rarely has direct contact with the client decision makers. The middle office often "enriches" the trade with key data/information and is responsible for transitioning the trade or activity from the front office's decision making and revenue producing activities to a form that can be handled by the back office "factories." These processes may include risk management and error resolution. In our auto analogy, middle office processes might include logistics, supply chain management and invoice production. Key components of post production also reside within the middle office. These may include certain testing and quality control functions – in the auto analogy, the middle office would own the test track!

16.2.3 The back office: functions and processes

The back office is the assembly line of a financial institution. Sometimes referred to as a factory, this is the group of functions and processes that maintain external relationships with exchanges and markets, process the transaction to "break it down" into its component accounting entries, to push the transaction through its lifecycle of events from inception to expiration/maturity. In the auto analogy, these processes are the mainstream manufacturing processes, the bolting together of the car and the roll-off process. The back office continues to interact with, primarily, the middle office throughout the lifecycle of the trade to manage and reconcile cash flows and product/trade updates.

As we proceed to dive deeper into our model and examine how transactions flow through the organization, we can anchor ourselves at each step in the applications and staff who have ownership of the trade and its related data.

16.3 The model

Our model is constructed with two constraints. First, this model is *generic*. We will describe and discuss data elements and financial information with names and terms that refer to its content or use. Every institution will have a unique and customized naming convention for data and the repositories that hold the data. The goal is to think more generically about the characteristics and use of the data and to have those characteristics, in fact, define the data itself. This generic model does not imply lack of granularity.

Second, the model is *logical* not physical. The specific architecture into which each financial institution organizes its processes, builds applications into an integrated processing structures and routes data though applications and databases is *not* a focus of this model. Processes and applications retrieve data for use or "consumption," enrich, advance or mature the data and return it for further use by other processes. While many high-performance organizations organize and maintain data centrally, the centralization of data is a question of efficiency. The expectation is that this generic, logical model can be applied to any specific institution's process and application architecture regardless of the degree of centralization achieved.

Let us begin by creating our generic financial institution in layers. At the highest level, every financial institution performs five master functions arranged in three key layers (see Figure 16.1).

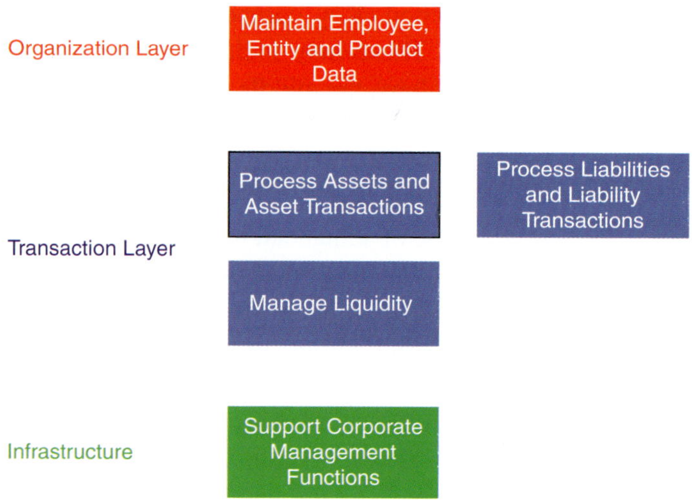

Figure 16.1 Layered model of a financial institution.

The organization layer

Every financial institution must maintain a repository of static information that defines its business. These data are central to the processing of the institution's daily business and transactions, define its organizational structure and the internal and external entities that participate in the organization's business. This is the layer where the basic data about the institution's foundational data are gathered, named, stored and controlled. These data can be organized by type.

Employee data: Define individuals within the institution. Employee data are utilized to provide employee services and corporate functions but also define the employee's role in the organization, entitlements, limits and relationships.

Entity data: Define the institution's internal legal structure and define containers, legal and organizational, that hold the institution's business. This category of data will include legal entities, trading desks, accounts, and books. In addition, for the purposes of our model, entity data will also refer to the external legal entities with which the institution does business. Each entity will be organized into a role in relation to the institution such as Client, Counterparty, Service Provider, and Exchange.

Product data: Organize and define the financial instruments and services that the institution provides to the marketplace or utilizes to generate revenue of any form.

The transaction layer

This layer defines the business activities conducted within the institution. Every business activity will reach into the organization layer "data repositories" to obtain information and the parameters or rules for the activity. At this highest level, transactions are grouped into three types:

Asset transactions: business activities that result in the creation of an institutional asset. Examples include:
- in banking – issuance of a loan,
- in insurance or trading businesses – purchase of a financial instrument,
- in investment management – provision of a client service resulting in a fee receivable.

Liability transaction: business activities resulting in the creation of an institutional liability. Examples include:
- in banking – deposit taking,
- in insurance – issuance of a life policy,
- in a trading business – issuance of a option or trade guarantee.

Liquidity transactions: business activities conducted to manage available cash and liquid assets. These transactions take many forms but can include securities lending, repo/reverse repo transaction, and short-term borrowings. Liquidity transactions, once completed or "booked," generate obligations that can be included in the institution's assets or liabilities.

These transactions are segregated, here, to distinguish the key difference in their purpose – to leverage/maximize the return on existing assets and liabilities. It is important to note, as well, that for risk analysis these transactions are two-tier transactions. For example, a funding transaction collateralized by a bank loan presents two layers of risk to be analyzed. The credit risk of the underlying collateral and the counterparty and credit risk of the trading counterparty on the funding transaction are critical points. As we proceed to assess the risk associated with the activities of a financial institution it is critical to recognize the complexity of these transactions. That complexity requires that we maintain and evaluate the risks associated with these layered transactions.

The infrastructure layer

For the purposes of any institution's day-to-day business activities, we could limit our model to the two layers described above. There are, of course, functions required to support all business activities. These include accounting, human processes, legal, compliance, financial reporting, risk management, regulatory reporting and technology infrastructure. These functions, although critical to the business, are predominantly consumers of financial data and risk information created in the primary course of the firm's business. This **infrastructure layer** is often a consumer of data created in the **organization layer** and the **transaction layer** but is fundamentally different – transaction data are not impacted or changed in this layer. These processes record data or utilize data to make calculations and report information internally and externally but do not conduct financial transactions. Shared functions often create financial obligations during the course of their activities. Two good examples of this concept are the payroll function, usually part of the human services organization, and the accounts payable function, usually part of the accounting organization. In any operating business these functions create liabilities (forward payroll liabilities or forward rent payments, for example). These "payables" are financial obligations critical to the efficient operations of the business. They are not, however, unique to a financial institution. For this analysis we will focus on the financial obligations and risks that are uniquely associated with conducting a financial operation. Of course a complete look at the organization's financial and risk profile would need to include all financial considerations. We will assume, here, that fundamental business operations, taxation and accounting principles common to all businesses are outside our analysis. A simple example will serve to illustrate these concepts.

Acme Financial is a global financial conglomerate with investment banking, investment management, retail and commercial banking operations. Acme also operates an insurance subsidiary providing life insurance and annuity products. To understand the internal organizational entity relationships it is important to understand the concept of a financial holding company. The financial holding company is the regulatory structure designed to provide oversight and transparency in a financial institution and its transactions.

Financial regulators have established the concept of a "holding company" as an organizational view into the relationships between legal entities, or operating entities, within a financial institution's "family." It is not unusual for a financial institution to maintain thousands of entities. Each entity is defined by its legal entity structure (corporation, partnership, special purpose vehicle), its geographical presence and by its relationships to other entity members of the family. A key purpose of this relationship structure is to provide a clear picture of the risk transactions of the institution to the outside world, regulators, trading partners, clients and shareholders.

At the highest level of a large financial institution is the "holding company." Every entity below this level in a large financial institution is subject to all the restrictions of the entities above it. Financial institutions gain protection, task advantage and trading benefits from linking organizations together but must also comply with capital usage restrictions and regulatory guidelines that accrue or consolidate upwards in the organization.

Let us use an example to illustrate this point. Acme Holdings (see below) is a financial holding company. As such, when its regulatory capital is assessed, the assets and liabilities of all the operating entities below it in the organizational chart are considered and regulated. Imagine that Acme were to organize an operating company called Acme Group to "own" Acme Holdings but to organize that as a parent of not only the financial holding company but another entity, say Acme Real Estate. Acme Real Estate could stand alongside Acme Holdings as a "subsidiary" of Acme Group. As such, its assets cannot be considered in calculating the capital position of Acme Holdings but its activities are not regulated as part of the holding company, either. This provides obvious advantages to certain global/diverse businesses in terms of regulatory or reach-of-claim circumstances.

For our modeling example, we will examine only the portions of Acme held within the financial holding company structure. Acme is organized as shown in Figure 16.2.

Each organizational entity within the Acme Holdings corporate chart will have a processing architecture within which its business transactions are conducted. At the beginning of the chapter we discussed the concepts of front, middle and back office. As we begin to examine the conduct of business transactions and how they flow through an organization it is critical that we transition away from the simplistic, flat "organizational model" represented in Figure 16.2 to looking at a functional

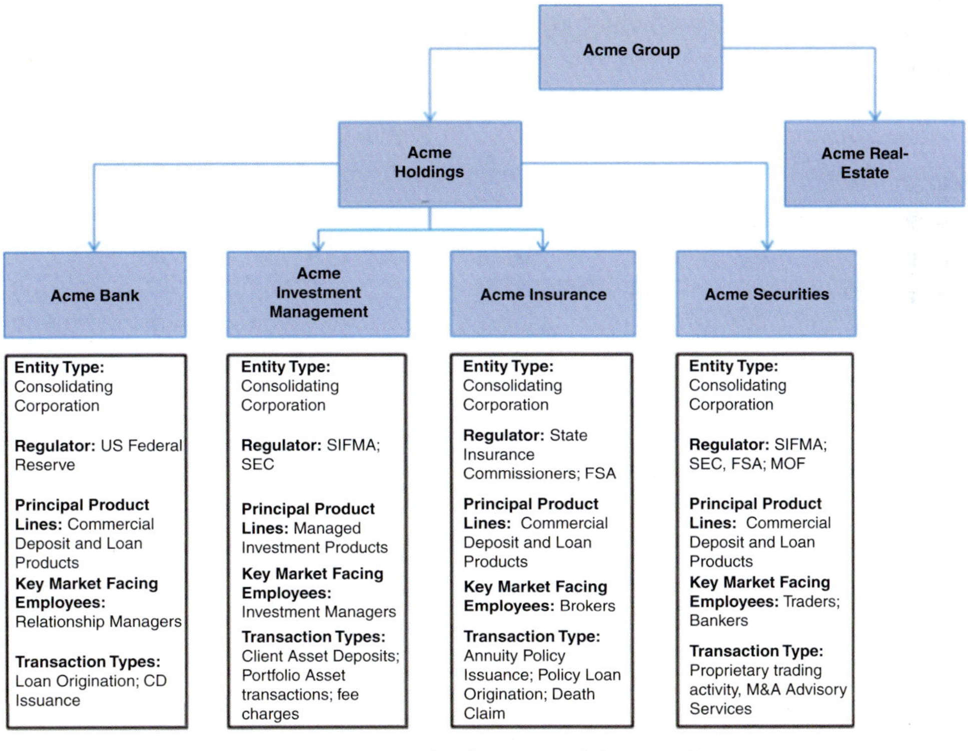

Figure 16.2 Organizational structure of the Acme Group.

operating model for Acme. We will utilize this functional application model and then apply it to each of Acme's subsidiary entities to demonstrate the flow of data and risk information through the architecture. We must understand however that the processing architecture in any actual financial institution will be far more complex and convoluted than in this generic model.

We now consider how business transactions flow through the organization's data processing architecture and functional architecture – the systems and people who process the business of the financial institution. Again it is helpful to think about an automobile manufacturer to provide context and contrast in function. If Acme were an automobile manufacturer it would start its production process by acquiring raw parts or raw materials – steel, leather, tires, preformed components. These materials would come into its manufacturing plant for assembly into an automobile. The assembly process would require many hands and steps along the assembly line to the test track and through the distribution process to the dealer's showroom.

We can use this manufacturing analogy to think about the financial process. Raw materials (cash) and preformed components (market securities) come into the financial institution. Design concepts are applied to shape "products" and the resulting trades are moved along an assembly line to deliver them through a sales organization to clients and trading counterparties. The processes that the front, middle and back office conduct along that financial assembly line consume and create data, just as the automobile assembly line consumes parts and materials to produce a fully formed automobile.

The functional operating model shown in Figure 16.3 is a representation of the financial assembly line that exists for processing financial transactions through each of Acme's manufacturing subsidiaries. The concept of layers introduced earlier is carried forth. The organizational layer owns the reference data about clients, entities, counterparties and products. This layer is also responsible for establishing the processing rules for the organization. It determines data entitlements, trading limits, access to information about business activities. That access and entitlement is most often determined by *who is accessing the data and for what purpose.*

The data of transaction information is, of course, owned within the transaction layer. For purposes of the model we will also assume that Acme maintains a repository of positions. This data can take many forms including a "stock record," position statements, client statements, and confirms. But in all cases, it represents *who is accountable for the performance, return, or profit/loss on each asset or liability.*

To further explore this concept, the "account" is the lowest level "transaction container" within any financial institution. Every transaction is conducted or "booked" into an account. An account is "owned" by a client/trading desk/counterparty. The account holds positions which represent the cumulative effect of all transactions conducted within that account. Buy/deposit/premium payment

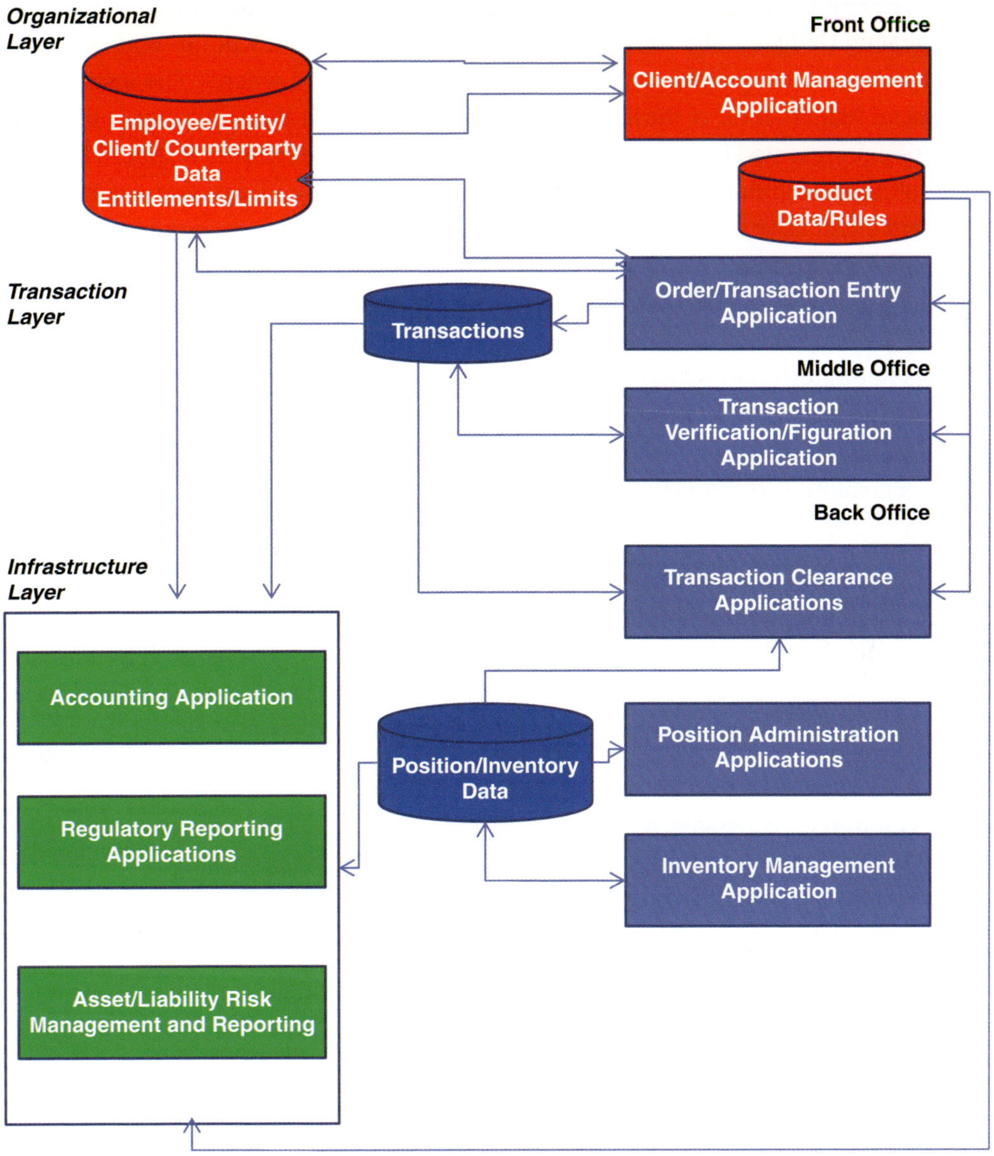

Figure 16.3 Functional model of a financial organization.

transactions (in securities firms, banks and insurance companies respectively) add to positions/balances; sell/withdrawal/loan transactions reduce positions/balances. The transactional data repository (or repositories) is the data record of the financial institution's accounts. Prices are applied to positions in accounts to determine point-in-time monetary equivalent value for each asset/liability.

Transactional data are distinctly different from the institution's financial ledgers, or general ledger. The differences are a matter of viewpoint and purpose. Transactional data provide a view to "who owns what and when." The institution's financial ledgers look at these same positions in terms of the institution's financial position represented by balance sheet and net worth, for example. It should be clear that although a financial institution holds an account of a client, that account will contain client assets/liabilities which are the mirror image when looked at from the financial perspective of the institution. A client asset held by the institution is a liability of the institution when assessed for balance sheet purposes. This transactional data repository is the key source of information for the positions/balances/assets/liabilities upon which all transactional risk is evaluated and measured.

Finally, it can be seen that the infrastructure layer is a consumer of data and a reporter of information, but that information is derived through access to the core data repositories of the institution.

16.4 Case study: transaction flows through the model

With this foundation of architecture, organizational relationships and model institution established, we can begin to step through specific transactions, or activities, and examine the data creation and consumption required for each process. First, let us look at Acme Bank and the key facts we know about this entity:

- **Entity type:** Consolidating corporation
- **Regulator:** US Federal Reserve
- **Principal product lines:** Commercial deposit and loan products
- **Key market facing employees:** Relationship managers
- **Transaction types:** e.g., Loan origination; CD issuance.

Utilizing the model first involves identifying the processing steps that must be made from start to end of the transaction. Our first transaction case study begins when the Acme Bank relationship manager receives a loan application from client company A. At this point the critical information contained in the application is entered into the client data repository of Acme Bank. These data will include all identifying information for the client as well as data required to determine creditworthiness, entity/parent relationships, size currency and pledged collateral descriptions. At this point the client is identified as a potential counterparty, the collateral is valued and any parent/child entity relationships are identified for the client. The loan approval is considered at several levels including the following.

- Is the loan a desirable risk when considered as a stand alone transaction?
- How should it be priced based upon creditworthiness of the borrower entity and the quality of the underlying collateral?

- When the borrower entity is related to other parent and related entities, what does this proposed new exposure do to overall credit exposure to the parent entity?
- What is the impact of the new asset (should the loan be made) to the Acme Bank's overall currency exposure?
- What is the impact of the new asset (should the loan be made) to the lending entity's portfolio aging, geographical exposure?

This analysis represents the pre-issuance considerations. To conduct this evaluation, the institution may also look at current exposures and do "what-if" analyses of the overall asset portfolio if the loan were made. Depending on the size and quality consideration of the loan, specific levels of approval will be required. Those levels are managed in the organizational layer based upon entitlements and employee relationships. At this point a loan approval decision is made. The decision is communicated to the applicant. If declined, the transaction stops. If approved, the money is moved and the booking transactions (assets/liabilities, income/expense) must be made.

Let us trace the narrative above through our model, shown in Figure 16.4. The heavy lines indicate data moving through the organization to facilitate the processes and decisioning required to complete the transaction.

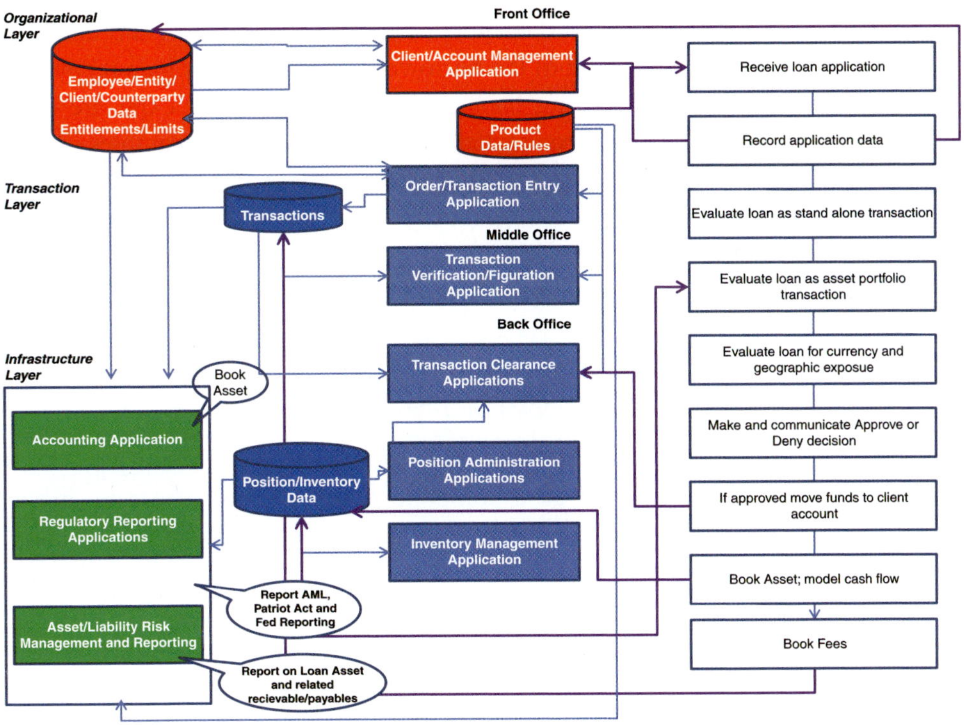

Figure 16.4 Processing a new loan.

Step 1: receive and record loan application

In Step 1, the prospective client provides information to the financial institution through verbal, written or electronic means. The data are recorded and stored. This is a very basic example of data creation. Product rules (i.e., features of the specific type of loan required – term, rate, repayment terms, collateral, etc.) are reference data to this step. This is a simple example of data consumption. The process cannot continue to completion until product rules are referenced and utilized. Once complete, the data are stored within the institution's architecture for reference by other processes. For example, that data will be referenced to provide electronic notification of approval or decline. The data may also be used for broad enterprise-wide processes beyond the boundaries of this specific transaction such as financial reporting as well as measuring and monitoring the overall risk of the bank.

Step 2: evaluate loan as a stand-alone transaction

In this step, the loan is considered on its own merits. Creditworthiness of the borrower, past experience, value and condition of collateral, and strategic fit for the transaction are all components of the analysis. It should be clear that significant amounts of external and internal data will be required to support the decision. Examples include the credit rating of the borrower (external), prior loan repayment experience (internal), compensating balances (internal), existing exposure (both external and internal). Data are also created during this process – examples include internal risk ratings for credit risk (e.g., Probability of Default and Loss Given Default) as well other kinds of risk ratings (such as those associated with anti-money laundering regulations). The data obtained, consumed and created during this process become part of the retained reference data for both the trade and the parties involved.

Step 3: evaluate loan in a "portfolio context"

The loan, if approved, will become part of the institution's asset/liability portfolio. As such, it must be considered as a component part of that whole to answer questions such as: Is the loan consistent with the current strategy of the bank? Does it overweight the portfolio for a specific economic, currency, geographical or credit risk?

This process, as well, is a significant consumer of data and rarely results in data creation. The data used in this case are primarily internal to the organization. A critical point to be made here is that portfolio risk analysis is a dynamic process and is sometimes constrained by product, regulatory or organizational barriers. These barriers may be determined by internal policies or external regulations. For example, let us consider the situation where this loan is being made to a client who is also an investment banking client of Acme Securities. There are regulatory restrictions on the data that can be shared between organizational entities about pending transactions. In some cases material and non-public information contained within one business unit will not be shared across entity lines. It is critical that the physical implementations and data sharing constructed within an organization recognize and respect these boundaries. Once a loan is evaluated in these two steps a decision is made and communicated to the prospective borrower in the next step.

Step 4: make and communicate approve or deny decision

In Step 5, reference data including contact information are accessed and the decision is communicated. Regulatory disclosures may also be required. This information is governed by the product rules maintained in the product data/rules repository contained in the organizational layer.

Step 5: move funds to client's account

In Step 5, the loan transaction is processed on the institution's trade processing architecture. Transactions are recorded to create the correct and balanced asset and liability entries. Income and expense are recorded and cash is moved. Step 5 is a data intensive step since it creates all the bookings used to manage and derive risk data and transaction information for the life of the loan. Asset and liability management, regulatory reporting, loan administration, client communication, sales management/CRM and financial accounting applications are all consumers of the loan information recorded in this step and in Step 1 for the remainder of the life of the loan.

We have discussed the concept of an account above. It is important to recognize that client and counterparty accounts may be segregated by product/activity type and are certainly separate where the client or counterparty has a relationship with Acme in different legal entities. A key component of effective risk analysis is for the financial institution to assess risk (market, currency, political, client, counterparty, credit, etc.) across accounts, entities and organizational structure. This is the true challenge for most financial institutions. Gathering, maintaining, consolidating and assessing risk across differing transactions is crucial. It is precisely why a generic view based upon a functional model can provide a cross-institution view and understanding of relationships and their intrinsic risks.

Steps 6 and 7: model cash flow and book fees

At the outset of a loan booking, as with many scheduled cash flow transactions like derivatives, the transaction's lifetime cash flows are projected. In the case of our simple loan transaction, these will include principal and interest payments and any periodic fees expected through the life of the loan. Most institutions will record these and "age" them as part of their financial reporting process. The value of this process, apart from the obvious need to do so to meet accounting rules and regulatory/capital requirements, is that the receivable risk is greater the further out in time the cash or fee income is expected.

Now that we have completed a narrative and a pictorial walk through of a sample transaction, let us analyze the process we went through to deconstruct the transaction and use the model to anchor that analysis. First, we identified a single transaction type. Although it seems intuitive and rather simple, this is a critical art form in the use of functional models. Often, modelers will mis-define transactions. To illustrate this concept let us remember *why* we use the functional model. It is a tool that allows us to identify similarities and differences in business processes. Often organizations will undertake this analysis governed by the unique organizational structure.

Within Acme Bank, undoubtedly loans will be made in various organizations:

- Auto finance division
- Home finance division
- Commercial lending
- Corporate banking
- Personal wealth management
- Others.

Our model can be used in these cases to deconstruct the process and document the physical architecture and data model. The physical implementation, data repositories, processing applications may be different for each line of business but the process and model will be identical. This is one of the key strengths of using a functional model to analyze a business process. *We are looking for similarities to construct the models and differences to identify distinctions in requirements.*

Transactions in other business lines: work assignment for interested readers

We have deconstructed the "loan origination" transaction for the case study above. To recap, the steps we took to conduct the analysis were:

- Define the business transaction
- Itemize the processing steps required to complete the transaction
- Align the processing steps to the functional model
- Document the data consumption and creation events for each step in the process.

In practical usage, our next steps would be to identify the physical IT components, data repositories and applications that would be impacted at each step of the process.

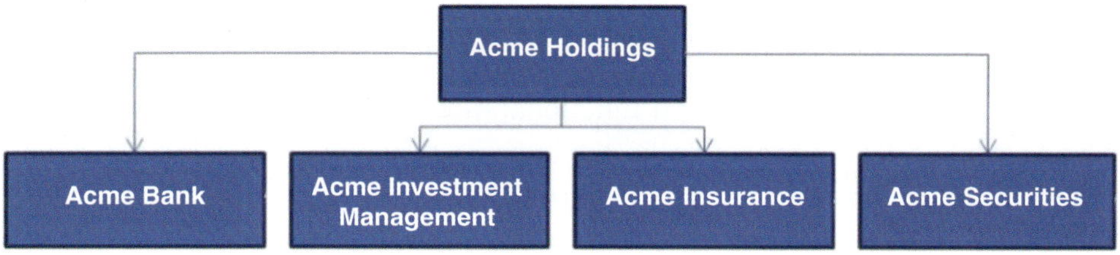

16.5 Practicing with the model

We have used Acme Bank for our case study. The reader can now identify and deconstruct a specific transaction type for each remaining entity. Practicing this skill of process deconstruction and alignment to a functional model is critical to using functional models for analysis and is only learned through repeated trial and error.

Suggested transactions are:

- **Acme Investment Management:** Buy securities in a managed portfolio. This is the process by which an Investment Manager constantly manages a pool of investments within specified rules. A good example is a mutual fund that invests only in tax exempt municipal bonds. Investment decisions must be made with the agreed constraints of risk/product type.
- **Acme Insurance:** Originate a life insurance policy. This is the process, often referred to as underwriting, whereby the front office "underwriter" works with an "agent" or "broker" to assess and price the risk acceptance of insuring a life. Key components of the analysis will include gathering demographic data, age, personal habits, occupational risks, health history
- **Acme Securities:** Execute an exchange-traded equity sale. This is the highly automated process by which a securities firm accepts an "order" from a client to sell a security on a public exchange. The securities firm takes an instruction to sell, and records whether there are limits within which the client will agree to sell. The institution then transmits that order to an exchange which matches buyers with sellers and reports back the agreed price. The financial institution then records that "trade" with price, counterparty and settlement date and fees in the client's account. On settlement date, cash flows and position entries are processed to remove the security from the client account and to credit that client with the net proceeds of the trade. Cash flow to the exchange/counterparty and to the revenue accounts of the firm is also made.

16.6 Practical uses of functional models

There are several invaluable uses for carefully constructed functional business models beyond the obvious process and flow documentation value.

1. **Architectural impact analysis:** Often we make changes to technical architectures without thought to the myriad processes that consume or share data. If an organization has a well-defined functional model of their business with a moderate level of detail for data flows, the impact of changing/replacing any component of the architecture is easily evaluated.
2. **Disaster recovery and business continuity planning:** Our model is very useful in understanding the impact of failure or any single component. The concept of "downstream data starvation" by which a failure of function in any one component renders further processes unable to complete, or worse, wrong, is easily identifiable if you can identify with surety the consumers of data further along the processing path from the failed process/application/hardware.
3. **Strategic planning:** Every modern organization is faced with questions of vendor dependency, application replacement, organizational spin-off, severability and isolation. It is important to remember that any institution can "exit a business" in a spectrum of ways. The business can be sold as a pool of assets and liabilities with no infrastructure. The business could be isolated as a stand-alone legal entity and perhaps subject to IPO. The functional business model is invaluable in identifying the what-if scenarios implied by these business critical events. With a well-constructed functional model, any organization can quickly identify alternatives for supporting every required function and data element in any architecture, outsourced solution or set of replacement applications.
4. **Living will analysis:** Under recent legislation in many jurisdictions, financial organizations are required to present a comprehensive plan for dissolution of the organization or for spin-off/divestiture of certain non-core businesses. The functional model is invaluable in conducting this planning. The planner can analyze, for each product line or entity, the functions performed and the application architecture required to support each function. It can then be determined, under selected scenarios for divestiture, how each function could be supported, outsourced or otherwise adapted to the stand-alone environment

16.7 Conclusion

In summary, we have shown that every financial institution can be broken down into an organizational chart of its entity relationships. We have seen the parent/child and sibling relationships between operating entities and we have shown that risk relationships may be established within and amongst these organization entities. Effective risk management requires a cross-entity view of all aspects of risk.

We have reviewed the high level groupings of functional responsibilities in any financial institution into front, middle and back office. We have aligned these groupings to the processes of an auto manufacturer. This analogy is helpful as the physical production of a car from raw steel and prefabricated parts is more easily visualized when compared with the esoteric concepts of financial instrument production.

We then constructed a functional process model that allows us to take *any* financial transaction and flow it through the processes and applications that support each process. This exercise, at the heart of our analysis, can be replicated by the reader for every type of financial transaction regardless of type of activity. It provides a valuable context for deconstruction of the processing requirements and is a container or anchor for the further examination of the specifics of the data and information created and consumed as these processes are conducted repeatedly in a financial institution.

17

Financial institutions' data requirements

Dianne Buonincontri and Nicholas Robson

17.1 Summary

Data are at the very heart of the financial industry both as a critical input to business processes and as an output from them. "Precise data content is the core factor of input into virtually every business process as well as the essential ingredient for the objectives of transparency and regulatory oversight."[1] Driven by regulations, compliance, client reporting, operational efficiency and a host of other imperatives, data management has become a strategic priority for Financial Institutions (FIs). The signing of the Dodd–Frank Wall Street Reform and Consumer Protection Act on July 21, 2010 in the USA and the emerging Basel Committee on Banking Supervision framework for liquidity risk management furthered the challenges FIs face and accelerated the need for strong data management governance across the enterprise. The regulations have created a fundamental shift towards enterprise-wide data where FIs now have to report business fundamentals, like cash flows, across all asset classes, geographies, legal entities and more.

Enterprise data comprises centralized, accurate and transparent datasets that are modeled to be consistent with all uses of the data, in all business processes. It should be based off industry standards to allow accurate communication with all regulators and counterparties. Standardized processes for the creation, enrichment and management of the dataset are the cornerstone of enterprise data. Robust data architecture to support distribution and integration is another pillar.

Enterprise data are often referred to as a "golden copy." It represents a strategic investment, not just because of the scale and significance, but also because enterprise data programs take many years to implement. Data alignment across a

[1] The Building Blocks of Enterprise Data Management, Lessons from the Financial Industry, EDM Council, July 13, 2011.

Handbook of Financial Data and Risk Information, ed. Margarita S. Brose, Mark D. Flood, Dilip Krishna and Bill Nichols. Published by Cambridge University Press.

dynamic business is a significant change from the historical situation. Traditionally, trading desks and business lines have been responsible for sourcing, normalizing and controlling their data needs, which has resulted in the unfortunate side-effect of multiple process and data silos that are the starting point and current reality for many FIs as they progress toward data management maturity.

This chapter covers the key data categories that make up enterprise data. This includes the reference datasets for legal entities, security master and pricing data as well as transactional data, commonly referred to as operational data. Starting with an understanding of each dataset in the context of the business processes that create, enrich and reference the data, the section will then overview the management process for the data categories. Finally, insights into implementation approaches, common problems and best practices will be reviewed.

17.2 Background: data processing

Before we get into the details of what constitutes enterprise data, it is worthwhile understanding how data are consumed within a financial institution. To do this, a generic securities process flow, Figure 17.1[2], is used to put context around the data that are required in advance of a customer placing an order and a financial institution executing and reporting on that order.

As orders are received by the front office through a number of means including electronic, phone, or email, they are matched up and executed using a number of internal applications and external (DTCC, NSCC, OTC, OCC) clearing and settlement services (see Chapter 15). The data required to capture the transaction details include (but are not limited to) customer information, security descriptions, transaction details,

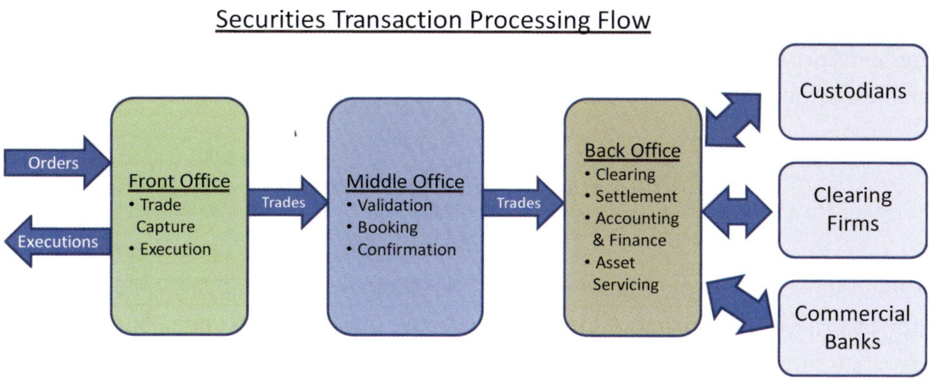

Figure 17.1 Securities transaction processing flow.

[2] Practical topics for IT professionals of Capital Markets – Khader Vali, `www.ksvali.com/securities-trade-life-cycle/`

and processing account information. The data are used to book a trade which flows from the front office to the middle office for further validation and processing.

The middle office which manages the institution's position-keeping is the division that analyzes the market and credit risks that are being taken onto the institution's balance sheet. Middle office staff evaluate the trade data captured to ensure that (1) economic risks are accurately recorded per agreement of commercial terms with the counterparty, (2) they are complete per standardized booking models in the appropriate system and (3) they are processed on time (often within 30 minutes of trade execution). To do this, middle office applications rely on data feeds from both internal and external sources for accurate and up to date information.

The back office is responsible for handling or overseeing the clearing, settlement and asset servicing processes. The process involves checking trades that have been conducted, ensuring that they are not erroneous and transacting the required transfers as illustrated in Figure 17.1. The data required to ensure the transaction can be processed without the need for re-keying or manual intervention, also known as straight-thru processing (STP), rely upon data that include customer name, settlement details, and a host of internal recordkeeping information. Historically, STP solutions were pursued to help financial markets firms move to one-day trade settlement of equity transactions to improve efficiencies and reduce costs as well as to meet the global demand resulting from the explosive growth of online trading. More recently the concepts of STP have also become attractive from the perspective of reducing systemic and operational risk and to improve certainty of settlement and minimize operational costs.[3] When fully realized, accurate and timely data contribute to financial institutions achieving straight through processing, reduced settlement and counterparty risks, and overall lowering of operating costs.

What are enterprise data and how can they be precisely designed for easy integration and effective retrieval and management? The following sections enumerate the core enterprise data components that are at the epicenter of data infrastructure transformations relating to systemic risk, regulatory and compliance discussions.

17.3 Legal entity reference data

17.3.1 Introduction

To accurately clear and settle transactions, financial institutions rely on data inherent to the transaction. Transaction details, such as descriptive data and accurate counterparty data, are required to ensure operational efficiencies. A critical component of the transaction detail is the Legal Entity (LE) reference data that describe all parties involved with the transaction. To efficiently process transaction flows, both

[3] See `http://en.wikipedia.org/wiki/Straight-through_processing`

internally and externally, financial institutions have come to realize that standardization of reusable or enterprise data, and the processes that reference these data, needs to be a priority. Data that are captured, validated and consistently used throughout the organization have proven to create operational efficiencies, reduced risks and are the essential ingredients for the objectives of transparency and regulatory oversight. This section defines what constitutes legal entity data, how they are created, enhanced and maintained, and draws compelling arguments for standardization of processes and data around a central data repository.

17.3.2 Definition of legal entity

Legal entities are counterparties with whom one negotiates agreements. The term counterparty can refer to either party or both, depending on the context. Any legal entity can be considered a counterparty. Examples of counterparties are partnerships, corporations, banks, government bodies, special purpose vehicles, funds (i.e., investment companies), and other entities of legal standing that have the capacity to negotiate contracts, assume financial obligations, and pay off debts. In contrast with natural persons, legal entities in the USA for example, are chartered by the states where they are organized. A corporation is a person in the eyes of the law, and it is responsible for its actions. For example, it can be sued if it fails to live up to its agreements.

Within financial institutions, the term *market counterparty* is used to refer to governments, national banks, national monetary authorities and international monetary organizations such as the World Bank Group, that act as the ultimate guarantor for loans and indemnities. The term may also be applied, in a more general sense, to companies acting in this role.

A counterparty can also be a broker, investment bank, and other securities dealer that serves as the contracting party when completing "over the counter" securities transactions. The term is generally used in this context in relation to "counterparty risk," which is the risk of monetary loss a firm may be exposed to if the counterparty to an over the counter securities trade encounters difficulty meeting its obligations under the terms of the transaction.

A legal entity is a single unit which the FI typically assigns a set of processing identifiers. These identifiers aid the FI in assigning counterparty credit limits, assessing counterparty risks, settling counterparty transactions and complying with regulatory reporting requirements.

> The ability of a financial institution to uniquely and precisely identify, define and link business entities is critical to a wide array of essential business and risk-monitoring processes. For example, business functions such as sales (that is, a holistic view of the client), compliance (for example, 'know your client' requirements), and risk management all rely on unique entity identification. Regulators rely on this degree of precision

as well, as they assess the financial health, systemic risk, antifraud, and other aspects of markets and their participants as part of their regulatory responsibilities.[4]

Legal entities in which jurisdictions or locations transactions are negotiated, executed and settled are subject to regulatory and statutory reporting requirements within those jurisdictions. FIs capture legal entity data attributes electronically in order to identify and appropriately classify the legal entity holdings and transactions. Legal entity hierarchies are assigned by the FI to identify the legal ownership of a counterparty transaction, including the direct parent and ultimate parent ownership. FIs collect and maintain subsidiary, affiliates and joint venture information along with shareholding percentage and inter-relationships to create legal entity hierarchies.

A sample legal entity hierarchy is shown in Figure 17.2.

FIs set counterparty limits at the legal entity parent or ultimate parent level to be spread across any LE in the hierarchy. Knowing the LE hierarchy helps FIs manage counterparty risks, and quickly identify financial exposures. By creating hierarchies, FIs can consolidate counterparty exposure from disparate source systems, support Know Your Client (KYC) and Anti-Money Laundering (AML)[5] functions, and

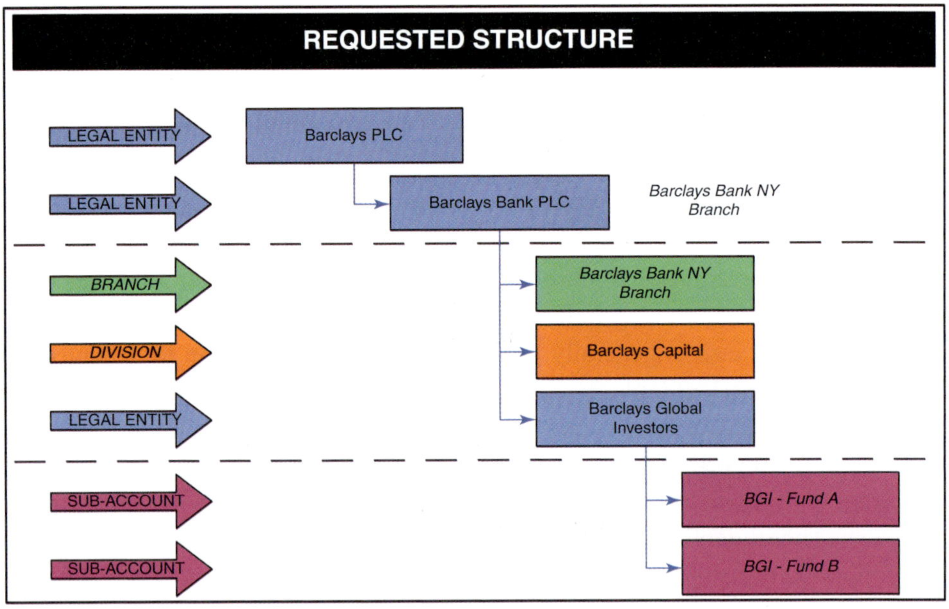

Figure 17.2 Sample legal entity hierarchy.

[4] Creating the Linchpin for Financial Data; The Need for a legal Entity Identifier, December 10, 2010, at `http://ssm.com/abstract=1723298`

[5] Definitions from Investopedia. Know Your Client (KYC), a standard form in the investment industry that ensures investment advisors know detailed information about their clients' risk tolerance, investment knowledge and financial position. Anti-Money Laundering (AML), a set of procedures, laws or regulations designed to stop the

expedite the processing of client on-boarding by leveraging common data elements, Standing Settlement Instructions (SSIs)[6] and internal and external credit ratings.

17.3.3 Client on-boarding

This section will outline the Client On-Boarding (CoB) process which is the key business process that creates legal entity data. Included within CoB are the important subprocesses of Know Your Client (KYC) and Anti-Money Laundering (AML).

The CoB process involves a thorough review of the institution, persons authorized to conduct business of the affiliated entities and internal departments that span the institution from front to back office. It may involve multiple applications required to facilitate, capture and maintain critical operational data that drive the transaction flow from initiation to settlement, through to financial and regulatory reporting.

For the CoB process to be fluid and error free, proper identification of legal entity data is paramount. This is a tall order in most globally complex, multinational financial institutions. It requires that details such as industry classification codes, standard settlement instructions, hierarchies that identify parents and subsidiaries, and many more data items have to be captured. Without proper identification, the flow from one process and application to another can be riddled with errors, causing the breakdown of legal and regulatory obligations such as AML and KYC, to the more onerous back office settlements. Core data, or enterprise data, that are shared across processes and inherent to the transaction, drive the legal and operational process flows of the institution. It is common for much of these data to be part of a paper document that is required for regulatory purposes, although increasingly FIs leverage electronic document technology and automated workflow to capture the many business approvals that are required in the process.

CoB and its subprocess KYC are both processes that involve the definition of corporate policies that meet the compliance requirements that frequently change. The implementation of these policies can involve manual processes and documents, or an application with the workflow capability to implement process management. This often includes an approval process from many different parts of an FI that needs to be captured and reproduced for audit and compliance purposes. A typical full CoB process may involve numerous teams, including front office/sales, KYC operations, client services, compliance, tax, account operations, credit and legal.

practice of generating income through illegal actions. In most cases money launderers hide their actions through a series of steps that make it look like money coming from illegal or unethical sources was earned legitimately.

[6] For recurring transfer of funds or securities, FIs set up standing settlement instructions within their applications that can be used each time a transfer request is made. For example, delivery versus payment, receive versus payment, Clearing Member Trade Agreement (CMTA), GUS for futures and futures options, advisor accounts for their own account settlements, etc. are pre-established and used based on the transaction initiated.

Functions in client on-boarding:

- Hierarchy definition and management
- Policy management
- Know your client
- Process management (workflow)
- Document management
- Data storage and management.

Facilitating the CoB process and ensuring that appropriate documentation and artifacts are obtained begins with the accurate identification and classification of the legal entity that is party to the transaction. Critical operations, such as trade confirmation, settlement and reporting, are reliant on the data captured and verified during the CoB process. Legal entity identification in conjunction with transaction details such as transaction type (outright, i.e., straight buy or sell, repo, or loan), provide essential information which is captured during this crucial process.

The mandate for stronger data governance and industry standardization has emerged in recent years through a plethora of industry and government initiatives. In 2009, the Committee to Establish the National Institute of Finance discussed ways in which improved data and analytics could help identify, minimize or perhaps prevent threats to the global financial system. In 2010, Federal Reserve Governor Daniel Tarullo testified to the banking subcommittee on the importance of creating data standards in finance to support improved transparency and risk analysis. In 2011, the Office of Financial Research (OFR) was established to identify and assess risks and potential treats to the stability of the US financial system with the intent to establish requirements for reporting data on financial contracts that include a standardized way of identifying counterparties.

With standardization, FIs are less prone to classification errors and are able to obtain operational efficiencies by straight through processing. In addition, standardization promotes the comparability of legal entities for measuring and monitoring risks. Specific risks, such as counterparty risk, are also very important. Until such time when a standard legal entity identification platform is established, FIs can meet their execution obligations and comply with compressed clearing and settlement cycles, and support customer compliance and KYC/AML requirements by leveraging a number of existing industry classifications and identifiers. Service providers such as Standard Industry Classification (SIC), the North American Industry Classification System (NAICS), and International Organization for Standardization (ISO) are relied upon to consolidate internal data and to measure counterparty risks.

17.3.4 Types of legal entities

FIs standardize counter parties by legal entity type, a typical example is shown in the table.

Legal entity		
Type	Definition	Example
Corporation	All corporate entities, insurance companies, branch and division entities, limited partnerships, colleges, foundations, other than those included below	Toyota Motor Corporation
Bank	Depository financial institutions, head offices, branch and division entities, savings institutions and credit unions	ABN AMRO N.V.
Non-depository financial institution	Non-depository institutions, fund managers, brokers and holding companies	Fidelity Investment Services Ltd.
Central government body	Country government bodies, departments and embassies (US and non-US)	Ministry of Finance, China
Central bank	Country central or federal reserve banks	Federal Reserve Bank Of New York (US, New York, NY)
US local government	US local government bodies including cities, towns and municipalities	County Of San Diego
US state government	US state bodies of government	State of North Carolina
Other US local	US local government departments, authorities and agencies	Dormitory Authority of the State of New York
Non-US local government	Non-US cities, towns, states, provinces, municipals	Province of Nova Scotia
Non-US government bodies	Non-US government bodies and departments including provinces and towns	Municipal Finance Authority of British Columbia
Supranational organization	Supranational organizations, including international and regional development banks	International Bank for Reconstruction and Development (US, Washington, DC)
FI owned entities	FI majority owned entities	ACME Inc.
Country central governments	Country central governments	Government of the French Republic
Private bank individuals	Private bank individuals who are customers of the FI	J. Smith
Special purpose vehicles	Special purpose vehicles created by the FI for transaction purposes	ACME plc.
Pension funds	Pension, employer, health and welfare funds and their management companies	Pimco BASF Corporation Pension Master Trust Fund
Funds and investment companies	Mutual, hedge, trusts and their management company	Wellington Management Company – Global Libor Alpha Fund

Once classified, additional attributes are collected to expedite operational efficiencies and risk management. At the time of this writing, efforts are underway to create a global standard identification to enable organizations to more effectively measure and manage counterparty exposure, while providing substantial operational efficiencies and customer service improvements to the industry.[7] The attributes collected serve to internally identify a legal entity for credit and legal purposes as well as to link transaction details for hierarchical reporting in measuring counterparty risks. Figure 17.3 depicts mandatory attributes for legal entities. Other legal entity types including branches and subaccounts will have slight variations.

Additional key data attributes used by FIs in their efforts to identify, classify and maintain legal entity data on a going forward basis include but are not limited to the following.

Legal entity attributes	
Attribute name	Definition
Credit identifier	Internal identifier used by credit, legal and regulatory departments for risk assessment and reporting
Legal entity identifier	A unique identifier used for maintaining legal entity data
Tax ID	A federal employer identification number (EIN) sometimes referred to as a tax identification number, a nine-digit code that businesses use to identify themselves for tax reporting, banking, and other purposes
Legal entity type	Internal code that classifies the entity by type, see list above
Legal name	The legal name of the entity under which it conducts its operation
Address	Legal mailing address
Country of domicile	Country where an organization is chartered or has its principal place of business
Customer name	Internal name given to the entity for transaction purposes; this may also be known as a short name
Date added	Date on which the legal entity data were first entered
SWIFT code	Society for Worldwide Interbank Financial Telecommunication code. SWIFT acts as the registration authority for BIC and national IBAN formats providing up-to-date and accurate information relating to transaction processing and routing
BIC	A bank identifier code (BIC) that is used in automated processing. BIC unambiguously identifies a financial institution, or an entity within a financial institution
IBAN	International bank account number (IBAN) is an international standard for identifying bank accounts across national boarders
SEDOL	A stock exchange daily official list (SEDOL) code is one of many codes used to identify UK listed securities

[7] Requirements for a Global Legal Entity Identifier (LEI) Solution, Trade Associations Global LEI Proposal, May 2011.

Ticker symbol	A short abbreviation used to identify publicly traded shares of a particular stock on a particular stock market
Bloomberg ID	A unique proprietary identification code for financial firms used for trade analysis and price discovery as well as settlement and transaction reporting.
DUNS number	The Dun & Bradstreet number is a unique nine-digit identification number for each physical location in which a legal entity does business; it enables customers, suppliers and lenders to find and learn more about a company
CUSIP	Committee on Uniform Security Identification Procedures (CUSIP) is a nine digit alpha-numeric code that provides a unique common language for identifying financial instruments across institutions, exchanges and nation for the purpose of facilitating clearing and settlement of trades
RIC code	The Reuters instrument code (RIC) is a ticker-like code used by Thomson Reuters to identify financial instruments and indices
SIC	The standard industry classification (SIC) classifies establishments by type of activity in which they are primarily engaged
NAICS	The North American industry classification system (NAICS) is an expansion of the SIC and is a six-digit code which classifies an entity by sector and subsector

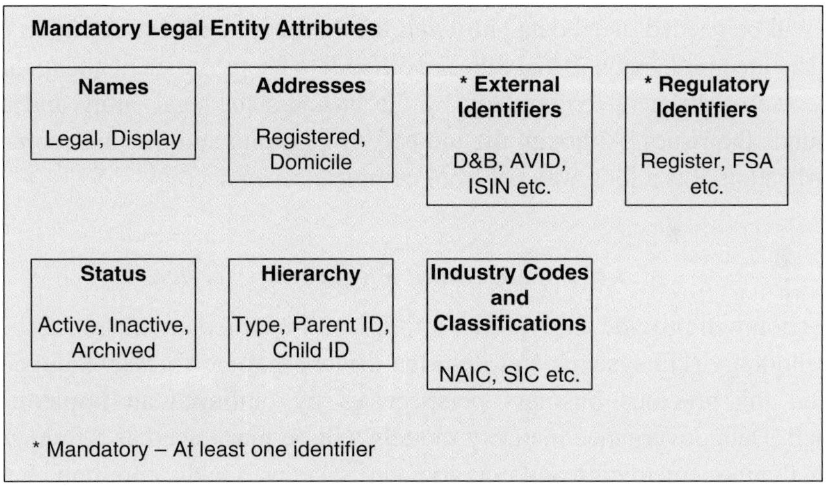

Figure 17.3 Mandatory legal entity attributes.

Other unique attributes specific to operational processes and/or risk management are more likely to be found within specialized applications supporting activities such as trade configuration, order comparisons, confirmations, clearing, settlements, etc. If linked properly with a common and unique identifier, the maintenance of data over time can be quite seamless. In contrast, segmented applications, or those processes that require dual keying, that is re-keying the same information into other applications and thereby increasing the chance of inconsistencies, create havoc for operational staff charged with these tasks. Information relating to problems, conflicts and resolution regarding enterprise master data can be found in subsequent chapters.

17.3.5 Operational and risk reporting

Functions within a financial institution rely on legal entity data to perform a number of operational and reporting processes. Front office functions, for example, are largely centered around trade execution, risk management and analysis and monitoring of real-time market data across the full range of markets and instruments. Middle office functions traditionally bridge the front office/back office divide and focus on allocation and are characterized by post-execution communication with counterparties. Back office functions are typically concentrated on fulfilling clearing and asset servicing, settlement obligations, reconciling data from different sources and producing metrics and reports to interested parties, in finance, audit or risk departments, or to portfolio managers and front office staff who need up-to-date positions and balances.

If the processes that create and enrich LE data are not updating one unique copy of the data, dual keying into more than one system to maintain one or more redundant copies will be needed. Dual data entry and local account database retention of legal entity data are problematic in that they are difficult to keep consistent and maintained, and the use of different terminology by the business for legal entity and account compounds the issues. Although the industry is maturing and moving more toward standardization, it is a long way off from complete acceptance.

17.3.6 Implementing a legal entity service

This section will provide an overview of typical implementations for LE information technology (IT) systems and describe drivers that are forcing change. It will build on the previous business perspectives by outlining an implementation approach. Data governance maturity models will be introduced as a framework to manage the implementation and operations of systems. Lastly, common challenges and solutions will be discussed.

Historically, processes and applications that create, enrich and reference LE data have often been implemented specific to a business function, business vertical or geography. This results in process and data silos that are not part of an enterprise-wide model, with operational inefficiencies and difficulties in creating an integrated holistic view. The causes of this inefficient implementation include the product-centric models in place in many FIs, business fragmentation resulting from mergers and acquisitions, lack of investment in enterprise-wide initiatives, reward structures that are tied to vertical performance that is business unit performance focused rather than enterprise focused, and the fragmented nature of the vendors who provide data and services in this area.

> For many organizations today, data are spread across multiple, complex silos that are isolated from each other. There are scores of redundant copies of data, and the business processes that use the data are just as redundant and tangled. There is little cross-organizational collaboration, with few defined governance and stewardship structures, roles and responsibilities.[8]

The diagram in Figure 17.4 shows a simplified view of this common current state. Each of the business function databases has large amounts of overlapping, redundant data as well as a subset of data for which it is the master source. Quite often the number of silo processes and data stores will be significantly more than shown in the figure. The key issue is that there is no such thing as "Equity LE data" and "FI LE data" as they should be one and the same.

The problems that result from an architecture, as in Figure 17.4, include the following.

- Inefficiencies of manual dual keying and reconciliations to maintain redundant data are common.
- Data chaining, resulting in the need to query all systems to get the complete dataset for any or all legal entities, that is, multiple data systems that capture specific data required by the enterprise for accurate reporting.
- Process errors on time sensitive business activities due to manual entries and cross reference failures.
- Data and process inconsistency that complicates integration with third parties such as settlement agencies.
- High cost of maintaining multiple repositories, technology stacks and reconciliation processes.
- Decentralized business rules create complexity, errors and maintenance problems.
- No single source of LE data to support the increasing number of enterprise-wide regulatory initiatives.

The implications of the above problems are significant. Four of the top seven reasons for failed settlements are linked to flawed counterparty data, according to some sources.[9] Failure to meet regulatory requirements can have a significant cost. In 2010 a major bank was fined $160 million for failing to monitor transactions and

[8] IBM Data Governance Council Maturity Model, IBM, October 2007.
[9] Avox website, `www.avox.info/challenges.shtml`

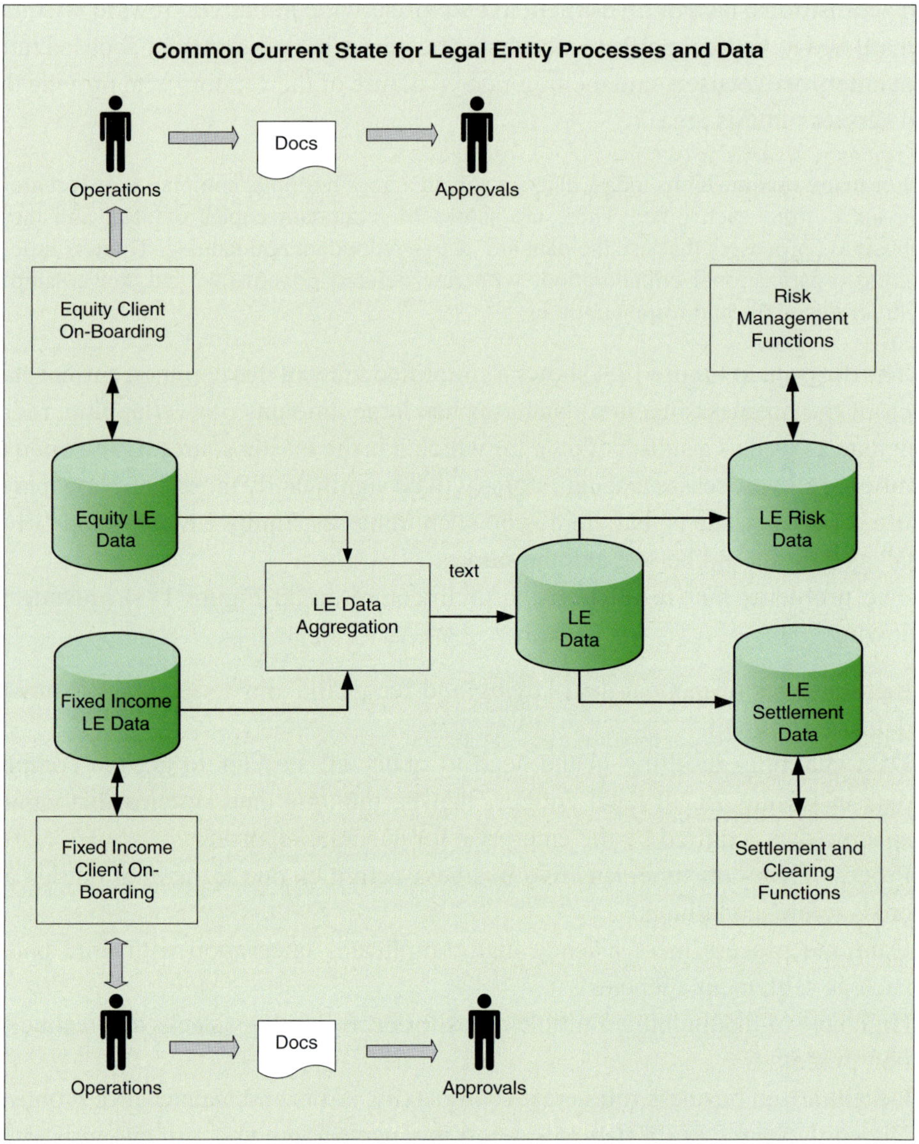

Figure 17.4 Common current state for legal entity processes and data.

maintain effective anti-money laundering involving drug traffickers.[10] Settlement exposure and market risks are increased as an understanding of the client's solvency and reliability is a foundation for risk assessment.

[10] Unlocking Data for Profitable Risk Managed Growth, Deloitte, December 2010.

Transformation of a complex architecture of process and data silos while maintaining current business is a significant challenge that will take several years. It requires a framework with an overall approach to breaking down the problem and creating structure to understand and communicate. This structure often includes implementation phases and processes within each phase that have inputs and outputs. Fortunately there are many data governance and data management models in the public domain that represent years of experience and learning that can be leveraged. The EDM Council[11] will add to this domain of knowledge with the release of the preliminary version of its Data Management Maturity Model, due for release in May 2012. These models can be customized to the specifics of the business requirements and current state if sufficient time and resources are dedicated before implementation starts. A key concept and a fundamental best practice is data governance (see Chapter 27).

> Data governance is a quality control discipline for adding new rigor and discipline to the process of managing, using, improving and protecting organizational information. Effective data governance can enhance the quality, availability and integrity of a company's data by fostering cross-organizational collaboration and structured policy-making. It balances factional silos with organizational interest, directly impacting the four factors any organization cares about most.[12]

In addition to a framework structure with processes, the models introduce other key data principles including data sourcing, data architecture, metadata, data quality, and data security. Explaining these in detail is beyond the scope of this chapter, but any team attempting to transform the systems architecture should be intimately aware of them. Understanding and best practice around each should be incorporated into the enterprise data management strategy. The strategy should also include principles that guide the implementation. Examples of principles that appear in many models and enterprise data strategies include:

- A coherent data architecture where functional systems are decoupled,
- Defining quality by outlining key aspects such as availability, performance and reliability,
- "Golden copy" data based on an integrated, centralized master which contains all data elements within the LE dataset,
- Leverage technical solutions to decrease manual processes such as dual keying data, paper based documentation and associated approvals.

[11] The Enterprise Data Management Council (EDM) is founded and owned by its members and is a non-profit industry association with goals to enable senior officers responsible for content management to share information on business strategies and practical implementation realities associated with achieving enterprise-wide control over data content, and to raise awareness of data management at the senior level.

[12] IBM Data Governance Council Maturity Model, IBM, October 2007.

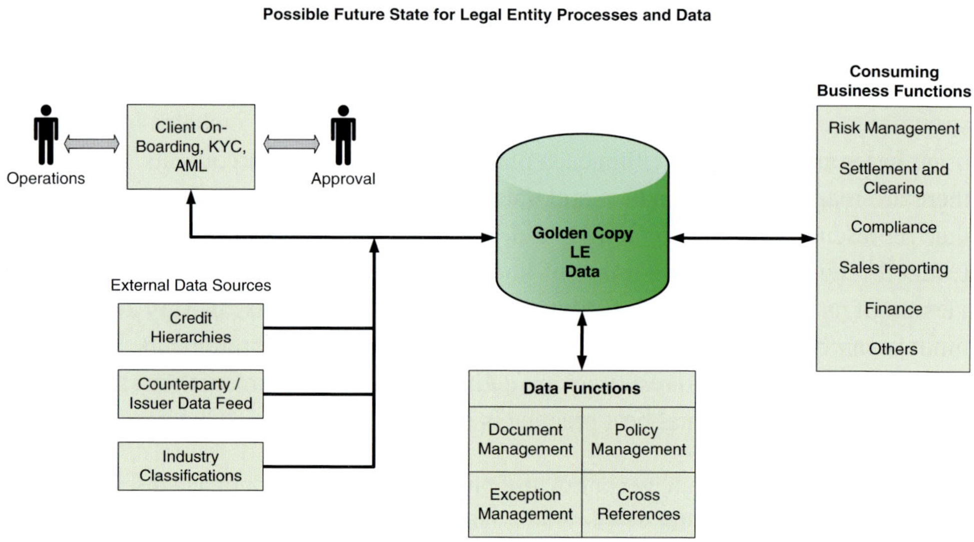

Figure 17.5 Possible future state for legal entity processes and data.

As Figure 17.5 shows, implementation on an enterprise information architecture that supports batch and real-time messaging data integration is an essential backbone of the implementation.

An enterprise data management strategy provides the crucial guidance on where to start. Typically this is an evaluation of the current situation that would identify business drivers and constraints by using a structured analysis process. Workshops, research and structured interviews would be completed along with external analysis of vendor products or services that could be part of the equation.

A phased approach is often recommended so that business value is delivered iteratively and more quickly. Tightly scoped iterations decrease complexity and risk. This phased, iterative approach leads to what many consider an inevitable intermediary state where some of the siloed process and data stores have been removed, but others remain. One major drawback of an intermediary architecture is that it still retains many points of failure and a high degree of complexity.

Vendor solutions exist for parts of the LE solution. The CoB functions that are responsible for the creation of the LE data are an area where the vendor solutions have reached a mature level. Leveraging pre-existing best practice for the CoB process, with preconfigured policies and experienced vendor service teams, can significantly decrease implementation time. Vendor solutions also exist for data sourcing that can be implemented as data feeds into the LE repository. One of the most commonly used LE data services is AVOX. This service "validates, corrects, enriches and maintains business entity reference data. This includes data such as corporate hierarchies, registered address information, industry sector codes and company identifiers."[13] Other data services for cleansing key data such as issuer information and counterparties can significantly improve data quality and decrease manual data management processes.

A key challenge to implementing an LE service is the lack of a universal, standard legal identifier (LEI) that can act as a lynchpin to link reference data across companies, organizations and firms. Although efforts are underway to establish a unique and precise FI identity, without a universal identifier that is consistent across all sectors, a patchwork of cross references has to be maintained that often leads to ambiguities and inconsistencies. To address this problem, the Dodd–Frank Act set a tight timeline for creating the rules for entity identification. At the time of writing, this complex collaborative process appears to be heading towards reusing the AVOX AVID as the basis for the LEI. This would reduce the complexity of implementation, as many large FIs already source and store this identifier. However, changing information architecture from the data sourcing, to coding the business analytics on live, business critical systems, is a pervasive and significant amount of work. Although the final requirements for the LEI are not complete, the time and risk to deliver this requirement can be significantly decreased by upfront investment in an enterprise data strategy. This would create a foundation of understanding, best practice and a delivery framework to make this change.

Other pitfalls to avoid in the implementation of an LE service include:

- Underestimating the time and effort it takes to analyze and understand operational systems,
- Selecting and locking into either a vendor or technology and limiting options in an area where regulatory requirements, vendor capabilities and technology solutions are dynamic,
- Having technology teams drive the program and not creating a governance and accountability model based on business alignment and ownership.

17.4 Security master reference data

17.4.1 Introduction

The rapid pace of financial innovation, particularly the recent explosion of derivatives and securitized products, has exponentially increased the amount of reference information that must be maintained by FIs. Changing market mechanisms, particularly in trade execution, have shifted the business toward highly automated and integrated processes that require enterprise-wide information architecture. In

[13] `www.avox.info`

parallel, technology innovation has facilitated new solutions. Historically, data silos that serve one or a few business processes have created unreliable data, low operational efficiency and high costs. To achieve high volume and low latency processing, financial instrument reference data, also known as security master data, must be pristine, centralized and managed as an enterprise service.

This section defines what security master reference data are, and how they are a critical factor in achieving operational efficiencies. It outlines how they can be implemented to create, modify and manage information. Lastly, it outlines some of the problems, solutions and best practice in implementation and operation.

17.4.2 Definition of security master

An advanced security master is a service that provides timely and accurate data to a diverse range of processes from front to back office within a FI; it includes the following:

- A database of securities and associated attributes such as descriptive data, issuer, ratings and classification data, and information relating to corporate actions;
- An application or Enterprise Data Management (EDM) system to manage the processes related to data events and implement advanced features such as security classification schemas;
- Data architecture to support ingestion and distribution of data.

These three elements require support from both IT and Operations teams who together turn the technology and data into a service for the various processes. Historically, IT and Operations teams within an FI refine data processes and have facilitated not just higher quality service, but also the ability to operate 24/7 and leverage off-shore resources for support functions.

Some firms, especially smaller ones, have not yet moved, or are still in the process of moving, to some of the advanced paradigms listed above. However, the evolution away from localized, data silos is gaining pace, as evidenced by the data management maturity models discussed later in this section.

17.4.3 The sec master's role

The sec master, as it is commonly termed, plays a pivotal role in operational processes such as transaction settlement, borrow and lending facilitation, and bank reconciliation, and allows for the aggregation of financial instruments for operation staffs to determine future cash flows, and trade settlement values. Increasingly, it is also a critical component to support regulatory reporting processes required by Dodd–Frank, Liquidity Risk Management and the like. The sec master reference data govern the economics of a position. For example, they drive the principal and interest information such as coupon, rate reset (for floating rates) and keep track of call/put prices. They also capture the coupon maturity date and include a universal data schema that accommodates all security asset types and associated data attributes. Real-time price data, however, are information that is not kept in a security master. As pricing data generally change daily, they are typically kept in the pricing service as discussed in the next section.

The sec master data are used for tracking compliance buckets including industry class, ratings, and issuer information and are the vehicle for capturing proprietary fields used for internal risk rankings, ratings, scores, and processing flags as well as the identification of security classification type such as cash (equities, fixed income) and non-cash (derivatives, futures, options). Inaccurate, inconsistent or incomplete reference data, held within a company's security master file are widely accepted as the single largest cause of settlement failures.

To initiate a transaction, front office staff requires security instrument data to be available and accurate in order to book and settle a trade. As new security instruments are traded, operations staff usually in middle office establishes the security set-up using a number of external sources available in the marketplace. Providers of security master data are extensive and as no one third party service provider can be relied upon for all instrument types, FIs typically subscribe to multiple data sources to augment and keep sec master data current.

To obtain processing efficiencies, a security master should include the following.

- Essential identification and settlement information, such as identifier, issuer name, exchange, ticker, currency, payment, frequency, coupon rate, maturity date, dated date, settlement date, first payment date and accrual method.
- Detailed terms and conditions data for debt (corporate, government, agency, municipal) and preferred and convertible stock, including call, put and sinking fund schedules, conversion details, coupon reset terms, payment-in-kind and step-up data, and default and credit enhancement data for municipal bonds.
- Tracking of corporate actions with comprehensive, timely information for numerous corporate action types, including mergers, rights offerings, tender offers, name changes, bankruptcies, recapitalizations, cash dividends, stock dividends, mutual fund payments, exchange listing changes and other shareholder notifications.
- Dividends, income payment details, earnings and amounts outstanding.
- Extensive classification and cross-reference information including a wide range of international security identifiers.
- Credit ratings from Moody's, Standard and Poor's and Fitch Ratings.
- US securities class actions information.

- Tax consequence and cost basis information for corporate actions.
- Business entity data, including issuer, parent entity and ultimate parent entity.
- Voting rights and entity debt outstanding information.
- Factor and related data for structured products.

Multiple identifiers for instruments exist and need to be planned for in a security master. Issue identifiers contain security IDs, such as CUSIP and ISIN, and the security master needs to be able to cross reference and match instruments by using the correct identifier. Maintaining cross references is a complex process that will be discussed further in the following section.

Furthermore, the sec master attributes should include issue type, name, identifier, identifier type (context type), issuer name, as-of date, market name, maturity date, and others. Providers of security master data include, but are not limited to the following vendors:

- Bloomberg Back Office Data
- Bloomberg Back Office Data Extended
- Bloomberg Per Security
- Reuters Datascope
- Reuters Datascope Select
- Telekurs Valordata Feed
- Dun & Bradstreet Worldbase
- Markit Red
- Moody's Rating Delivery Service
- Standard and Poor's ISID Plus.

For a large FI the costs associated with licensing all data from the above vendors for all data types discussed in this chapter will be of the magnitude of tens of millions of dollars per year. There are, however, ways to reduce data costs to meet an FI's needs. The three sources of Bloomberg data listed above, for example, all have significantly different price points. Per Security data, which is a real-time query for one or a batch of securities, and includes more complex features such as schedules, is exponentially more expensive than Bloomberg Back Office Data. One central security master combined with a well-managed and cost effective data licensing model can significantly lower operational costs. However, this requires an understanding of data requirements and associated licensing costs that often receives insufficient planning.

17.4.4 Data lifecycle and business processes

Sec master data go through a lifecycle that involves creation, updates and archiving. Typically, the lifecycle for sec master data begins with a daily load from one or more of the data providers listed above. This is followed by intraday updates from these providers. This represents the vast majority of the data creation and updates. However, the operations team that supports the security master may also provide some manual updates to the data.

Perhaps the most important business process that results in updates to the sec master data is corporate actions. Corporate actions are events that are agreed upon by the board of directors and authorized by the shareholders. Such events include stock splits, dividends, mergers and acquisitions, rights issues and spin-offs. These events have the potential to cause a change in the company's share price. Depending on the event, whether it is mandatory (participation by shareholders is mandatory, i.e., stock split), voluntary (shareholders elect to participate, i.e., tender offer), or mandatory with choice (shareholders have several options to choose from, i.e., cash or stock dividend option), there is a level of risk associated with them. When a publicly traded company issues a corporate action, it is initiating a process that brings material change to the company and affects its shareholders.

- A *stock split* is a corporate action in which the company's shares are divided into multiple shares. Although the number of shares outstanding increases by a specific multiple, the total dollar value of the shares remains the same compared to the pre-split amounts, because no real value has been added as a result of the split.
- A *dividend* is a corporate action in which a portion of a company's earnings, decided by the board of directors, goes to a class of its shareholders. The dividend is most often quoted in terms of the dollar amount each share received (dividend per share). It can also be quoted in terms of a percentage of the current market price.
- A *merger* is the combining of two or more companies, generally by offering the stockholders of one company securities in the acquiring company for the surrender of their stock. The decision is usually mutual between both firms.
- An *acquisition* is a corporate action in which a company buys most, if not all, of the target company's ownership stakes in order to assume control of the target firm. Acquisitions are often made as part of a company's growth strategy whereby it is more beneficial to take over an existing firm's operations and market share compared to expanding on its own. Acquisitions are often paid in cash, the acquiring company's stock or a combination of both and often at a premium on the target company's market price in order to entice shareholders to sell.
- *Rights issues* are a corporate action in which a company issues rights to existing shareholders to buy a proportional number of additional shares at a given price,

usually at a discount and within a fixed period of time in order to raise capital. With the issued rights, existing shareholders have the right to buy a specified number of new shares from the firm. Companies usually opt for a rights issue either when they are having problems raising capital through traditional means or to avoid interest charges on loans.
- *Spin-off* is a corporate action in which an independent company is created through the sale or distribution of new shares of an existing business/division of a parent company. A spin-off is a type of divestiture. Companies wishing to "streamline" their operations often sell less productive or unrelated subsidiary businesses as spin-offs. The spun-off companies are expected to be worth more as independent entities than as parts of a larger business.

There are three primary reasons for companies to create a corporate action: (1) return profits to shareholders in the form of cash dividends; (2) influence the share price to make it more attractive to potential investors; (3) corporate restructuring in order to increase profitability.

In a Cutter Associate white paper which reported that "there are many different kinds of corporate actions events – as of early 2010, ISO 15022 (an industry messaging standard used for the transmission of corporate action events) defined between 50 and 60 different corporate action event types." They went on to say "the root of the issue is that there is no industry wide-accepted definitive listing of corporate action event types."[14]

Given that these corporate actions result in fundamental changes to their associated securities, it is important that the update process is accurate and timely. Typically, corporate action updates are supplied by data vendors into the security master, but operations teams perform manual checks to make sure the data are correct. FIs receive data regarding corporate action events from a multitude of sources, directly from the issuer, the issuer's agents, transfer agents, exchanges, custodians, central securities depositories and third party data suppliers. Often the same event will be reported by two or more sources. Inconsistencies can lead to differences in how the event is classified and consequently, significant financial losses to the institution and or their clients.

Another key process that affects the lifecycle of the sec master data is the "end dating" of matured securities. Securities such as bonds have a maturity date that represents the date at which it can no longer be traded. As such, the sec master data have to be updated to represent this and the data archived so that they can be referenced if needed, but are no longer distributed.

[14] Corporate Action Processing: Phases and Functionalities, March 24, 2010, ID:G00174720, Mary Knox.

17.4.5 Implementing a security master

A security master can be implemented either as a proprietary build, or as the installation of a vendor product. A product implementation allows a firm to leverage vendor expertise, existing functionality and ongoing upgrades for a price typically lower than the cost to build. Vendor products can be a very attractive proposition if there is a good alignment between requirements and vendor functionality. A procurement process with a disciplined Request for Proposal (RFP) process and Proof of Concept (POC) are essential. The downsides of less flexibility and possible constraints of a product implementation should be assessed carefully.

The following section will outline the basic components of a security master and how data are created, modified and managed. The following high level components are common to most proprietary software developments and products (see Figure 17.6).

- **Data Providers** – A security master will have one or many third party data providers. Often the data provider will deliver data as an end of day file, intraday file and real-time messages.
- **Golden Copy Database** – Central to the storage, retrieval and data integrity of the security master is an enterprise quality database. This requires the integrity of a well-modeled and normalized transactional system, with the data volume processing speed of a data warehouse. Finding the balance between these two requires high levels of database skill and experience. A central part of the data processing is the mapping from one data representation to another, for example from the data provider's file layout into the relational system. This is then followed by another mapping from the database to the message format for publishing.
- **Data Management GUI** – There are many data management processes that are required to be performed by the operations team who support the service. A GUI application that encapsulates these processes is therefore essential. Example management processes include end-dating matured securities so that they are managed according to this date driven characteristic. Securities creation capabilities are required for securities that need to be active and tradable quicker than they can be received from a data provider.
- **Load Scheduler** – Intelligent file and message processing with the capabilities to parallel load large data volumes, manage exceptions and contention with publishing is a core capability. Exception reports that are customizable based on error type and severity facilitate quick resolution of data loading issues.
- **Publishing Service** – The distribution of securities via files or messages is referred to as publishing. Monitoring tools to report on delays and associated data latency

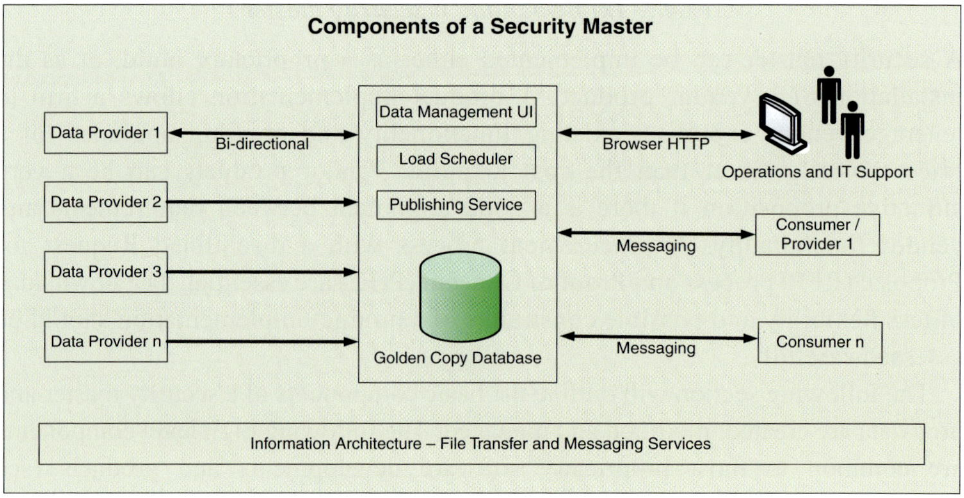

Figure 17.6 Components of a security master.

issues are an important part in what is often the slowest area of the information flow of the sec master.

- **Operations Support Team** – A specialist support team is vital to the operation of the service and the execution of data management tasks, data auditing and data cleansing.
- **IT Support Team** – Performs both monitoring and support as well as ongoing enhancement efforts for initiatives such as adding consumers, adding products and additional data providers.
- **Internal Consumers and Suppliers** – Consumers receive data by processing files as they are sent, listening to messages on topics they subscribe to and initiating message based requests. Typically a consumer will have a local store of security information, but most often this would be a read only copy of the security master. Front office systems can be the exception. Here, the flexibility and speed needed demands they have the capability to create securities. In this instance the information flow to the security master would be bi-directional. Often, the consumers will only be interested in and receive a subset of the security master. The definition of this subset is controlled with a Securities-of-Interest list per consumer and associated functionality that references this list.
- **Information Architecture** – To support the file transfer and messaging, an enterprise quality technology foundation is required. Real-time messaging requires more than just an investment in technology components such as a messaging bus, message queues and adaptors – it requires components such as the establishment of messaging standards and a corporate XML schema.

With the above components in place, the operation of the sec master is driven by incoming data events, monitoring and management activities from the support teams, and data queries from consuming applications.

A common data management feature in a sec master is chronologically recording the historical changes to all data attributes. An example of this is the capture of credit ratings data from a rating agency such as Standard and Poor's or Moody's. Each rating has a start date to display when the rating became effective and – in the case that a new version (i.e., new rating value) of the same rating has been received – an end date to display when this rating was replaced. The change history is kept for the life of the record. The logic for determining whether to update the existing record or end-date and create a new record is controlled within the business rules. This time stamp functionality can handle current, retrospective as well as future dated changes.

FIs look for a consolidated data store containing data from all data sources. The consolidated data store, or golden copy as it is commonly referred to, merges overlapping data from different vendors into a single record per instrument. To create a single record a hierarchy, typically by asset class, is specified to rank the data sources. This ranking is based on the staff's evaluation of different vendors' competence, and changes over time. For example, the hierarchy for fixed income asset class may rank Bloomberg ahead of Reuters for issue descriptions and Reuters ahead of Bloomberg for issue market capitalization.

To ensure the accuracy of the security master data, most large FIs staff an operational team dedicated to maintaining the security master. Security master operations usually operate on a 24/7 timeframe. Primary operational sites span the globe and are established to service the majority of trades where efficient executions can only occur by maintaining a local presence and leveraging local expertise. Such sites or "centers of excellence" include the USA, UK, Japan, Hong Kong, and Brazil. These sites ensure that local area expertise is available to set up new products and resolve issues through local market contacts as issues arise. Typically, the sites operate from one consolidated procedures manual and operating system, and by passing the operations on as the local day draws to a close.

17.4.6 Problems, solutions and best practice

The downside of a centralized sec master supporting a wide variety of business processes is complexity, and the potential for a single point of failure. This section will start by highlighting a few of the common problems. It will finish by outlining some best practices and methodologies that are available to support the development and operation of a security master.

Organizational ownership of the security master has historically been with the IT department, but has gradually changed to a cooperative ownership with the operations team. "For the most part, data management programs continue to move out of technology departments into an integrated IT and operations function."[15] Both teams provide essential roles in the operation and quality of the service and clear alignment and role definition is important.

A key problem that has yet to be solved is the lack of one universal standard identifier for instruments that guarantees uniqueness. Although the securities industry is littered with identifiers such as CUSIP, ISIN and Bloomberg Unique ID, not one of them provides uniqueness across all securities. A CUSIP is unique for most fixed income securities, but not for equities where the market identifier and currency are required in addition to the equity identifier; an example is IBM trading on the London Stock Exchange (IBM.L) versus IBM trading on the New York Stock Exchange (IBM.N). The Bloomberg Unique ID is unique, but for securities not covered by Bloomberg this solution does not work. When building large data stores this is a fundamental problem and often requires complex look-up logic to check the asset type and based on this determine which identifier to use and any additional attributes to reference. Currently this solution is not elegant or easy to implement.

Mapping errors from one format to another are the most common problem in security masters. In part this is because the instruments are evolving and new attributes are being added and these are not always consistent between data vendors. Any sec master database schema and corporate XML schema has to be adapted to handle this, as do all the downstream consuming applications.

One of the most common problems is a human resource knowledge gap due to the highly complicated and specialized nature of sec masters, and the many business processes they support. The natural polarization between technologists and business oriented IT staff is a problem in the world of reference data. Finding resources that have the ability and experience to bridge these worlds is not easy, and a thorough program of cross training and resource management is essential to avoid high costs of hiring consultants to fill the gap, or worse still, the potential to lose control of this strategic and central asset.

At a strategic level, any EDM program requires an effective governance model which can be problematic. Because of the difficulties associated with providing a clear business case on the benefits of data management, the implementation of an effective EDM strategy can be challenging. The core of the challenge is due to the

[15] EDM Council's Verdict: Making Progress But No Cigar for Enterprise Data Management, April 2012, www.issmag.com

fact that data quality in itself has little intrinsic value, rather it is an enabler of other processes. The true benefits of an effective data management are systematic and intertwined with other processes. This makes it difficult for data management professionals to justify the budget required to enhance downstream applications to create upstream improvements.

Solutions, best practices and methodologies to help with the many data management issues and complexities are widely available, but require time and experience to utilize fully. The one key organization is the Enterprise Data Management (EDM) Council which has a focus on resolving key issues such as standardized terms and definitions to create a common language. Other areas of focus for the EDM include the Data Management Maturity Model, Standards, Data Quality and Industry Benchmarking. The goal of these efforts is to standardize the terms and definitions of all reference data attributes that are stored in the master files of financial institutions and that are passed among supply chain partners. Creating precise nomenclature translates into a common language between systems and sources, reduces the cost of doing business, and promotes confidence in data among business users. Data Management Maturity Model (DMM), launched by the EDM and currently in the content model phase of development, is the essential reference framework that provides a roadmap for program implementation. This structure provides a definition of the key components at the specific business process level for assessment and adoption of best practice. Below are some of the most important and general best practices, but all companies should take time to review the models that are appropriate.

- **Requirements** – A disciplined and thorough approach to understanding organizational needs and building them into your plans from design and delivery to Service Level Agreement (SLA) definition.
- **Procedures** – Operating policies and procedures should be documented and include change management and key areas of data management such as security, normalization and cleansing.
- **Technology Foundation** – A comprehensive data model, robust information architecture, established technology partners and proven technologies are important cornerstones.

17.5 Pricing reference data

17.5.1 Introduction

Pricing data are one of the key elements required by FIs to perform asset valuation and support operations functions and analytics. Regulators place a significant importance on the valuations functions that use these data to determine the price of an asset, because errors will affect many key functions such as daily financial positions and profit and loss. Mis-pricing of assets has been well documented as a key part of a number of high profile failures and frauds. FIs have to create stringent and transparent pricing policies and be able to show methods used to reach a price. These valuations functions often leverage many different pricing data sources based on the asset in question, and use sophisticated modeling techniques to determine the price. Pricing data are the primary input to the valuations function, but not the function itself. The valuations function is the application that performs the calculations and maintains the business rules. Various independent sources are required to ensure the integrity of the valuation. Therefore, an enterprise quality pricing data service is required to support this and other functions that consume pricing data.

Fortunately, and with the exception of model driven pricing, the lifecycle of pricing data is not as complex as that of legal entity or security master data. Typically, data are created from various internal and external sources and then distributed. In contrast to KYC for LE data and corporate actions for sec master, there are few processes that update and maintain the pricing data.

This section describes key price data elements, how they are used, and establishes a framework to determine the sources by asset class that are required to properly support FIs business objectives. Lastly, it will outline the fundamentals of implementing a pricing data service, challenges and best practice solutions.

17.5.2 Categories of required pricing data

Pricing data required to support global markets analysis include real-time, clean "execution" prices that are firm and non-negotiable for operations and customer trading, intraday, end of day, benchmark or evaluated prices. End of day and indicative prices found in a security master are several hours old and therefore would only be used for processes where this level of data latency is not an issue. Price attributes include not only price type, but also price source, data provider, quote type and pricing method, and manually entered valuations.

Pricing data can come from many sources and be based on a number of algorithms. Prices can be exchange based, based on spreads, or determined via complex formulations. Critical aspects in determining specific data needs are the ability to identify sources and maintain price attribution to source, along with the ability to provide the selected prices depending on the business requirements. Data sources need to be integrated and compared based on rules determined by the pricing policies. Pricing outputs are then integrated with various other functions with an FI for consumption and use.

17.5.3 Use of pricing data

Pricing data are used by analytical staff to perform financial market analysis and to monitor global markets for arbitrage and other economic opportunities. End of day price data across an array of financial instruments provide both breadth and depth of market analysis to predict daily price movements, as well as to identify mid-term and long-term trends. Trading desks reference end of day prices, but typically set real-time prices for execution on market conditions and trends. The following represents some common use cases for price data as required by FIs analytical and operational support staff. The number of functions within the list, as well as the increased scrutiny by the regulators on these functions, is the primary driver for investment in enterprise quality pricing data services.

- The valuation function within an FI is a key user of pricing data. Outputs from valuations feed into both risk management and finance functions to create risk calculations such as VaR, daily financial positions, profit and loss, performance, attribution and more.
- Foreign Exchange staff responsible for monitoring and analyzing foreign exchange markets and liquidity, as well as identifing global financial market developments, utilize near real-time FX Rates and FX Options pricing data.
- Investment managers calculating the firm's foreign and domestic reserves invested in US dollar, sterling, euro, yen, and other currencies, utilize real-time market prices and/or evaluated prices to evaluate foreign securities and foreign non-marketable securities (i.e., time deposits).
- Operational staff performing various back office functions for bank transactions: settlement activities, monitoring collateral balances for repo operations, collateralizing securities lending and borrowing transactions loans, confirming clearing information for securities, and other authorizing procedures. Staff performing these activities for proprietary and customer accounts, must have clean "execution" prices that are reliable and that are centrally managed to ensure accuracy and completeness throughout the organization.
- Operations staff performing repurchase agreement (Repo) market analysis and repo operation execution also perform liquidity analysis and rely on real-time price data for these activities.
- Operations staff performing foreign and domestic cash projections and validating prior day's forecasts rely on historical price data.
- Analytics staff managing and monitoring portfolio holdings in terms of allocation and performance require real-time pricing in asset classes including equities, fixed income, commodities, derivatives, etc.

- Staff performing operational activities, including settlements and securities lending and borrowing operations, must have access to real-time clean "executed" price data to ensure settlements are facilitated.

17.5.4 Pricing data sources

FIs can keep their market price data current by subscribing to a number of service providers including, but not limited to, Bloomberg, Reuters, Dealer websites, MARKIT just to name a few. Defined by business users once vendor analysis is complete, data sources are then prioritized in a hierarchical order to be assigned based on asset class rules. A large FI will have dozens of overlapping data providers and commonly would use a handful of sources, assigned by asset class, to set a price. Internal or "house prices" will often also be a source with feeds from front office systems.

There are a significant number of vendors that are available in the marketplace, those that specialize in one asset class and those that provide broader coverage. Below is a sampling of those sources.

- **BG Cantor Market Data** supplies real-time, tradable, indicative, end of day, and historical fixed income and derivatives data. Their services are used for price discovery, analysis, risk management, portfolio monitoring and valuation, and research. Coverage includes fixed income, foreign exchange and foreign exchange options, credit derivatives, and structured products.
- **GFI** maintains brokerage operations in credit, equity, FX, energy, commodities and emerging markets, and is an information hub for cash and derivative products. GFI provides the capital markets with data that refer to actual executable and executed market prices and trade information, reflecting market sentiment rather than indications gleaned through consensus pricing.
- **ICAP BrokerTec** manages an electronic trading platform for the fixed income markets. BrokerTec delivers anonymous and orderly electronic fixed income trading for market participants for Canadian Government bonds, EU repos, European Government bonds, UK gilts, mortgage backed securities, US agency securities, US credit default swaps, US Treasury and US repos.
- **ICAP EBS** provides access to global markets for spot FX, precious metals and non-deliverable foreign currencies. EBS provides anonymous and highly liquid trading opportunities, as well as authoritative real-time and historical market data.
- **MarketAxess Corporate BondTicker** data service combines the FINRA TRACE data with MarketAxess tools to provide market participants with corporate bond price information. The data include actual execution prices as well as

MarketAxess estimated spread-to-Treasuries and asset swap spreads for approximately 30,000 investment grade and high-yield bond trades disseminated by FINRA.

- **Tullet Prebon** is an inter-dealer broker with positions in global fixed income securities, money markets, capital markets, equities, energy, credit and associated derivative products. Tullet utilizes the price flows generated by these brokerage areas as its principal source of market information and uses its own team of price reporters to enhance and add value to the information provided by the desks.
- **TradeWeb** market data, sourced directly from over 30 leading dealers, provides streaming, real-time price updates on a broad range of markets, for example US Treasuries TBA-MBS, interest rate swaps (US dollar, euro, yen and sterling).

Additional data source providers include exchanges, such as the New York Stock Exchange, Nasdaq and alike, as well as the following.

- **Depository Trust & Clearing Corporation**, through its subsidiaries, provides clearing, settlement and information services for equities, corporate and municipal bonds, government and mortgage-backed securities, money market instruments and over-the-counter derivatives. In addition, DTCC is a leading processor of mutual funds and insurance transactions, linking funds and carriers with their distribution networks.
- **IFR Markets** provides real-time financial markets commentary and analysis of market sentiment and deal flow.
- **Markit** is a global financial information services company providing independent data, valuations and trade processing across all asset classes.
- **MTS** delivers market data services for fixed income market participants to set best practice for price discovery, verification and validation.

Through additional analysis "best of breed" pricing sources can be identified and once identified automated to create a hierarchy of preferred vendor data by asset class.

17.5.5 Implementing a pricing data service

Given the extensive and increasingly important use of pricing data within FIs and the high costs of data licensing, there are strong drivers to implement enterprise quality data management services. Manual processes for managing pricing data are prone to error and no longer meet the strict regulator standards for transparency, or the needs of the increasingly complex pricing policies. This section will provide an overview of the core functions that a pricing data service performs and provide insight into implementation issues and best practices. The below section is written for end of day or end of month pricing data used for functions such as valuations.

Real-time pricing data implementations would be significantly different and specific to an asset class and marketplace, making them difficult to define in a general overview such as this.

The current state within the industry regarding the implementation of many of the below features varies widely. However, the one consistent thing is that most FIs have started the journey towards fulfilling many of the key pricing data service functions listed below.

- **Load** – Support the loading of data from dozens of sources in a variety of formats with a robust exception handling process that resolves specific data issues, without overall load delays. This intelligent workflow process needs to support regular changes to sources, formats and timings.
- **Data model** – Map and store prices in an enterprise quality data model that supports the distribution and query requirements of all the consuming business function applications.
- **Cross reference** – Store all required instrument codes, such as ISIN, so that the pricing information can be linked to all the details on the underlying instruments that are stored in the security master.
- **Attribution** – Maintain attribution to the source to meet the transparency required by the regulators regarding pricing policies and data sources used per asset class.
- **Price history** – Provide current day as well as historical pricing views for any asset class.
- **Distribution** – Support both the batch and real-time queries with an information architecture that meets the needs of all consuming applications.

It is important to differentiate between the pricing data service and the valuations service. The former provides the required data and the latter implements all the business logic around the valuations function. It is best practice to separate these. Although the valuations service is often the driver behind the creation of the pricing data service, it is not the only function that will reference pricing data. As such, it is important that they are implemented separately, because coupling them will complicate future enhancements. Also important to understand is the difference between pricing and valuation as terms. Pricing is a dynamic process executed by the front office and on exchanges where a price is determined for an instrument based not just on asset valuations but also market conditions and other variables. This price is used as an execution price on transactions. Valuations are a more scientific assessment of an asset's value, often using models applied against various independent sources to calculate the value of a position.

The most common implementation problem for pricing services is the mapping of data fields from the many sources and formats to the appropriate data fields within the enterprise data model. The lack of consistency regarding data formats requires a skilled team of analysts who understand standard data formats such as FINXML and FIX, and can quickly learn and map new formats. Semantic differences will also create mapping errors, so iterative implementations where mappings are frequently validated by the functions that use them are more efficient. The implementation plan should anticipate some remapping to achieve realistic timelines[16]

17.6 Trade and position data

17.6.1 Introduction

Trade data are the operational data from business activity. They are sourced from the internal systems of a financial institution. This internal origin sets it apart from reference data previously discussed, which was largely sourced from outside an FI. A position is an investment holding that could be the result of one or many trades, can be long or short, open or closed. Trades therefore are the building blocks of positions.

The creation of a single view of positions is becoming an increasingly critical goal for FIs. Numerous drivers that will be outlined in more detail below include risk management, cash management, collateral management, regulatory compliance, client reporting and more. Yet building a consistent view is extremely demanding, as an FI will have to understand and integrate dozens of products, traded on many markets, often in several time zones, and executed on a variety of complex, idiosyncratic IT applications.

This section will provide a brief insight into the drivers that make trade and position data so important. It will provide an overview of the specifics of the dataset and how it is created and managed.

17.6.2 Drivers for enterprise trade and position data

The following list shows examples of drivers for enterprise-wide trade and position data, highlighting not only its importance, but also how central this dataset is to the financial services industry.

- **Risk** – Position data identify a firm's exposure and counterparty risk. Liquidity risk requires an understanding of cash flows from positions.
- **Operational efficiency** – Decreases numerous point to point integrations and associated reconciliation; provides clear, unified reporting of data quality issues for user action.

[16] FinXML, www.integral.com/resources/finxml/

- **Compliance** – Pre-trade and post-trade compliance that evaluates accounts and positions across asset types and business verticals.
- **Client reporting** – Accurate, timely and functional client reporting is a differentiating feature in investment management and trade; position details and trends are a central part of this.

17.6.3 Trade and position data management

The principles of enterprise data management outlined in the reference data sections apply to trade and position data management equally. The goal is a centralized golden copy trade and position service that can provide timely and accurate data to meet the needs of the business drivers. The importance of executive commitment to the initiative cannot be understated, because engaged and proactive front office business and IT functions are crucial to understanding the instrument trade cycles and data.

Probably the biggest challenge with building a trade and position data service is the variety that exists in the operational systems that create the data. It is common for an FI to use financial products from multiple vendors for front, middle and back office operations. This often creates inconsistency in key areas such as trade flow, order status or product definitions. Some systems are built for a subset of the trade lifecycle such as order management, inventory control, operations, accounting, funding, profit and loss calculation (P&L), risk management or tax. Some source systems do many or all of these functions. So bringing the data from these different systems and products together is no small undertaking. The starting point is to create a generic asset class definition for each product.

For a large FI active in many asset classes, the list of trade and position data attributes would be in the thousands. Below are a few key fields, but it is important to realize that the challenge is not to list them all. The difficulty lies in understanding them, resolving semantic differences between sources and products, and mapping source specific fields to generic asset class definitions for products so they can be modeled, standardized and communicated.

- **Economics** – buy or sell, price, currency, sales credits, markups, broker fees etc.
- **Counterparty information** starting with the identifier link to cross reference further legal entity data
- **Instrument information** usually consisting of cross referenced IDs linking back to security master
- **Execution information** such as exchange, source system, trader, broker, trade status

- **Processing information** including settlement date, book, GL account number, position ID.

Integration of the trade and position service with the other reference data services discussed in this chapter is essential. Trade information would be enriched by cross references to these repositories. Standard component development to support data provisioning that seamlessly spans all of these data services would enhance ease of integration for data consumers.

17.7 The Marketplace

To help manage the daunting task of building and maintaining high quality enterprise data, several vendors have emerged in the marketplace with products and services to increase time to market, and decrease implementation risk. Vendors and products change frequently, but information on the dynamic list of offerings is widely available on the Internet. Below is a sample of vendors intended to provide some initial insight and a starting point for further analysis.

Avox validates, corrects and enriches business entity data. The firm has established a shared pool of business entity data regularly contributed to by their international clients. These industry data are combined with up-to-date content from other authoritative sources including regulators, business registries and company filings. Changes made by an Avox client to a data record are immediately revalidated by an Avox analyst and, if proven correct, are broadcast, complete with a full audit trail, to all other clients holding the same record. Data in Avox include corporate hierarchies, addresses, industry codes, identifiers, regulatory content and links to issued securities.

GoldenSource provides an integrated enterprise data management (EDM) platform for the securities and investment management industry. Their software and expertise help firms to better manage, store and distribute data company-wide. Key product coverage includes security master, pricing data service and client reference data service.

Interactive Data provides reference data on over 10 million financial instruments to improve risk management, and maintain compliance with regulatory mandates to maximize operational efficiency.

Kingland Systems is a provider of enterprise risk, compliance, and Master Data Management (MDM) solutions. They provide enterprise software, professional services, and global databases, with expertise in the banking, brokerage, insurance, accounting, government, and healthcare industries.

Markit is a global financial information services company providing independent data, valuations and trade processing across all asset classes required to enhance transparency, reduce risk and improve operational efficiency. Its client base covers a broad range of institutional participants in the financial marketplace.

SIX Telekurs specializes in procuring, processing and distributing international financial information. Their market specialists gather information from all the world's major trading venues, directly and in real time. The SIX Telekurs database captures more than 5.6 million financial instruments.

SunGard's StreamCore is designed to help firms leverage data from across the enterprise. The solution normalizes and consolidates data from disparate applications across the enterprise into a single repository of accounts, market data, transactions, positions and balances.

18

US residential-mortgage transfer systems: a data-management crisis

John Patrick Hunt, Richard Stanton and Nancy Wallace

18.1 Introduction

The residential mortgage and mortgage-backed security (RMBS) markets in the USA are very large,[1] and their operational practices have been deeply implicated in the continuing financial crisis.[2] The systems in place to manage, monitor, store, and analyze the huge volumes of data associated with these markets have not kept pace with the rapid financial developments that have occurred in the last few decades, leaving the legal status of the market increasingly unclear, and leading to many data-related problems that have exacerbated the situation for millions of borrowers who either have lost or are about to lose their homes. Levitin (2010) discusses several examples, including questions regarding the following points.

- Numerous failures of financial institutions with a primary focus on residential lending.
- Widespread allegations of mispricing of mortgages and mortgage-backed securities.
- The validity of the mortgage chain of title in the securitization process, which could potentially cause MBS holders to be unable to go after collateral in the event of a default (see also Hunt et al., 2012).
- The effect of the use of the Mortgage Electronic Registration System (MERS) on the legal status of the mortgages underlying MBS (see also Hunt et al., 2012).

[1] The outstanding stock of the US residential mortgage market was $10.290 trillion at the end of the fourth quarter of 2011 (see *Federal Reserve Statistical Release*, Z.1, Flow of Funds Data, Board of Governors of the Federal Reserve System, Washington, DC, 20551, `www.federalreserve.gov/releases/z1/Current`). The outstanding stock of RMBS was $6.437 trillion at the end of the fourth quarter of 2011 (see *Inside MBS & ABS*, March 16, 2012).

[2] See, for example, Acharya and Richardson (2010) and Mian and Sufi (2009).

Handbook of Financial Data and Risk Information, ed. Margarita S. Brose, Mark D. Flood, Dilip Krishna and Bill Nichols. Published by Cambridge University Press.

- Whether investors will be able to "put-back" to banks securitized mortgages on the basis of breaches of representations and warranties about the quality of the mortgages.

Combined with widespread allegations (both true and fraudulent) of lost documents, these problems strongly suggest that data management in the whole-loan and mortgage securitization industry has reached a crisis.

In this chapter, we outline the central role of data transfer and management in US mortgage and mortgage-backed security markets. We point out numerous inadequacies in the availability of mortgage and mortgage-backed securities data and in the management of existing data, inadequacies that place lenders, borrowers, and MBS investors at risk. We conclude with some recommendations for improvement.

18.2 Whole-loan mortgage data structures

18.2.1 Contract design and asymmetric information

In the US residential-mortgage market, mortgage lenders set the menu of mortgage-contract features that are offered to borrowers from websites, internal loan officers, and independent mortgage brokers. The mortgage menus, or wholesale-rate sheets, are refreshed frequently, usually daily, and they define the types of loans (e.g., fixed rate versus adjustable rate) that can be offered to borrowers on a given date as well as identifying the available combinations of mortgage contract rates, discount points payable at origination, amortization and payment structures, maturities, loan-to-value ratios, and permissible FICO scores, among many other contract features. The wholesale-rate sheets reflect the business decisions of the lender concerning the types of borrowers they are seeking to attract and the types of products that they wish to originate. During the run-up to the financial crisis, there is convincing evidence that lenders intentionally designed their mortgage menus to enable lending to borrowers who were willing to pay high mortgage coupons in exchange for greater risk metrics such as having no documentation, having a higher loan-to-value ratio, or having a low FICO score (see Berndt et al., 2012).

As an illustration of the choices available to borrowers on a typical mortgage menu, Table 18.1 shows the wholesale-rate schedules for four of the largest residential-mortgage lenders in the USA on January 25, 2012. The schedules reported in Table 18.1 are for borrowers seeking conventional, conforming loans (i.e., maximum loan amounts of $417,750 with borrower credit – FICO – scores of at least 620) and who seek to lock in their quoted rate for 30 days. The schedule presents to the borrower the current menu of coupon rate and discount points (both positive and negative, or credits to lower settlement costs). The choices presented in Table 18.1 are for 30-year maturity, fixed-rate mortgages. Although not reported

Table 18.1 *Prime wholesale-rate sheet offered on January 25, 2012 by four of the largest US residential mortgage lenders for conventional conforming fixed rate mortgages with a 30 day lock-in period*

	Discount points			
Contract Rates	Lender 1 Points	Lender 2 Points	Lender 3 Points	Lender 4 Points
5.500	(5.500)			
5.375	(5.500)			
5.250	(5.500)			
5.125	(5.237)	(5.788)		
5.000	(4.905)	(5.467)		(5.000)
4.999				(4.875)
4.875	(4.520)	(4.999)	(5.539)	(4.375)
4.750	(4.263)	(4.766)	(5.539)	(4.250)
4.625	(3.953)	(4.467)	(5.539)	(4.000)
4.500	(3.551)	(4.096)	(5.539)	(3.875)
4.375	(2.996)	(3.511)	(3.515)	(3.750)
4.250	(2.429)	(2.894)	(3.063)	(2.625)
4.125	(2.073)	(2.492)	(2.680)	(2.250)
4.000	(1.610)	(2.051)	(2.261)	(1.875)
3.950			(2.050)	(1.250)
3.875	(0.954)	(1.407)	(1.560)	(0.375)
3.750	(0.204)	(0.587)	(0.830)	
3.625	0.25	(0.021)	(0.394)	
3.500	0.818	0.467	0.083	
3.375	1.536	1.143	0.880	
3.250	2.438	2.122	1.730	

here, the rate schedules provide similar menus for 15-year maturity, fixed rate loans and for adjustable rate mortgages with differing resets and maturities.

As shown in Table 18.1, if, on January, 25, 2012, a hypothetical borrower sought a 30-year fixed rate mortgage to finance a balance of \$417,000 with a guarantee to lock-in the offer for 30 days, and a contract rate of 3.375%, then Lender 1 would require an additional charge of 1.536% for an up-front interest payment (discount points) to be paid by the borrower. The discount points are paid up-front because the lender originates a loan for \$417,000, but only distributes \$410,594.88 of that principal.[3] Lender 2 would charge discount points of 0.587% (0.01143 × \$417,000 = \$4,766.31) for the same contract and coupon, Lender 3 would charge

[3] Given the required discount points, the cash equivalent owed to the lender on the origination day is \$6,405.12 (0.01536 × \$417, 000 = \$6, 405.12). Thus, the lender decreases the principal paid out to the borrower to \$410,594.88 (\$417,000 – \$6,405:12 = \$410,594:88).

0.83% (0.0088 × \$417,000 = \$3,699.6), and Lender 4 will not offer a 3.375% contract rate for such a loan on this date. Similarly, if the same borrower agrees to pay a rate of 3.250% to Lender 1, the borrower would have to pay discount points at origination of 2.438% (0.02438 × \$417,000 = \$10,166.46) for the lender to price the loan at par. If, on the other hand, the borrower agreed to pay Lender 1 a coupon of 4.00% on January 25, the lender would pay "negative points" of \$6,713.7, which is cash paid by the lender to the borrower to defray other origination costs.

In addition to the menu choices over maturity, coupon interest rates, and points, the rate schedules also delineate the borrowers' options on downpayment and mortgage coupon combinations. As shown in Table 18.2, on January 25, 2012, the wholesale-rate schedule for this lender required that a borrower with a 680 FICO and a loan-to-value ratio of exactly 75% would have to add an additional 125 basis points to the contract rate.[4] The same borrower would have to add only 50 basis points if he or she was willing to lower the loan-to-value ratio to 65%.

The key intuition behind these menus is that they provide lenders with a way to resolve the well-known information asymmetries between what borrowers know about themselves and what they truthfully (and credibly) reveal to lenders. The menus provide the borrowers with incentives to select combinations of mortgage characteristics that best suit their (often hidden) preferences and risk characteristics. Empirical evidence suggests that, for a given coupon rate, mortgages with low points tend to be prepaid more rapidly than mortgages with high points (see Brueckner, 1994; Hayre and Rajan, 1995; Downing et al., 2009), suggesting that differences among the behavioral characteristics of borrowers may be associated with the interest rate/points trade-off. Dunn and Spatt (2005) and Stanton and Wallace (1998) find that the points versus coupon trade-off sorts borrowers by type. Borrowers who plan to move soon ought to take out loans with a high periodic interest rate and low points, whereas those who plan not to prepay (except possibly for interest-rate-related reasons) should take out loans with higher points and a lower periodic interest rate. Following this logic, a borrower's choice of contract serves as a self-selection device (see Rothschild and Stiglitz, 1976), allowing the lender to learn private information about potential borrowers' mobility.[5]

Similarly, it seems likely that the menu choices over loan-to-value ratio and mortgage coupon are additional self-selection mechanisms, which reveal

[4] The add-ons over an 80% loan-to-value fall due to additional charges associated with the mortgage insurance coverage required for high loan-to-value charges. These charges would be 87.5 additional basis points of coupon interest for an 85% loan-to-value loan and 225 basis points of addition coupon interest if the loan-to-value ratio was greater than 90%.

[5] Chari and Jagannathan (1989) propose a model in which individuals face an (uninsurable) risk of moving, and their expected income, conditional on moving, is higher than if they do not move. If they take out a loan with points and a below market interest rate, their average payment is high if they move and lower if they do not move. The contract thus provides partial insurance against moving and its associated income shock. However, the resulting correlation is counter to what is observed in practice.

Table 18.2 *Prime wholesale-rate sheet offered on January 25, 2012 by large US residential mortgage lenders for conventional conforming fixed rate mortgages with a 30 day lock-in period*

	Loan-to-value percentages							
	<=60%	60.01% – 70%	70.01% – 75%	75.01 % – 80%	80.01% – 85%	85.01% – 90%	90.01% – 95%	95.01% – 97%
	All fixed rate and adjustable rate products with terms greater than 15 years							
FICO score								
Greater than 740	(0.250)	0.000	0.000	0.250	0.250	0.250	0.250	*0.250
720–739	(0.250)	0.000	0.250	0.500	0.500	0.500	*0.500	*0.500
700–719	(0.250)	0.000	0.750	1.000	*1.000	*1.000	*1.000	*1.000
680–699	0.000	0.500	1.250	1.750	*1.500	*1.250	*1.250	*1.125
660–679	0.000	1.000	2.125	2.625	*2.750	*2.250	*2.250	*2.000
640–659	0.500	1.250	2.625	3.000	NA	NA	NA	NA
620–639	0.500	1.500	3.000	3.000	NA	NA	NA	NA
Less than 620	NA	NA	NA	NA	NA	NA	NA	NA

* Only eligible and applicable to FNMA HOMEPATH Loans.

borrowers' likely preferences about the exercise of their default options (higher-coupon-paying borrowers with higher loan-to-value ratios would be expected to be more likely to default). These borrowers should thus either be screened out or charged higher discount points at origination, in compensation for this greater risk.

The existence of these self-selection mechanisms in the mortgage origination process also presents potential difficulties in the ex post application of statistical risk analytics to evaluate the expected performance, or valuation, of mortgage contracts. Although standard statistical techniques exist (see Heckman, 1979) to correct for sample-selection biases, a significant problem in the mortgage market is that loan-specific information on the magnitude of the discount points paid at origination, a crucial factor in the selection process on the mortgage menu, is not retained either for securitized loans or for loans held in the lender's portfolio. Thus, with currently available mortgage origination and performance data, the analyst does not know for a given mortgage coupon rate whether the borrower paid high or low discount points at origination. Even though the mortgage menu was specifically designed to have borrowers reveal private information about their likely mobility preferences or, possibly, their default-risk preferences through their points, coupon, and loan-to-value ratio choices, the current state of available data for mortgage analytics does not include the level of discount points that were paid by the borrower at origination. As a result, it is likely that fitted forecasts of mortgage performance suffer from uncorrected sample-selection biases arising from the correlations between the residual structure of these forecasts and an important omitted variable, the magnitude of the discount points, related to the endogenous self-selection rule.

The ex ante design of the origination menu is also never retained for ex post analytics. For this reason, it is not possible to identify the effects of errors in the relative structure of the menu either across lenders or across mortgage products for a given lender. Analysts thus cannot determine whether the menu structure itself was badly designed (e.g., mortgages were priced incorrectly relative to the choices on the menu) or whether specific contract types (e.g., option ARMs) are simply inherently bad products and attract poorly performing borrowers. These policy-relevant questions concerning the ex post performance of loan contracts are impossible to address given the current lack of information concerning the ex ante mortgage menu structure that led to the origination of specific loans and the types of borrowers that chose them. This, again, is another form of sample-selection bias, because the analyst cannot control for the endogenous choices of borrowers over the menus, since the needed information is simply not available. In fact, generally in the US mortgage market, the wholesale-rate schedules are considered to be proprietary information by the banks.

18.2.2 Origination data

Mortgage origination data comprise the set of static information related to the mortgage at the time of the loan origination. As shown at the top of Table 18.3, the loan record is identified by an *internal* loan identification number. Currently, in the USA there is no permanent, unique, and verifiable loan identifier (like the CUSIP number in the bond market) attached to each loan at origination. Instead, loan identification numbers are recreated by the different owners and managers (such as servicers and pool trustees) of the loan origination and performance datasets. Nearly always, the loan IDs are changed as the loans travel though the mortgage supply chain (which will be described in subsequent sections of this chapter), making it all but impossible to track a unique loan through the supply chain from its originators, via its servicers, to its securitized pool. Private-sector data providers such as Bloomberg, Corelogic, Lender Processing Services (LPS), CTSLink, and ABSNet, and public data sources such as the Home Mortgage Disclosure Act (HMDA) also usually assign their own identification system to the loans, which again makes it difficult to compare information that purports to represent the same universe of loans. Without unique and permanent identifiers, the only way to track loans is through complicated, and often erroneous, computer matching schemes that link the information by common loan elements such as zip code, loan amount, and contract features.

The data reported in Table 18.3 are fully available internally to the analysts of the loan originator, the loan servicer, the GSEs, and their regulator, the Federal Housing Finance Administration (FHFA). However, only subsets of the data are available through the private data vendors who represent the primary data source to investors and analysts in the securitized mortgage bond market and to the regulatory institutions, such as the Federal Reserve. The private data providers and all of the trustees expunge nearly all of the borrower and co-borrower identification information reported in Table 18.3. Usually, the only remaining borrower-specific information is the original FICO score, the borrower's age, and income at origination. Similarly, the information about the property characteristics at origination is usually limited to the city, state, and zip code of the property. It is also important to reiterate that these data include no information about the mortgage menu or about the discount points paid at origination.

Because the origination data are static, many of the data fields, such as those representing underwriting characteristics like loan-to-value ratio and the debt-service-coverage ratio, and those representing the lien and servicer status, may become stale over time. Often, risk modelers rely heavily on static information such as the original FICO score, the original loan-to-value ratio, and original debt-service coverage in modeling mortgage prepayment, default, and valuations, because updated information is not available. Another common practice is to

Table 18.3 *Typical data fields for residential mortgage borrower, property, underwriting, and lien characteristics at origination*

The data fields reported in this table summarize the data fields that are available from a variety of sources including Freddie Mac, Loan Variable Disclosure, Bloomberg, CTSLink, and ABSNet Lewtan.

Loan identifiers	
Internal loan identification number	
Borrower information	
Borrower/co-borrower name	
Borrower/co-borrower age	
Borrower/co-borrower income	
Address	Street, city, county, zip code, state
Borrower FICO score	
First-time home buyer flag	
Number of borrowers	
Property characteristics	
Address	Street, city, county, zip code, state
Number of units	
Occupancy status	Investment property, primary or secondary home
Original appraised value	
Purchase price	
Underwriting characteristics	
Documentation flag	Extent of income and asset verification
Documentation type	Income, assets, employment
Original loan-to-value ratio	
Original combined loan-to-value ratio	Sum of all indebtedness positions
Debt-to-income ratio	
Credit/risk grade	
Private mortgage insurance flag	
PMI provider	
Underwriting exception	Non-standard underwriting flag.
Third party originator flag	
Originator name	
Originator address	
Lien and servicer status	
Lien position	
Servicer	
Silent second flag	
Senior lien amount	
Servicer contact	
Originator	

simulate the expected path of loan-to-value ratios and property prices using house-price indices such as those of Case Shiller S&P and the FHFA. Often, these indices were constructed using geographical aggregates of transaction data for sales that do not precisely match the available loan address on the property.

There are also significant problems with access to residential house price and characteristic data in the USA. These data are usually available in the records of the Assessor's Office in the county in which a given property is located. They are archived by Assessors Parcel Number (APN) and are difficult and time consuming to search or aggregate into larger geographical units such as census tracts or cities. Large private data vendors, such as Corelogic and Dataquick, collect the assessors' data (which are public), aggregate them into proprietary datasets, and then sell them to the capital markets. The vendor data include information on the exact property addresses, transaction prices, the names of the owner(s)/borrower(s), and other physical characteristics of each property.

Instead of working with raw house-price data, however, most financial institutions monitor expected house-price levels, or expected loan-to-value ratios, for mortgages or mortgage-backed securities using either automated valuation models (AVMs) or house-price indices. AVMs estimate property values at a specific point in time using proprietary statistical models and data samples that correspond to a subject property at either the state, Metropolitan Statistical Area (MSA), county, city, zip code or zip-code-plus-four level of geographical specificity, depending on the AVM vendor. The two most commonly used repeat-sales price indices, the Federal Housing Finance Administration[6] (FHFA) price index and the S&P-Case-Shiller price index, (see Case and Shiller, 1987, 1989), measure the rate of house price growth at either the state or MSA levels. The available AVMs and price indices are proprietary so there is no transparency concerning the details of the statistical models or the data used to calibrate the models.[7]

Table 18.4 shows the typical data available on the origination characteristics of the loan. These data are again static, and are primarily designed to provide information on the contractual rules for setting up the amortization and payment schedules for the loans. To generate the actual expected cash-flow structure of a particular mortgage using these data entails: (1) a careful assembly of the contractual rules of the loan; (2) acquisition of information not directly provided in the origination data fields, such as the actual, or forecasted, time series for the designated ARM index type; and (3) a great deal of customized computer code to generate the loan-level principal and interest payments. Additionally, since residential mortgages in the

[6] The FHFA is the regulator for Fannie Mae and Freddie Mac.

[7] A large academic literature has evaluated the strengths and weaknesses of these indices (see Meese and Wallace, 1991; Hwang and Quigley, 2003; Gatzlaff and Haurin, 1997, 1998; Goetzmann and Peng, 2006; Stanton and Wallace, 2009, among others).

USA always contain embedded default options and usually contain embedded prepayment options, the mechanical representation of the cash flows that is possible using only the data from Table 18.4 would generate very inaccurate representation of the actual expected cash-flow performance of the mortgages. To get an actuarially accurate representation of the expected cash flows of any given mortgage requires additional detailed information about the delinquency, default, and prepayment performance of representative samples of mortgages. These type of data are called performance data.

18.2.3 Performance data

As shown in Table 18.4, the mortgage-performance data include data fields that are updated monthly. These data fields provide information on the time-series performance of the loans with respect to delinquency, default, and prepayment. Since the financial crisis, these loan-performance data fields have been expanded to include information on the loan-modification status of the affected loans. Overall, these data provide a comprehensive picture of the current status of the loans. Within the financial institutions, these data may be updated contemporaneously, though because these data come from servicers or pool trustees they are more commonly updated with one- or two-month lags. The primary use of these data is to develop forecast models for borrowers' exercise of their delinquency, default, and prepayment options. The determinants of the exercise of these options are usually associated with dynamic information about the current value of the property, the current amortization and payment structure of the loan contract, and the current characteristics of the borrower such as his/her current FICO score, income, and employment status. As previously discussed, the origination data can be used to construct the amortization and cash flow structure of each mortgage up to its stated maturity date. Mortgage valuation modeling, however, requires accurate data on the realized default and prepayment performance of similar loans in order to develop forecasts for the probability that any given loan will terminate before the contractual maturity date. In addition, accurate mortgage valuation forecasts require time-varying information concerning the loan-to-value ratio of the loan, the value of the index (for adjustable-rate mortgages), and the FICO score of the borrower. This information is only available at origination, and for that reason it must be estimated using information from other sources.

18.2.4 Financial statement and regulatory uses

Banks and savings and loan institutions (S&Ls) are required to make an ex ante choice as to whether specific mortgage loans are to be held in the "Hold-for-Investment"

Table 18.4 *Typical data fields for residential mortgage contract characteristics at origination*

The data fields reported in this table summarize the data fields that are available from a variety of sources including Freddie Mac, Loan Variable Disclosure, Bloomberg, CTSLink, and ABSNet Lewtan.

Contract characteristics	
Origination date	
Original interest rate	
Original loan amount	
Maturity date	
First payment date	Date first loan payment was due
Product type	Fixed, ARM
Original amortization term	
Prepayment penalty flag	
Prepayment penalty term	
Periodic rate cap %	
Periodic rate floor %	
Periodic payment cap %	
Periodic payment floor %	
Lifetime payment cap %	
Maximum life of loan cap %	
Minimum life of loan cap %	
Rate adjustment period	
Rate adjustment frequency	
Negative amortization flag	
Maximum negative amortization %	
Initial fixed rate period	For hybrid ARMS
Gross margin %	Number of percentage points added to the index
Look-back period	Number of days from the index publication to reset
Interest only term	
Balloon flag	
ARM converting flag	
HELOC flag	
Interest only flag	
Pledged asset flag	Other assets pledged as security
ARM index type	
Balloon date	
ARM conversion date	
Payment frequency indicator	
ARM lockout period	
Initial interest rate reset date	
Initial payment reset date	
Draw period	For home equity lines of credit
Credit limit	For home equity lines of credit
Interest only end date	

(HFI) portfolio or the "Hold-for Sale" portfolio. The accounting treatment for mortgage loans held in these two portfolios is different. Loans held in the "Hold-for-Sale" portfolio can be held in two forms: "Trading Securities" (TS) for loans that will be actively traded, or "Securities-Available-for-Sale" (AFS) for loans that will not immediately (but will eventually) be traded. The reporting method used for TS loans is fair value with unrealized holding gains and losses included in net income.[8] The reporting method used for the AFS loans is also fair value. However, unrealized holding gains and losses are excluded from net income.

Loans held in the HFI portfolio are usually the largest position on bank and S&L balance sheets.[9] They are reported at amortized cost and the holding gains and losses associated with these loans are unrealized for financial reporting purposes (see Spiceland et al., 2011). Instead, banks and S&Ls are required to hold reserves, called *Allowances for Loan and Lease Losses* (ALLLs), on their balance sheets in anticipation of credit losses for individually evaluated mortgage loans that are determined to be impaired and for groups of loans that are not identified as impaired. Additions to ALLLs are recorded as an accrued expense on the bank's and/or S&L's income statement and are recognized as accounting transactions indicating a likely reduction in the institution's cash flows since expected loan repayments would be anticipated to be less (see Barth and Landsmen, 2010; Laux and Leuz, 2009; Huizinga and Vaeven, 2009).

The most important required financial statement, or regulatory use, of mortgage origination and performance data, is the quarterly (or more frequent, if warranted) calculation and reporting of the ALLLs. According to the *Interagency Policy Statement on the Allowance for Loan and Lease Losses (ALLL)*,

> The ALLL represents one of the most significant estimates in an institution's financial statements and regulatory reports. Because of its significance, each institution has a responsibility for developing, maintaining, and documenting a comprehensive, systematic, and consistently applied process for determining the amounts of ALLL and the provision for loan and lease losses (PLLL). To fulfill this responsibility, each institution should ensure controls are in place to consistently determine the ALLL in accordance with GAAP, the institutions's stated policies and procedures, management's best judgment and relevant supervisory guidance. (pages 2–3)[10]

[8] The Financial Accounting Standards Board (FASB) provides a hierarchy that prioritizes the inputs financial institutions should use when determining the fair value for loans. Level 1 is market prices. Level 2 is inputs other than market prices that are observable or are quoted for similar assets. Level 3 is unobservable inputs that reflect the entity's own assumptions about the assumptions the market would use to price the asset (see Spiceland et al., 2011).

[9] See Huizinga and Vaeven (2009), Laux and Leuz (2009) and Barth and Landsmen (2010).

[10] See *Interagency Policy Statement on the Allowance for Loan and Lease Losses* Office of the Comptroller of the Currency, Board of Governors of the Federal Reserve System, Federal Deposit Insurance Corporation, National Credit Union Administration, Office of Thrift Supervision, `www.federalreserve.gov/boarddocs/sfletters/e006/SR0617a1.pdf`

The managers and board of directors of banks and S&Ls are responsible for maintaining the ALLLs at appropriate levels and for documenting the analyses used to derive them. As part of these responsibilities, management is required to adopt and adhere to written policies and procedures that are appropriate to the size and risk of the institution and the "nature, scope, and risk of its lending activities."[11] Usually, the ALLLs are obtained using empirical models to estimate the probability of default and the losses given default for loans and/or groups of loans for a specific evaluation date and for a specific holding period, usually 36 months. The statistical estimates needed to compute the ALLLs are obtained using the mortgage and performance data for the loans in the HFI portfolio. The data that are typically used to estimate the ALLLs are the underwriting characteristics reported in Table 18.3, such as as the loan-to-value ratio, the borrower FICO score, and the geographical location of the property, and the contract characteristics reported in Table 18.4 that classify the loans by type. The loan performance data, reported in Table 18.5, is required to identify the impaired and unimpaired loans using the data on the delinquency and default status of the loans.

The ALLL estimation process requires that the institution has effective loan review systems, loan classification systems, and/or credit grading systems that can accurately identify, monitor, and address asset quality problems in a timely fashion. Institutions are also required to have adequate data capture and reporting systems to supply the information necessary to support and document the ALLL estimates. The management of regulated financial institutions are required to document their evaluations and conclusions regarding the appropriateness of their estimated credit losses and the appropriateness of ALLL forecasting models. Institutions are also required to validate the ALLL statistical methodology, and this validation process must include parties who are independent of the institution's credit approval and ALLL estimation process. Overall, the ALLL estimation process requires highly functional data storage and processing systems. These systems must allow for the verification of the accuracy and completeness of newly originated mortgage data, of mortgage data migrations from real time to historical warehousing facilities, and of mortgage data transformation processes either within different business units or in the aggregate measurement of enterprise-wide risk.

As is clear from Tables 18.3, 18.4 and 18.5, mortgage loan-origination and performance data are voluminous and can present significant data management and storage challenges for financial institutions and agents in the economy, as well as in regulatory institutions that require actuarially sound mortgage-risk analytics. Currently, the management and storage of origination and performance

[11] See *Interagency Policy Statement on the Allowance for Loan and Lease Losses*, `www.federalreserve.gov/boarddocs/sfletters/e006/SR0617a1.pdf`, p. 5.

Table 18.5 *Typical data fields for residential mortgage monthly performance*

The data fields reported in this table summarize the data fields that are available from a variety of sources including Freddie Mac, *Loan Variable Disclosure*, Bloomberg, CTSLink, and ABSNet Lewtan.

	Field description
Loan identifiers	
Internal loan ID	
Monthly performance	
Beginning balance	
Ending balance	
Delinquency status	30, 60, 90, 120, 240 days
Bankruptcy flag	
Foreclosure flag	
REO flag	
Prepayment date	First record date
Liquidation date	First record date
Repurchase date	First record date
Bankruptcy date	First record date
Foreclosure date	First record date
REO date	First record date
Post modification contract	
Original loan balance	
Mod balance	
Mod principal and interest	
Mod rate	
Mod capitalized amount	
Mod interest forgiven	
Mod principal forgiven	
Mod deferred amount	
Mod balloon date	
Mod maturity	
Mod first payment date	
Mod next due date	
Mod margin	

mortgage data, and the ALLL and risk-modeling analytics that have been customized to them, are highly labor-, time-, and resource-intensive activities. They also present opportunities for inaccuracies in mapping contracting structures to cash-flow generation and then accurately accounting for expected credit losses associated with borrower exercise of their default options. These problems are likely to lead to imprecision in expected credit losses and poor quality ALLLs.[12] The purpose of

[12] Ng and Rusticus (2011) find that the quality of loan loss provisions and earnings is increasing in the strength of the mapping between loan loss provisions and current and future net charge-offs.

ALLLs is to allow the institution to cover "expected losses," that is, loan losses that occur on average. The overall effectiveness of regulatory capital as a buffer against "unexpected shocks" is predicated on the appropriateness of the subsidiary buffer of the reserves created through the ALLLs.

18.3 Mortgage data transfer over the supply chain

Prior to the financial crisis, mortgages in the USA traveled along two distinct supply chains: one for Freddie Mac and Fannie Mae[13] and a second for the private-label market, which usually (but not always) securitized mortgages that were ineligible for securitization by Fannie Mae and Freddie Mac. Currently, with the collapse of private-label mortgage securitization, Fannie Mae and Freddie Mac securitize nearly all of the newly originated mortgage-backed securities in the USA.[14]

18.3.1 Mortgage data transfer through the GSE supply chain

Figure 18.1 presents a diagram of the two-stage process used to create multi-class Real Estate Mortgage Investment Conduit (REMIC) residential mortgage back securities.[15] Single-class mortgage-pass-through securities are created in the first stage of the process. These bonds pay out a pro rata share of the cash flows from a pool of conventional, conforming residential mortgages.[16] The single-class pass-through bonds are issued by a special-purpose vehicle (SPV),[17] which is organized as a grantor trust[18] and is insured against default losses by either Fannie Mae or Freddie Mac, the government sponsored enterprises (GSEs). As shown in Figure 18.1, there are two possible channels in the first stage of the process. In one of these channels, shown as a hashed section beneath the financial institution on the left-hand side, the financial institution accumulates a pool of newly originated

[13] Until September 2008, Fannie Mae and Freddie Mac were privately owned but publicly chartered financial institutions that provided default-risk insurance for pools of securitized conventional, conforming mortgages. They are known collectively as the government sponsored enterprises (GSEs). During the financial crisis, defaults in the large portfolio holding of the GSEs led to a US government bail-out and the GSEs were put into conservatorship in September 2008.

[14] In the first quarter of 2012, Fannie Mae and Freddie Mac accounted for 77.97% of all residential mortgage-backed security production in the United States (see *Inside MBS & ABS, April 6, 2012*).

[15] The REMIC designation is defined in the Internal Revenue Code, see 26 U.S.C. §§860A–860G.

[16] Conventional, conforming loans are loans with balance of $417,000 or less that meet the underwriting guidelines of Freddie Mac and Fannie Mae.

[17] An SPV can be defined as a "finite-lived, standalone financing vehicle" that "de-link[s] the credit risk of the collateral pool from the credit risk of the originator." (Bank for International Settlements, 2005, The Role Of Ratings In Structured Finance: Issues And Implications 5).

[18] Under the Internal Revenue Code, a non-REMIC mortgage securitization vehicle, including a non-REMIC grantor trust, will be taxed as a corporation and not treated as a pass-through entity if it issues debt obligations with more than one maturity. 26 U.S.C. §7701(i). Moreover, a grantor trust may be recharacterized as a taxable entity if it can purchase new receivables, substitute receivables, or reinvest assets. Treas. Reg. § 301.7701–4(c); Rev. Rul. 73–460; Rev. Rul. 78–149.

conventional, conforming mortgages and then "converts" them into an agency pass-through of equivalent principal balance as either Freddie Mac Participation Certificates (PCs) or Fannie Mae Mortgage-Backed Securities (MBS).[19] This process of converting mortgages into MBS/PCs shifts all of the borrower default risk to either Fannie Mae or Freddie Mac, thereby transforming default risk into the risk of early return of principal to the financial institution, who is now the MBS/PC investor.[20]

Following the MBS/PC conversion process, the financial institution can either retain the MBS/PCs in its portfolio or it can sell them on the TBA (to-be-announced) forward-contract market. The TBA market is a highly organized and liquid market for the forward delivery of newly created MBS/PCs. It is the primary market channel to sell MBS/PCs, with an average of $302 billion of agency MBS traded each day as of the second quarter of 2011.[21] The TBA transaction process is shown in the second hashed box in the center of Figure 18.1. Originators use the TBA market to lock in a price for forthcoming MBS/PCs as they are originating

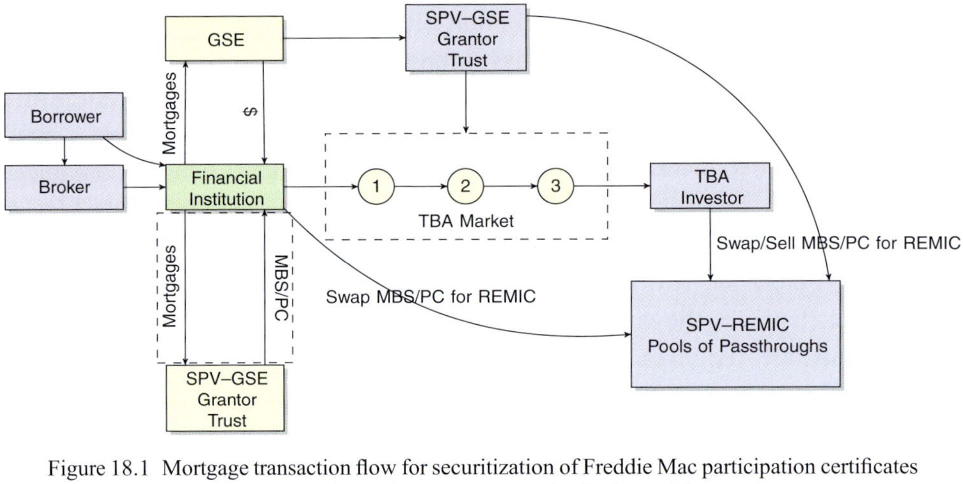

Figure 18.1 Mortgage transaction flow for securitization of Freddie Mac participation certificates (PCs) and Fannie Mae mortgage-backed securities.

[19] Although not shown here, there is a similar conversion, or swap, process that is used for the securitization of federally insured Federal Housing Administration loans originated by private sector financial institutions. These loans are then converted into government-insured Government National Mortgage Administration (GNMA) mortgage backed securities. GNMA guarantees the timely payment of interest and principal, and the insurance is backed by the full faith and credit of the US Federal government. The GNMA bonds are then traded on the TBA market.

[20] This conversion occurs because the GSE insurance fully reimburses all losses associated with any defaults on the underlying mortgages. Thus, when defaults occur the pass-through investors receive the full principal balance as if the borrower had instead prepaid the mortgage for other reasons. For that reason, default and prepayment are cash-flow equivalent in these pools.

[21] See SIFMA, *TBA Fact Sheet*, 2011.

mortgages, thus allowing lenders a mechanism to hedge or forward fund their origination pipeline.[22] Although not shown in Figure 18.1, there is also another very much smaller STIP (stipulated pool) market for MBS/PCs that operates over the counter for specific pools, however, the trading volume is very limited.

As shown in the hashed box in the center of Figure 18.1, there are three key dates in a TBA securities trade: (1) the trade date; (2) the notification date (48-hour day); and (3) the settlement date. On the trade date there is an agreement reached on six criteria for the securities that are to be delivered: issue, maturity, coupon, face value, price and the settlement date. SIFMA has established a monthly schedule for trade date settlements. The notification date occurs two business days before the contractual settlement date and at this point the forward-contract seller, the financial institution, will communicate to the buyer the details for the MBS pools that will be delivered on the contract. Two days later, on the settlement date, the specified securities are delivered and the buyer pays the seller the agreed upon price.

Trading in the TBA market follows standards called "Good-Delivery Guidelines," that detail the rules covering quality and allowable variance between the delivered pools and the trade date criteria. The market functions, however, on the fundamental assumption that the MBS pools are homogeneous. Given the prior discussion in this chapter on the sample selection biases that are likely to be induced by the mortgage menu, mortgages with the same maturity and contract rates could have been chosen by borrowers who paid very different levels of discount points at origination and thus have very different preferences concerning their expected holding period for the loan. Stanton (1996) and Downing et al.(2009) find evidence that mortgage pools exhibit significant heterogeneity in their prepayment speeds, even for pools that appear homogeneous in other contract features such as the contract rate, loan-to-value ration, and origination date. This evidence is, of course, inconsistent with the fundamental homogeneity assumption underlying the TBA market.

The second channel to the creation of the single-class pass-through is a direct sale for cash of the mortgages to a GSE. The GSE then securitizes the mortgages through a grantor trust SPV that funds the mortgage purchase by selling the MBS/PCs forward on the TBA market. The TBA trades made through this channel follow the same trade sequence as discussed above. Here again, the forward-contract trade occurs only on the basis of the six criteria under the presumption of homogeneity in the pool performance.

A second important feature of the TBA market is that it is the only segment of the mortgage market in which there are publicly available pricing data. Daily trading

[22] Another feature of the TBA market, is that originators can also carry out a closing sale transaction, basically a cash settlement of the contract, or they can roll the delivery commitment to a later TBA contract.

prices are available for a wide range of coupons and maturities for GNMA, Freddie Mac, and Fannie Mae TBA forward contracts. These prices are widely used as a benchmark for other segments of the mortgage market for which no price data are available. Other than the TBA market, all the other mortgage trading occurs within proprietary over-the-counter markets. An important limitation of the TBA prices is that they cannot be matched to the true underlying characteristics of the mortgages in the pools. These characteristics are unobservable due to the rules for trading forward contracts and the fundamental assumption of the TBA trading platform that one MBS pool can be considered interchangeable with another. The cheapest-to-deliver option that is embedded in the forward-contract trading rules and the Downing et al. (2009) empirical results suggest, instead, that the TBA market is a market for lemons. This evidence suggests that the benefits for the liquidity of the TBA market is intrinsically linked to a tacit market acknowledgement that TBA trading prices reflect the likelihood that, on average, only the worst pools will be delivered.

In the final stage of the two-stage REMIC securitization process, market participants deliver MBS pools to the GSE in return for a pro rata share of each REMIC tranche, a swap, or through a direct sale for cash. Because market participants very often exchange equivalent *principal* positions of MBS pools for *principal* positions in the REMIC bonds, the REMIC origination market also embeds a cheapest to deliver option. As in the TBA market, market participants maximize the value of the REMIC cheapest-to-deliver option by contributing low value pools, and they expect all other market participants to do the same. The assets of the REMIC SPVs are pools of the pass-through securities from numerous grantor trust pools.

The REMIC trusts, different from the Grantor Trust organizational structures, are allowed to issue multiple classes of securities with, often highly, customized cash flow structures. The primary advantage of the REMIC trusts is the diversification potential of pooling the pass-through pools and the opportunity to customize the payout structure of the REMIC bonds to meet the objectives of a broader investor base. Since the pass-through pools must issue bonds that are essentially a fractional interest in a long-term, 15 or 30 year, mortgage security, their payout structures are not attractive to investors with portfolios of short-term liabilities such as banks. The REMIC bonds can be easily customized to create both shorter-term and longer-term bonds from the cash flows of the underlying pass-through pools and thus they appeal to a broader set of investors.

Pass-through and REMIC mortgage data availability

Fannie Mae and Freddie Mac provide access to the underlying mortgage data for their pass-through and REMIC bonds and these data are organized by pool CUSIP. Through their website, `www.freddiemac.com/`, Freddie Mac provides the loan-level origination and performance data for each pool. The REMIC prospectuses also provide the CUSIPs that make up the REMIC pools, so that the disaggregate loan-level information associated with each CUSIP can be assembled. Through their website, `www.fanniemae.com`, Fannie Mae also provides loan-level origination and performance data for the whole loans that underlie their REMIC pools. These data are organized in a manner similar to that described in Tables 18.3 and 18.4. The agency loan-level data again have no unique loan ID but instead use an ID system unique to each agency. The agency data also do not include the actual loan origination date, and instead only have the "note signed date" which is the date the mortgage note was transferred into the pool. Finally, there is no information about the discount points that the borrower paid at origination.

The GSE websites also provide access to the prospectuses for each REMIC. These documents provide the legally binding rules for the payouts to the REMIC bonds, identify the pass-through CUSIPs that make up the collateral of the REMIC, and provide summary statistics for the underlying loan collateral, among other information concerning the management of the trust and the pooling and servicing agreements with the trustee. For analysts to evaluate the performance of a specific REMIC bond, they would have to: (1) assemble the loan-level origination and performance data for each pass-through CUSIP that is included in the REMIC; (2) write and run customized computer code to generate the contractual loan-by-loan principal and interest payments for all of the loans in the deal; (3) use the prospectus to determine the cash flow allocation rules for the various classes of bond that make up the REMIC capital structure; and (4) write and run customized computer code to generate the contractual bond-level principal and interest allocations. All of these data preparation steps precede the actual analytical procedures required to determine the "expected," or risk adjusted, cash flow structure of the bonds conditional on the performance of the embedded prepayment and default options that are exercised by borrowers.

An additional type of data that is available on the GSE websites is pool-level origination and performance data by CUSIP for both the pass-through and REMIC bonds. The pool-level data are reported as bond-level summary statistics. They are often used to model the performance of forward contracts since the underlying mortgage data are limited to the six TBA market criteria. The pool-level data are also frequently used to determine the mortgage "option adjusted spreads" which are correction factors for the spread between observed and modeled TBA forward yields. The OAS is often used as a proxy measure for the degree of MBS market illiquidity, perhaps due to supply and demand disequilibria.

Prior to the financial crisis, the pool-level data were the only information available on agency pools. As shown in Table 18.6, the pool performance data are quite limited, providing information only about the mortgage balance weighted averages

Table 18.6 *Typical data fields for agency pool-level origination and performance*
The data fields reported in this table summarize the data fields that are available from the Freddie Mac and Fannie Mae websites.

	Field description
CUSIP number	
Pool number	
Pool type	
Original pool balance	
Current date	
Current balance	
Maturity date	Last payout date on the pool
Original weighted average maturity (WAC)	
Current weighted average maturity (WAC)	
Original weighted average coupon (WAM)	
Current weighted average coupon (WAM)	
Weighted average loan age (WALA)	
Current factor	A measure of early principal terminations
Pass-through rate	The WAC minus the servicing fee
Issue date	CUSIP issuance date
Seller name	Seller of the mortgages
Servicer address	
State concentrations	
Loan types	Fixed or adjustable

of the coupon and maturities of the underlying loans, some information about the loan originators and the state locations of the loans, and information about termination speeds, which are the performance "factors."

Overall, data availability has improved significantly in the agency pass-through and REMIC bond markets. Nevertheless, the data are not at all easy to assemble and they must be downloaded CUSIP by CUSIP from the agency websites. There is also no publicly available standardized clearing source for the data. Instead, the data must be downloaded from the two agency websites where they are available in text files without comma delimiters.

18.3.2 Mortgage data transfer through the private-label supply chain

As shown in Figure 18.2, the private-label mortgage securitization supply chain begins with a borrower who takes out a mortgage loan on a residential property working with an agent of the financial institution, such as a loan officer, or with a mortgage broker who receives a fee from the financial institution for their underwriting services. Both the loan officer and the independent broker make their loan offers to borrowers based upon the originator's wholesale-rate sheets, or mortgage

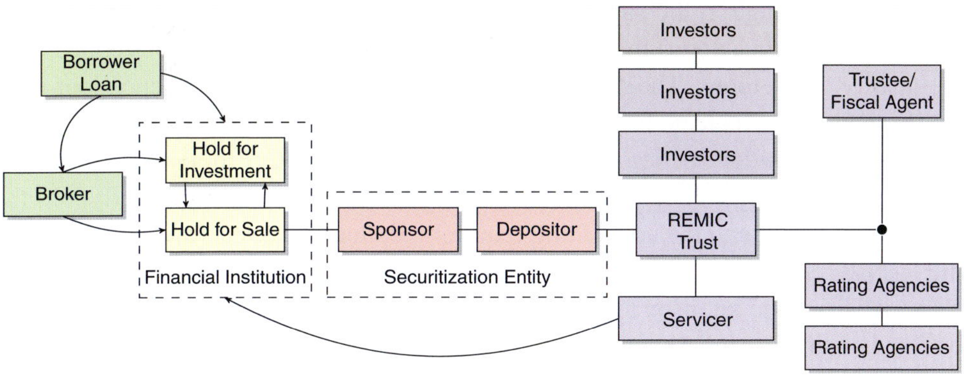

Figure 18.2 Mortgage transaction flow for private-label securitization.

menu, as discussed above. As discussed above, if the financial institution, shown in Figure 18.2 is a bank, a credit union, or a savings and loan institution, the institution has a choice of holding the mortgage on their balance sheet in their "Hold-for-Investment" portfolio or, alternatively, if they intend to actively trade or eventually sell the loan, they can hold the loan in their "Hold-for-Sale" portfolio. If the originator is a mortgage bank or a finance company, the loans will be sold to an aggregator, who is either an investment bank or a large commercial bank, and the aggregator will initiate the rest of the securitization process with a securitization entity as shown in the second hashed box in Figure 18.2.

In the next stage in the securitization supply chain, the originators sell the loans to a sponsor, as shown in Figure 18.2.[23] The sponsor is a special-purpose stand-alone entity, often affiliated with a large financial institution or investment bank. For this reason, the sponsor is shown in the figure as part of the securitization entity. The sponsor purchases the mortgages and receives the loan-by-loan origination and performance data. The sponsor structures the securitization by devising the bond payout rules and the capital structure of the bonds under the advice of two rating agencies and the underwriter who will sell the bonds.[24] The sponsor often retains the "first-loss" bond or "over-collateralization (OC)" bond.[25]

[23] The originator and the sponsor could also be the same entity. In this case, the first transfer would be from sponsor to depositor.

[24] In the typical private-label structuring there would be 18–20 bond classes rated from AAA to below investment grade. There would also be one or more residual classes making up the equity position in the trust and bearing all of the tax liability. The residual class is held by the depositor.

[25] These bonds are structured to pay high yields if the pool performs well. The "first-loss" bond is the first bond to bear exposure to default risk if the pool experiences principal losses due to mortgage defaults. The OC bond receives a proportion of the cumulated excess interest (a sum aggregated from the spread differential between the coupon interest paid by the mortgages and the coupon interest earned on the senior bonds). The OC bond functions as an alternative form of subordination support to protect the senior bonds from losses.

In the third stage in the supply chain, the sponsor initiates the securitization by selling the loans to a depositor along with warranties and representations concerning the quality of the loans. The depositor is an entity designed to be independent from the sponsor and to have no liabilities or risk of bankruptcy. The depositor then issues the certificates under Section 15 of the Securities Act, USC 77b(a)(4).

Prior to January 1, 2002, the sponsor would have sold the loans directly into the SPV. However, due to the accounting requirement that the assets (the mortgages) of the REMIC SPV must achieve legal isolation from the seller of the loans, this one-step sale from the sponsor to the REMIC became problematic when the sponsor retained positions such as a retained interest in the "reserve fund" or credit enhancements of the SPV.[26] For this reason, starting January 1, 2002, the accounting authorities affirmed that using a two-step process, with an additional sale from sponsor to depositor, presumptively created "legal isolation."[27]

In the final stage of the chain, the depositor sells the mortgages to a REMIC trust. The depositor has ongoing responsibilities in conjunction with the trustee to appoint a successor servicer and/or to appoint a successor trustee. Usually, the depositor is a corporate affiliate of the sponsor, so it is shown in Figure 18.2 inside the securitization entity, along with the sponsor. The certificates of the REMIC SPV are sold to investors, who receive payments of principal and interest following the allocation rules defined in the REMIC prospectus.

The prospectus and the prospectus supplement: defining the certificate payment rules

The Federal National Mortgage Association Charter Act provides that securities issued or guaranteed by Fannie Mae will be considered exempt securities. The Federal Home Loan Mortgage Corporation Act contains a similar provision for securities issued by Freddie Mac.[28] These exemptions allow Fannie Mae and Freddie Mac pass-through and REMIC securities to be offered and sold without registration under the Securities Act. Although the GSEs do not file any MBS offering materials with the SEC, they do prepare offering documents that are similar in form to the core prospectuses filed by private-label issuers in registered offerings. They then make the deal-specific information available through either a final prospectus supplement or website disclosures.

[26] Deloitte and Touche, *Learning the Norwalk Two Step*, Heads Up, April 25, 2001, at 4. Our survey of securitization agreements suggests that it is quite common for the sponsor to retain interests in the mortgages through its ownership of certain classes of certificates. Jim Johnson and Jim Mountain, 2005, *Securitization Accounting: The Ins and Outs (And Some Do's and Don'ts) of FASB 140, FIN 46R, IAS 39 and More.*

[27] See Financial Accounting Standards Board, Statement of Financial Accounting Standards No. 140 §9 (2000) and FASB, Technical Bulletin No. 01–1 (2008).

[28] See 12 U.S.C. §1723c.; 12 U.S.C. §1455g.

In contrast, the REMIC securities of private-label MBS are subject to the registration requirements of the federal securities laws. To offer and sell these securities, the sponsor must file a registration statement with the SEC and this statement must meet the disclosure, content, and procedural requirements of the Securities Act. When private-label issuers file a registration statement to register an issuance of a REMIC security, they typically use a "shelf registration."[29] The sponsor first files a disclosure document, known as the "core" or "base" prospectus, that outlines the parameters of the various types of REMIC securities offerings that will be conducted in the future through the sponsor's shelf registration. The registration statement will also contain the form of a prospectus supplement outlining the format of deal-specific information that will be disclosed when a shelf offering actually occurs.[30] REMIC SPVs could also choose not to register the securities offering and thus limit the eligible set of investors in the certificates to 144A investors.[31]

Fannie Mae and Freddie Mac post their base prospectuses and the prospectus supplements for GSE REMICs and pass-through securities on their websites. The prospectus supplements for private-label securitization can be obtained from the Security and Exchange (SEC) website (`www.sec.gov`),[32] or by subscription access to Bloomberg, CTSlink.com, or ABSnet.com. As shown in Table 18.7, the disclosures found in the prospectus supplements of private-label securitizations are very detailed and they are not in standardized formats. The purpose of the document is to reveal all the material characteristics of the securities offering, including the cash-flow structure of the bonds, possible risk factors, prepayment and yield considerations, as well as detailed information on the duties of the various agents involved in maintaining the smooth operation of the trust for the benefit of the certificate holders.

[29] At the time of enactment of the Secondary Mortgage Market Enhancement Act (SMMEA), the SEC amended Rule 415 of the Securities Act, known as the shelf rule, to allow SMMEA-eligible mortgage-related securities to use the shelf-offering process (see Simplification of Registration Procedures for Primary Securities Offerings, Release No. 33–6964, Oct. 22, 1992, and SEC Staff Report: Enhancing Disclosure in the Mortgage-Backed Securities Markets, January, 2003, `www.sec.gov/news/studies/mortgagebacked.htm#secii`).

[30] Prospectus supplements are filed under Rule 424(b) of Regulation C. Under Rule 430b and Rule 430C. (See *Frequently Asked Questions about Shelf Offerings*, Lloyd S. Harmetz and Nilene R. Evans, Capital Markets Group, Morrison & Foerster LLP, 2011, `www.mofo.com/files/uploads/documents/faqshelfofferings.pdf`).

[31] The Rule 144A is a non-exclusive safe harbor from the registration requirements of the Securities Act and it permits resales to institutional investors that meet the criteria for "qualified institutional buyer" (QIB) of certain privately placed securities (see 17 C.F.R. §230.144A, and SEC Staff Report: Enhancing Disclosure in the Mortgage-Backed Securities Markets, January, 2003, `www.sec.gov/news/studies/mortgagebacked.htm#secii`).

[32] For the full search, click on "Search for Company Filings" under "Filing & Forms(EDGAR)." Under "General-Purpose Searches" click on "Companies & other filers" and then, in the "Enter your search information" dialog box, give the issuing entity name next to "Company name" and click on "Find Companies."

Table 18.7 *Disclosures in the prospectus supplement to a private-label REMIC prospectus*

The data fields reported in this table summarize the usual table of contents for private-label prospectus supplements.

	Field description
Front page	Identification of total deal principal offered, offered bond classes: Initial principal by class; Calculation rules for the offered bond pass-through rates; Offered bond types; Offered bond ratings; Assets of the issuing entity; Credit enhancement; and Interest rate support.
Risk factors	Prepayment, default, swap counterparty risk, risky loans, house price risks, insufficient bond overcollateralization.
Mortgage loan pool	Aggregate loan summary statistics, loan underwriting guidelines.
Sponsor	Described above.
Depositor	Described above.
Servicer	Services and administers the mortgage loans on behalf of the issuing entity for the benefit of the certificates.
Master servicer	Aggregates monthly servicer reports and service oversight.
Static pool information	Sponsor's past securitization performance history.
Issuing entity	Legal name of the REMIC SPV.
Securities administrator	Preparation and distribution of the monthly service reports.
Trustee	Role defined by the pooling and serving agreement.
Custodian	Responsible for administering the loans on behalf of the REMIC and safeguarding the mortgage notes and mortgage files on behalf of the certificate holders.
Interest rate cap and swap counterparty	
Description of the certificates	Principal and interest distribution rules by bond class.
Pooling and servicing agreement	Defines responsibilities and rights of the servicers, trustee, among others over principal and interest collections and foreclosure processes.
Prepayment and yield considerations	"Prepayment assumption" used to compute quoted bond yields.
Legal aspects of the mortgage loans	General information about state real property law.
Federal income taxes, state and local taxes	Information on REMIC tax treatment among other tax considerations.
ERISA considerations	General information on fiduciary responsibilities under the Employee Retirement Income Security Act (ERISA).
Legal investment	Bond eligibility as "mortgage related securities" under Secondary Mortgage Market Enhancement Act (SMMEA).
Methods of distribution	Allowed methods for selling the certificates.
Ratings	Information on on-going rating requirements.
Reports to certificate holders	Requirements for monthly pool performance reports.
Glossary of terms	

Table 18.7 presents the primary disclosure categories found in most prospectus supplements. Overall, the prospectuses are very difficult to read and are usually lengthy (often more than 300 pages of text and tables). The first page of every prospectus identifies the corporate names of the depositor, issuing entity, sponsor, servicer, and shelf name. It also provides the total pool offering amount and information on the offered bonds such as the initial principal of each class, the pass-through rates, the bond types, and the bond ratings. There is also information on the assets of the SPV, the credit enhancement for the senior bonds, and the interest-rate risk support, usually in the form of an interest-rate swap, and the identity of the swap counterparty.

As shown in Table 18.7, numerous risk factors are outlined for the investor. The effect of these factors on bond performance is summarized generally. The supplement also provides general information on the effects of the prepayment assumptions used to structure the deal, State and Federal tax issues, and aspects of state real-property law that may affect the performance of the bond if foreclosure or default occurs. The prospectus supplement also identifies the functions of the agents involved in the operation of the SPV on behalf of the certificate holders. These entities include the servicers, trustee, sponsor, depositor, securities administrator, custodian, and any swap counterparties. The information provided about the underlying mortgage collateral is in the form of aggregate summary statistics and in some cases summary statistics are also provided for subsets of the loan data such as adjustable rate and fixed rate mortgage products. The summary statistics focus on characteristics of the loans at origination such as the distribution of origination coupons in the pools, the maturity distributions, the geographical coverage of the pool, or pools, the loan-to-value ratio distributions, the debt-to-income ratio distributions, and the FICO score distributions. The summary statistics for the adjustable rate mortgages include information on the reset periods, the indices, average interest rate caps and floors, life-of-loan caps, margin distributions, and negative amortization distributions.

The prospectus supplement also provides very precise principal and interest distribution rules for all of the offered bond classes.[33] The bond distribution rules are often highly customized. The expected cash flows for the bonds are contingent on the subordination structure of the pool and on the performance of the underlying collateral, including the degree to which the loans in the pool are well diversified over key characteristics such as coupon, product type, house-price dynamics, and regional business cycles. However, because the underlying mortgage origination data are not available to analysts at the time of the offering, it is very difficult to

[33] Because the first-loss-bond, the OC bond, and the residual bonds are often not offered for sale, their class-flow distribution rules are not provided in the prospectus supplement.

estimate values for the offered bonds. The only other data available in the prospectus supplement are an analysis of the expected yields on the bonds if they are evaluated using the deal-structuring prepayment assumptions. The prepayment sensitivity analyses provided in the supplement are usually very stylized, and thus unlikely to be sufficient to make actuarially sound investment decisions.

The pooling and servicing agreements (PSA) govern the rights and responsibilities of the parties responsible for administering the REMIC SPV. Most of the focus of the PSA is on the responsibilities of the servicer, who is required by the agreement to use the same care as it customarily employs in servicing and administering similar mortgage loans for its own account, in accordance with customary and standard mortgage-servicing practices of mortgage lenders and loan servicers. The servicers of REMIC SPVs are also required to cover temporary shortfalls in the receipt of principal and interest payments from the mortgage borrowers if the servicer determines that such advances would be recoverable. The PSAs usually stipulate that the servicer is required to deliver to the securities administrator and the depositor a servicer remittance report every month to enable the securities administrator to make the distributions to the certificate holders. The agreements also define the servicing oversight duties of the master servicer, the securities administrator, and the custodian, whose responsibilities include safeguarding the mortgage and note files.

In the wake of the financial crisis, the PSAs have generated significant criticism because of their default and loan-modification provisions. Although there is considerable variety in the wording of the PSAs across REMICs, in general the agreements require the servicer to make reasonable efforts to collect all payments called for under the contractual requirements of the individual mortgages. Under the PSAs, the servicers are often allowed to: (1) waive any late payment charge or, if applicable, any penalty interest; or (2) extend the due dates for the monthly payments for a period of not more than 180 days. The PSAs often allow considerable latitude in the servicer's choices concerning the resolution of default, or the expectation of default. These choices often include: (1) foreclosing on the mortgage loan; (2) accepting the deed to the related mortgaged property in lieu of foreclosure; (3) granting the borrower under the mortgage loan a modification or forbearance, which may consist of waiving, modifying or varying any term of such mortgage loan (including modifications that would change the mortgage interest rate, forgive the payment of principal or interest, or extend the final maturity date of the mortgage); and (4) accepting payment from the borrower of an amount less than the principal balance of the mortgage loan in final satisfaction of the mortgage loan. In all instances, the final maturity date of any mortgage loan is not extendible beyond the final scheduled distribution date for the certificates.

REMIC data reporting

As previously discussed, both prior to the crisis and currently, there is no loan-level information available for the mortgage collateral held as assets in the REMIC SPVs at the date of the issuance of the prospectus supplement or the date of the initial offering of the certificates. Instead, the supplement provides information on the aggregate principal balance at origination, the number of underlying loans, and often the average geographical composition of the pool(s). In addition, the mortgage-origination data are summarized using principal-weighted averages for: scheduled loan principal balance; gross interest rate; interest rate net of servicing fee; original FICO score; original loan-to-value ratio; combined original loan-to-value ratio; combined original loan-to-value ratio with silent seconds; stated remaining term (months); seasoning (months); months to roll from fixed to floating; gross margin; initial rate cap; periodic rate cap; gross maximum lifetime rate; percentage of mortgage loans with silent seconds; debt to income ratio at origination; and percentage of mortgage loans with mortgage insurance. Because many of the REMIC SPVs were composed of more than one distinct pool of mortgages, often the summary statistics would be provided for each of the sub-pools rather than for the collateral aggregates. Of course, mortgage analytics based solely on this information would be challenging, because the full distributional effects of the loan characteristics on the payments of principal and interest could not be specified.

The prospectus supplement also contains very limited information concerning the expected effects of the embedded prepayment and default options on the bonds' cash flows and yields. As previously discussed, the supplement only provides information on the "assumed prepayment levels" that were used to determine the initial subordination structure of the bonds and to determine expected yields on each of the bond classes *without providing any information on the expected sales prices of the bond classes*.[34] The prospectus supplements are usually much less analytical concerning default. Instead the prospectuses tend to include only a general discussion of the expected risks of default among the other listed risk factors for the bond offerings.

Actually modeling the effects of the embedded default and prepayment options, of course, would require performance data. However, at the date of the offering there would only be about three to four months of the pool performance data available. With only information on the weighted averages of the loan collateral, applying risk analysis metrics would be akin to modeling the performance of one mortgage, or a

[34] One significantly missing piece of information on the front page of the prospectus is the offered price for each of the bond classes. As previously discussed, due to the over-the-counter nature of this market pricing information is considered proprietary.

defined number of identical mortgages, rather than a diversified pool of loans. These significant analysis limitations existed even though it was, and is, well known that actuarially accurate pricing of these instruments requires information concerning the covariance of contractual elements of the underlying mortgages, interest rates, and house prices and about higher moments of the relevant distributions, such as the variance of the contract elements, interest rates, and house prices.

Table 18.7 presents the monthly remittance data that would be available from a REMIC trustee. Most of the fields are self-evident. As shown, the reports focus on the cash flow performance of the certificates, the performance of the principal subordination available for the outstanding bonds, the aggregate and loan-level delinquency, default, foreclosure, and prepayment status at the end of the accrual month. The remittance reports also provide information on the possible contingent payout structures for the OC bonds, which as previously discussed are primarily held by the sponsors. OC flag fields are indicators as to whether or not the various triggers that could to lead to payouts to the OC bond class are true or false. The logic of these triggers is specified in the prospectus supplement and must be accounted for to understand the cash supports that exist for the more senior offered bonds. Finally, the remittance report provides updated information concerning the level of advances paid by the servicer to bondholders for shortfalls in the receipt of interest payments (or information concerning the servicer's decisions not to make such advances) as well as the periodic fees that have been paid out to the interest rate swap provider, the interest cap provider, the guaranty provider, and the administrative fees.

During the run-up to the crisis, the only data that were available to analyze the loan origination and loan performance data for securitized mortgages were the data generated as the result of the PSA data management and reporting requirements stipulated in the prospectus supplement. These activities were carried out by the servicers and trustees of the REMIC SPVs. Interestingly, the prospectus supplements never require that the monthly remittance statistics for the principal and interest payouts on the loans, the loan balances, and current loan delinquency or prepayment status be subject to external verification by accountants. Access to the remittance data is possible through subscriptions to private vendors such as ABSNet Lewtan and Bloomberg, the servicers, such as LPS and LoanPerformance (now Corelogic), and the trustees, such as CTSlink. Since the vendors source their data differently, the data that they maintain and sell are in part unique from, and in part overlap, data available from other sources. Because there was, and is, no unique loan identifier, and because only some of the sources include data on the securitization status of the loans, it was and remains nearly impossible to obtain a consistent aggregate of securitized-loan characteristics and performance in the USA.

18.3.3 Legal foundations of transfer: notes and mortgage/deeds of trust

"Mortgages" in the USA consist of two contracts. The first is the promissory note, which establishes the borrower's legal obligation to repay the loan principal and interest, stipulates the periodic payment structure, defines the contractual rules for exercising the prepayment option, and identifies the conditions that would trigger default and foreclosure. States do not require that the promissory note be recorded for it to be enforceable.

The second contract, the mortgage (or deed of trust, security deed, or other similar instrument), grants a lien or other security interest in the borrower's real property to the lender (or to a trustee, for the lender's benefit) to secure the contractual obligations of the promissory note.

Although the law is unclear and varies from state to state, recording rules may be relevant to determining priority of ownership claims to the mortgage.[35] At least in some states, for any of the mortgage originator, the sponsor, the depositor, or the REMIC SPV trust to have clear first priority among competing claims to own the mortgage, it must have been the first to record the mortgage. Even if an unrecorded mortgage is enforceable against the borrower, it is exposed to the risk that it would lose priority to a junior lien that was created later but recorded promptly. And even if the mortgage itself is recorded, failure to record an assignment along the chain exposes the non-recording assignee to additional risk of losing priority to a subsequent assignee, particularly one who does record. For these reasons, the recipients of mortgage transfers anywhere in the mortgage transfer supply chain should have a strong incentive to record their mortgages as quickly as possible.

Generally speaking, mortgage recording is carried out at the office of the county recorder or equivalent official in the county where the collateral is located. The recorders' offices maintain records on who owns each tax parcel in the county and records the existence of liens on these properties in the form of mortgages and trust deeds, among others. County recorders typically are elected officials and recorders' offices usually charge a fee for each document that is recorded. For the two-step private-label mortgage securitization process, as discussed above, the mortgage and the promissory note must be sold at least twice to achieve legal isolation. Under the mortgage recording system, each subsequent owner of the mortgage in the mortgage transfer supply chain would need to re-record its ownership of the mortgage at the appropriate recording office for the property. Since the two-step process shown in Figure 18.2 usually was completed within four months after the loans were originated, the rapid growth of private-label securitization

[35] See discussion in Section 18.6. Hunt et al. (2012) discuss how recording protects mortgage owners from subsequently arising claims of ownership, and how failure to record exposes mortgage owners to such risks.

apparently put significant pressure on the processing capacity of recorders' offices (see Peterson, 2010a; Levitin 2010).

18.4 MERS and mortgage recording

In 1995, in what apparently was at least in part a response to the recording backlogs in recorders' offices, 28 mortgage industry companies and organizations including the Mortgage Bankers Association, Fannie Mae, Freddie Mac, First American Title Insurance Corporation, and large commercial lenders such as Wells Fargo Bank, Bank of America, Citimortgage, Chase, and Washington Mutual became shareholders of a closely held private corporation, called MERSCORP, Inc.[36] In 1998, a subsidiary of MERSCORP, Inc., called Mortgage Electronic Registration Systems, Inc. (MERS, Inc.) was incorporated in Delaware.[37] These entities are responsible for the Mortgage Electronic Registration System (MERS).

The purpose of MERS was to serve as the "mortgagee" in the county land records for mortgages registered on the MERS system. The corporate members of MERS have entered into a membership agreement with MERS, Inc. and MERSCORP, Inc. in which each member agrees that MERS, Inc. shall serve as the member's nominee as the mortgagee in the land records in exchange for the Member registering the mortgage on the MERS system.[38] As shown in Figure 18.3, MERS was designed on the assumption that as long as all the mortgage transfers within the two-step process occur within the MERS membership list, no further recording is required because MERS remains the owner of record at all times. Thus, the county recording system prior to MERS, where at least three assignments of the mortgage would have occurred in the typical two-step securitization process shown in Figure 18.2, has now collapsed into a system with only one recording of the mortgage at the recorder's office, as shown in Figure 18.3. In addition, under the MERS system there is only one fee payment to the recorder, whereas internally MERS charges a two-part tariff that includes an annual membership fee for its 5,643 members and a payment for each mortgage "e-registry" ($6.95) and each mortgage transfer ($2.00). Under the new one-time MERS recording structure, the borrower pays the recording fee at origination.[39]

As shown in Figure 18.3, the mortgage remains recorded in the name of MERS, Inc. as nominee for the originator and its successors in interest through all of the transactions within the mortgage securitization supply chain. If the mortgagor

[36] `www.mersinc.org`

[37] See Certificate of Incorporation of Mortgage Electronic Registration Systems, Inc. State of Delaware Secretary of State, Division of Corporations, Filed 03:01 pm 12/30/1998, 981509524–2990193. For brevity, we refer to this entity as "MERS, Inc."

[38] See MERS Terms and Conditions §2, `www.mersinc.org`.

[39] `www.mersinc.org`

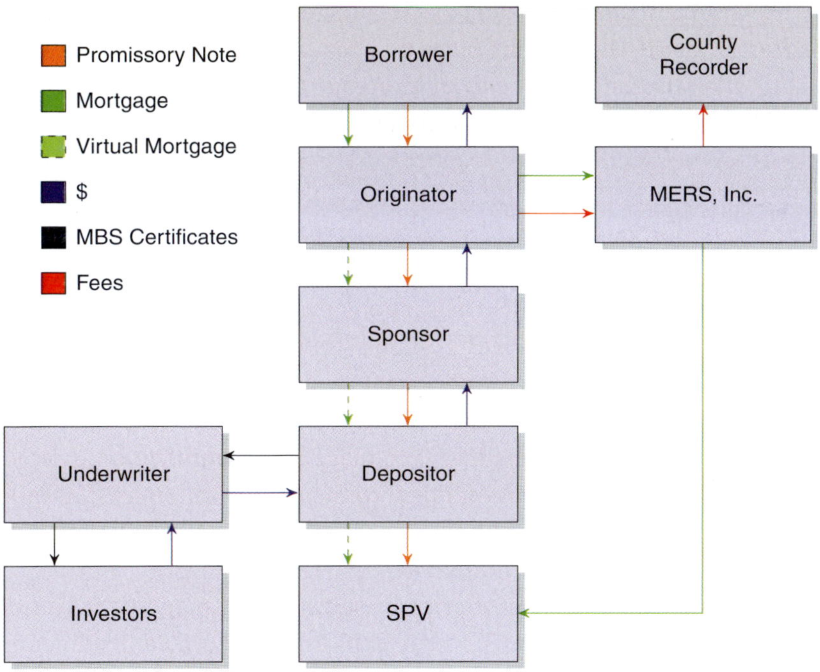

Figure 18.3 Mortgage recording with MERS.

defaults, MERS, Inc. may assign the mortgage to the securitization trustee or its servicer for foreclosure and record that assignment, but up until that time the mortgage remains recorded in the name of MERS, Inc. as nominee.[40]

18.4.1 State laws may require owners to record mortgage assignments to protect their interests

As discussed in the appendix in Section 18.6, all 50 states and the District of Columbia have real property recording statutes. Recording statutes conventionally are divided into three categories: "pure race," "race-notice," and "pure notice." A purchaser who fails to record an interest in real property may under certain circumstances lose that interest to a subsequent purchaser under all three types of statutes. The specifics differ from category to category; in particular, the different types of statutes give differing effects to what the second purchaser knows and when the second purchaser records (Powell and Rohan, 2011).

[40] Although MERS, Inc. is described as a "nominee" in many documents, other documents from MERS, Inc. and MERS' members suggest that MERS, Inc. is not just a nominee (at least if "nominee" means something different from "agent") (see Hunt et al., 2012).

States' laws are not just heterogeneous but unclear. A mortgage conventionally is described as a conveyance of an interest in real estate (American Law Institute, 1997), suggesting that one might expect mortgage assignments to be covered by the recording statutes. However, it is not always clear just what follows if mortgage assignments are not recorded. In Section 18.6, we discuss this question under the laws of California and Florida, the two states with the largest numbers of private-label securitized mortgages.

To the extent that state law requires recording of mortgage assignments for full protection of the assignee, MERS transactions may have problems. In the typical MERS transaction, at least up until foreclosure the only public record of the mortgage states that MERS, Inc. is the legal owner of the mortgage on behalf of the originator and its successors. There is no explicit public record of the several subsequent mortgage assignments that take place in a securitization. For example, there is no public record of any assignment to the securitization trustee, at least for mortgages that are not in foreclosure. The securitization trustee's interest is unrecorded, and therefore potentially vulnerable.

The heterogeneity and lack of clarity in states' laws suggest that an effort to standardize and clarify the law is worth exploring. Indeed, amendments to the Uniform Commercial Code adopted early in the last decade seem to move in this direction by purporting to give full protection to unrecorded mortgage assignments in all states. As we discuss in Section 18.6, it is not clear how these amendments interact with other law, so it is not clear how successful they have been. Moreover, the Code's approach may be criticized normatively because it does not provide an incentive to maintain public records of mortgage ownership.

Further efforts to bring further clarity and uniformity to state mortgage assignment law could face challenges, however. To be sure, uniform acts in some areas have been quite successful (examples include the Model Penal Code, Model State Administrative Procedure Act, Model Unfair Trade Practices Act, Uniform Anatomical Gift Act, and most parts of the Uniform Commercial Code) (Rosinia, 2012). But real property is a special case. It is linked to the territorial aspect of state sovereignty (Florey, 2009) and states' unique interest in regulation of land continues to be recognized by the Supreme Court (US Supreme Court, 1997). Coordinating a common set of clearly applicable mortgage-assignment rules will present political as well as technical hurdles given longstanding differences in state real-property statutes and thus in the legal foundations of transfer from state to state.

18.4.2 Data risks to the property registration system

The creation of MERS apparently responded to a perception that county real-property records systems were dated, expensive, and sometimes cumbersome.

The county recorders' (or equivalent county offices') systems for real property recording are designed to: (1) verify the "recordability" of the documents that are submitted either in paper form or as scanned electronic copies from trusted document preparers such as title insurance companies and financial institutions;[41] (2) scan, organize, and store the recorded documents;[42] and (3) apply Optical Character Recognition (OCR) technology to capture some fields of the scanned recorded documents to allow for electronic search, permit aggregation and evaluation of key fields, and allow for on-line display at least internally in the recorder's office.[43] The putative benefit of the MERS registration system is that their members could sell mortgage loans to others within their system without having to record each transfer. The MERS system thus eliminated the delays and costs associated with frequently recording assignments of the mortgages and the deeds of trust through the county recorders' offices.

A significant potential cost of the MERS system identified by legal scholars (see Peterson, 2010a; Levitin, 2010) concerns the significant reductions in the transparency of country records concerning the actual ownership of the real property liens and inaccuracies related to the true identities of the grantees of beneficial interests under the Deeds of Trust in the event of a Deed upon Sale.[44] Nationwide about 66 million loans have been registered and tracked on the MERS system, and about 31 million of these loans are still active.[45] Coincident with the massive growth in MERS lien registrations, the loss of fee revenues from lien assignments at the local county recorders' offices, concerns, especially in California, about the lack of oversight in the expedited non-judicial foreclosure process and the large number of foreclosures that are currently taking place under this process, it appears that the informational quality of county lien registration records has degraded considerably.[46] Although MERS claims that the transfers between MERS members have been internally documented, these transfers are often unknown, and in some cases unknowable to those outside the MERS system. A second set of problems is that MERS apparently lacks internal verification mechanisms such as an audit trail, has not enforced any requirement that its members update assignments on its database, and does not keep digital or hard copies of assignments (see Peterson, 2010b).

[41] The process of verifying that the documents are properly signed and notarized. Verification of the accuracy of the recorded documents is left to law enforcement and the courts.

[42] For real property, these include: deed of trust; mortgage; notice of default; notice of trustee sale; trustee's deed; and deed in lieu of foreclosure

[43] Peterson (2010b, p. 9) claims that over "... 480 jurisdictions that are now offering fully electronic recording of mortgages and mortgage assignments."

[44] This document provides evidence that a foreclosure sale will be conducted by a trustee in exercise of power of sale.

[45] See Testimony of R. K. Arnold, President and CEO of MERSCORP, Inc. before the Subcommittee on Housing and Community Opportunity, House Financial Services Committee, November 18, 2010, at 11.

[46] There is some dispute about how useful county records systems were in identifying mortgage owners before the advent of MERS (see Hunt et al., 2012).

In light of these concerns, the San Francisco County Recorder's office commissioned an analysis of a sample of records, such as assignments and or Notice of Trustee's Sales, associated with foreclosures (see Pizante et al., 2012). In this study, the authors compared the foreclosing beneficial interest under the Deed of Trust that was named in the Trustee's Deed Upon Sale to the investor information found on the internal MERS system. The study results indicated that 112 of 192 subject loans had incorrect investor information on the MERS system. In addition, they found that the MERS information error rates were significantly higher than for non-MERS recorded Trust Deeds in all categories evaluated.

This new evidence on the relative quality of the information supposedly maintained by MERS in its registration system and its current role as the supposed custodian of accurate lien and assignment information is very sobering. These results suggest that significant damage may already have occurred in publicly available real property records at the local level in the USA. The cost of recovering from this damage and of designing modern electronic and verifiable lien assignment and transfer systems that are consistent with the real property laws of states presents a significant challenge in terms of time and cost. Ben Bernanke recently called for a federal title recording system that would accept electronic filings.[47] However, he did not indicate how this new system would be achieved or paid for. Solving this data management problem is a significant hurdle with broad public-policy implications for both the US housing and mortgage markets.

18.5 Conclusions

Despite the size and importance of the mortgage market in the overall US economy, current data management practices make it difficult or impossible for borrowers, lenders, investors and government regulators to perform the oversight and analysis functions necessary to maintain an orderly market and ensure fair pricing of securities backed by those mortgages. In this chapter, we have outlined the data available to the various participants, pointed out where these data are significantly lacking, and where it is most important that improvements be made.

18.6 Appendix: State recording statutes

All 50 states and the District of Columbia have real property recording statutes,[48] and how these statutes apply to recording of assignments of mortgages or similar

[47] See Board of Governors of the Federal Reserve System, 2012, The U.S. Housing Market: Current Conditions and Policy Considerations, 24–25, January 4.

[48] See LEXIS/NEXIS, *LEXIS/NEXIS 50 State Comparative Legislation/Regulations, Real Property Records and Mortgage Recording* (August 2011).

instruments is not always clear. We focus on two sample states, California and Florida. These are the two states with the largest numbers of mortgages securitized in private-label transactions. We discuss specific aspects of the law applicable to recording assignments for each state below. The issues we discuss are as follows.

Leading instrument We identify the legal form of the leading real property security device used in the state: the deed of trust in California and the mortgage in Florida.

Interest in real property Whether the instrument creates an "interest in real property" under state law is a basic question. It can become an important question in the event of a bankruptcy where the debtor has some kind of interest in mortgages, because the bankruptcy trustee has certain powers over "interests in real property" under Section 544(a)(3) of the Bankruptcy Code.[49]

Applicable recording law We cite the provision or provisions of state law that do or may relate to recording of assignments of mortgages or deeds of trust.

Consequences of failure to record We focus on the consequences of failure to record assignments of mortgages or deeds of trust for contests over ownership of the instruments and/or proceeds from the instruments. We do not discuss any potential consequences of failure to record for disputes over enforceability of the mortgage or deed of trust, for example in foreclosure litigation.

Ability to transfer note along with mortgage or deed of trust The *Restatement (Third) of Property: Mortgages* provides that an assignment of a mortgage or deed of trust may take the note along with it unless the parties manifest a contrary intention or the Uniform Commercial Code prevents transfer of the note.[50] In other words, according to the *Restatement*, the note may follow the mortgage when the mortgage is transferred. It is not clear that this formulation reflects current law in all states. We address this issue with respect to each state's law. This issue could become important in bankruptcy, as a bankruptcy trustee's powers with respect to the mortgage or deed of trust may relate to whether the debtor could transfer the note along with the mortgage or deed of trust.[51]

Interaction of recording statutes with Uniform Commercial Code The Uniform Commercial Code contains provisions, proposed in 1998 and adopted in all states by the beginning of 2002, that are relevant to mortgage transfer and ownership.[52] In general, these provisions may purport to make recording of

[49] 11 U.S.C. §544(a)(3). We discuss this issue in detail in Hunt et al. (2012).

[50] Restatement (Third) of Property: Mortgages §5.4(a)-(b) (1997).

[51] See Hunt et al. (2012).

[52] See, e.g., U.C.C. § 9–203(b) (creation of security interest in promissory note); U.C.C. § 9–203(g) (security interest in mortgage or other lien on real property attaches when security interest in note attaches); U.C.C. § 9–308(e) (security interest in mortgage or other lien on real property is perfected when security interest in note is perfected).

mortgage assignments irrelevant to contests over mortgage ownership.[53] However, it is not clear how the Code provisions interact with real-property recording law. These provisions may contradict some states' recording laws, and not all states have changed their mortgage recording law to acknowledge the primacy of the Code. We discuss the interaction of the Code and real-property recording law with respect to each state's law.

California

Leading instrument

The deed of trust is the leading instrument for security interests in real property in California.[54] Under a deed of trust, the owner of property (trustor) conveys title to the property, at least in form, to a trustee who holds title on behalf of a beneficiary.[55] Typically, the borrower is the trustor.[56] The trustee may be a bank or title company.[57] Historically, the lender typically has been the named beneficiary.[58] Despite use of the name "trustee," the trustee of a deed of trust is sometimes described not as a true trustee but as a "ministerial statutory actor"[59] whose function is limited to selling the property in foreclosure or returning title to the trustor upon payment of the debt.

Interest in real property

Whether the beneficial interest under a deed of trust is treated as an interest in real property under California law is unclear and appears to be context specific.

A federal bankruptcy court applying California law recently found that "deeds of trust. . . are interests in real property,"[60] so that a provision of the Bankruptcy Code governing "transfer[s] of real property"[61] applied to the assignment of deeds of trust.[62] As described in more detail below, this court held that unrecorded mortgage assignments before it were vulnerable to avoidance by the bankruptcy trustee.

Courts frequently are called upon to describe the beneficial interest in a deed of trust without making a holding as to its status, and when they do so they routinely

[53] See generally Permanent Editorial Board for the Uniform Commercial Code, Application of the Uniform Commercial Code to Selected Issues Relating to Mortgage Notes (November 14, 2011).

[54] Harry D. Miller et al., 10 Miller & Starr California Real Estate § 10:1 (2003 & Supp. 2011); Roger Bernhardt, 1 California Mortgages, Deeds of Trust, and Foreclosure Litigation § 1.35 (2012) ("In California, the deed of trust has completely eclipsed the mortgage as the lending and title industries' preferred security instrument.").

[55] Bernhardt, note 54, §1.35. [56] Bernhardt, note 54, §1.39 (noting that trustor need not be borrower).

[57] Bernhardt, note 54, §1.40.

[58] Bernhardt, note 54, §1.43. The use of MERS purports to change this practice. §1.25B.

[59] Bernhardt, note 54, §1.40 (citing Pro Value Props. v. Quality Loan Serv. Corp., 88 Cal. Rptr. 3d 281 (Cal. Ct. App. 2009)).

[60] See In re Cedar Funding, 2012 WL 1346365, at *4 (Bankr. N.D. Cal. April 5, 2010).

[61] 11 U.S.C. §547(e)(1). [62] See Cedar Funding, note 60, at *4.

describe the beneficial interest as a security interest in the underlying property.[63] Statutes governing specific aspects of deeds of trust state that a deed of trust creates an interest in real property. California's Marketable Record Title Act provides that a deed of trust creates "a security interest of record in real property,"[64] while California's attachment statute refers to an "interest in real property arising from . . . a deed of trust."[65]

On the other hand, a California court applying a different statute suggested that a deed of trust does not involve any conveyance of an interest in real property, thus implying that the beneficial interest under a deed of trust is not an interest in real property. In *Secrest v. Security Nat'l Mortgage*,[66] the court was called upon to decide whether the Statute of Frauds covered an agreement to modify a deed of trust. The Statute of Frauds generally provides that agreements it covers must be evidenced by a writing to be enforceable,[67] and one type of agreement it covers is an agreement to sell an interest in real property.[68] The court distinguished cases from other states holding that mortgages are interests in land and therefore covered by the statute, noting that in those states "mortgages are considered conveyances of legal title."[69] The court stated that "[i]n contrast, under California law, a mortgage or deed of trust is a lien on property."[70]

This decision, with its distinction between a "conveyance of title" and a "lien" thus reflects the view that California embraces the "lien theory" and not the "title theory" for mortgages and deeds of trust.[71] It is not totally clear why the choice of title theory as opposed to lien theory should matter to whether the beneficial interest under a deed of trust is an interest in real property, as a lien generally is held to be a security interest in the underlying property.[72]

[63] See, e.g., Cadlerock Joint Venture v. Lobel, 143 Cal. Rptr. 3d 96, 99 (Cal. Ct. App. 2012) (when two separate loans are secured via separate deeds of trust by the same real property, senior creditor's nonjudicial foreclosure "extinguishes the junior lienholder's security interest"); Lona v. Citibank, 134 Cal. Rptr. 3d 622, 637 (Cal. Ct. App. 2011) ("deeds of trust granted [lender] a security interest, with the power of sale, in residential property"); 1538 Cahuenga Partners, LLC v. Turmeko Props., 97 Cal. Rptr. 3d 552, 553 (Cal. Ct. App. 2009) (lien purported "to have priority over the security interests of [lender] through its deeds of trust.").

[64] Cal. Civ. Code, §882.020.

[65] Cal. Code Civ. Proc., §483.010(b); Bank of America, N.A. v. Stonehaven Manor, LLC, 113 Cal. Rptr. 3d 57, 59 (Cal. Ct. App. 2010).

[66] 84 Cal. Rptr. 3d 275 (Cal. Ct. App. 2008). [67] See Cal. Civ. Code, §1624(a).

[68] Cal. Civ. Code, §1624(a)(3). [69] *Secrest*, note 66, at 282–83. [70] *Secrest*, note 66, at 283.

[71] Under the lien theory, the security instrument grants a lien, defined as a charge imposed on specific property by which the property is security for the performance of the obligation. Miller et al., note 54, §10:2 (citing Cal. Civ. Code, §2872). Under the title theory, the security instrument grants the lender at least some attributes of legal ownership of the property. See Bernhardt, note 54, §1.35 (2012); Miller et al., note 54, §10:2; Michael Alan Wolf et al., 4 Powell on Real Property, §37.03 (2012).

[72] See Los Angeles Fed. Credit Union v. Madatyan, 147 Cal. Rptr. 2d 768, 772 (Cal. Ct. App. 2012) ("An equitable lien is a property interest that can be converted."); Aviel v. Ng, 74 Cal. Rptr. 200, 205 (Cal. Ct. App. 2008) ("A subordination agreement is a contract by which a party holding a senior lien or other real property interest agrees to lower its priority in relation to that of another holding an interest in the same property."); Berry v. Hannigan, 9 Cal. Rptr. 2d 213, 216 (Cal. Ct. App. 1992) (towing company operator's "property interest" in towed vehicles "is a lien").

But even assuming that the difference between the theories is relevant, and that the beneficial interest in a deed of trust can be a real property interest only under the "title theory," the trend in recent foreclosure cases is to find that deeds of trust, but not mortgages, are in fact governed by the title theory. The issue comes up frequently in cases evaluating Section 2932.5 of the California Civil Code, which provides that non-judicial foreclosure is available to the assignee of a "mortgagee or other encumbrancer" if "the assignment is duly recorded." It is not clear whether this statute covers deeds of trust or only mortgages, and the question is important because many foreclosure cases involve deeds of trust that were transferred without recorded assignments. Most courts have found that the provision does not cover deeds of trust, so that assignments of beneficial interests in deeds of trust do not have to be recorded as a prerequisite to foreclosure, because Section 2932.5 applies only to mortgages and not deeds of trust. A stated basis for these decisions is that a deed of trust involves a conveyance of title to the underlying property to a trustee while a mortgage does not.[73] Although embraced by most courts, this interpretation does seem to depart from a previous consensus that mortgages and deeds of trust were functionally identical[74] and governed by the lien theory.[75]

Courts' decisions about whether beneficial interests in deeds of trust are real property interests seem to be heavily influenced by the context in which they arise and do not seem consistent across statutory contexts. This seems to reflect the fact

[73] See Calvo v. HSBC Bank USA, 130 Cal. Rptr. 3d 815, 818 (Cal. Ct. App. 2011) (requirement of recorded assignment for enforceability applied to mortgage and not deed of trust on grounds that deed of trust passes title to trustee with the power to transfer marketable title to a purchaser). Later decisions of the California state courts have followed Calvo. See Haynes v. EMC Mortg. Corp., 140 Cal. Rptr. 3d 32 (Cal. Ct. App. 2012); Herrera v. Federal Nat'l Mortg. Ass'n, 141 Cal. Rptr. 3d 326 (Cal. Ct. App. 2012). Federal courts applying California law generally have reached the same conclusion. See, e.g., In re Salazar. 470 B.R. 557, 562 (S.D. Cal. 2012); Myers v. Encore Credit, 2012 WL 4511033 (E.D. Cal. Sept. 30, 2012) (adopting reasoning of Calvo); Pedersen v. Greenpoint Mortg. Funding, Inc., 2012 WL 4510854, at *10 (E.D. Cal. Sept. 30, 2012) (same); Wadhwa v. Aurora Loan Servs., LLC, 2012 WL 762020, *11-*16 (E.D. Cal. March 8, 2012) (same). But see Tamburri v. SunTrust Mortg., Inc., 2011 WL 2654093, at *5 (N.D. Cal. July 6, 2011) (finding "serious questions" about whether assignments of beneficial interest in deed of trust had to be recorded for foreclosure in part because of possibility that the "formalistic distinctions between mortgages and deeds of trust are outdated") (internal quotation omitted); In re Cruz, 457 B.R. 806, 816 (Bankr. S.D. Cal. 2011) (distinction between mortgages and deeds of trust for purpose of recording statute rests on "archaic title theory of deeds of trust rather than the modern lien theory").

[74] See, e.g., Bernhardt, note 54, §1.35 (2012); Miller et al., note 54, §10:1 ("The mortgage and deed of trust are practically identical"). Although the California Supreme Court said in 1933 that "deeds of trust, except for the passage of title for the purpose of the trust, are practically and substantially only mortgages with a power of sale," Bank of Italy v. Bentley, 20 P.2d 940, 944 (Cal. 1933), recent cases have emphasized that the Court "did not obliterate the distinction" between mortgages and deeds of trust in Bank of Italy. See Wadhwa v. Aurora Loan Servs., Inc., 2012 WL 762020, at *12 n.5; see also Calvo, note 73, at 819 ("[T]he court in *Bank of Italy* did not hold that a mortgage is the same as a deed of trust. Far from it; the *Bank of Italy* court recognized that the distinction between a mortgage, which creates only a lien, and the deed of trust, which passes title to the trustee, "has become well settled in our law and cannot now be disturbed.") (quoting *Bank of Italy*, 20 P.2d at 944).

[75] Miller et al., note 54, §10:2 ("For trust deeds, courts generally arrive at the same conclusion as the 'lien theory' traditionally applicable to the mortgage. In practical effect, if not in legal parlance, a deed of trust is a lien on the property."); Bernhardt, note 54, §1.35 ("Both instruments now have the same effect, or lack of it, on the debtor's title.").

that the deed of trust has an "anomalous"[76] nature under California law, as title to the underlying property is conveyed to a trustee, but is conveyed "only so far as may be necessary to the execution of the trust."[77] For example, courts apparently have reached different conclusions about the status of the beneficial interest when applying the Statute of Frauds than when applying the assignment recording statute for foreclosures. It seems likely that a court called upon to determine whether a deed of trust is an interest in real property under California law, for example when required to do so by the Bankruptcy Code, would apply an analysis specific to the statute in question.

Applicable recording law

A California recording statute, California Civil Code Section 2934, explicitly covers assignments of beneficial interests under deeds of trust:

> Any assignment of a mortgage and any assignment of the beneficial interest under a deed of trust may be recorded, and from the time the same is filed for record operates as constructive notice of the contents thereof to all persons; and any instrument by which any mortgage or deed of trust of, lien upon or interest in real property, (or by which any mortgage of, lien upon or interest in personal property a document evidencing or creating which is required or permitted by law to be recorded), is subordinated or waived as to priority may be recorded, and from the time the same is filed for record operates as constructive notice of the contents thereof, to all persons.[78]

In addition, it is possible that a separate, general recording statute, Civil Code Section 1214, applies to the assignment of the beneficial interest under a deed of trust. Section 1214 applies to "[e]very conveyance of real property,"[79] and a "conveyance" includes "every instrument in writing by which any estate or interest in real property is created, aliened, mortgaged, or incumbered, or by which the title to any real property may be affected, except wills."[80] As an assignment of the beneficial interest under a deed of trust may "affect" the title to real property, Section 1214 may cover such an assignment.

We have located no case holding that this general provision governs the assignment of a beneficial interest in a deed of trust,[81] although the provision has been treated as applicable to protect the party who is first to record the initial grant of a

[76] Monterey S.P. Partnership v. W.L. Bangham, Inc., 777 P.2d 623, 626 (Cal. 1989). [77] See note 76.

[78] Cal. Civ. Code, §2934.

[79] Cal. Civ. Code, §1214; see also Cal. Civ. Code, §1107 (grant of an estate in real property is conclusive against the grantor and against those claiming under a grantor, "except a purchaser or incumbrancer who in good faith and for a valuable consideration acquires a title or lien by an instrument that is first duly recorded").

[80] Cal. Civ. Code, §1215.

[81] Compare Security Mortg. Co. v. Delfs, 47 Cal. App. 599, 603–04 (Cal. Ct. App. 1920) (assuming for the sake of argument that Section 1214 governs mortgage assignments, but holding that Section 2934 controls to the extent the provisions are in conflict).

deed of trust from subsequent claims to the underlying property.[82] Moreover, as discussed above, recent decisions in foreclosure cases focus on the deed of trust as a conveyance of real property, although these cases have stressed the conveyance to the trustee rather than to the beneficiary. The recent case law therefore may suggest that the general recording statute comes into play when the trustee is substituted rather than when the beneficial interest is assigned.

In sum, a specific recording statute governs assignments of beneficial interests under deeds of trust, and it is unclear whether the general recording statute covers the assignment of beneficial interests in deeds of trust.

Consequences of failure to record

Although Civil Code Section 2934 expressly provides for recording mortgage assignments, it is not clear what follows if the assignment is not recorded. Specifically, it is unclear whether an assignee's unrecorded beneficial interest under a deed of trust is vulnerable to competing claims.

Some decisions state or suggest that a deed of trust cannot be transferred apart from the note it secures.[83] It might be argued based on these cases that recording is irrelevant to ownership contests and that all that matters is possession of the note. At least one older California Supreme Court case seems to support this proposition.[84]

However, commentators do state that it is the "better practice" to record assignments of mortgages and deeds of trust.[85] One reason given for this recommendation is that "an unrecorded assignment is not effective against the assignor's successors who do not have notice of the assignment."[86] It appears that recording is relevant to contests between the assignee and a successor of the assignor.

Courts have found that recording is relevant to contests between earlier and later assignees. In *Security Mortgage v. Delfs*,[87] a bank that had been assigned a mortgage released the mortgage and note back to the assignor, who then assigned the mortgage to another party.[88] The court held that the bank could not claim ownership

[82] See In re Cortez, 191 B.R. 174, 178 (B.A.P. 9th Cir. 1995)("In California, the deed of trust is an instrument providing security or collateral which must be perfected by recordation to bind subsequent purchasers."); In re Cedar Funding, 2012 WL 1346365, at *4 (Bankr. N.D. Cal. April 5, 2010); *In re* Planned Protective Servs., 130 B.R. 94, 98 (Bankr. C.D. Cal. 1991) (stating, with reference to a deed of trust, "[p]rior to recordation, a interest in real property is not effective against intervening creditors"); Powell v. Goldsmith, 199 Cal. Rptr. 554, 556–57 (Cal. Ct. App. 1984) (finding deed of trust owner with recorded interest not protected against competing deed of trust by this statute because of owner's agent's fraud).

[83] See, e.g., Kelley v. Upshaw, 39 Cal. 2d 179, 192 (1952) ("[P]urported assignment of the mortgage without an assignment of the debt which it secured was a legal nullity."); Domarad v. Fisher & Burke, 76 Cal. Rptr. 529, 535 (Cal. Ct. App. 1969).

[84] See Adler v. Newell, 109 Cal. 42, 48 (1895) ("one in possession of a mere instrument of mortgage, which purports on its face to be security for a certain note, is bound to know that if the note had been assigned to another the mortgage is of no legal value to him."). The loser in Adler took a forged instrument and the case may be understood as being about forgeries rather than recording generally.

[85] See Miller et al., note 54, §10:38. [86] Miller et al., note 54, §10:39; Cal. Civ. Code, §2934.

[87] 47 Cal. App. 599 (Cal. Ct. App. 1920) [88] 47 Cal. App. at 600–01.

of the mortgage because it was negligent in releasing the note and mortgage with no indication of its claim of ownership,[89] noting "[i]f [the bank] had recorded its assignment . . . its rights could have been protected."[90]

A federal bankruptcy court applying California law recently found that recording is relevant in contests between the assignee and a bankruptcy trustee. In In re *Cedar Funding, Inc.*, a bankrupt debtor had assigned interests in notes and deeds of trust to investors, first sending the investors copies of the notes and deeds of trust with executed endorsements of the notes, and then later recording assignments of the deeds of trust.[91] The bankruptcy trustee sought to bring the notes and deeds of trust into the bankruptcy estate. The specific issue presented was when the investors' interest in the deeds of trust was "perfected," that is, good against claimants other than the assignor.[92] The court held that under Civil Code Section 2934, the mortgage assignment recording statute, the investors' interests were not perfected until they were recorded: "only upon recording" is there "constructive notice to third parties to prevent them from gaining a superior interest."[93]

Thus, despite some authority suggesting that mortgage recording does not matter and courts should look only to possession of the note to determine ownership of the note and mortgage, courts have in some circumstances found that failure to record can imperil mortgage ownership. The losing party did fail to protect itself by maintaining physical possession of the note in the cases discussed here. But, as discussed below, in at least one case a California court has found that even a party that does not give up possession of the note can lose out in an ownership contest if it fails to record its interest.

Ability to transfer ownership of note along with mortgage

California courts have not discussed the note-follows-the-mortgage rule of the *Restatement (Third)* described above, although federal courts applying California law have approvingly cited other mortgage-transfer provisions of the *Restatement*.[94]

In cases decided before the *Restatement (Third)*, California courts have held that "a deed of trust has no assignable quality apart from the note"[95] and that a deed of trust is "inseparable"[96] from the note, but this does not explain what happens if the

[89] 47 Cal. App. at 602. [90] 47 Cal. App. at 602–03.

[91] In re Cedar Funding, Inc., 2012 WL 1346265, at *2 (Bankr. N.D. Cal. April 5, 2010).

[92] Cedar Funding, note 91, at *4. The trustee proceeded under Section 547(e) of the Bankruptcy Code, which allows avoidance of transfers made in a 90-day preference period. Whether a transfer falls within the 90-day period depends on when it is perfected. 11 U.S.C., §547(e).

[93] Cedar Funding, note 91, at *4.

[94] See In re Vargas, 396 B.R. 511, 516 (Bankr. C.D. Cal. 2008); Davidson v. Countrywide Home Loans, 2010 WL 962712, at *5 (S.D. Cal. March 16, 2010).

[95] Domarad v. Fisher & Burke, 76 Cal. Rptr. 529, 535 (Cal. Ct. App. 1969); see also Kelley, note 83, at 30 ("purported assignment of the mortgage without an assignment of the debt which [it] secured was a legal nullity")

[96] Domarad, note 95, at 535.

record owner of a deed of trust purports to transfer the deed of trust and the note together, but does not transfer possession of the note.[97] Is the transfer ineffective because the deed of trust is subsidiary to the note? Or does the transfer carry the note along with it?

Domarad v. Fisher & Burke, Inc. held that a record owner of a deed of trust could transfer the deed of trust and note free and clear of a third party's security interest, even though the record owner did not possess and therefore did not transfer possession of the promissory note.[98] In *Domarad*, a trust deed company borrowed from a finance company to finance its purchase of notes and deeds of trust for resale.[99] In one particular transaction, the trust deed company received the physical deeds of trust, as well as assignments of the notes and deeds of trust.[100] The trust deed company recorded the assignments.[101] The associated notes, which were pledged to the finance company as security for the funds it loaned to the trust deed company to buy the notes and deeds of trust, were held in escrow.[102] The trust deed company then purported to sell the notes and deeds of trust.[103] The court held that the finance company was estopped to assert its security interest in the notes because it was negligent in allowing the trust deed company to appear to be the sole owner of the notes and deeds of trust when the finance company knew that the trust deed company intended to sell the instruments.[104] Notably, the court held that the finance company was negligent, and that its unrecorded interest was vulnerable, even though it never surrendered possession of the note to the deed of trust company. In effect, the court held that under the circumstances of the case, the record owner of the deed of trust could transfer the deed of trust, with the note following, even without delivering the note.

Interaction of recording statutes with Uniform Commercial Code

It does not appear that California amended its recording statutes specifically to make them consistent with the 1998 revisions to the Uniform Commercial Code. However, a provision originally enacted in 1963, Civil Code Section 2944, provides, "None of the provisions of this chapter applies to any transaction or security interest governed by the Uniform Commercial Code . . . ."[105] "This chapter" includes Civil Code Section 2934, the statute covering recording of assignments

[97] In Kelley v. Upshaw, where the court found that a purported assignment of a chattel mortgage on fixtures without the underlying debt was a "nullity," Kelley, note 83, at 30, it seems that all parties understood that the intent was to split mortgage and note, rather than to transfer both together by means of a mortgage assignment. The assignor apparently was seeking to induce the assignee to become indebted to the assignor on the promissory note by giving the assignee the associated mortgage, *id.*, an arrangement that would make no sense if the assignee was going to take the promissory note as well.

[98] Domarad, note 95, at 538–39. [99] Domarad, note 95, at 532–33. [100] Domarad, note 95, at 536.

[101] Domarad, note 95. [102] Domarad, note 95, at 533. [103] Domarad, note 95.

[104] Domarad, note 95, at 537. [105] Cal. Civ. Code, §2944.

of interests in mortgages and deeds of trust.[106] Section 2944 has apparently never been interpreted by a court, and it apparently was intended to exclude security interests in fixtures or personal property from the rules for mortgages and deeds of trust rather than to change the rules for recording these instruments. Nevertheless, its text may encompass sales of mortgages after the 1998 revisions to the Code described above, under which such sales arguably are governed by the Uniform Commercial Code.[107] Section 2944 may now cover mortgage sales because, after 1998, the buyer's interest in a deed of trust apparently is a "security interest" under the Commercial Code.[108]

Thus, it is unclear how California's mortgage recording statute may interact with the 1998 revisions to the Uniform Commercial Code, although an older statute may be interpreted to make the recording statute give way.

Summary

Due in part to uncertainty about the status of a deed of trust as compared to a mortgage, California law is unclear on almost all the points we discuss. There is uncertainty, or at least variation across statutory contexts, about whether a deed of trust is an interest in real property.

Although it seems fairly clear that at least one California recording statute covers assignments of beneficial interests in deeds of trust, it is not clear what consequences follow from failing to record, and in particular when the assignee's ownership interest is vulnerable to competing claims because the assignee failed to record. For example, it is unclear whether the record owner's assignment of a deed of trust can also transfer the associated promissory note even without delivery of the note. When the record owner can effect such a transfer, that makes the unrecorded assignee's interest vulnerable even if the assignee retains the physical promissory note.

Furthermore, it is not clear whether California's recording statutes are in conflict with the Uniform Commercial Code or cede primacy to the Code. It seems the legislature should seriously consider action to clarify California law.

Florida

Leading instrument

The mortgage is the leading instrument for security interests in real property in Florida.[109]

106 Both the mortgage assignment recording provision (Section 2934) and Section 2944 appear in Division 3, Part 4, Title 14, Chapter 2 of the California Civil Code.

107 See U.C.C. provisions,note 52.

108 See Permanent Editorial Board for the Uniform Commercial Code, note 53, at 8–9.

109 Thomas E. Baynes, Jr., Florida Mortgages, §1–1 ("[A]ny instrument which uses the property for security will be deemed a mortgage") (collecting authorities).

Interest in real property

Florida courts have held that a mortgage assignment "is not a 'conveyance' or 'transfer' of any interest in land covered by the mortgage," so that Florida's general real property recording statute apparently is not applicable to mortgage assignments.[110] However, mortgages are covered by a special recording statute.

1. An assignment of a mortgage upon real property or of any interest therein, is not good or effectual in law or equity, against creditors or subsequent purchasers, for a valuable consideration, and without notice, unless the assignment is contained in a document that, in its title, indicates an assignment of mortgage and is recorded according to law.
2. This section also applies to assignments of mortgages resulting from transfers of all or any part or parts of the debt, note or notes secured by mortgage, and none of same is effectual in law or in equity against creditors or subsequent purchasers for a valuable consideration without notice, unless a duly executed assignment be recorded according to law.
3. Any assignment of a mortgage, duly executed and recorded according to law, purporting to assign the principal of the mortgage debt or the unpaid balance of such principal, shall, as against subsequent purchasers and creditors for value and without notice, be held and deemed to assign any and all accrued and unpaid interest secured by such mortgage, unless such interest is specifically and affirmatively reserved in such an assignment by the assignor, and a reservation of such interest or any part thereof may not be implied.
4. Notwithstanding subsections (1), (2), and (3) governing the assignment of mortgages, chapters 670–680 of the Uniform Commercial Code of this state govern the attachment and perfection of a security interest in a mortgage upon real property and in a promissory note or other right to payment or performance secured by that mortgage. The assignment of such a mortgage need not be recorded under this section for purposes of attachment or perfection of a security interest in the mortgage under the Uniform Commercial Code.
5. Notwithstanding subsection (4), a creditor or subsequent purchaser of real property or any interest therein, for valuable consideration and without notice, is entitled to rely on a full or partial release, discharge, consent, joinder, subordination, satisfaction, or assignment of a mortgage upon such property made by the mortgagee of record, without regard to the filing of any Uniform

[110] Garrett v. Fernauld, 57 So. 671, 672 (Fla. 1912); Free v. Free, 936 So. 2d 699, 703 (Fla. Dist. Ct. App. 2006) ("A mortgage creates a lien under Florida law, and . . . does not convey an interest in real property"); Baynes, note 109, §7–2; Note, *Mortgages: Effect of Failure to Record a Mortgage Assignment in Florida*, 7 FLA. L. Rev. 93, 93 (1954).

Commercial Code financing statement that purports to perfect a security interest in the mortgage or in a promissory note or other right to payment or performance secured by the mortgage, and the filing of any such financing statement does not constitute notice for the purposes of this section. For the purposes of this subsection, the term "mortgagee of record" means the person named as the mortgagee in the recorded mortgage or, if an assignment of the mortgage has been recorded in accordance with this section, the term "mortgagee of record" means the assignee named in the recorded assignment.[111]

Consequences of failure to record

It appears that recording mortgage assignments can help protect the assignee's interests under Florida law. For example, courts applying Florida law have held that "if the original mortgagee assigns the mortgage to Entity A and Entity A fails to record that assignment, Entity A cannot claim priority over a latter assignee of the same mortgage (Entity B)."[112]

Ability to transfer ownership of note along with mortgage

Although Florida courts have not cited the *Restatement* provision that transfer of the mortgage can also transfer the note, there is some indirect support for this proposition in Florida cases,[113] and at least one Florida commentator has embraced it.[114]

Interaction of recording statutes with Uniform Commercial Code

As paragraph (4) of the statute quoted above indicates, Florida apparently amended its recording statute in 2005,[115] so that the recording statute now defers to the Uniform Commercial Code's provisions relating to mortgage transfer. This provision seems to be in tension with the recent decisions that recording is important for protecting the assignee's interest. It appears that no Florida court has discussed the tension.

[111] Fla. Stat. Ann., §701.02.

[112] JPMorgan Chase Bank v. New Millennial, LLC, 6 So. 3d 381 (Fla. Dist. Ct. App. 2009) (citing In re Halabi, 184 F.3d 1335, 1338 (11th Cir. 1999) (§701.02 recording requirement "is applicable . . . to (and enforceable by) competing creditors or subsequent bona fide purchasers of the mortgagee"). See also Rhodes v. JPMorgan Chase Bank, 2012 WL 2504043, at *3 n.4 (S.D. Fla. June 28, 2012) (same). But see American Bank of the South v. Rothenberg, 598 So. 2d 289, 290 (Fla. Dist. Ct. App. 1993) (section 701.02 irrelevant to dispute between successive assignees of mortgage and note, which "is governed by negotiable instruments law"); Note 110, at 96–98 (1954) (concluding, in "absence of Florida case law directly in point," that recording statute is "probably inapplicable" to contests between earlier and later assignees of same mortgage).

[113] See WM Specialty Mortg. Co. v. Salomon, 874 So. 2d 680, 682 (Fla. Dist. Ct. App. 2004) (holding that it was a question of fact whether physical transfer of mortgage document transferred right to enforce mortgage when transferee did not "own or hold" note).

[114] Baynes, note 109, §7–2 ("[I]f it is the intent of the parties that an assignment of the mortgage instrument shall operate as an assignment of the note, then the assignment will still be effective.").

[115] See Fla. Stat. Ann., §701 (Historical and Statutory Notes) (citing Laws 2005 c. 2005–241, § 20).

Summary

The mortgage is the leading real-property security instrument in Florida. A mortgage apparently is not considered to be a grant of a real-property interest, but a special statute governs recording of mortgage assignments. Although this statute apparently defers to the UCC rules on the relationship between mortgage and note, Florida courts continue to state that recording a mortgage assignment is relevant to protecting the assignee's interest. There is at least indirect support for the proposition that transferring a mortgage in Florida may transfer the note as well.

Acknowledgements

The authors gratefully acknowledge helpful contributions from Rob Freund, Patrick Greenfield, and Dwight Jaffee.

Bibliography

Acharya, V. and M. Richardson, 2010, Causes of the financial crisis, in *Critical Review*, Routledge, Volume **21**, 195–210.

American Law Institute, 1997, *Restatement (Third) of Property: Mortgages*.

Barth, M. E. and W. R. Landsmen, 2010, How did Financial Reporting Contribute to the Financial Crisis, Working Paper, Stanford Graduate School of Business, forthcoming, *European Accounting Review*.

Berndt, A., B. Hollifield and P. Sandas, 2012, The Role of Mortgage Brokers in the Subprime Crisis, Working paper, Carnegie Mellon University and NBER.

Brueckner, J. K., 1994, Borrower mobility, adverse selection and mortgage points, *Journal of Financial Intermediation*, **3**, 416–441.

Case, K. E. and R. J. Shiller, 1987, Prices of single family homes since 1970: New indexes for four cities, *New England Economic Review* September/October, 45–56.

Case, K. E. and R. J. Shiller, 1989, The efficiency of the market for single family homes, *American Economic Review*, **79**, 125–137.

Chari, V. V. and R. Jagannathan, 1989, Adverse selection in a model of real estate lending, *Journal of Finance*, **44**, 499–508.

Downing, C., D. Jaffee and N. Wallace, 2009, Is the market for mortgage-backed securities a market for lemons? *Review of Financial Studies*, **22**, 2457–2494.

Dunn, K. B. and C. S. Spatt, 2005, The effect of refinancing costs and market imperfections on the optimal call policy and the pricing of debt contracts, *Real Estate Economics*, **33**, 595–617.

Florey, K. J., 2009, State courts, state territory, state power: Reflections on the extraterritoriality principle in choice of law and legislation, *Notre Dame Law Review*, **84**, 1058–1134.

Gatzlaff, D. H. and D. R. Haurin, 1997, Sample selection bias and repeat-sales index estimates, *Journal of Real Estate Finance and Economics*, **14**, 33–50.

Gatzlaff, D. H. and D. R. Haurin, 1998, Sample selection and biases in local house value indexes, *Journal of Urban Economics*, **43**, 199–222.

Goetzmann, W. and L. Peng, 2006, Estimating house price indexes in the presence of seller reservation prices, *Review of Economics and Statistics*, **88**, 100–112.

Hayre, L. and A. Rajan, 1995, Anatomy of Prepayments: The Salomon Brothers Prepayment Model, Working paper, Salomon Brothers.

Heckman, J., 1979, Sample selection bias as a specification error, *Econometrica*, **47**, 153–162.

Huizinga, H. and L. Vaeven, 2009, Accounting Discretion of Banks During a Financial Crisis, Working Paper, International Monetary Fund.

Hunt, J. P., R. Stanton and N. Wallace, 2012, All in One Basket: The Bankruptcy Risk of a National Agent-Based Mortgage Recording System, Working Paper, University of California at Berkeley, forthcoming, *U.C. Davis Law Review*, **46**.

Hwang, M. and J. Quigley, 2003, Selectivity, quality adjustment and mean reversion in the measurement of house values, *Journal of Real Estate Finance and Economics*, **28**, 161–178.

Laux, C. and C. Leuz, 2009, Did Fair-Value Accounting Contribute to the Financial Crisis, Working Paper no. 266/2009, ECGI, forthcoming, *Journal of Economic Perspectives*.

Levitin, A. J., 2010, Robo-signing, chain of title, loss mitigation, and other issues in mortgage servicing, Hearing Before the Subcommittee on Housing, and Community Opportunity of the House Financial Services Committee, 111th Congress, November 18, 2010.

Meese, R. and N. Wallace, 1991, Nonparametric estimation of dynamic hedonic price models and the construction of residential housing price indices, *AREUEA Journal*, **19**, 308–332.

Mian, A. and A. Sufi, 2009, The consequences of mortgage credit expansion: Evidence from the U.S. mortgage default crisis, *Quarterly Journal of Economics*, **124**, 127–145.

Ng, J. and T. O. Rusticus, 2011, Banks' Survival During the Financial Crisis: The Role of Financial Reporting Transparency, Working Paper, MIT Sloan School of Management.

Peterson, C. L., 2010a, Foreclosure, subprime mortgage lending, and the Mortgage Electronic Registration System, Inc., *University of Cincinnati Law Review*, **78**, 1359–1407.

Peterson, C. L., 2010b, Two faces: Demystifying the Mortgage Electronic Registration System's land title theory, Working Paper, University of Utah School of Law, forthcoming, *Real Property Probate & Trust Law Journal*.

Pizante, L., M. Rappaport, J. Patterson and A. Sheffield, 2012, Foreclosure in California: A crisis of compliance, Working Paper, Aequitas, forthcoming, *Real Property Probate & Trust Law Journal*.

Powell, R. and P. J. Rohan, 2011, *Powell on Real Property*, Matthew Bender and Company, New York.

Rosinia, N., 2012, How "reasonable" has become unreasonable: A proposal for rewriting the lasting legacy of Jackson v. Indiana, *Washington University Law Review*, **89**, 673–703.

Rothschild, M. and J. Stiglitz, 1976, Equilibrium in competitive insurance markets: An essay on the economics of imperfect information, *Quarterly Journal of Economics*, **90**, 629–650.

Spiceland, J. D., J. F. Sepe and M. W. Nelson, 2011, *Intermediate Accounting*, McGraw-Hill Irwin, New York.

Stanton, R., 1996, Unobservable heterogeneity and rational learning: Pool-specific versus generic mortgage-backed security prices, *Journal of Real Estate Finance and Economics*, **12**, 243–263.

Stanton, R. and N. Wallace, 1998, Mortgage choice: What's the point? *Real Estate Economics*, **26**, 173–205.

Stanton, R. and N. Wallace, 2009, Dynamic House Price Indices: Foundations for a New Estimator, Working Paper, University of California at Berkeley.

United States Supreme Court, 1997, *Idaho v. Coeur d'Alene Tribe of Idaho*.

PART V

DATA MANAGEMENT TECHNOLOGIES

EDITORS' INTRODUCTION TO PART V

Martijn Groot

In this section we will survey the technologies and methodologies used to process financial markets data. The importance of a streamlined and controlled process for sourcing, quality-proofing and channeling information steadily increased even before the crisis due to the explosion of incoming information on the one hand and the rapid rise in reporting requirements to investors, clients, regulators and for internal purposes on the other hand. Frequently, IT departments as well as business users find themselves stuck in the middle trying to meet the resulting information processing challenge.

You could make the case that all financial services is about information processing. We can abstract the whole infrastructure of financial services as sets of information and sets of business logic. A bundle of business logic is an application, a set of rules that takes various inputs (e.g., equity prices and volumes) and that produces new information (e.g., trading signals, risk scores, client portfolio statements). But these activities can only produce meaningful results when fed with correct, timely and sufficient information. Bad information leads to bad decisions. Paraphrasing a well-known tagline we can summarize by saying "the data is the application."

Faced with the incoming floods of data, many financial institutions found themselves singularly ill equipped to deal with the increased reporting requirements in the wake of the financial crisis. IT infrastructures typically last far longer than anticipated, and large firms frequently have systems dating back decades that suddenly also have to cope with, use and produce a vastly increased and much more diverse set of content that has to permeate the whole organization. Pervasive data needs, increased variety of sources, mounting volumes, a sometimes forbidding internal data morass and escalating external reporting requirements all add to the data challenges. In this section we cover some of the data management considerations, technologies and best practices that have emerged over the last decade.

The financial content space and the technology to deal with it, to source, process, quality control, deliver and monitor financial information is undergoing rapid change. Casting the challenge as a supply chain problem, we are looking at finding the most effective way to move data from A to B, or, in other words, to get it unadulterated from the point of creation to the point of consumption. Keeping this picture in mind, the financial services content industry is plagued by confusion along the different steps in this supply chain as to the different bits of information needed, how to separate or even identify them, how to unambiguously define them, what formats or representations to use and very different information requirements and desired trade-offs between, for example, quality and speed for different users. Some of the recent developments in the financial data management landscape are as follows.

- Data consumers face an ever mounting task with keeping up a sufficient information supply to customers, investors and regulators. The number of direct information sources from exchanges, alternative trading venues, brokers, index providers, rating agencies and third party administrators is exploding and the costs of simply getting the right data from A to B and of ensuring that quality information arrives in time are huge. Part of the problem is a lack of standards, part sheer volume, part product complexity and the enormous amount of terms, part changing reporting requirements. Part of it is also that more understanding is needed by suppliers of data and software of their customers' use cases. The job of data management technology is to take away friction along this supply chain and to allow quality information to reach business users and applications in a timely, secure and cost-effective way.
- Regulators are much more active, not only in demanding more information, but also in being much more actively involved in the syntax and scope of data standards. Regulators realized that to prevent drowning in information they need standards, if only for themselves. They have understood that disclosure is not the same as transparency. Disclosure without standards, consolidation and access could mean the disclosed information is inaccessible and essentially useless. Data collection does not necessarily lead to actionable information. Similarly, data quality is not universally defined and there is a recognition that bad data is one of the biggest operational risks to a business. Many of the techniques and methods discussed in this section are there to take that risk out of the data.
- A realization that the gap between front office and back and middle office technologies in terms of information provision has become too large is driving the data management agenda. The industry has lived with silo-based businesses that pushed reconciliation, transparency and reporting requirements down the

trade lifecycle for too long. Even within risk there were separate silos to handle credit, liquidity, market and operational risk.

Data management technology has matured dramatically in the past decade. While many of these advances have been used in an ad hoc nature on risk information management implementations, there has not been a universal, strategic application of these new technologies. Utilizing these techniques in a comprehensive manner will go a long way to helping financial organizations implement a comprehensive, long-lasting and strategic risk management asset.

This section will address the various technologies applied to different aspects of risk information management. We start off with Bill Nichols, Jim Northey and Karla McKenna discussing the benefits of standards, some of the current standards in use, the challenges to be overcome in implementing standards and the road ahead. Dilip Krishna then discusses approaches to addressing the storage challenges involved in coping with these vast amounts of data. Martijn Groot addresses the role of data management technologies in sourcing and integrating information and combining different sources so that value and insights can be extracted. After this, Victoria Lemieux, Brian Fisher and Thomas Dang cover approaches to visualizing both raw data and the results from analysis in an appropriate manner in order to make more sense of the data, spot patterns more easily and use visualization techniques in an overall workflow. This chapter is followed by Marc Alvarez who addresses the technology that is required to maintain the integrity of data such as master data, data quality and metadata tools. Finally, Michael McCabe discusses technology for data security, an often overlooked area but still a major source of operational risk.

In short, in this section we seek to bridge the all too common gap between technology and business practices. Rapid developments take place in technologies for visualization, storage, data integration, data and metadata standards and quality management. The service models through which these technologies are shifting too – with a broader range of supply options available to firms. The amounts of information and the information needs of firms and their clients, shareholders, business partners and regulators will only increase. To effectively address current and future information challenges, financial institutions are well advised to take advantage of these developments in shaping a new financial services landscape.

19

Financial data interchange standards

Karla McKenna, Jim Northey and Bill Nichols

19.1 Introduction

Standards make modern commerce possible. In fact, modern commerce is wholly dependent upon a multitude of standards. It is not unreasonable to see the function of commerce and finance as an exercise in the creation of standard definitions for valuation and subsequent exchange of value Ferguson, 2008, p. 7). The dependency on and the necessity of standards is likely not fully understood by most people. We all see glimpses of these issues when consumers are confronted with vendors vying to become a de facto consumer standard, such as the recent competition between Blu-Ray and High Definition technology, or the oft used example of VHS versus Beta as the videotape standard in the 1970s.

This chapter will focus on the relevant current and emerging standards that the risk management practitioner should be aware of with respect to the production, dissemination, and aggregation of financial information. We will also provide an overview of the environment of standards and a framework for standards selection and usage. This framework is being provided as a form of risk mitigation in and of itself, as there is an inherent risk in every standards adoption decision.

19.1.1 Context for standards in finance

The expansion and sophistication of modern finance has followed the trajectory of, among others, three primary concurrent trends:

1. relaxation of laws governing finance in response to ascendant political philosophies stressing deregulation and "efficient market" economic theories coupled with the massive globalization of commerce and resultant extraterritoriality and corresponding opportunities for regulatory arbitrage by financial firms;

Handbook of Financial Data and Risk Information, ed. Margarita S. Brose, Mark D. Flood, Dilip Krishna and Bill Nichols. Published by Cambridge University Press.

2. the application of advanced mathematics to finance, especially in the area of options pricing theory; and
3. the information technology and telecommunications revolution.

The vast majority of this expansion of finance and, more importantly, the speed with which it has occurred, has been built upon standardized methods to exchange data.

19.1.2 Vertical versus horizontal standards

There are two separate dimensions used to describe the ways in which standards fit into the overall information landscape. Every industry has standards that are specific to the business domains they deal with. Manufacturing, medical, construction, aerospace and finance, of course, are examples of industry sectors that maintain their own standards. In the language of standards, each of these areas creates *vertical standards*. These vertical standards often overlap in areas of applicability, especially around data representation and information technology, resulting in standards that have a similar purpose but are quite different in terms of implementation and the terminology used to specify the standard.

Information technology standards are generally applicable across multiple industry sectors. Because of their general applicability, these standards are often referred to as *horizontal standards* or *horizontal technology standards*. There is a natural tension and conflict between vertical standards and horizontal standards. The trend that has occurred over the past few decades sees vertical standards being modified or replaced with horizontal technology standards, in place of vertical industry-specific technology implementations, albeit with the specific requirements for the vertical being addressed by specific horizontal technology methods.

The information technology revolution has resulted in the majority of the important horizontal technology standards falling within the broad category of Information and Communication Technology (ICT). In recent years there has been a push to implement industry-specific vertical standards using existing horizontal technology standards. An example of this is the use of the Tagged Image File Format (TIFF) standard to encode and exchange check images. A more nuanced example would be the construction of XML-based vertical dialects such as XBRL or FIXML. In that case, the horizontal syntax standard XML is used to create specific vertical applications for financial and accounting data.

This chapter will look at the impact of horizontal standards – technologies and methods applicable to multiple application domains – on the financial industry, an industry that has not yet adopted a vertical supply chain framework.

19.1.3 Historical context of standards

Commerce has always been the most important driver behind standardization. The early historical focus of standards was on weights and measures. Today we take for granted and would have a difficult time conceiving of a world without standardized weights and measures (Spivak and Brenner, 2001, p. 7). A small anecdote from England at the time of the creation of the Magna Carta serves to provide some historical context on the genesis and general importance to society of standards.

King John and the Magna Carta

On June 15, 1215, King John was forced to sign the Magna Carta at Runnymeade by the English Barons in an attempt to recover some of the rights granted by earlier rulers or established feudal customs long in effect prior to John. John had violated the earlier rules for his personal benefit or gain. Among the some 70 articles in the Magna Carta was Chapter 35, which dealt with "measures." It read: "Let there be one measure of wine through our whole realm; and one measure of ale; and one measure of corn, to wit, 'the London quarter' and one width of cloth (whether dyed or russet, or 'halberget'), to wit, two ells within the selvedges; of weights also let it be as of measures."

Prior to 1215 various ordinances (assizes) were issued to regulate the sale of commodities. Assizes were issued to cover beer, wine, and bread, also weights and measures. Richard the Lionheart published the Assize of Cloth (1197), which tried to overcome the inconvenience experienced by traders who met with varying standards as they moved their goods from place to place. Of more importance, this assize sought to prevent frauds being perpetuated on buyers under shelter of ambiguous weights and measures; thus the requirement for "one measure" for each of the commodities, ale, corn, etc. Dyed cloths were required to be uniform all across the fabric. Merchants were prohibited from hanging over their windows "cloth whether red or black, or shields so as to deceive the sight of buyers seeking to choose good cloth."

One of the complaints against John was that he did not attempt to enforce the regulations but was content to fine the violators with the fines going to the Royal Treasury! Clearly, these references reflect the concerns for the potential of abuse of weights and measures that had already been established as "standards of the realm" (Spivak and Brenner, 2001, pp. 9–10).

19.1.4 Units of monetary value

While weights and measures for trading of commodities are often discussed when talking about standards related to trade, less attention is generally given to the necessity for, or the characteristics of, the standardization of units of value necessary to communicate monetary value. A currency is a standardized unit of value. A share of stock in a company is a highly standardized unit of measure and, when coupled with a price (yet another measure of value), is a representation of a monetary value assignment for an ownership unit of a legal entity. When a firm decides to raise money in the public markets by "going public," much of the issuance process which is required for the public listing is devoted to assuring that the accounting practices of the company, used to assess a comparable and standard measure of value, comply with existing standards and practices. These standards and practices are compulsory based upon the jurisdiction within which the legal entity is based and those of the market venue that lists the issuance of the stock.

Compulsory standards are most commonly referred to as *de jure* standards. When a practice becomes a common practice such that it dominates a particular niche for a business function, it is usually referred to as a *de facto* standard. While the difference may at times appear negligible, the distinction is of critical importance when evaluating public policy approaches – or when assessing technology or data strategies for private sector firms.

19.2 Types of standards

Spivak and Brenner (2001, p. 2) list five general types of standards:

1. Physical standards or units of measure;
2. Terms, definitions, classes, grades, ratings, or symbols;
3. Test methods, recommended practices, guides, and other application to products and processes;
4. Standards for systems and services, in particular quality standardization and related aspects of management system standards for quality and the environment; and
5. Standards for health, safety, consumers, and the environment.

The discussion following will focus on those standards covered in the third topic above, especially the application to products and processes. Specifically with regard to financial services, there are two broad types of standards supporting financial transactions – data content standards and data technology standards, collectively financial technology standards. Additionally, there are accounting and risk management practice standards that you should be familiar with as a risk management practitioner that are outside the scope of this chapter on data interchange.

19.2.1 Context for viewing different types of financial technology standards

In discussing financial technology standards, the important distinction between the content and technology viewpoints – what does a piece of information mean or

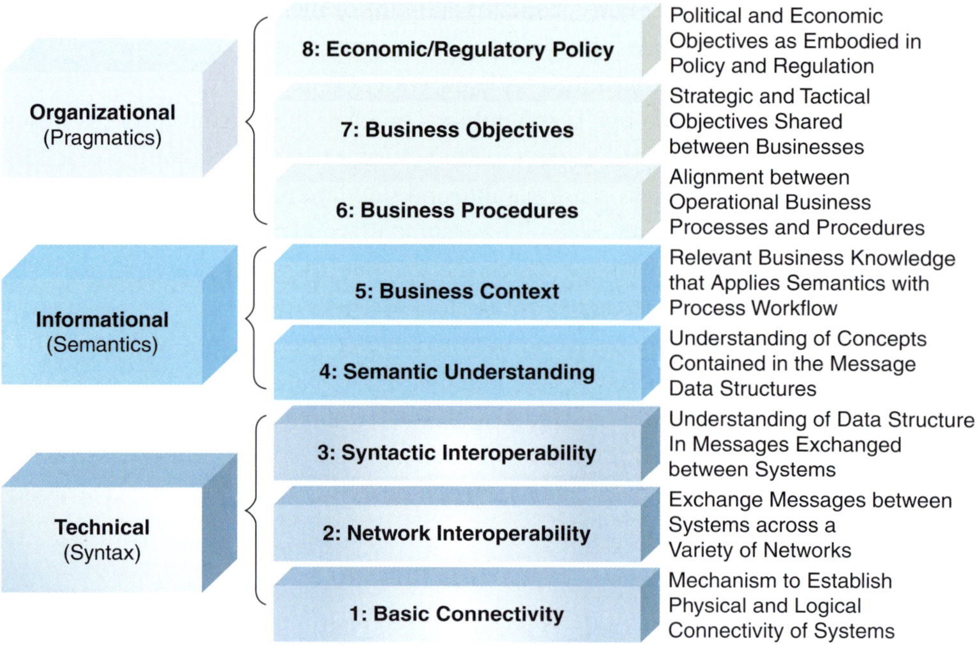

Figure 19.1 Interoperability stack.

represent versus how is the information stored, moved, etc. – is context dependent based upon which "level" is being discussed. Figure 19.1 (Gridwise Architecture Council Interoperability Framework Team, 2007) shows an interoperability "stack" used to describe these viewpoints in some recent efforts related to energy markets.

As used in the diagram and in the discussion here, interoperability means "being able to accomplish end-user applications using different types of computer systems, operating systems, and application software, interconnected by different types of local and wide area networks" (O'Brien and Marakas, 2000). Interoperability may be assessed in at least four dimensions:

1. Basic technical – common tools, interfaces, etc., providing uniformity in navigation and access;
2. Syntactic – allowing the interchange of metadata and protocol elements;
3. Functional – share service definitions, or functional primitives;
4. Semantic – allowing access to similar classes of objects or services across multiple systems, including multilingual functionality.

For our purposes, we are concerned primarily with those standards used in the Informational and Technical areas shown in the diagram. While there are emerging horizontal standards in the Organizational areas, these are generally immature with respect to financial services and are dependent upon the availability of clean, accurate data with established and shared definitions, provenance, and quality characteristics. Achieving this type of data is in turn dependent upon standards at each of the technical and information levels.

19.3 Organizational frameworks for enterprise data

Large organizations, also known as enterprises, typically have relatively common structures with regard to horizontal technologies. There are generally vertical areas of specialization "riding on top" of more generic networking and storage components. *Enterprise architecture* is the discipline that is used in large organizations to manage the overall inventory of technologies used to support operations. It is worthwhile to spend a moment providing some background on how these approaches have evolved, as it sheds light on the interdependencies among systems, processes, and people that characterize interoperability and standards issues.

19.3.1 The evolution of technology organization approaches

Early (1960s–1970s) large-scale technology systems development projects such as the original US NASA "moon-shot" and other large government exercises (primarily defense and statistics oriented) were pursued in an environment characterized by "glass rooms" and raised flooring (which served to both hide and protect the cabling required to connect hardware components). Organizational processes were managed around "big iron" requirements, and workflows were structured to coordinate the setup of computationally difficult parts of processes so that calculation time on the machines was optimized. As hardware costs dropped and large commercial firms began using computing power in early accounting systems, business process design (though not called this yet) focused on marshaling the required input information to ensure that data entry operators had everything prepared on "business forms" prior to entering the data. The data entry screens and underlying software were prepared based on the premise that the information required would be gathered by hand and organized in order to expedite key-entry. Software development was primarily focused on screen interfaces that transmitted form data (typically a screen-full at a time) to data storage mechanisms that were closely bound to physical storage requirements of early disk technologies. Competing proprietary methods for organizing stored data concentrated on their ability to access data "randomly" (i.e., to be able to find individual records directly, without having to sequentially read through large numbers of records).

The advent of "mini-computers" in the late 1970s and the development of relational database theory and related Relational Database Management Systems (RDBMS) in the early 1980s, along with the emergence of networked architectures tying multiple machines together, provided the genesis for early efforts in organizing the design and planning of more sophisticated workflows crossing wider geographical boundaries. Ongoing developments in government and academic circles fed the maturation of the Internet, which led to early commercial networking infrastructures that began addressing "network of network" requirements. The explosion in computing power driven by successful microprocessor architectures, and the resulting proliferation of competing protocols and standards, led to a significant increase in the level of complexity required to assemble, maintain, and evolve business systems, somewhat mitigated by the increasing standardization of SQL-based RDBMs as the primary method for data storage and retrieval.

In 1987, John Zachman, an IBM employee, created the Zachman Framework, the first attempt at organizing the complete set of information required to manage and maintain systems planning for large organizations (see Figure 19.2 below) Others in government circles (primarily Department of Defense-related at the time) were pursuing similar paths, in conjunction with standards organizations such as the Institute of Electrical and Electronics Engineers (IEEE) and its development of the Portable Operating System Interface (POSIX).[1] The Technical Architecture Framework for Information Management (TAFIM) was first published in 1991, with an emphasis on non-proprietary, open systems architectures then emerging in commercial and academic circles. From these came the Open Group Architecture Framework (TOGAF) system approach, which has significantly influenced government and defense-related Enterprise Architecture (EA) approaches.[2] Earlier POSIX-based US Department of Defense architectures have tended to merge with TOGAF-based frameworks over the past two decades.

19.3.2 Enterprise architecture framework

The seminal academic text articulating an approach for Enterprise Architecture was Spewak and Hill's 1993 publication, *Enterprise Architecture Planning – Developing a Blueprint for Data Applications and Technology*. Long a fairly obscure niche area outside of very large-scale business and government environments, the changes in systems technologies driven by the rapid adoption of Internet internetworking technologies, especially since the introduction and the growth of the World Wide Web, has led to a "mainstreaming" of EA in business and government circles. Large organizations have struggled to keep up with rapidly evolving technologies and the proliferation of computing power resulting from advances in microprocessor and storage-related manufacturing that has been accompanied by global network capabilities. Today, every organization can (indeed, must) assume a

[1] See `http://standards.ieee.org/develop/wg/POSIX.html` [2] `www.opengroup.org/togaf/`

low-cost network connection to the Internet; the ability to interface with customers, suppliers, regulators, and other constituencies across the Internet using Web interfaces is a requisite core competency. This has significantly increased the complexity of planning and management for every component, service, and function that organizations must perform. EA frameworks represent current state-of-the-art for large-scale efforts in organizing and managing this complexity.

An enterprise architecture framework can be thought of as a "proposed blueprint" for designing and building an EA for a specific organization or enterprise. In other words, an EA framework serves as the starting point to create the "to do list" necessary to implement a program management process for documenting the current state of an organization's system(s), defining the desired target state, and describing the set of documentation and artifacts necessary to enable the related change management and development processes.

19.3.3 Defining data standards: the Zachman framework

The Zachman framework is perhaps the simplest to use in terms of defining interoperability and standards for data.[3] The Zachman framework can be nicknamed "the reporter's approach," since it seeks to frame the problem domain by describing "Who, How, What, Where, When, and Why?" These questions are laid out along the x-axis of a grid, with the y-axis representing the party responsible for the information required in "filling in the box" created by intersecting columns (see Figure 19.2).[4]

Context and perspective are either explicit or implicit in any modeling system. Enterprise architectures try to frame context in such a way as to make the creation of documentary artifacts (models, descriptive documents) that describe a system form a coherent collection. In the Zachman framework, the rows of the grid refer to the perspective of the actors listed in the "Who" column *as seen by* the practitioners in the leftmost column – those tasked with building and maintaining the system.

In the diagram, the practitioner designation is attached to the scope of the activity. Accordingly, a business sponsor is responsible for defining the overall scope of the effort, beginning with designing or designating the core business concepts, how they should impact the existing business functions, what types of changes will be required of existing operations, where these changes will be made, who will implement them, the sequencing and timing of the new business processes, and the reasons for the changes in operational behavior.

[3] The other commonly used framework, used mostly by technology teams, is The Open Group Architecture Framework (TOGAF). The Federal Enterprise Architecture (FEA) (`www.whitehouse.gov/omb/e-gov/fea`) is also worth a look.

[4] `http://en.wikipedia.org/wiki/File:The_Zachman_Framework_of_Enterprise_Architecture.jpg`

	Why	How	What	Who	Where	When
Contextual	Goal List	Process List	Material List	Organizational Unit & Role List	Geographical Locations List	Event List
Conceptual	Goal Relationship	Process Model	Entity Relationship Model	Organizational Unit & Role Relationship Model	Locations Model	Event Model
Logical	Rules Diagram	Process Diagram	Data Model Diagram	Role Relationship Diagram	Locations Diagram	Event Diagram
Physical	Rules Specification	Process Function Specification	Data Entity Specification	Role Specification	Locations Specification	Event Specification
Detailed	Rules Details	Process Details	Data Details	Role Details	Location Details	Event Details

Figure 19.2 The Zachman framework.

The succeeding rows, from the top down, describe those tasks and models necessary to understand what must be done to specify, document, and manage the operational and IT business systems. The final row refers to the actual operating enterprise itself.

By design, the Zachman framework is agnostic to standards for describing and maintaining the design and documentation artifacts it specifies. While various devotees have emphasized different toolkits and approaches, it is only within the past few years that commercial products seeking to provide a "platform approach" to EA documentation have come to market, and none have yet achieved significant dominance. From a standards perspective, however, we can cross-reference each row and column of the Zachman framework with the categories from the interoperability stack (Figure 19.1) to provide a "landscape view" of where horizontal standards will typically be integrated with, or overridden by, vertical standards for a specific industry application.

19.4 Financial industry landscape – industry supply chain overview

Each organization within the financial services supply chain can be viewed internally from an enterprise architecture viewpoint to analyze the creation and management of information, but interoperability between organizations is the primary requirement to ensure that data used within an organization are of known provenance and quality. This requires a view of the industry supply chain that cuts across multiple organizations. While the use of advanced supply chain analysis and management approaches is common in many industries, relatively little work has been done to date to utilize these approaches for analyzing financial services supply chains.[5]

Understanding risks – of all types – within and among financial market participants is typically done with approaches that have a primary emphasis on data and algorithms. With the development of ICT systems and their dominant role in business operations, the creation, collection, integration, and interpretation of data within a firm and across a supply chain has become a core competency requirement for players. Sharing data electronically between market participants – especially with regard to products, prices, customers, order processing and fulfillment, payments, and related transactional information – is a requirement to participate in any industry supply chain in the modern world. Supply chain analysis has become a sophisticated set of management practices that include things like integrating product catalogs and shipping mechanisms to create "just in time" manufacturing and distribution systems that are completely electronic and require minimal human interaction.

The ability to manage and integrate electronic supply chains is dependent on standards for defining and managing data. From syntax – How is it represented? What character set(s) are used to implement grammars and lexical organization? – to semantics – What does it mean? How does it relate to the data around it? How are messages constructed and recognized? – to transmission – What connection(s) are created and managed how? – there are standards for each component, appropriate for the physical or logical level with which the ICT systems are being used to accomplish a specific task. From the range at which a current moving through a wire will fluctuate in order to represent 1s and 0s in an electronic circuit to the syntactical representations of those 1s and 0s as character sets, all the way "up the stack" to highly structured information such as accounting data, or semi-structured data such as text files; at each step of the way, there are both vertical and horizontal, de jure and de facto standards.

19.4.1 Current environment in financial services

Over the past 20–30 years, the revolution in supply chain management has gone largely unnoticed within the financial services industry. Basic data management

[5] A notable exception is found in Groot (2008).

issues addressed in other industries – examples include shared identification and classification systems – have been generally relegated to third or fourth tier considerations. This is in part due to the entrenched and expanded industry focus on the immediate profitability of transactions as the metric for compensation. As in any industry, those who generate sales tend to be the most highly compensated; unlike most industries, however, trading firms have long applied an extreme discount to the value of any other functions outside of creating profitable transactions. It is not possible to ignore the costs of inventory and assembly in a manufacturing environment, but financial services have minimal visible "manufacturing friction" compared to most other industries. Additionally, the product manufacturing process in finance has become the specialized purview of those who can create increasingly complex mathematical models that can be used to create transactions dependent on pre-existing transactions, driving both revenue and risk into increasingly complex interdependent transaction streams. In other words, within financial markets the immediate profits are sourced at the trading desk – the traders call the shots.

This situation has led to one where the vast majority of investment has been in the systems and processes that can create transactions, with relatively minimal investment in the people, processes, or systems used to process the outputs of the transaction creating systems or the integration of the information created. It is relatively easy to create many transactions, but clearing, settlement, and payment systems suffer from minimal investment and a perception that they are not a place for top performers. This has, unfortunately, left financial services as an industry badly out of line with modern supply chain practices and with a data environment that may be best characterized as chaotic.

To complicate the environment even further, the financial services industry has had a rapid increase in the number and type of intermediaries over the past 20 years.[6] Figure 19.3 is drawn from the International Standards Organization (ISO) Financial Instrument Business Information Model (FIBIM)[7] describing some of the intermediary roles of firms involved in the creation and distribution of equity, debt or fund securities, and provides a hint of the types of players, the complexity of the supply chain, and the resulting fragmentation of information sources.

As is apparent from Figure 19.3, there are multiple players with multiple roles and objectives that are responsible for creating and sharing different pieces of interrelated information for any given transaction. Today, the actors in the financial system have to compare, aggregate and integrate millions of records created daily by multiple different computer systems, using different identifiers, different naming

[6] See, for example, The Kay Review of UK Equity Markets and Long-Term Decision Making (2012).

[7] The ISO 19312 Financial Instrument Business Information Model work, 2007, has since been incorporated into the ISO 20022 standard.

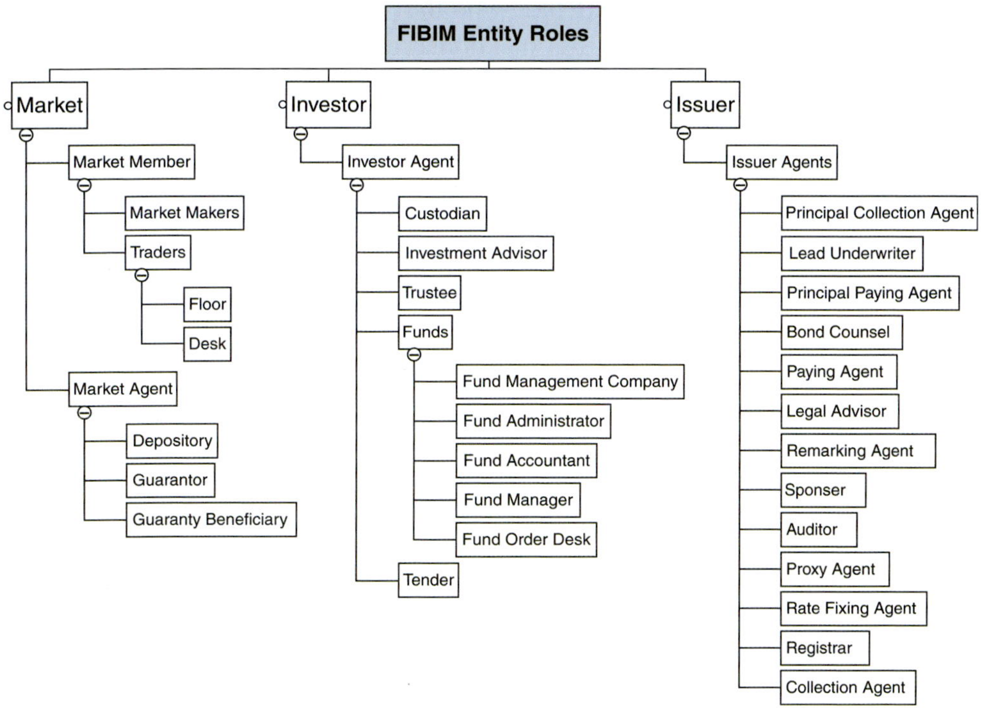

Figure 19.3 Financial instrument business information model.

conventions and different definitions. Absent multiple data standards and shared definitions at each level of the "stack," and the predictable result is that each firm will spend significant resources just to maintain "directionally correct" information related to overall interrelated risks and exposures. There is also recent research indicating that the overall levels of complexity in the technical implementations of multiple firms in the financial services supply chain(s) may in and of themselves be a source of systemic risks and create an almost impossible task for those within large enterprises who are responsible for assessing related risks (Sommerville et al., 2012).

Later we will provide a more detailed framework for integrating the different information created by capital markets activities amongst various players.

19.4.2 Building a standard model of the financial system

In order to describe and measure financial information, one is faced with a mix of technology, content, and business dimensions that must be combined within an information framework to create a meaningful representation of the system. In order to assess risks within and across organizations, one must be able to understand what the landscape of the system looks like, who the players are, and how they create, manage, and share information. As mentioned above, this type of supply chain analysis does not really exist for capital markets. Here we propose a broad framework for describing the global capital markets system. We will use the framework below to organize the presentation of relevant financial standards related to securities transactions.

The common background

An organizing observation is to acknowledge the delineation between the different combinations of technology and operational analyses and models emerging from the evolutionary forces of the past 20–30 years. Chief among these are (a) distributed networks, (b) consumer-focused technologies, and (c) enterprise management strategies. These can each be regarded as dimensions along which one can measure or model the ways in which firms and people create information, what can be done with that information (especially the constraints imposed by the actual mechanics involved in managing the information), and what changes need to be made to increase visibility into the system.

Distributed networks: The Internet is the obvious example of a large distributed network, but large firms today maintain and operate distributed network environments that share many of the structural forms used in designing the Internet. These are characterized by their use of *registries* to maintain authoritative sources for defining entities, the attributes that describe those entities, and the services available to manage the related information.

Consumer-focused technologies: The rapid evolution in physical form factors, interfaces, and applications for mobile technologies is the most visible factor of the evolution of consumer-focused technologies, but the corresponding changes in the ways in which software is created and information is managed are possibly more relevant when thinking about implementing and maintaining changes in the systems of information management used to examine and measure capital markets related activities. The intersection of consumer interaction and the Web has had multiple impacts on the way in which information is created, stored, and maintained, many of which are driven into the enterprise through employee insistence on using these more flexible and more integrated toolsets. In addition, much of this innovation has been fundamentally driven by the intersection of open source software practices, the sheer scale of the Internet, and the resulting "rain forest" ecosystem, where ideas and applications appear and disappear like clothing styles, but many leave behind some novel approach to a particular problem. This open ecosystem has fundamentally changed the nature of systems development and maintenance, and continues to do so at an accelerating pace.

Enterprise management strategies: Enterprise management strategies can be looked at along two primary dimensions – the technical platform dimension and corresponding developments in IT and data management practices that are specifically designed to address enterprise issues and the simultaneous development of management and operational approaches designed to either control or take advantage of these practices. The evolution of business process design and associated modeling and specification languages (Business Process Execution Language (BPEL), for example) is a very visible example of where emerging horizontal standards are moving up the abstraction ladder, encompassing increasingly complex areas of functionality. In other words, business practices for analyzing enterprise workflow processes have led to a variety of tools and software that promise to bring automation to more complex areas of activity.

From a horizontal technologies perspective, enterprise organizations today can be described as "closed" or "walled garden" entities: they run an information management system that they own and control, along with strict "cross border" definitions, rules, and controls on what is and what is not "internal" versus external information. These "garden" enterprises all utilize the open Internet, but also communicate and connect over private and semi-private networks. Many, if not most, of the "private" networks also run on top of the Internet; even in the case where the connections are strictly private (i.e., there is no direct network connection to the Internet), the common Internet standards for networking (TCP/IP, etc.) are almost always the underlying mechanisms.

A generic vertical industry landscape is characterized by multiple different "walled garden" organizations exchanging different types of information along different pathways, all the while connecting to (and often through, for purposes of connecting to each other) the larger economic network which is a subset of the global Internet. (While this definition lacks the specificity necessary for describing certain types of entity relationships, it is sufficient for dealing with coarse-grained economic structures such as firms and transactions.)

19.4.3 Financial markets framework

Abstractly, the capital markets can be viewed as composed of vertical, industry-specific, specialized organizations that interact with each other to engage in specific (vertical) sets of activities. For our purposes, the primary activities can be divided amongst types of players: market participation, regulatory, and infrastructure,

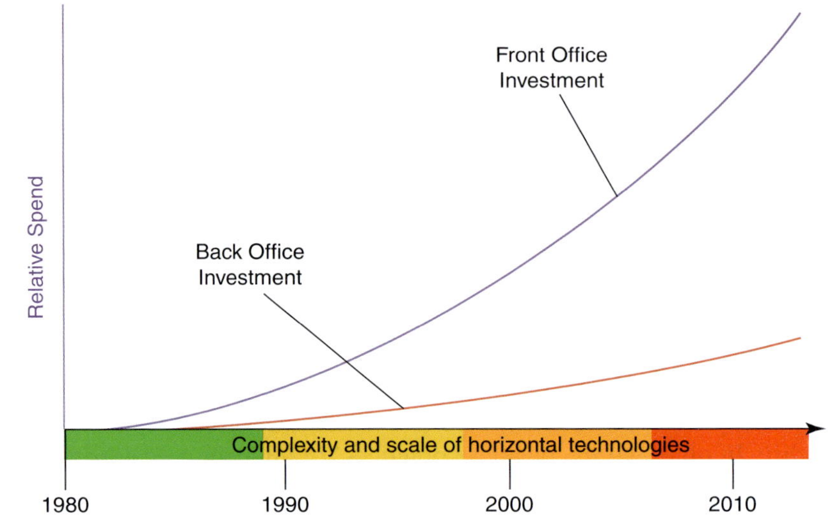

Figure 19.4 The relative rate of investment in front office resources versus back office resources.

collectively "capital markets organizations." These firms all operate "on top of" the evolving "network cloud" (i.e., the Internet, not to be confused with "cloud computing") that is encompassing the services economy in particular and the overall economy from a larger perspective.

One aspect of the past few decades of interdependent evolution of horizontal and vertical technologies, with related vertical business products and practice in the financial industry, has been a significant impact on the quality and availability of information available for risk management. Specifically, the relative disparity in resources allocated between the front office and back office cited above has been a key factor in the dearth of adequate, timely and quality information for risk management. This is illustrated in Figure 19.4.

Briefly, the diagram illustrates the relative rate of investment in front office resources (e.g., sales and trading) versus back office (e.g., settlement and clearing) resources over the past few decades, while the bars on the *x*-axis represent the increases in complexity – and risk – of the underlying horizontal technologies.[8] Put another way, the capability to create transactions grows faster than the capabilities required to manage them after the sale. It should be noted that this is nothing new – the original development of identification numbers for securities in the US markets

[8] The diagram is only meant to be illustrative, and it should not be interpreted to imply that industry back office spending is small. The diagram is based on personal experience and many conversations with industry players over the past decade, and has been shown to hundreds of industry participants – where the general consensus is that it understates the issue.

was brought about by the breakdown of back office systems in the late 1960s: paper-based trading systems fell apart when the New York Stock Exchange share volume hit approximately 13 million shares per day. (In 2009, volumes regularly exceeded 2 billion shares per day.) But while not new, the technologies used and global scale of the industry makes the impact of associated data issues a key source of complexity and a significant barrier to transparency for either firm management or regulators.

A framework for describing the information created and maintained within the industry must be precise enough to describe the actual system while general enough to enable the data to be organized in a way that permits the information to be useful and accessible to multiple players with differing goals. While a top-down global view of the entire industry will necessarily have to work through multiple abstraction layers to be useful, the information relevant to risk analysis for individual firms has to be sourced from activities related to the acquisition and disposal of securities – the investment process, and especially a lower level view of the trade lifecycle.

19.4.4 Modeling the system

Organizing an abstract model of a system in such a way that it can be used to create a digital representation of the financial system is constrained by the boundaries of how the available technologies implement the storage and management of the underlying information. Here we present a generic outline of such a model that is agnostic with regard to software methods – in other words, specific implementations will be done at a lower level based upon appropriate standards. Keeping this type of model in mind is useful primarily to make sense of the way different messaging schemes and database structures are constructed. When you are faced with a mountain of data, start by drawing a map of what it is supposed to represent (Figure 19.5).

The first task in building such a model is to specify the boundaries of the system, formally the Domain. As the diagram illustrates, within the Domain there are Things – some of them more important than others, typically the Things the system is designed to represent, some of them necessary but not the primary focus. There

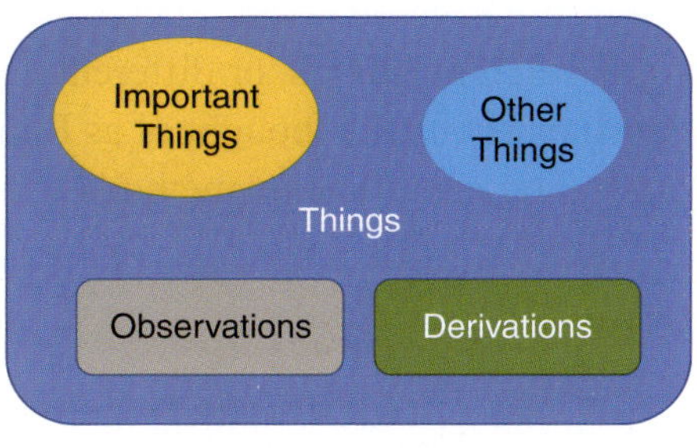

Figure 19.5 Data mapping.

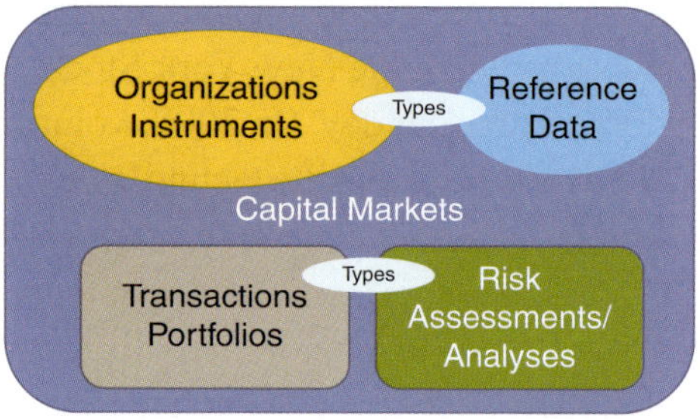

Figure 19.6 Data mapping – capital markets.

are observations about these Things and their interactions, and derivations based on these observations. While not specifically shown in the first diagram, note that each Thing has a Type.

Applying this type of modeling abstraction approach to the capital markets domain, we end up with something like Figure 19.6. The primary entities that are tracked and measured are Firms, Instruments, Portfolios, and Transactions.

There are four basic aspects of each Thing, or Entity, that must be implemented in software (horizontal or vertical applications) for the resulting system to be useful, as follows.

1. Entities – or Things – the primary focus of the underlying organizational mission – what are the important things being described by the applications? There is a top-level classification here that creates the organizing taxonomy for the overall system. This is reflected in a master taxonomy of Entity Types.
2. For each Entity – there must be an Identification Method. You have to be able to distinguish two things of the same type from each other by some sort of uniqueness criterion. This is typically done through some form of identifier, but there are many variations and new methods emerging rapidly in the horizontal systems space.
3. Groups – what categories can things be grouped into? This involves creating and maintaining multiple purpose specific taxonomies.
4. Relationships – how are different Things, or different groups of Things, related to each other? Relationships should be thought of as Entities, or Things, in their own right, with corresponding identification methods and classification schemes.

19.4.5 The trade lifecycle

Using the framework just described (or something similar), we can eventually come up with a more pragmatic representation of the system based on observations and measurements of actual entities and their interactions. In order to analyze the trade lifecycle, we need to answer the question(s), "Who Trades What, Where, How?" While not definitive, the outline below gives an overview of the Things or Entities involved.

- Who – buyers and sellers. At the initial level we look only at descriptive attributes, not intentions or strategies
- Trades – what types of transactions, defined by what characteristics?
- What – types of securities, instruments, or contracts, fungible or not.
- Where – organized or informal venues, listed or OTC.
- How – the mechanics of the trade(s), including supply chain requirements such as clearing and settlement or asset servicing.

Who – The Bank of International Settlements (BIS) defines the types of "financial markets firms"[9] as:

- Central Banks
- Deposit-taking corporations except the central bank
- Money market (MM) funds
- Non-MM funds
- Other financial intermediaries except insurance and pension funds
- Financial auxiliaries
- Captive financial institutions and money lenders
- Insurance corporations
- Pension Funds.

In examining the trade lifecycle, one will see each of these firms acting either directly or through intermediaries, depending on the asset class and type of transaction. It is important to note that there are currently no formal standards for classifying firm types based on their roles in various market transactions.

Trades – There are literally hundreds of documented trade types, usually divided into three stages based on their position in the overall sequence – pre-trade, trade, and post-trade. The FIX Protocol currently lists 50 trade types, each of which can be combined with other message attributes to define variants.[10] Structures of trade transactions also vary by asset type and market venue.

What – Securities and contract types. Classification of securities is an area where prior efforts at standardization (see ISO Classification of Financial Instruments, below) have been overtaken by the changes in different markets and the introduction

[9] BIS, May 2009.

[10] Financial Information eXchange (FIX) Protocol is a series of messaging specifications for the electronic communication of trade-related messages, see `http://fixwiki.fixprotocol.org/fixwiki/TrdType`.

of new types of assets and related securities. Since instrument supply chains vary across asset classes and instrument types, the complexity inherent in the system shows up clearly. Classification of securities – and clean identification schemes that permit accurate tracking of instruments from issuance or contract writing – are important areas for standardization. Absent continued efforts to standardize, coupled with development of standards for types of firms engaged in the transaction lifecycles, there is little chance that market transparency will increase.

Where – Venues for financial transactions typically include public exchanges, dealer markets, electronic communication networks (ECNs), and private markets (dark pools, etc.).

How – This describes the different sequences of events and inputs to them that are the result of the combination of asset classes, market participant types, and trade venue.

From an individual security or contract perspective, the trade lifecycle can be seen as a fairly linear progression (Figure 19.7).

When looked at from the perspective of the different players in the supply chain, however, the cyclical nature of the overall set of activities is more apparent (Figure 19.8).

When we look at Figure 19.8 in light of the above analysis, we can see that all of the specialized organizations involved in the supply chain are using similar horizontal technology stacks, i.e., TCP/IP based internetworking, relational database management systems, etc. The corresponding internal business processes are

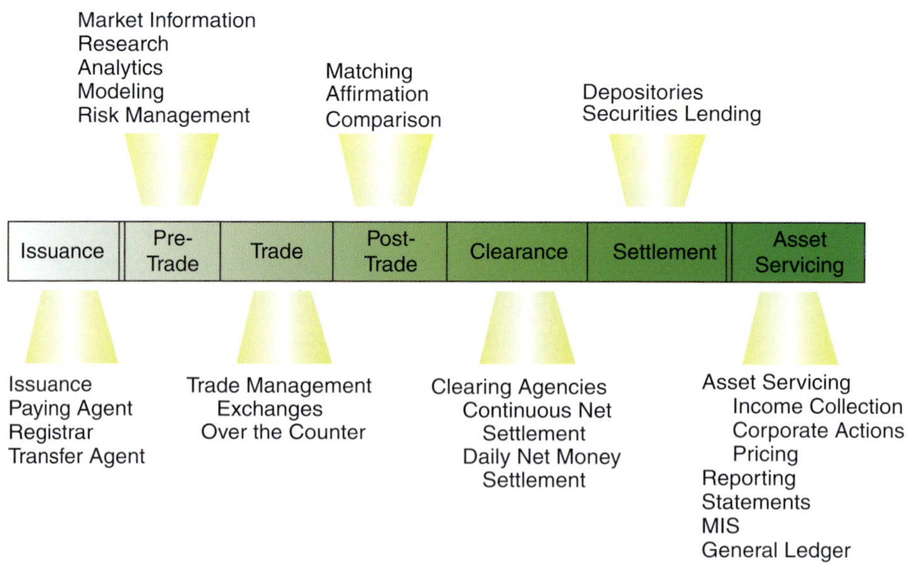

Figure 19.7 The trade lifecycle. (Diagram courtesy of ISITC.)

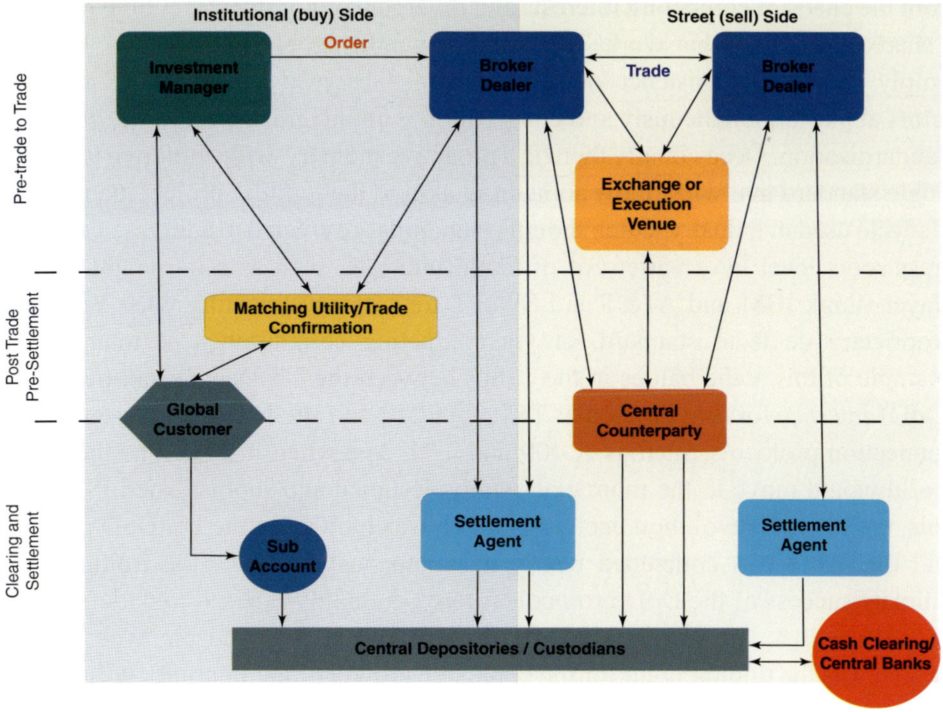

Figure 19.8 Trade supply chain. (Bill Nichols, 2010.)

shaped by the requirements of these horizontal toolsets – data must be arranged in normalized table formats for databases, etc. As the horizontal technologies mature and newer ones evolve, the different market participants (in their vertical businesses) will adapt at different rates, leading at times to "frozen pockets" of technologies and operational processes. These technologies require standardized interfaces to the newer, more powerful approaches before they can be adopted, and all of the players face a possible first mover penalty – or advantage. The following sections describe in detail how and why standards are key to the financial services trade lifecycle.

19.5 How standards are created

One of the answers from Gerald Weinberger's (1986) book *The Secrets of Consulting* to the age old question of "Why are things the way they are?" is disarmingly straightforward. Weinberg's response is "They got that way." Similarly, standards just get that way.[11] More often than not, standards emerge

[11] For a more academic explanation, see Puffert (2009).

from the chaos of competing interests and divergent requirements, evolve rapidly to a shared view of "what works" to solve a specific problem, and often appear to simply spring into existence out of nowhere. Observed closely, there are many actors at play simultaneously and competition is an integral aspect of the process of standardization. Occasionally there is a proactive industry-wide initiative to create a single standard that will follow something akin to the familiar project management lifecycle used in industry. Often though, standards are created to address needs at the grass roots level by a variety of different interests, such as a dominant industry player (think IBM and AT&T and SWIFT for instance) creating what becomes a proprietary de facto standard, or via competing communities of interest. One example of this is the battles in the 1980s between the US Department of Defense (DoD) internetworking protocols: TCP, UDP, IP and the ISO open systems interconnection protocols, such as X.400 and TP4. Just when it looked as though the world would move to the more elaborately designed and sophisticated ISO protocols, the Internet revolution occurred, which was built upon the US DoD protocols and the battle was concluded nearly instantaneously. One of the ironies of the ultimate success of the DoD protocols (more commonly referred to colloquially as TCP/IP) is that Tim Berners-Lee, the inventor of HTML and the Web Browser (which was the tipping point for the explosive growth of the Internet), was working for one of the strongest proponents and adopters of the ISO-OSI protocols, the European Organization for Nuclear Research (CERN) in Switzerland.

19.6 Types of organizations that create standards

The number and types of organizations that create standards can appear quite daunting to comprehend – the online "Essential Guide to Standards" claims there are over 500 standards-setting organizations that create technology standards.[12] There are major international standards setting bodies: ISO, the International Electrotechnical Commission (IEC), and the International Telecommunications Union (ITU). Participating in these organizations are representatives from national bodies from each participating country. Within the last few decades, largely as a result of the explosion in information technology, the *consortium model* for standards has emerged in response to actual or perceived rigidity and special-interest entrenchment found in the established international standards bodies. Consortiums are made up of participants with a common interest, often aligned within a specific industry or for a specific technology. Professional societies, such as the Institute of Electrical and Electronics Engineers (IEEE) also create standards. Sometimes dominant market participants, such as The Bell System during the AT&T monopoly

[12] www.consortiuminfo.org/essentialguide/

years, IBM during the mainframe computer era, and Microsoft during the initial PC wave, create proprietary standards due in large part to their market share and sometimes due to their marketing prowess.

We will introduce important examples of each type of standards organization and then specifically define the important standard setting organizations that will likely impact risk practitioners.

19.6.1 International standards

Most people only slightly familiar with standards will be aware of ISO – the International Organization for Standardization. There is nearly no aspect of modern life that is not somehow impacted by an ISO standard. The inclusion of the ITU and IEC is not as readily clear, as these organizations are referred to as *information and communications technology* (ICT) standards organizations. Modern financial systems are wholly dependent on ICT standards. The horizontal technology standards created by standards organizations have major implications in terms of the continuing revolution of modern finance, where technology-centric development provides the primary avenue for product and practice development.

International Organization for Standardization (ISO)

The International Organization for Standardization (ISO), located in Geneva, Switzerland is the premier standards body responsible for the development and management of international standards. Currently, the number of ISO standards exceeds 19,000.[13] Standards work is performed within ISO's over 260 Technical Committees (TCs),[14] their subcommittees (SCs) and working groups (WGs).[15]

ISO was created in 1947, formed from a previous organization, the International Federation of the National Standardizing Associations (ISA). ISO was revitalized to provide international standards as much of the world was being rebuilt following World War II. The organization draws its membership from a world-wide network of national standards bodies which are the standards-setting entities in each of their local markets. Examples are the American National Standards Institute (ANSI) in the USA, the British Standards Institute (BSI) in the UK, and the Standardization Administration of the People's Republic of China (SAC) in China.

In the ISO structure, member countries can join committees as participating (P) members or observer countries (O) members (see Figure 19.9).[16] Participating

[13] www.iso.org/iso/home/about.htm
[14] www.iso.org/iso/home/standards_development/list_of_iso_technical_committees.htm
[15] www.iso.org/iso/home/about/about_governance.htm, see also Craig N. Murphy and JoAnne Yates (2009).
[16] www.iso.org/iso/home/about.htm

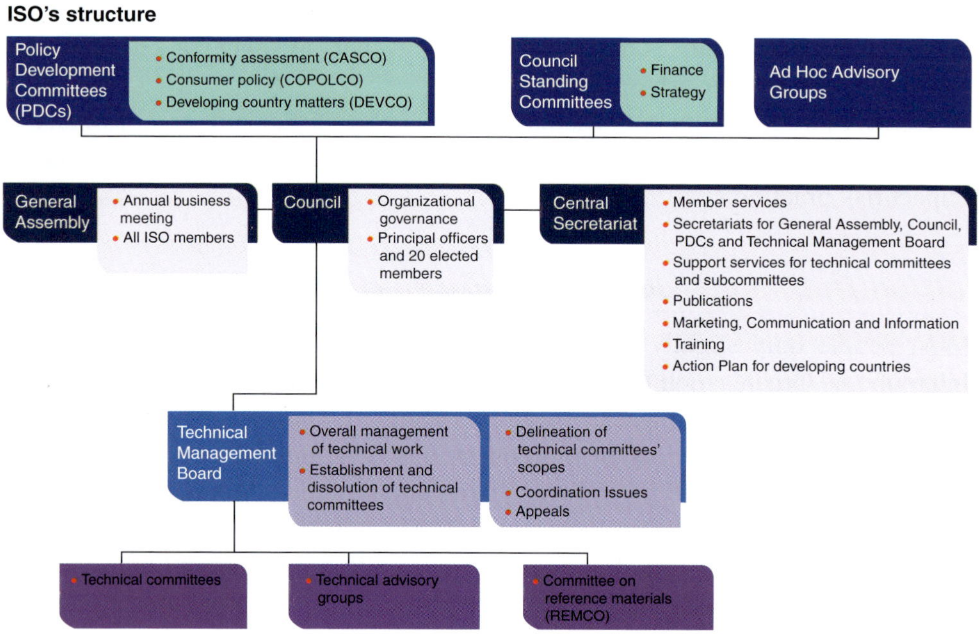

Figure 19.9 ISO structure.

members are active participants in the committees that they choose to join and are eligible and expected to contribute expert members to standards working groups. Observer countries are kept up-to-date on the work of the committees but do not actively participate in the work.

ISO standards also benefit from the participation of liaison organizations. These are international or broadly based regional organizations working or interested in similar or related fields.

Cooperation with industry-led standards groups Considerable attention was focused on a battle over network standards in the late 1980s and into the mid-1990s. The battle focused on a proposed standard network architecture called the ISO Open Systems Interconnect (ISO-OSI) Network model and the US Department of Defense network architecture, referred to colloquially as either the Internet protocols or the TCP/IP protocol. The technology revolution had a significant impact on the conservative, process-oriented ISO standardization system. The trend in recent years has been for ISO to work with other standards bodies in adopting de facto standards and industry-created standards as ISO standards. There are now well-established agreements whereby widely adopted and successful consortia created standards can become ISO standards.

International Electrotechnical Commission (IEC)

The IEC was formed in 1906 to support standards setting in the electrotechnical industries. It works with industry, government, academia and end-users to develop consensus-based international standards, and manages conformity assessment systems for electric and electronic products, systems and services.[17] The conformity assessment establishes that a product is reliable and meets expectations in terms of performance, safety, efficiency, durability and other criteria.

International Telecommunication Union (ITU)

ITU was founded in Paris in 1865 as the International Telegraph Union. It took its present name in 1934, and in 1947 became a specialized agency of the United Nations. Although its first area of expertise was the telegraph, the work of ITU now covers standards from digital broadcasting to the Internet, and from mobile technologies to 3D TV. An organization of public-private partnership since its inception, ITU currently has a membership of 193 countries and some 700 private-sector entities. ITU is headquartered in Geneva, Switzerland, and has 12 regional and area offices around the world. One of the divisions of the ITU that was formed in 1956 was the International Telegraph and Telephone Consultative Committee (CCITT, from French: Comité Consultatif International Téléphonique et Télégraphique) which was responsible for a large number of telecommunications standards. The CCITT was renamed ITU-T in 1993.

Coordination of the three organizations

The IEC, ITU, and ISO work together to coordinate consistent global technology standards. As technology has evolved beyond bespoke, vertical foci, the IEC, ITU, and ISO all are playing a much more important role in helping to provide consistent horizontal technology standards that can be adopted across industries.

19.6.2 National standards organizations

National standards bodies, in addition to consortia and liaison organizations, are the primary source of labor for international standards. Each country has different approaches to standards creation. There can be multiple governmental and non-governmental organizations (NGOs) involved in creating standards.

The USA relies on the American National Standards Institute (ANSI), an NGO, to create national standards and to participate in international standards within ISO. In turn ANSI relies on a variety of organizations who are accredited to create ANSI standards and participate in ISO organizations.

[17] www.iec.ch/about/history

The Accredited Standards Committee (ASC) X9 is the accredited standards body that represents the USA in all financial services standards within ISO and also creates and adopts national (ANSI) standards on behalf of ANSI.[18] It is easiest to think of accredited standards organizations as franchisees of the national body. Similar to a franchisee, the accredited standards organization undergoes periodic reviews by the franchiser (the national standards body). The British Standards Institute (BSI) works in a similar manner. For example, via a contractual agreement with BSI, UK Payments is responsible for the majority of financial services standards in the UK and in representing the UK in ISO standards development and decision making.[19]

The USA also has the National Institute of Standards and Technology (NIST), which is a governmental organization. Part of the Department of Commerce, NIST was created as a physical science laboratory to improve measurement technology in manufacturing. One of the major services provided by NIST is the measurement of time. This service[20] is especially important in situations where multiple systems at multiple locations must synchronize transaction timestamps, a common issue in the securities industry.

19.6.3 Professional societies

Another source of industry standards are professional societies, such as the Institute for Electrical and Electronics Engineers (IEEE), the American Institute of Certified Public Accountants (AICPA), and ASTM International (formerly the American Society for Testing and Materials). The IEEE provides important standards for computer networking and data representation; the AICPA sets standards for accounting and audit practices in the USA.

19.6.4 Industry consortiums

Early information and communications technology was largely dominated by proprietary technology solutions, such as the Bell System – AT&T Monopoly in the USA, that because of market size and structure became de facto standards. Largely due to advances in semiconductor manufacturing and the resultant information technology revolution, the need for standardization far outpaced the ability of large entrenched market participants to control and evolve platforms. One of the fundamental requirements for information and communication technology is interconnection and interoperability. The need to develop standard interfaces and

[18] www.x9.org/about/ [19] www.ukpayments.org.uk/about_ukpayments/
[20] www.nist.gov/pml/div688/grp40/its.cfm

Table 19.1 *Consortiums in the financial services industry*

Consortium	Year initiated	Objective	Website
FIX Protocol Ltd.	1994	Securities processing – pre-trade, trade, post-trade, derivatives clearing	www.fixprotocol.org
Financial Products Markup Language	1999	OTC derivatives	www.fpml.org
Interactive Financial eXchange	1997	Banking, payments, ATM/POS, card management	www.ifxforum.org
Open Financial Exchange	1997	Communicate financial data between financial institutions and business and consumers	www.ofx.net

communication protocols behind which multiple companies could compete in terms of service offering spawned an entire industry of standards organizations focused on special purpose, highly focused processes. These special purpose standards organizations are referred to as consortiums. There are estimated to be roughly 500 industry standards consortiums.[21] Consortiums are often specific purposed with a specific time horizon, while some continue on and have quite a long time horizon. Table 19.1 presents several examples of consortiums in the financial services industry.

19.6.5 The Internet standards organizations

The collection of standards referred to as the Internet standards are managed by a group of related organizations that are reflective of the culture that created the Internet, as described below.

Internet Engineering Task Force (IETF)

Arguably, the Internet Engineering Task Force (IETF), the first of this group of organizations, redefined the standards process. The traditional standards development organizations were largely driven by governmental and corporate entities. The IETF from its inception was formed to be the anti-organization and the

[21] Updegrove, Gesmer, The Essential Guide to Standards, www.consortiuminfo.org/essentialguide/whatisansso.php

anti-standards organization. Membership was not official and it was at the individual level not the organization level. The IETF prides itself in being a meritocracy where the only barrier to entrance is competence. Despite the success of the Internet, the IETF has maintained its anti-organizational bias and culture. Structure and multiple organizations have arisen, given the size and scope of the Internet itself, yet the IETF ethos remains at the heart of the organization.

Internet Society (ISOC)

The Internet Society (ISOC) was formed in 1992 to provide a formal organizational structure to address the Internet. The Internet Society provides support for the IETF and the Internet Engineering Steering Group provides technical management of the IETF.

Internet Architecture Board (IAB)

The Internet Architecture Board (IAB) is a committee of the Internet Engineering Task Force that oversees the architecture, the standards process, and provides advice to the Internet Society. The long-term issues and direction for the Internet were moved from the IETF to the Internet Research Task Force (IRTF). The IETF was left to address short- to medium-term issues.[22]

Internet Corporation for Assigned Numbers and Names (ICANN)

Prior to 1998, the US Government, via the Internet Assigned Numbers Authority (IANA), was responsible for issuing internet address and domain names. This function was privatized to the Internet Corporation for Assigned Numbers and Names (ICANN). IANA became a department of ICANN upon successful transition of responsibilities to ICANN. ICANN operates under a Memorandum of Understanding with the US Department of Commerce. The Domain Name System, the assignment of Internet addresses (IP addresses), and the management of competing distributed Internet registrars, is one of the primary examples (and certainly the largest) of the power and scalability of a distributed registry.

19.6.6 Horizontal technology standard consortiums

Three organizations were formed to provide horizontal information technology solutions in response to technical innovations. While the World Wide Web Consortium (W3C) has stayed to its core mission of supporting XML horizontal technologies, both the Organization for the Advancement of Structured Information Standards (OASIS) and the Object Management Group (OMG) have evolved since their origin.

[22] RFC 4440 IAB Thoughts on the Role of the Internet Research Task Force (IRTF).

World Wide Web Consortium (W3C)

The W3C was created to support open standards for the World Wide Web in 1994 by Tim Berners-Lee, the inventor of the World Wide Web. Incredibly significant contributions from the W3C include: HTML (HyperText Markup Language) in 1995; The Extensible Markup Language (XML) in 1996; the Semantic Web in 2001; and Web Services in 2002. Many of the most important financial data interchange formats are based upon XML (ISO 20022, XBRL, FpML, FIXML).[23]

The Object Management Group (OMG)

The Object Management Group (OMG) was created by a group of computer vendors and American Airlines in 1989 as part of the open systems movement. The OMG was initially responsible for creating a heterogeneous distributed object standard.[24] The mission of the OMG has evolved along with the industry; some of OMG's latest initiatives are focused on the financial services industry and semantics. The Financial Domain Task Force (FDTF) is a partnership between the OMG and the EDM Council.

OASIS (Organization for the Advancement of Structured Information Standards)

OASIS is a not for profit consortium that began in 1993 as SGML. The Standard Generalized Markup Language (SGML) was the precursor to both HTML and XML. With the creation of XML, the OASIS name was adopted. OASIS maintains a very diverse set of both horizontal and vertical standards based upon XML. OASIS is also home to the AMQP messaging standard that originated within the financial services industry.[25]

19.7 ISO TC68 the International Standard for Financial Services

The International Standards Organization Technical Committee 68 (ISO TC68), responsible for standards for international financial services, was established in 1948. ISO TC68 has developed and manages international standards used by financial organizations to support financial transactions covering the broad business categories of securities and payments, with the requisite security standards to support financial services transactions (see Figure 19.10).

ISO TC68[26] supports financial services through standards that apply to the broad financial services community (the Legal Entity Identifier (LEI) (ISO 17442) standard and ISO 20022) and through standards for the broad business categories

[23] www.w3.org/Consortium/facts
[24] http://en.wikipedia.org/wiki/Object_Management_Group [25] www.oasis-open.org/org
[26] Source: Karla McKenna, ISO TC68 Structure and the Legal Entity Identifier (LEI) Standard, June, 2011

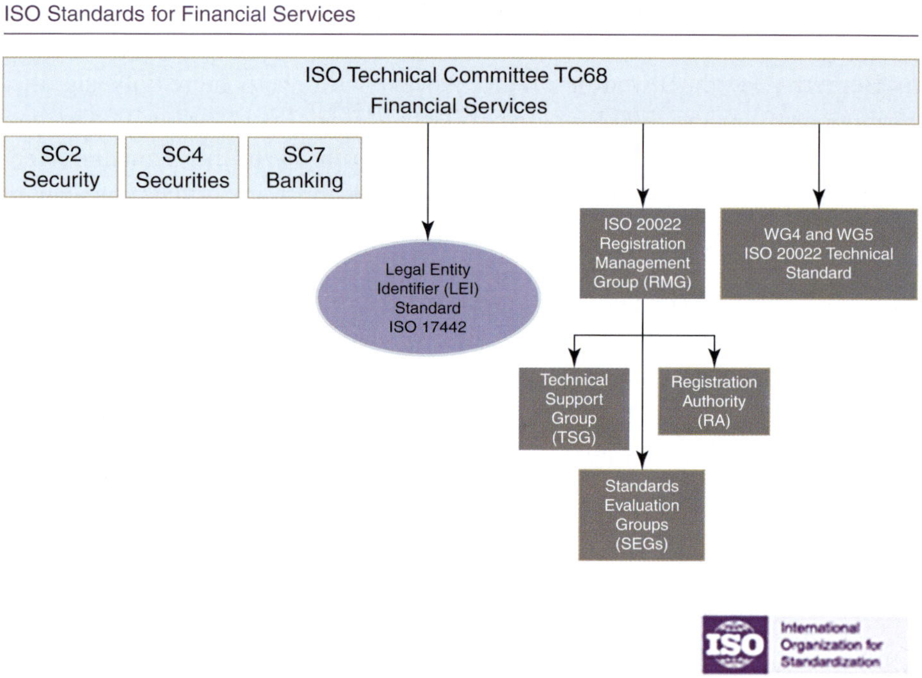

Figure 19.10 ISO standards for financial services.

of security (SC2), securities (SC4) and payments (SC7). Examples of standards developed and managed by these subcommittees include:

- SC2 Financial Services, Security: PIN, biometrics, PKI, encryption algorithms
- SC4 Securities and Related Financial Instruments: ISO 15022, International Securities Identification Number (ISIN, ISO 6166), Classification of Financial Instruments (CFI, ISO 10962), Market Identifier Codes (MIC, ISO 10383)
- SC7 Financial Services, Core Banking: Business Identifier Code (BIC), ISO currency codes, International Banking Account Number (IBAN)
- TC68: ISO 20022 standard, Universal Financial Industry Message Scheme.

Of the broad categories of standards, the majority of ISO TC68's standards are data content standards. We will see in the next section that one of TC68's newest standards also extends past data content and utilizes data technology standards to deliver value to the financial services standards users.

19.8 Data content standards

Data content standards support the *What*, the information that is needed to describe financial instruments and execute financial transactions. These can include key elements such as identifiers, identification and classification codes, as well as lists of key information that describe financial instruments (an algorithmic representation of a credit default swap, for example) or messages to complete the settlement of a financial transaction. These standards ensure consistent data elements within and across business areas.

19.9 Standards for data representation

The need to define a standard encoding for numeric and character data is by now taken for granted. Surprisingly, this area still draws attention when not properly considered. Data representation should be explicitly agreed upon, and not assumed, when exchanging information. While a bit technical and somewhat detailed, the discussion below is used as an example of the level of granularity necessary to create data representation standards. It is important to understand that working systems are built up out of multiple layers of such detailed specification. When evaluating operational risks inside an organization, it is important to be able to understand where the "handoffs" are between horizontal standards implementations ("pretty much everyone in every industry does it this way") and vertical standards applications that are built for industry specific business requirements and that implement standards specific to that vertical industry.

19.9.1 Character data standards

You are likely aware that computers represent information as binary digits or bits: 1s and 0s are strung together to represent the characters and digits we use in our natural language alphabets. The representation of character data as a string of on and off values predates electronic computing machines by a couple of centuries, going back to Francis Bacon. Most readers will probably be familiar with a more recent coding set from telegraphy, Morse Code. Today's predominant standard and others you should be aware of are set forth below.

ASCII standard

The ASCII standard from the USA and its extensions to represent European character sets has become a dominant standard for data representation for data interchange. There are two main categories of ASCII character sets. The original 7 bit version contains "printable" alpha-numeric and punctuation, as well as non-printable control characters. The non-printable control characters served as rudimentary data transmission and data formatting commands for printers and terminals when character based devices dominated prior to the introduction of modern graphical device interfaces. The ASCII character system was limited to use 7 bits to permit the 8th bit of an 8-bit byte (octet) to be used for parity checking, when computer communications were much less reliable than today. The original 7 bit ASCII format was extended to use all 8 bits of an octet to represent additional characters required to represent additional European language characters.

Table 19.2 *The printable range of the character encoding from the ASCII character set*

Binary digits	Hexadecimal	Character	Description
100000	0x20		Space
100001	0x21	!	Exclamation mark
100010	0x22	“	Double quotes (or speech marks)
100011	0x23	#	Number
100100	0x24	$	Dollar
100101	0x25	%	Percentage
100110	0x26	&	Ampersand
100111	0x27	‘	Apostrophe
101000	0x28	(	Open parenthesis (or open bracket)
101001	0x29	)	Close parenthesis (or close bracket)
101010	0x2A	*	Asterisk
101011	0x2B	+	Plus
101100	0x2C	,	Comma
101101	0x2D	-	Hyphen
101110	0x2E	.	Period, dot or full stop
101111	0x2F	/	Slash or divide
110000	0x30	0	Zero
110001	0x31	1	One
110010	0x32	2	Two
110011	0x33	3	Three
110100	0x34	4	Four
110101	0x35	5	Five
110110	0x36	6	Six
110111	0x37	7	Seven
111000	0x38	8	Eight
111001	0x39	9	Nine
111010	0x3A	:	Colon
111011	0x3B	;	Semicolon
111100	0x3C	<	Less than (or open angled bracket)
111101	0x3D	=	Equals
111110	0x3E	>	Greater than (or close angled bracket)
111111	0x3F	?	Question mark
1000000	0x40	@	At symbol
1000001	0x41	A	Uppercase A
1000010	0x42	B	Uppercase B
1000011	0x43	C	Uppercase C
1000100	0x44	D	Uppercase D

Table 19.2 (*cont.*)

Binary digits	Hexadecimal	Character	Description
1000101	0x45	E	Uppercase E
1000110	0x46	F	Uppercase F
1000111	0x47	G	Uppercase G
1001000	0x48	H	Uppercase H
1001001	0x49	I	Uppercase I
1001010	0x4A	J	Uppercase J
1001011	0x4B	K	Uppercase K
1001100	0x4C	L	Uppercase L
1001101	0x4D	M	Uppercase M
1001110	0x4E	N	Uppercase N
1001111	0x4F	O	Uppercase O
1010000	0x50	P	Uppercase P
1010001	0x51	Q	Uppercase Q
1010010	0x52	R	Uppercase R
1010011	0x53	S	Uppercase S
1010100	0x54	T	Uppercase T
1010101	0x55	U	Uppercase U
1010110	0x56	V	Uppercase V
1010111	0x57	W	Uppercase W
1011000	0x58	X	Uppercase X
1011001	0x59	Y	Uppercase Y
1011010	0x5A	Z	Uppercase Z
1011011	0x5B	[	Opening bracket
1011100	0x5C	\	Backslash
1011101	0x5D	]	Closing bracket
1011110	0x5E	^	Caret – circumflex
1011111	0x5F	_	Underscore
1100000	0x60	`	Grave accent
1100001	0x61	a	Lowercase a
1100010	0x62	b	Lowercase b
1100011	0x63	c	Lowercase c
1100100	0x64	d	Lowercase d
1100101	0x65	e	Lowercase e
1100110	0x66	f	Lowercase f
1100111	0x67	g	Lowercase g
1101000	0x68	h	Lowercase h
1101001	0x69	i	Lowercase i
1101010	0x6A	j	Lowercase j
1101011	0x6B	k	Lowercase k
1101100	0x6C	l	Lowercase l
1101101	0x6D	m	Lowercase m
1101110	0x6E	n	Lowercase n
1101111	0x6F	o	Lowercase o
1110000	0x70	p	Lowercase p
1110001	0x71	q	Lowercase q
1110010	0x72	r	Lowercase r

Table 19.2 (*cont.*)

Binary digits	Hexadecimal	Character	Description
1110011	0x73	s	Lowercase s
1110100	0x74	t	Lowercase t
1110101	0x75	u	Lowercase u
1110110	0x76	v	Lowercase v
1110111	0x77	w	Lowercase w
1111000	0x78	x	Lowercase x
1111001	0x79	y	Lowercase y
1111010	0x7A	z	Lowercase z
1111011	0x7B	{	Opening brace
1111100	0x7C	|	Vertical bar
1111101	0x7D	}	Closing brace
1111110	0x7E	~	Equivalency sign – tilde
1111111	0x7F		Delete

The ASCII table also included non-printable control characters for communicating between different computer devices and supporting printer and terminal equipment based upon typewriters. ASCII has special character values to indicate that there should be a line feed. This character has survived and is used by many operating systems as the break in a row of data (or a record). In an example of where a widely held de facto standard from the UNIX operating system, the use of a line feed to indicate a new record, Microsoft operating systems used the more verbose CR (character number 13 in ASCII) followed by LF (character 10). If you have ever edited a file on a PC and experienced having a continuous string of characters when you expected several lines of characters, this is the reason.

EBCDIC

IBM's dominance in computing created an environment where many of their offerings became de facto standards. Still today it is likely that somewhere in the processing centers of large financial institutions are systems that use IBM's character set EBCDIC, Extended Binary Coded Decimal Interchange Code. EBCDIC was based upon an early 6 bit encoding, BCD, derived from the IBM Card, which predates electronic computers.

International support

As computing technology expanded globally, the ability to represent characters from other languages (such as, Greek, Arabic and Cyrillic) became a requirement. Multiple extensions to the ASCII code were introduced by using the previously unused 8 bit range, thus doubling the range of values above 127 (0x7F). Several

Table 19.3 *Important non-printable characters from the ASCII character set*

Binary digits	Hexadecimal	Character	Description
0000000	0x00	NUL	Null
0000001	0x01	SOH	Start of heading
0000010	0x02	STX	Start of text
0000011	0x03	ETX	End of text
0000100	0x04	EOT	End of transmission
0000101	0x05	ENQ	Enquiry
0000110	0x06	ACK	Acknowledge
0000111	0x07	BEL	Bell (beep)
0001000	0x08	BS	Backspace
0001001	0x09	HT	Horizontal tab
0001010	0x0A	LF	Line feed
0001011	0x0B	VT	Vertical tab
0001100	0x0C	FF	Form feed
0001101	0x0D	CR	Carriage return
0001110	0x0E	SO	Shift out
0001111	0x0F	SI	Shift in

conflicting extended character sets emerged using the values 128–255 now available using full 8 bits. On the IBM PC there were multiple "pages" available for various languages. Eight bits was not sufficient however to represent Asian languages, which had more than 255 characters.

The need to represent Asian characters, such as Chinese, Korean, and Japanese, led to the creation of a variety of double byte and multi-byte character sets. There were two varieties of multi-byte character sets – fixed length and variable length. Double byte character sets can represent up to 65,536 characters. To maintain compatibility with ASCII, many of the double byte characters were actually multiple bytes, sometimes one or two bytes. The two dominant multi-byte character sets are Shift-JIS (Japanese) and Big5 (Chinese).

Unicode

To address the difficulties in representing the various characters from all the existing languages and to overcome interoperability problems caused by different encodings, Joe Becker of Xerox and Mark Davis of Apple started an effort to come up with one common character set. The initial version of their proposal was named "Unicode" and presented a 16 bit (two byte) character encoding. Seeing the feasibility of the approach, other companies soon joined the initiative and an industry consortium, The Unicode Consortium, was created. It is important to understand that Unicode separated the identification of characters from the actual encoding as a string of bits. Unicode characters are represented by Unicode **code**

Table 19.4 *Character codes of importance for financial technology*

Character set	Description	Number of bits	Year created	Type of standard
ASCII	American Standard Code for Information Interchange	7	1960	ANSI
BCD	Binary Coded Decimal	6	1928	Proprietary
EBCDIC	Extended Binary Coded Decimal Interchange Code	8	1963	Proprietary
ISO/IEC 8859	8 bit extended character sets			ISO/IEC
UNICODE		Varies	1991	

Table 19.5 *Encodings for Unicode of importance for financial technology*

UCS-2	Original two-byte encoding for Unicode
USC-4	Four-byte encoding for every character
UTF-8	The standard ASCII characters 0–127 are represented as one byte, other characters are represented by two or more bytes
UTF-16	Replaced UCS-2 which only supported big endian encoding
UTF-16BE	Big endian version of UTF-16
UTF-16LE	Little endian version of UTF-16
UTF-32	Four-byte encoding with enhanced functionality compared with UCS-4

points. A code point is represented by a hexadecimal value between 0x and 110FFFFx. This permits 1,114,112 code points. To support backward compatibility with existing computing devices and peripherals and to provide flexibility for the future, multiple encodings are defined. Conversion of Unicode characters into a computer readable format is done using Unicode Transformation Formats (UTF). Because of byte order differences, as discussed below in Section 19.10, representing characters in multiple bytes creates interoperability issues between different computer architectures.

There are several encodings defined by Unicode. The list shown in Table 19.5 is a summary of the most common.

Unicode as an ISO Standard

The success of Unicode had led to a parallel international standard, ISO 10646, which is defined by the ISO/IEC. The standard is overseen by a joint technical committee JTC1/SC2/WG2. The W3C in its role in developing standards for the Internet also has embraced and supports Unicode.

Base64

Base64 has nothing to do with encoding Unicode characters. Base64 is used to encode binary data (image files, PDF documents, etc.) so they can be carried safely along with character data, specifically US-ASCII character data. Two places where Base64 encodings are in use are email and XML. Many existing standards for data interchange, such as FpML and FIX, are being extended to support attachments using Base64 encoding. As with one of the common themes in this chapter, care must be taken when adopting "Base64" encodings to guarantee interoperability. The IETF issued RFC 4648 "The Base16, Base32, and Base64 Data Encodings"[27] which replaces RFC 3548, which is intended to improve the level of specificity around what is "Base64."

19.9.2 Numeric data standards

The representation of numeric data is another area where a variety of formats is available. In the previous section we discussed representing characters. Representing numbers as a series of characters in a character set is essential for human interaction with a computer; however, these are of little useful value within the computer. The string "2183646" is represented in the ASCII character code as a series of bytes: 0x32, 0x31, 0x38, 0x33, 0x36, 0x34, 0x36 is not useful for internal computation in the computer. Computers represent values in various binary formats. A useful way to represent this particular value would be as an integer. An integer is a whole number that does not have a fractional component. The binary value of the number 2183646 is 0x2151DE. Given the high volume of transactions and information, increasingly financial information is being sent in a binary format instead of character representation so it is important to be familiar with a basic introduction to numeric formats.

Binary number representations

One may assume that all computers would store a numeric value, say the number 2183646, in exactly the same way. There used to be a myriad of encodings for numbers in different computers. The encoding of numeric data, as with the rest of information technology, has standardized and there are now two primary ways of storing the bytes that make up an integer number. These are little endian and big endian numbers. Big endian numbers have their value listed left to right with the left containing the most significant byte. A little endian number would store the least significant byte first.

[27] `http://tools.ietf.org/html/rfc4648`

Table 19.6 *Big endian versus little endian numbers*

Endian-ness	Representation of the number 2183646 (0x2151DE hexadecimal)
Big endian	0x21 0x51 0xDE
Little endian	0xDE 0x51 0x21

Table 19.7 *Operating system time*

Operating system	Time unit (tick)	Epoch	Maximum time representation
Unix, Linux 32 bit	1 second	January 1, 1970	January 19, 2038
Unix, Linux 64 bit	1 second	January 1, 1970	AD 292,277,026,596
Microsoft Windows	1 millisecond	January 1, 1601	AD 30,828

Network byte order

Early in the evolution of Internet standards, the choice was made to use big endian representation. This was formalized in RFC 1700.[28]

Floating point numbers

The last area to be standardized in terms of binary representation within a computer was floating point numbers. Each CPU architecture had its own representation and precision in terms of floating point numbers. The IEEE created the IEEE-754 numeric representation in response to the lack of interoperability in this most complex of data representation areas.

Standards for time and date

There are two types of time representation that should be understood. The first is the internal representation used to perform date and type calculations, and the second is the external representation that is used to format a time format into a human readable format.

Internally, computer operating systems store the time as an integer value that is some number of time units since an epoch. The epoch is the set starting date and the value stored in the time value is the number of time units.

Microsoft Excel date and time representation

Microsoft Excel has its own date and time representation. Excel uses a decimal number to store date and time. The date is stored in the integer part of the number

[28] `http://tools.ietf.org/html/rfc1700`

Table 19.8 *Examples of ISO 8601 formatted dates and times*[29]

Format	Example	Description
YYYY-MM-DD	2012-09-07	Standard date format
YYYYMMDD	20120907	Lesser known, but fully supported, date format
YYYY-**W**nn	2012-W01	First week of 2012
YYYY**W**nn	2012W01	First week of 2012
YYYY-DDD	2012-251	2012–09–07 represented as day of the year
YYYYDDD	2012251	2012–09–07 represented as a day of the year
hh:mm:ss	11:59:46	Time representation with punctuation
hhmmss	115946	Time without punctuation
YYYY-MM-DD**T**hh:mm:ss	2012–09–07T12:59:46	Date and time in the same field with punctuation Notice the "T" separator
YYYYMMDD**T**hhmmss	20120907T125946	Date and time in thesame field without punctuation
A time value with no additional information than hh:mm:ss or hhmmss is considered to be in local time. To define a time as being in Universal Time Coordinated (UTC) a "Z" is appended or an offset in hours and minutes is appended to the time value.		
hh:mm:ssZ	12:59:46Z	12:59:46 UTC (aka "Zulu")
hhmmssZ	125946Z	12:59:46 UTC (aka "Zulu") without punctuation
hh:mm:ss[-+]hh:mm	12:59:46–05:00	12:59:46 US Eastern Time
hh:mm:ss[-+]hh:mm	12:59:46+01:00	12:59:46 Central European Time
ISO 8601 supports a variable sized fractional second.		
hh:mm:ss.sss	12:59:46.125	12:59:46 and 125 milliseconds
hh:mm:ss.ssss	12:59:46.1254	12:59:46 and 125.4 milliseconds
hh:mm:ss.ssssss	12:59:46.125418	12:59:46 and 125418 microseconds
hhmmss.ssssss	125946.125418	12:59:46 and 125418 microseconds without punctuation
hh:mm:ss.sssssssss	12:59:46.125418321	12:59:46 and 125418321 nanoseconds
YYYY-MM-DDThh:mm:ss.sssZ	2012–09–07T12:59:46.125Z	Example of a fully expanded UTC date time field

and the time is stored in the fractional portion. With this system, Excel can add, subtract, or compare dates and times just like any other numbers, and all dates are manipulated by using this system. The value 1.0 represents 1/1/1900 12:00:00 a.m. Times are stored as fraction numbers between 0.0 and 0.99999, where 0.0 is

[29] `www.cl.cam.ac.uk/~mgk25/iso-time.html`

00:00:00 and 0.99999 is 23:59:59. The range of the Excel date is December 31, 9999. The largest time value that can be stored is 9999:59:59.

The human readable display of date and time has increasingly followed a comprehensive ISO standard: Data elements and interchange formats – Information interchange – Representation of dates and times (ISO 8601). ISO 8601 provides considerable options for date and time representation.

Standards for time synchronization

The dramatic increase in message rates of "ultra low latency" finance has created a requirement for much lower tolerances between the differences in system clocks, known as drift, on the various computers and network components used in the financial markets. Computer messaging and processing for ultra low latency finance is now being measured in nanoseconds (billionths of a second) with the most sophisticated players attempting picosecond (trillionth of a second) measurement. Commercial products exist that claim to perform 20 pre-trade risk decisions in less than 90 nanoseconds; these are being marketed and sold as of Fall 2011.

Various open standards exist for synchronizing times between computers. These protocols attempt to account for the time it takes to transmit and process a time synchronization message. The synchronization messages originate from computers synchronized to highly accurate atomic clocks. Newer protocols rely on the time synchronization signals sent by Geographical Positioning Satellites

Table 19.9 *Time synchronization protocols*

Protocol	Name	Standards	Description
NTP	Network time protocol	IETF RFC 5905,[30] RFC 5906,[31] RFC 5907,[32] RFC 5908[33]	The original protocol for Internet based time synchronization. The de facto standard.
PTP	Precision time protocol	IEEE 1588–2002: standard for a precision clock synchronization protocol for networked measurement and control systems.[34]	Newer, much more precise in terms of managing drift than NTP. Firms requiring higher message rates are adopting PTP.
GPS	Global Positioning System		Both smart grid and ultra low latency financial messaging are using GPS for time synchronizaiton.

[30] www.ietf.org/rfc/rfc5905.txt [31] www.ietf.org/rfc/rfc5906.txt
[32] www.ietf.org/rfc/rfc5907.txt [33] www.ietf.org/rfc/rfc5908.txt
[34] http://en.wikipedia.org/wiki/Precision_Time_Protocol

(GPS). Several vendors now provide and sell highly accurate time synchronization hardware components.

19.10 Standards for data interchange

With some very basic understanding of how data itself are represented, we can then next look at how information is exchanged.

19.10.1 Comma Separated Variable (CSV) format

Comma separated variable files represent information in a character format. Each data item (also known as a field) is separated by a delimiter character. This delimiter character is often the comma (ASCII 0x2C), giving the file format its name. The CSV format was created in the very early days following the introduction of spreadsheets during the early days of the Microcomputer revolution. CSV files were used as a minimal data interchange format between competing spreadsheet products, such as Visicalc, Microsoft Excel, and Lotus 1–2–3. The simplicity of the file format and the ubiquitous support from spreadsheet software have made CSV formats ubiquitous and by far the simplest interchange format available. Industry standards have been created for the CSV, such as the IETF RFC 4180.[35]

The CSV format, given its spreadsheet origins, is generally used to represent two-dimensional tabular data organized as rows and columns. A row corresponds to a record in a file and a column represents a data item or field in a record. Records are separated either using a carriage return (CR ASCII 0x0D) or a combination of the carriage return and line feed (CR-LF) characters (CR-LF ASCII 0x0D 0x0A).

Internationalization issues

The CSV format reveals its US origins by the use of the comma character as the delimiter. Unfortunately, the comma is used in several countries as a separator for the whole number and the fractional (decimal) portions of a number. Therefore, the comma delimiter is not an optimal approach for global data interchange. Alternative characters, such as the tab character (ASCII 0x09), have been proposed as international alternatives. While the tab character resolves the issue created by the choice of comma, it is not without its own failings. The tab character is a non-printable character making the character non-displayable in many computer applications. The tab character, a remnant from mechanical typewriters, was used to align information at specified tabstops which can make the display and editing of tab delimited data confusing and somewhat difficult.

[35] `www.ietf.org/rfc/rfc4180.txt`

Optional header row

A common convention to specify the metadata used to identify each column in the CSV file is to create a single header row with the name of the column as the first record (row) in the file. The use of a header row permits the data to be self-describing and is the dominant practice in use today when using CSV files for data interchange.

Example:

```
GS,859430–1,100753.00,345,"Main account"
  JPMC,1045A3,9439.42,94,"Customer ABC"
  DB,ABC1379–01,34576.03,1203,"Mutual Fund ABC"
```

Example with header row:

```
BIC,Account,Balance,TransactionCount,Memo
  GS,859430–1,100753.00,345,"Main account"
  JPMC,1045A3,9439.42,94,"Customer ABC"
  DB,ABC1379–01,34576.03,1203,"Mutual Fund ABC"
```

Despite all of the elaborate XML standards and alternative data representations, the CSV format is still a very widely adopted and efficient standard for data interchange. It should be considered the lowest common denominator for a self-describing format for the exchange of financial information. The data are largely two dimensional, although this is certainly not always the case and should not be assumed. Data types are implicitly defined. Character data can optionally be delimited using a double quote or single quote character. Fields consisting only of numeric characters that do not contain a decimal place indicator (remember the "." is just one possibility for this delimiter) is considered to be an integer. Numeric character fields with a decimal place indicator ("." or ",") are considered to be a decimal data type.

19.10.2 Microsoft Excel

The exchange of financial information using the Microsoft Excel file formats is an alternative to the rudimentary CSV family of file formats. The Microsoft Excel format provides considerable advantages to CSV files. Datatypes can be specified and formulas (calculations) can be provided. When looking for standards in terms of data interchange, the reliance on a proprietary product from one vendor, even (or especially) such a ubiquitous vendor as Microsoft, has long been a subject of controversy within the software community.

Microsoft re-engineered its office suite to use an XML format. The new format is named Office Open XML (OOXML). At the same time the firm strongly advocated the international standardization of OOXML. As a result of its efforts, OOXML is the new international standard ISO/IEC 29500. This opening of the Microsoft document formats has fostered considerable innovation and competition in a variety of specialized areas.

The primary disadvantage of spreadsheets and a significant operational risk to organizations is the lack of information control and security in the use of data. Likely every major financial institution is vulnerable from an information integrity and information security perspective due to vital information being stored within spreadsheets and not in a controlled corporate database.

19.10.3 Open Protocol Enabling Risk Aggregation (OPERA)

The OPERA standard is being presented here as an example of a spreadsheet based standard. A group of hedge funds created an informal consortium to create a standard for "reporting procedures for collection, collation and conveying hedge fund risk information."[36] An interesting, and oft overlooked, purpose of a standard is to create "consistent practices," not necessarily "best practices." The concept of a standard focusing more on consistency versus an optimized or best practice is an important consideration when determining the purpose and value of an existing or new standardization effort.

19.11 Standards for document interchange

Despite our attempts to fully quantize all information, there are times when prose or documents need to be interchanged. Both FpML and FIXML have been extended to include attachments that further document financial instruments and trades. The risk practitioner should be aware of common formats used for document interchange.

19.11.1 Adobe Portable Document Format (PDF)

Adobe revolutionized document interchange with the creation of the Portable Document Format (PDF).[37] The ubiquity of the PDF format today belies its slow adoption rate following its introduction in 1993. PDF was finally released as an open standard in 2008 as ISO 32000–1. PDF includes a subset of an earlier

[36] www.theopenprotocol.org/top/home
[37] http://en.wikipedia.org/wiki/Portable_Document_Format

standard for device independent printing, PostScript, which is still used for portable document interchange, though its use has been greatly eclipsed by PDF.

19.11.2 Microsoft Word (OOXML)

Many documents exist in the proprietary formats of Microsoft Word prior to the introduction of Microsoft Word 2007.

Microsoft worked with the European Computer Manufacturer's Association (ECMA) to standardize the format of their Microsoft Office product suite named Office Open XML (OOXML)[38]. Once standardized as ECMA-376, it was then approved as ISO/IEC 29500. At first met with great political skepticism, the availability of an open standard, even one dominated by a single major vendor, has been beneficial in terms of interoperability, automation, document automation, and document interchange. Microsoft's opening of their format has permitted a variety of office suite products to compete.

19.11.3 Microsoft RTF (Rich Text Format)

Microsoft created a proprietary document file format designed for document interchange named Rich Text Format (RTF) in 1987. The format was used as a de facto standard for exchanging word documents between different word processing applications, with a certain loss of fidelity in terms of how the documents were rendered. RTF is still in use, although the use of OOXML is fast replacing RTF for new document creation.

19.11.4 XHTML/HTML

The original markup language for the World Wide Web was the HyperText Markup Language (HTML). The origins of HTML were the Standard Generalized Markup Language (SGML) (ISO 8879). SGML was a comprehensive documentation and publishing system. HTML was a very simple subset. After the eXtensible Markup Language (XML) was created, an XML compliant version of HTML was developed. XHTML and HTML can serve as a valid format for document interchange.

19.12 Standards for Identification

As this chapter is being written, the entire financial services industry is focusing on the Legal Entity Identifier (LEI), the usage and impact of which will not be decided

[38] `http://en.wikipedia.org/wiki/Office_Open_XML`

when this book goes to print.[39] There is also considerable controversy around business practices by firms providing both standard and proprietary identifiers, with the European Commission conducting anti-trust investigations and actions. This section is intended to explain the standard financial identifiers that a risk management practitioner should have familiarity with at this current time given the confusion and myriad of overlapping identifiers.

19.12.1 Fundamental identifiers

Identifiers for currencies and countries are fundamental to the global financial markets.

Currency codes

The Codes for Representation of Currencies and Funds (ISO 4217) defines the three character currency codes that are used to identify currencies as a financial instrument both for trading, clearing, and settlement. These currency codes are used as attributes in the definition of other financial instruments.

Country codes

The Codes for the Representation of Names of Countries and their Subdivisions (ISO 3166) standard provides standard codes for countries and regions. ISO 3166 is a collection of standards.

Table 9.10 *Example currency codes*

Currency code	Currency
CHF	Swiss Franc
JPY	Japanese Yen
BRL	Brazilian Real
USD	US Dollar
EUR	Euro
CNY	Chinese Yuan Renminbi
GBP	British Pound
AUD	Australian dollar
CAD	Canadian dollar
SGD	Singapore Dollar
RUB	Russian Ruble

[39] The LEI is also discussed in other chapters in the Handbook, including Chapters 5, 6 and 11.

Table 9.11 *ISO 3166 standard sections*

Title	Standard	Description
Part 1: Country codes	ISO 3166–1 alpha-2	Two-character country codes
	ISO 3166–1 alpha-3	Three-character country codes
	ISO 3166–1 numeric	Numeric country codes assigned by the UN
Part 2: Country subdivision code	ISO 3166–2	Subdivisions, provinces of countries defined in ISO-3166–1
Part 3: Code for formerly used names of countries	ISO 3166–3	Deleted code values

Table 9.12 *Example two-character alpha country codes*

Alpha-2 value	Country
AR	Argentina
AU	Australia
BR	Brazil
CH	Switzerland
CM	Cameroon
CN	China
DE	Germany
FR	France
GB	UK
LI	Liechtenstein
NL	Netherlands
US	USA
ZA	South Africa

Market identifier codes

The Codes for Exchanges and Market Identification (MIC) (ISO 10383) standard provides standard codes for markets and venues.[40] During the last revision of the standard, MICs were extended to provide additional values for segments of a market, for instance a dark pool at an exchange.

19.12.2 Identification of financial instruments

As described above, unlike the supply chains found in manufacturing or publishing, the financial industry is a complicated and costly entanglement of

[40] www.iso10383.org/

Table 9.13 *Example MICs (ISO 10383)*

MIC	Market
XMOC	Montreal Climate Exchange
XSWX	Swiss Exchange
BATE	BATS Europe
SWAP	SWAPSTREAM
XISE	International Securities Exchange
XCBO	Chicago Board Options Exchange
XNAS	NASDAQ
XNYS	New York Stock Exchange, Inc.
XJSE	Johannesburg Stock Exchange
XSTO	NASDAQ OMX Nordic

proprietary identification systems that are de facto standards and de jure standard identification systems that are essentially proprietary and costly. Redundant identifiers abound. Even the smallest of financial market participants will likely be faced with the need to maintain cross references of securities and pay multiple licensing fees for identifier databases. The financial services industry tolerates the practice where not only does the broker have to pay for the use of standard identifiers but there are charges for ultimate end investors as well. It is within this bleak context that we will introduce a necessarily incomplete list of financial instrument identifiers with which the risk management practitioner should have familiarity.

In an industry where liquid securities trade in extremely efficient markets, that employs some of the best educated, most competent technologists available, one would be likely to conclude that the most efficient processes would be available for managing the identification of the products that the industry creates and trades. Unfortunately, this is not the case. Some would even argue that industry resources are at times devoted to making sure that keeping track of the securities and derivatives traded is opaque. While we do not have documentation for (and tend to discount) this view, a close examination of the way in which product identification is managed across the industry will not refute the argument.

Supply chain management is largely dependent upon the maturity of the identification processes across market participants – if the identification systems do not work, the supply chain is less optimized and/or transparent. The differences with regard to identifier systems between the financial industry and the manufacturing, distribution, or publishing industries border on shocking. The financial industry is a complicated and costly entanglement of mainly proprietary identification systems, some of which are either de facto or de jure standards. These are combined with a level of complexity in IP constraints and technical specifications that ensures that tracking identifiers will be a requirement for anyone trying to track much of any level of market activity. We can add to this additional confusion caused by exchanges that do not use standards for identification and often have both a ticker symbol and an internal identifier that are exposed to the users of the market. This entire labyrinth is sometimes colloquially referred to as "symbology" and at times referred to as the "problem of symbology."

International Securities Identification Numbering System (ISIN)

Securities and Related Financial Instruments – International Securities Identification Numbering System (ISIN) (ISO 6166).

The ISIN is a standard beneath ISO TC68. The standard is administered by the Association of National Numbering Agencies (ANNA). ANNA is the Registration Authority (RA) for ISINs globally. The ISIN was created in response to the globalization of markets where firms traded and held positions in securities across borders. The ISIN provides a country prefix in front of existing national identifiers (defined within ANNA as the National Security Identification Number). The ISIN is a federated identifier. Each National Numbering Agency (NNA) is responsible for assigning its own identifiers. The prefix by country prevents collisions in identifiers across NNAs.

The ISIN consists of:

- A **prefix** which is either an ISO 3166–1 alpha-2 country code or "XS" for international securities deposited in the International Central Securities Depositories (ICSD), or other prefixes defined by the Registration Authority.
- A **Basic Number**, which is nine alpha-numeric characters. If the basic number (also known as the National Security Identifier Number) is less than nine characters in length, it should be right justified with leading zeroes.
- A **Check Digit**, which is calculated to ensure the code has not been changed or altered. The algorithm for calculating the check digit is available in the standard.

National Security Identification Numbers (NSIN)

NSINs are a nine position identifier assigned by a National Number Agency (NNA). The concept of an NSIN is implied by the ISO 6166 standard. The name and concept are more formally recognized within the Association of National Numbering Agencies (ANNA). Each NNA decides on the standard for its own NSIN.

Ticker symbols

Ticker symbols were developed to provide short easily writable and easily human learnable unique symbols to represent securities being traded on a market. Within

Table 19.14 *Example ISINs*

Type	Issue description	Issuer or origin	Issuer's country	ISIN Prefix	ISIN
Equities Shares	Com stk	Hitachi Ltd., Tokyo	JP	JP	JP3788600009
Equities Depositary receipts	BDRS(ISS Barcl Bank-Units 1/20th sh)	Motors Liquidation Co	GB	GB	GB0003658282
Equities Units	Inhaber-Anteile	Union-Investmentgesellschaft GmbH, Frankfurt a.M.	DE	DE	DE0008491002
Debt instruments All	EO-Bonds 2006(11/Und.).	Allianz Finance II B.V.	NL	DE	DE000A0GNPZ3
Debt Instruments ICSD	EURO issue bond EUR 0 183300000	Residential mortgage Sec//A2C/ REGS	GB	XS	XS0203542088
Debt Instruments Private placements, no CSD/ICSD	Tranche # TR 00528 Dtd03/20/02	General Elec Cap Corp MediumTerm Nts Book Entry	US	US	US36962GXZ26
Debt instruments Official Strips	Principal Certificate on Stripped Gvmt Bonds OAT 0%250423 PPMT 8–5%23 EUR 1	France Emprunt d'Etat	FR	FR	FR0000571077

Table 19.15 *National security identification numbers*

Security Identifier	Name	Standard	Format	Country	Organization
CUSIP®	Committee on Uniform Security Identification Procedures	ANSI X.9.6	Six position number, with two digit suffix, and a check digit	US, CA	American Banker's Association CUSIP Global Services division of Standard and Poor's
SEDOL	Stock Exchange Daily Official List		Six number identifier with one digit check digit	GB	London Stock Exchange
WKN	Wertpapier kennnummer		Six digits or capital letters (excluding I and O), no check digit	DE	WM Datenservice
VALOR	Valor		Numeric opaque identifier used in conjunction with a MIC (ISO 10383) or Currency code (ISO 4217)	CH	SIX Financial Information
Sicovam (Full ISIN now used)	Société Interprofessionnelle pour la Compensation des Valeurs Mobilières		Six digit identifier	FR	Euroclear France

Table 19.16 *Example ticker symbols*

MIC	Market	Ticker symbol	Company
XNYS	NYSE	T	AT&T
XNAS	NASDAQ	AAPL	Apple Inc.
BATE	NASDAQ	WDFC	WD-40 Company
XNYS	NYSE	TTM	Tata Motors Ltd.
XNYS	NYSE	WFC	Wells Fargo & Company
XNYS	NYSE	CAT	Caterpillar, Inc.

Table 19.17 *Proprietary identifiers*

Security identifier	Name	Standard	Format	Country	Organization
CLIP	CUSIP-linked MIP Code	Proprietary		Global	S&P CUSIP Bureau and Markit
LXID	LoanX ID	Proprietary		Global	Markit LXID is a unique identifier applied to syndicated loans.
BBUD	Bloomberg Unique ID	Proprietary		Global	Bloomberg
MEI Loan Markit	Markit Entity Identifier	Proprietary		Global	Markit

the USA, the original stock ticker symbol methodology was created by Standard and Poor's. Equity securities and listed options on equities still use the stock ticker symbol for exchange based trading.

Proprietary identifiers

Two of the dominant security identifiers are from data vendors that consolidate market data and redistribute it to market participants. Both Bloomberg and Thomson-Reuters consolidate data across the majority of market venues. Financial instruments are consolidated and normalized and assigned proprietary identifiers.

A subclass of proprietary identifiers is the commodity and futures contracts that are proprietary to an exchange. Futures and commodity contracts are in general non-fungible. Since the contract is only traded on one market venue, the symbology is proprietary to that venue as well.

Table 19.18 *Proprietary identifiers that are now open*

Security identifier	Name	Standard	Format	Country	Organization
BBGID	Bloomberg Global ID	Proprietary	12 digit alphanumeric "BBG" prefix 8 alpha-numeric characters randomly assigned, one check digit	Global	Bloomberg
RIC	Reuters Instrument Code	Proprietary	Six number identifier with one digit check digit	Global	London Stock Exchange

New open symbology initiatives

Bloomberg and Thomson-Reuters both created free offerings in response to European Commission anti-trust investigations. Bloomberg and Thomson-Reuters are two of the leading suppliers of financial instrument reference data and market data. Their identifiers span asset classes and markets to provide global coverage for financial instruments. With these new freely available identifiers and the restrictive nature in terms of licensing and use of certain national standards, the distinction between open and closed and proprietary and standard are quite blurred.

19.12.3 Legal entities and parties

When compared to the identification of the legal entities involved in the trading process, the state of financial instrument identification looks rather simple and straightforward. Especially for risk managers responsible for aggregating risk by counterparty, the task of entity identification is challenging. The Legal Entity Identifier was intended to mitigate the problem with disparate identification of entities involved in transactions to permit views into aggregate risk on a systemic level. At the time of this writing, a new Legal Entity Identifier (LEI) standard has been approved as an ISO standard. The ISO 17442 LEI will be governed by a Regulatory Oversight Committee, the creation of which is being managed by the Financial Stability Board (FSB) of the G20 countries.[41] The exact scope and impact of the LEI is yet to be determined. If the industry embraces the LEI, if a sensible governance structure is created, if a sensible, open, standardized technology implementation is achieved, then the LEI could be an incredible benefit in terms of cost

[41] www.financialstabilityboard.org/

reduction and improvements in efficiency, accuracy, and transparency throughout the entire supply chain of modern finance.

The Business (formerly, Bank) Identifier Code (BIC) (ISO 9362), for which SWIFT is the ISO Registration Authority, is a fundamental identifier in bank transactions. The use of the BIC was extended to trading with its adoption by the FIX Community around 2000. Originally implemented as a network address, the BIC has evolved into a hybrid with significant "legacy overhang." Nonetheless, it is a significant piece of the puzzle. BICs are now also issued for non-banking institutions.

Table 19.19 lists some of the primary entity identification implementations used in the securities markets.

19.12.4 Concordances

A concordance provides a cross reference between two or more identification schemes or standards. A typical example would occur where you have acquired Reuters reference and market data, which contain the ticker symbol and the RIC (Reuters Instrument Code); however, for FINRA TRACE reporting you are required to use the CUSIP. The most common way of handling this is to use a concordance that will provide a cross reference map between the RIC database and the CUSIP.

19.13 Standards for classification of financial instruments

Securities industry participants regularly need to manage information by general asset class, as well as look at subsets of specific asset classes for a wide variety of business purposes including risk management, collateralization and mark to market requirements. The current ISO Standard for categorizing securities instruments is ISO 10962 Classification of Financial Instruments or "CFI" for short. Given that the pace of industry evolution is considerably more rapid than the rate of change that has been supported to date by the formal Standards process, the result has been a proliferation of both general purpose and custom proprietary categorization schemes among market participants. The number of asset classes and the composition of instruments within asset classes are both evolving. Further, new instruments consisting of existing asset types, pieces of existing asset types, custom indicators/instruments, and in some circumstances based on specific trading venue or customer group requirements, have no coherent representation in most classification or categorization approaches across industry participants.

New regulatory requirements require careful attention to asset class correlations and the supporting data beneath them. The current state of the CFI and related

Table 19.19 *Financial industry focused entity identifiers*

Entity identifier	Name	Standard	Country	Organization
BIC	Business Identifier Code	ISO 9362	Global	SWIFT is the registration authority for the BIC. A BIC consists of either eight (BIC8) or 11 (BIC11) contiguous characters. These characters comprise either the first three, or all four, of the following components: bank code, country code, location code, and branch code.
FRN	FSA Registration Number	Public	UK	The FSA Registration Number or FRA is a unique identifier assigned to every firm, individual or other body regulated by the Financial Service Authority of the UK.
CNPJ	Cadastro Nacional de Pessoas Jurídicas	Public	Brazil	Secretaria da Receita Federal Brazilian company identifier
FINID	Bankers Almanac Bank Identifier	Proprietary		Banker's Accuity www.bankersaccuity.com Bankers Almanac offers reference data on financial institutions for payments, compliance and risk assessment.
UKREG	UK Registration Number		UK	UK Companies House. It also has introduced a simple Uniform Resource Identifier (URI) for all companies on its register. The URI is permanent and represents the company. When used as an Internet address, it will return basic company details in a format appropriate for the requesting system.
CICI	CFTC Interim Compliant Identifiers	Public	USA/ Global	DTCC and SWIFT on behalf of the Commodity Futures Trading Commission (CFTC). The CICI is intended to be an interim identifier until the launch of the LEI.
ZC	Zengin Code	Public	Japan	Japanese Bankers Association. The Zengin Code is a unique number used in Japan to identify banks and branches for

Table 19.19 (*cont.*)

Entity identifier	Name	Standard	Country	Organization
				domestic bankwiring, mass transfer, Paypal transfer and Google Adsense payments. The Zengin code for a bank company comprises four digits, with three more digits added to identify the branch.
RSSDID	Replication Server System Database ID	Public	USA	US Federal Reserve System. A unique identifier is assigned to financial institutions by the Federal Reserve.
CIB	Code interbancaire	Public	France	Banque de France. CIB is assigned by the Bank of France to credit institutions under French law and branches of foreign credit banking businesses in France or Monaco.
FIID	Fitch Issuer Identification Number	Proprietary	Global	Fitch Ratings. FIID is a unique proprietary 12-digit tracking number applied to each issuer of debt rated by Fitch Ratings.
MIN	Moody's Issuer Number	Proprietary	Global	Moody's Investor Services. MIN is a unique ten-digit identifier applied to each issuer of debt rated by Moody's Investor Services.
PID	Moody's KMV Identifier	Proprietary	Global	Moody's Analytics. The Moody's KMV unique permanent company identifier, PID or MKMV ID, is a six-digit number applied to companies covered by Moody's CreditEdge Plus™ service, notably used by the EDF™ (Expected Default Frequency) credit measure.
RED "RED Code"	Markit Reference Entity Database identifier	Proprietary	North America and Europe	Markit. Markit RED (Reference Entity Database) is the market standard that confirms the legal relationship between reference entities that trade in the credit default swap market and their associated reference obligations, known as "pairs." Each entity is identified with a

Table 19.19 (*cont.*)

Entity identifier	Name	Standard	Country	Organization
				six-digit alpha-numeric code, and a nine-digit code identifies the pair.
AVID	Avox International Business Entity Identifier	Proprietary	Global	AVOX, a division of DTCC. The Avox International Business Entity Identifier or AVID is an identifier assigned by Avox Limited. Avox assigns a unique identifier to each business entity it validates for its clients.
DUNS	Data Universal Numbering System	Proprietary	Global	Dun & Bradstreet Inc. (D&B). The Data Universal Numbering System, DUNS® or D-U-N-S®, is a system developed and regulated by Dun & Bradstreet (D&B), that assigns a unique nine-digit numeric identifier, referred to as a "DUNS number" to a single business entity.
CABRE	CUSIP Avox Business Reference Entity Identifier	Proprietary	Global	CUSIP Global Services and Avox, a division of DTCC. The CUSIP Avox Business Reference Entity Identifier, or CABRE, is a combined effort by CUSIP Global Services and Depository Trust & Clearing Corporation/Avox to provide a universal identification system for global business entities.
CPLID	Counterparty Link ID	Proprietary	Global	Counterparty Link. CounterpartyLink was formed in 2005 with the sole remit to collect, maintain and deliver accurate and auditable legal entity reference data from around the world.
GVKEY	Compustat ID	Proprietary	Global	S&P Capital IQ. Compustat GVKEY is a unique six-digit number key assigned to each company (issue, currency, index) in the Capital IQ Compustat database. It is a company (issue, currency, index) identifier similar to a

Table 19.19 (*cont.*)

Entity identifier	Name	Standard	Country	Organization
				ticker symbol. It represents the primary key for a company that is an index constituent.
FDS	Factiva Data Symbol	Proprietary	Global	Dow Jones Factiva. FDS is a code applied to quoted and unquoted companies mentioned in Factiva's database of business information.
DTCPA	Depositary Trust Corporation Participant Account	Proprietary	Global	The Depositary Trust and Clearing Corporation. The Depositary Trust Corporation Participant Account code is a four-digit number used by the Depositary Trust and Clearing Corporation to identify participating depositary institutions.
BBCID	Bloomberg Company ID	Proprietary	Global	Bloomberg. The Bloomberg Company ID is an eight-digit number that uniquely identifies a company on the Bloomberg system.
LEI	Legal Entity Identifier	ISO 17442	Global	ISO TC 68 maintains the 17442 standard; implementation is overseen by a group of international regulators (ROC).
IBAN	International Bank Account Number	ISO 13616	Global	SWIFT is the Registration Authority, the IBAN consists of up to 34 alpha-numeric characters: first the two-letter ISO 3166–1 alpha-2 country code, then two check digits, and finally a country-specific Basic Bank Account Number (BBAN).
MPID	Market Participant Identifier	Public	US	FINRA. MPID is a four character alpha code required of market participants trading and reporting trades.

categorization coding schemes is not adequate for consistently identifying variations in or across asset classes or enabling the level of rigor required for the calculations of non-diversifiable risk in Basel II measurements of economic capital. From an operational standpoint, the CFI is regarded as both too complex and not comprehensive enough, depending on the process for which this information is used. For example, in the USA, for the purpose of classifying financial instruments for clearing and settlement, the external code scheme developed and maintained by ISITC (International Securities Association for Institutional Trade) is widely used.

In response to this situation, several industry organizations initiated a joint effort to create a workable categorization scheme that incorporates the current market practices regarding variations within asset classes that derive from the instrument lifecycle from issuance through custody and possible retirement or expiration. Overall, the group looked at schemes from the CFI and ISIN (ISO 6166, International Securities Identifying Number) standards, those used by FIX Protocol Limited (FPL), ISITC, International Securities Market Advisory Group (ISMAG), Depository Trust and Clearing Corporation (DTCC), Financial Products Markup Language (FpML), the EDM Council's semantics repository, and Market Data Definition Language (MDDL).

Feedback and experience from this effort has found its way into the formal Standards process as part of the current revision work for the CFI.

19.14 Standards for external code lists

To understand the role of code lists in data and when controlled vocabularies need to be managed and versioned separately, one needs to understand something about the rate of change of information. Static data do not change as they are primarily the structure of information not the information itself; these are the classes, data items, objects that define the business domain. Transactional data are constantly being created and are dynamic. Between the static structure and dynamic data are data items that change at a slower rate. These data items are at the same time data, but also have a structural role.

Let us examine currency codes, something familiar to most readers. A trade occurs and its price is denominated in a currency. The currency is a structural reference from the transaction. Yet currencies do change periodically. Country codes, financial instrument types, settlement types, are all examples of lists of data items that make up the transaction. Code values, or enumerations, are usually assigned to these data items to simplify conveyance of the data item. For instance "USD" is the standard code for the US Dollar. "CS" is the code for the FIX Security Type (financial instrument type) for Common Stock. Lists of these data items are Code Lists. These code lists each have their own lifecycle. When we are fortunate, these code lists are standardized and shared across business users and vertical financial standards (such as security types shared by FIX Protocol, ISITC, and as an external code list to ISO 15022). With the move to taxonomies and semantics, the current term for defining a set of specific enumerations for the values associated with a structural data item or concept is called a controlled vocabulary, that is, "closed" as opposed to "open" or "uncontrolled."

Since these code lists have their own lifecycle, a mechanism for the management, versioning, and dissemination of code lists is required. The area of code list management was neglected until recently, as the evolution of financial technology and automation first addressed data stores and message flows. Code lists are the connection between the two, and the realization of the need for the standard was discovered mid-process. At this point in time the management of code lists is inconsistent. The use of Excel spreadsheets and CSV files is still common. Two industry standards do exist that have been adopted within the financial industry.

19.14.1 Genericode

Genericode[42] is an OASIS standard first released in 2007 by OASIS, based upon an original proposal delivered in 2004 by Anthony Coates at the XML 2004 conference.[43] Genericode is an XML based standard that provides a very simple and extensible structure for managing code lists. FpML was the first protocol to adopt and fully embrace Genericode for managing code lists. The following is the list of asset classes supported within FpML.

```
<?xml version="1.0" encoding="UTF-8"?>
<gcl:CodeList xmlns:gcl="http://xml.genericode.org/2004/ns/CodeList/0.2/"
xmlns:doc="http://www.fpml.org/coding-scheme/documentation"
xmlns:xsi="http://www.w3.org/2001/XMLSchema-instance"
xsi:schemaLocation="http://xml.genericode.org/2004/ns/CodeList/0.2/ CodeList.xsd">
   <Annotation>
      <Description>
         <doc:definition>Defines a simple asset class categorization. Used for
classification of the risk class of the trade.</doc:definition>
      </Description>
   </Annotation>
   <Identification>
      <ShortName>assetClassScheme</ShortName>
      <Version>2-0</Version>
      <CanonicalUri>http://www.fpml.org/coding-scheme/asset-class</CanonicalUri>
      <CanonicalVersionUri>http://www.fpml.org/coding-scheme/asset-class-2-
0</CanonicalVersionUri>
      <LocationUri>http://www.fpml.org/coding-scheme/asset-class-20.xml</LocationUri>
   </Identification>
   <ColumnSet>
      <Column Id="Code" Use="required">
         <ShortName>Code</ShortName>
         <Data Type="token">
            <Parameter ShortName="maxLength">63</Parameter>
         </Data>
      </Column>
```

[42] www.genericode.org/ [43] www.genericode.org/2004/presentations/xml-2004/codelists.pdf

```
    <Column Id="Source" Use="optional">
      <ShortName>Source</ShortName>
      <Data Type="string"/>
    </Column>
    <Column Id="Description" Use="optional">
      <ShortName>Description</ShortName>
      <Data Type="string"/>
    </Column>
    <Key Id="PrimaryKey">
      <ShortName>key</ShortName>
      <ColumnRef Ref="Code"/>
    </Key>
  </ColumnSet>
  <SimpleCodeList>
    <Row>
      <Value>
        <SimpleValue>Credit</SimpleValue>
      </Value>
      <Value>
        <SimpleValue>FpML</SimpleValue>
      </Value>
      <Value>
        <SimpleValue>Credit.</SimpleValue>
      </Value>
    </Row>
    <Row>
      <Value>
        <SimpleValue>InterestRate</SimpleValue>
      </Value>
      <Value>
        <SimpleValue>FpML</SimpleValue>
      </Value>
      <Value>
        <SimpleValue>InterestRate.</SimpleValue>
      </Value>
    </Row>
    <Row>
      <Value>
        <SimpleValue>ForeignExchange</SimpleValue>
      </Value>
      <Value>
        <SimpleValue>FpML</SimpleValue>
      </Value>
      <Value>
        <SimpleValue>ForeignExchange.</SimpleValue>
      </Value>
    </Row>
    <Row>
      <Value>
        <SimpleValue>Equity</SimpleValue>
      </Value>
      <Value>
        <SimpleValue>FpML</SimpleValue>
      </Value>
      <Value>
        <SimpleValue>Equity.</SimpleValue>
      </Value>
    </Row>
```

```
        <Row>
           <Value>
              <SimpleValue>Commodity</SimpleValue>
           </Value>
           <Value>
              <SimpleValue>FpML</SimpleValue>
           </Value>
           <Value>
              <SimpleValue>Commodity.</SimpleValue>
           </Value>
        </Row>
    </SimpleCodeList>
</gcl:CodeList>
```

19.14.2 SDMX

Statistical Data and Metadata eXchange (SDMX) is a standard for exchanging statistical data.[44] While broadly suited for most types of statistics, it was initially developed by central bankers and has been largely used for economic statistics. The original sponsoring institutions were the Bank for International Settlements, the European Central Bank, Eurostat (the statistical office of the European Union), the International Monetary Fund (IMF), the Organisation for Economic Co-operation and Development (OECD), the United Nations Statistics Division, and the World Bank. With the release of v2.1 in 2011, the information model of SDMX can be viewed as becoming a horizontal technology for managing statistical time series of any type, and is an excellent example of abstracting structure and format across content domains. The logical models for maintaining code lists, hierarchies, and taxonomies are applicable to many content sets, including some of those used for financial instruments or institutional ownership hierarchies. The previous version of SDMX is an ISO standard ISO 17369:2005. A revision to ISO 17369 is in process for v2.1.

SDMX is much more sophisticated, and as a result a bit more difficult to master than Genericode, however, there are some important capabilities that practitioners now require, most importantly hierarchical code lists. The ability to derive and extend a code list from another list is a common requirement. SDMX supports simple code lists, hierarchical code lists, and hybrid code lists.

The following excerpt shows the code list for frequencies maintained by the ECB.

```
<Structure xmlns:="http://www.SDMX.org/resources/SDMXML/schemas/v1_0/message"
...
<Header/>
<CodeLists>
<structure:CodeList id="CL_FREQ"agency="ECB">
<structure:Name xml:lang="en">"Frequency code list"</structure:Name>
<structure:Code value="A">
```

[44] www.sdmx.org

```
<structure:Description xml:lang="en">Annual</structure:Description>
</structure:Code>
<structure:Code value="B">
<structure:Description xml:lang="en">Business (not supported)</structure:Description>
</structure:Code>
<structure:Code value="D">
<structure:Description xml:lang="en">Daily</structure:Description>
</structure:Code>
<structure:Code value="E">
<structure:Description xml:lang="en">Event (not supported)</structure:Description>
</structure:Code>
<structure:Code value="H">
<structure:Description xml:lang="en">Half-yearly</structure:Description>
</structure:Code>
<structure:Code value="M">
<structure:Description xml:lang="en">Monthly</structure:Description>
</structure:Code>
<structure:Code value="Q">
<structure:Description xml:lang="en">Quarterly</structure:Description>
</structure:Code>
<structure:Code value="W">
<structure:Description xml:lang="en">Weekly</structure:Description>
</structure:Code>
</structure:CodeList>
...
```

19.14.3 XBRL

The use of the controlled vocabularies of XBRL for managing code lists has been proposed as an alternative standard.

19.15 Financial technology standards

Data technology standards support how information is collected, aggregated and transported. An example of a standard is the FIX Protocol for trading, eXtensible Business Reporting Language (XBRL), which defines methods to allow machine readable data on financial statements that can then be transported, exchanged, and stored in a consistent manner. Sometimes confusion can arise when people speak of different "levels" of standards. Standards exist both for the basic data element level and as lists and messages to represent financial instruments and financial transactions. Standards also exist for the syntaxes or physical representation of the elements. There is room and value for all of these levels of standards. The key is having an appreciation for the convergence on the meaning and definitions of the elements themselves.

19.15.1 The investment roadmap

Alan Dean of HSBC, in his elected role as the Global Technical Committee Co-chair of FIX Protocol Ltd., raised an important issue from the perspective of a large investment bank. With so many standards, many overlapping, the question rightfully asked by executive management at HSBC was, what standards should we use where? Out of this simple question was formed the Standards Coordination Group. The Standards Coordination Group is an informal and purposefully unofficially organized group of representatives from the various standards setting organizations (SSOs). The group is organized around maintaining the investment roadmap, which was ultimately the answer to the question raised by HSBC, but a question shared by any financial institution. Participants include SIIA's Financial Information Services Division (FISD), FIX Protocol Ltd., FpML, ISITC, SWIFT, and XBRL US, with the Chair of ISO TC68 having an observer seat. The investment roadmap is shown in Figure 19.11.

The summary rationale from the group laying out the goals and approach of the roadmap is as follows.

Executive summary of the investment roadmap

Within the financial services industry, there are multiple messaging standards being used, and the purpose of the Standards Coordination Group has been to come up with an approach that leverages and includes these standards into a broader framework without reinventing and creating redundant messages that increase implementation costs and cause confusion for the industry.

This investment roadmap affirms the commitment of each organization (FIX, FpML, SWIFT, XBRL US, ISITC and FISD) to the ISO 20022 business model by laying the groundwork for defining a common underlying financial model and ensuring some level of interoperability by producing a consistent direction for utilization of messaging standards and communicating that direction clearly to the industry.

The respective business processes included in the roadmap are or will be incorporated within the ISO 20022 business model and the model allows for ISO 20022 XML based messages to be created to support the business processes, while at the same time provides, in certain circumstances, for existing domain specific syntaxes and protocols to be maintained in order to protect the investments of market participants.

The organizations are committed to meeting on a consistent basis to ensure the roadmap continues to accurately depict the standards environment.

19.15.2 ISO 15022

ISO 15022 is the current standard in use primarily on the SWIFT network. ISO 15022 replaced the previous ISO 7775 and ISO 11521 standards in 2007. ISO 15022 was one of the initial standards that removed the actual messages and data elements from the normative text of the standard. Instead, the standard set forth a process and the standards for designing messages, the message flows, and the data dictionary. Approved messages are published on the ISO 15022 website and are freely available for use.

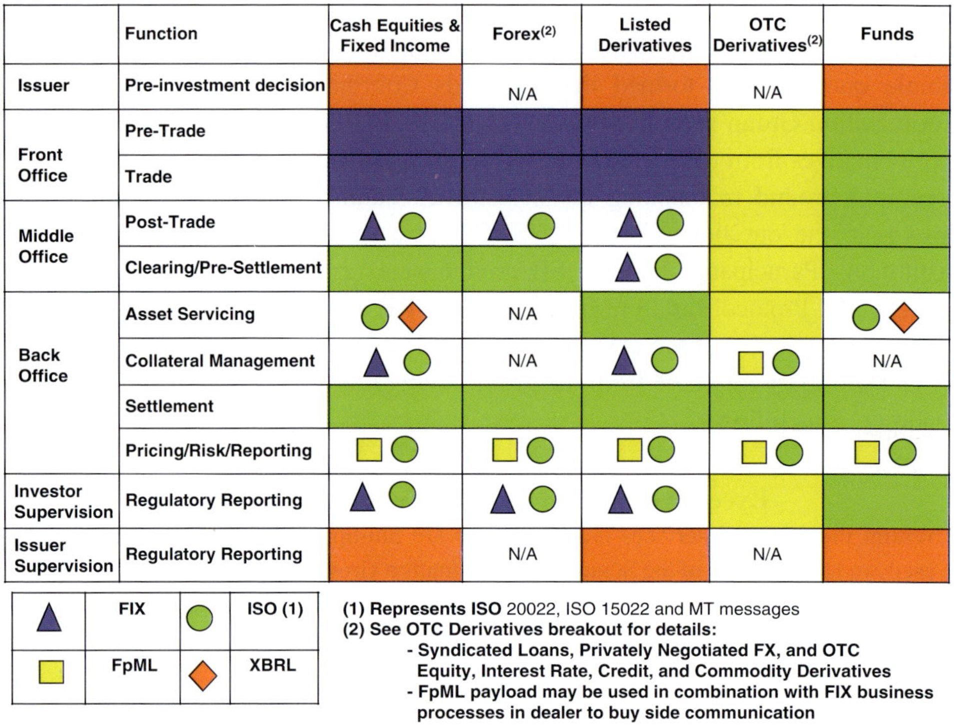

Figure 19.11 The investment roadmap.

ISO 15022 is another example of a vertical technology standard. The syntax and format for ISO 15022 messages are not based upon a horizontal technology standard, such as ASN.1 or XML for instance, instead the format was created by SWIFT as part of the earlier ISO 7775 standard.

ISO 15022 has been designed to incorporate and be upwards compatible with the previous securities message standards ISO 7775 and ISO 11521, as updated in ISO TR 7775. As a result, the initial Data Field Dictionary and Catalogue of Messages accommodate ISO TR 7775 data fields and messages. However, some of the previous fields and messages are not fully compliant with the Enhanced ISO 7775 syntax. In addition the initial Data Field Dictionary incorporates the International Securities Association for Institutional Trade Communications (ISITC) DSTU 1/1995 and the Securities Standards Advisory Board (SSAB) data dictionaries.

The ISO 15022 standard is described in the document ISO 15022 Securities – Scheme for Messages (Data Field Dictionary), Part 1 – Data Field and Message Design Rules and Guidelines and Part 2 – Maintenance of the Data Field Dictionary and Catalogue of Messages.[45]

[45] www.iso15022.org/

19.15.3 ISO 20022

ISO 20022 is the agreed methodology used by the financial industry to create consistent message standards across the industry.[46] As the scope of this standard covers the global financial services industry, this allows coverage across business areas (securities, payments, foreign exchange, for example), across asset classes and internationally.

The ISO 20022 method starts with the modeling of a particular business process, for example, the execution of an equity securities transaction. The activities that are required to complete this particular process are identified along with the actors and their roles needed to accomplish the process. The relationships and interactions between the actors, along with the information that they must share to execute the process also are identified. Once all this is done, the result is organized into a formal business model using UML (Unified Modeling Language[47]).

The formal business model of the process and the business information needed to support this particular process are organized into "business components" and placed into a repository. It is this repository that represents the core of the ISO 20022 standard. Once a business component is modeled and present in the ISO 20022 repository, the component can be reused in applicable transactions within other separate, but related business processes.

To illustrate by example, let us keep with our example of the execution of an equity securities transaction. Within this process, the business concept of a trade date exists. Further along in the processing chain is the process of actually settling this trade transaction. In the settlement business process, the concept of trade date can be reused as this concept and its definition and meaning already exist in the ISO 20022 repository.

The UML model forms the basis for the next step in the ISO 20022 process. Although the business model and its business components are an important foundation of the ISO 20022 standard, those undertaking developments using ISO 20022 most often are looking to develop messages that will allow for the transfer and transport of information to support the electronic processing of transactions.

So in the next step, the business components identified during the modeling process are ordered into one or more messages needed to support the business process. The elements that are ordered within these messages are called "message components." The logical messages also are part of the ISO 20022 repository.

When developing messages for the transfer and transport of information to support the electronic processing of transactions, the users also must agree on the syntax of the messages, the physical representation of the elements of the message.

[46] `www.iso20022.org/`
[47] An OMG standard for modeling technical systems. UML has multiple flavors or variants used to address modeling issues at varying levels of granularity.

ISO 20022 originally was introduced with an XML syntax, with the original version of the ISO 20022 having rules for transformation from the models to messages only for the XML syntax. ISO 20022 since has evolved to be able to support transformation of the message model into other syntaxes. In the business area that we have been using as our example, the FIX syntax of FIX Protocol Ltd. (FPL) will be used for messages to execute equity securities trades.

There are several excellent sources for additional detail regarding the ISO 20022 standard. First, the ISO 20022 official website, iso20022.org, contains information on the completed and ongoing ISO 20022 developments and includes details of the processes, committees and groups that support governance of the ISO 20022 standard and its developments. Also, in 2010, The SWIFT Standards Team published *ISO 20022 for Dummies*, as a useful source of information for current and potential users of ISO 20022.

The following is an example of a payment initiation message.


```
<?xml version="1.0" encoding="UTF-8"?>
<?Standards Smart Test Messages for SEPA Credit Transfer Scheme product version 3.0?>
<?valid true?>
<?description ABC Corporation, New York has received three invoices:

1. An invoice with number 4562, dated 08 September 2012 from DEF Electronics, London:
10 million JPY needs to be paid to DEF Electronics account 23683707994215 with AAAA
Bank, London (AAAAGB2L). ABC Corporation assigns reference ABC/4562/2012-09-08 to the
payment. Payment transaction
charges are shared between ABC Corporation and DEF Electronics.
2. An invoice with number ABC-13679, dated 15 September 2012 from GHI Semiconductors,
Brussels: 500,000 EUR needs to be paid to GHI Semiconductors account E30001216371411
with DDDD Bank, Belgium (DDDDBEBB). ABC Corporation assigns reference ABC/ABC-
13679/2012-09-15 to the payment. The accounts receivable department of GHI
Semiconductors needs to be advised when the funds have been credited on the account on
telephone number +32/2/2222222. GHI Semiconductors will bear all
payment transaction charges.
3. An invoice with number 987-AC, dated 27 September 2012, from their branch ABC
Corporation, California: 1 million USD needs to be paid to the branch account 4895623
with BBBB Bank, San Francisco (BBBBUS66). ABC assigns a reference ABC/987-AC/2012-09-
27 to the payment. Payment transaction charges are shared.
ABC Corporation holds an account 00125574999 with BBBB Bank, New York (BBBBUS33) and
instructs its bank to execute payment of the invoices with a
CustomerCreditTransferInitiation
message.?>
<Document xmlns:xsi="http://www.w3.org/2001/XMLSchema-instance"
xmlns="urn:iso:std:iso:20022:tech:xsd:pain.001.001.04">
      <CstmrCdtTrfInitn>
          <GrpHdr>
              <MsgId>ABC/120928/CCT001</MsgId>
              <CreDtTm>2012-09-28T14:07:00</CreDtTm>
              <NbOfTxs>3</NbOfTxs>
              <CtrlSum>11500000</CtrlSum>
              <InitgPty>
                  <Nm>ABC Corporation</Nm>
                  <PstlAdr>
                      <StrtNm>Times Square</StrtNm>
                      <BldgNb>7</BldgNb>
```

```
            <PstCd>NY 10036</PstCd>
            <TwnNm>New York</TwnNm>
            <Ctry>US</Ctry>
         </PstlAdr>
      </InitgPty>
   </GrpHdr>
   <PmtInf>
      <PmtInfId>ABC/086</PmtInfId>
      <PmtMtd>TRF</PmtMtd>
      <BtchBookg>false</BtchBookg>
      <ReqdExctnDt>2012-09-29</ReqdExctnDt>
      <Dbtr>
             <Nm>ABC Corporation</Nm>
             <PstlAdr>
                    <StrtNm>Times Square</StrtNm>
                    <BldgNb>7</BldgNb>
                    <PstCd>NY 10036</PstCd>
                    <TwnNm>New York</TwnNm>
                    <Ctry>US</Ctry>
             </PstlAdr>
      </Dbtr>
      <DbtrAcct>
         <Id>
            <Othr>
               <Id>00125574999</Id>
         </Othr>
      </Id>
   </DbtrAcct>
   <DbtrAgt>
      <FinInstnId>
         <BICFI>BBBBUS33</BICFI>
      </FinInstnId>
   </DbtrAgt>
<CdtTrfTxInf>
      <PmtId>
            <InstrId>ABC/120928/CCT001/01</InstrId>
            <EndToEndId>ABC/4562/2012-09-08</EndToEndId>
      </PmtId>
      <Amt>
            <InstdAmt Ccy="JPY">10000000</InstdAmt>
   </Amt>
   <ChrgBr>SHAR</ChrgBr>
   <CdtrAgt>
      <FinInstnId>
            <BICFI>AAAAGB2L</BICFI>
      </FinInstnId>
   </CdtrAgt>
   <Cdtr>
      <Nm>DEF Electronics</Nm>
      <PstlAdr>
         <AdrLine>Corn Exchange 5th Floor</AdrLine>
         <AdrLine>Mark Lane 55</AdrLine>
         <AdrLine>EC3R7NE London</AdrLine>
         <AdrLine>GB</AdrLine>
      </PstlAdr>
   </Cdtr>
   <CdtrAcct>
      <Id>
         <Othr>
```

```
                 <Id>23683707994125</Id>
           </Othr>
        </Id>
     </CdtrAcct>
     <Purp>
           <Cd>CINV</Cd>
     </Purp>
     <RmtInf>
            <Strd>
                  <RfrdDocInf>
                        <Nb>4562</Nb>
                        <RltdDt>2012-09-08</RltdDt>
                  </RfrdDocInf>
            </Strd>
     </RmtInf>
</CdtTrfTxInf>
<CdtTrfTxInf>
     <PmtId>
           <InstrId>ABC/120928/CCT001/2</InstrId>
           <EndToEndId>ABC/ABC-13679/2012-09-15</EndToEndId>
     </PmtId>
     <Amt>
           <InstdAmt Ccy="EUR">500000</InstdAmt>
     </Amt>
  <ChrgBr>CRED</ChrgBr>
  <CdtrAgt>
        <FinInstnId>
              <BICFI>DDDDBEBB</BICFI>
        </FinInstnId>
  </CdtrAgt>
  <Cdtr>
        <Nm>GHI Semiconductors</Nm>
        <PstlAdr>
              <StrtNm>Avenue Brugmann</StrtNm>
              <BldgNb>415</BldgNb>
              <PstCd>1180</PstCd>
              <TwnNm>Brussels</TwnNm>
              <Ctry>BE</Ctry>
        </PstlAdr>
  </Cdtr>
  <CdtrAcct>
        <Id>
              <IBAN>BE30001216371411</IBAN>
        </Id>
  </CdtrAcct>
  <InstrForCdtrAgt>
        <Cd>PHOB</Cd>
        <InstrInf>+32/2/2222222</InstrInf>
  </InstrForCdtrAgt>
  <Purp>
        <Cd>GDDS</Cd>
  </Purp>
  <RmtInf>
        <Strd>
              <RfrdDocInf>
                    <Tp>
                          <CdOrPrtry>
                                <Cd>CINV</Cd>
                          </CdOrPrtry>
```

```
                    </Tp>
                    <Nb>ABC-13679</Nb>
                    <RltdDt>2012-09-15</RltdDt>
                </RfrdDocInf>
            </Strd>
        </RmtInf>
    </CdtTrfTxInf>
    <CdtTrfTxInf>
        <PmtId>
            <InstrId>ABC/120928/CCT001/3</InstrId>
            <EndToEndId>ABC/987-AC/2012-09-27</EndToEndId>
        </PmtId>
        <Amt>
            <InstdAmt Ccy="USD">1000000</InstdAmt>
        </Amt>
        <ChrgBr>SHAR</ChrgBr>
        <CdtrAgt>
            <FinInstnId>
                <BICFI>BBBBUS66</BICFI>
            </FinInstnId>
        </CdtrAgt>
          <Cdtr>
              <Nm>ABC Corporation</Nm>
              <PstlAdr>
                  <Dept>Treasury department</Dept>
                  <StrtNm>Bush Street</StrtNm>
                  <BldgNb>13</BldgNb>
                  <PstCd>CA 94108</PstCd>
                  <TwnNm>San Francisco</TwnNm>
                  <Ctry>US</Ctry>
              </PstlAdr>
          </Cdtr>
          <CdtrAcct>
              <Id>
                  <Othr>
                      <Id>4895623</Id>
                  </Othr>
              </Id>
          </CdtrAcct>
          <Purp>
              <Cd>INTC</Cd>
          </Purp>
          <RmtInf>
              <Strd>
                  <RfrdDocInf>
                      <Tp>
                          <CdOrPrtry>
                              <Cd>CINV</Cd>
                          </CdOrPrtry>
                      </Tp>
                      <Nb>987-AC</Nb>
                      <RltdDt>2012-09-27</RltdDt>
                  </RfrdDocInf>
              </Strd>
          </RmtInf>
        </CdtTrfTxInf>
      </PmtInf>
    </CstmrCdtTrfInitn>
</Document>
```

19.15.4 The Financial Information Exchange Protocol (FIX)

The FIX Protocol was created beginning in 1992 by Bob Lamoureux, Jim Leman, Chris Morstatt, and Jacques Perold of Fidelity Investments and Salomon Brothers. The goal was simple: create an open protocol for communication of stock trading between buy-side and sell-side firms. At the time, large brokerages were dominated by IBM computer architectures, the sell side to a large extent DEC VAX. However, emerging were Internet standard networking protocols in the form of TCP/IP and open systems based upon UNIX.

The founders followed an open approach in the adoption of TCP/IP as one of the primary transports. The implementation of FIX is quite simple. All messages are composed of fields that are represented in a tag=value format. An integer number is used to represent the field. The original FIX protocol combined a reliable session layer for bidirectional ordered delivery of messages with session recovery in the event of a loss of the underlying communication session. Although widely adopted using TCP/IP, the actual FIX specification itself is independent of the underlying transport. The timing of the creation of FIX was coincident with the growth of securities and commodities markets globally. By the mid-1990s FIX had been extended to support equities, fixed income, options, futures, and FX. The basic set of business processes, which spanned pre-trade through allocations, were expanded considerably to include the majority of business processes up through post-trade processing, pre-settlement. FIX is used throughout the entire securities trading infrastructure (see Figure 19.12).

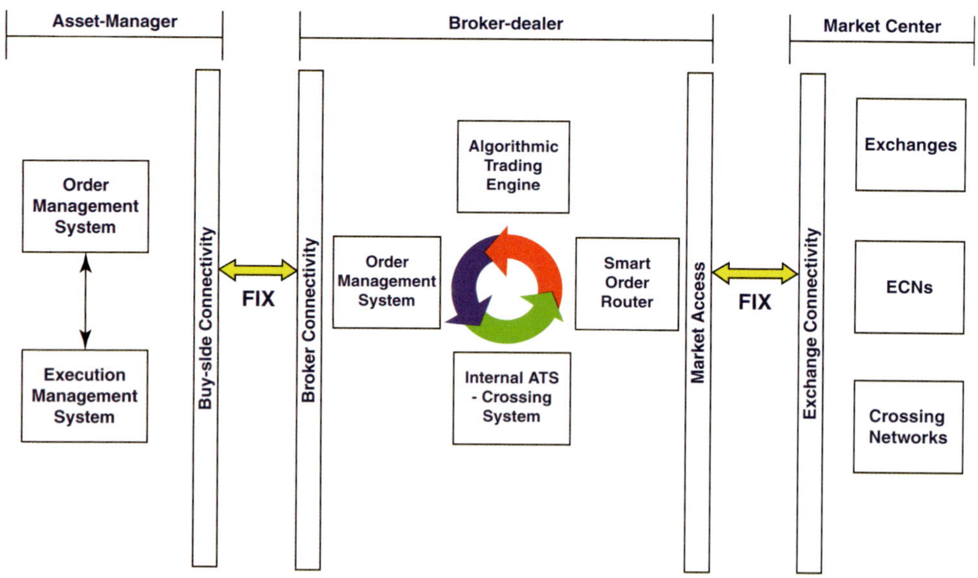

Figure 19.12 FIX protocol.

```
8=FIX.4.1^9=0235^35=D^34=10^43=N^49=VENDOR^50=CUTOMER^56=BROKER^52=1
9980930-09:25:58^1=XQCCFUND^11=10^21=1^55=EK^48=277461109^22=1^54=1^
38=10000^40=2^44=76.750000^59=0^10=165
```

Header		*Body*	
8 = FIX.4.1	**BeginString**	1 = XQCCFUND	Account
9 = 235	**BodyLength**	11 = 10	ClOrdID
35 = D	**MSGType**	21 = 1	HandInst
34 = 10	MsgSeqNum	55 = EK	Symbol
43 = N	PossDupFlag	48 = 277461109	SecurityID
49 = VENDOR	SenderCompID	22 = 1	IDSource
115 = CUSTOMER	OnBehalfofCompID	54 = 1	Side
56 = BROKER	TargetCompID	38 = 10000	OrderQty
52 = 19980930-09:25:58	SendingTime	40 = 2	OrdType
		44 = 76.750000	Price
		59 = 0	TimeInForce
Trailer			
	Checksum		

Figure 19.13 FIX message example.

The original version of FIX used a very simple tag=value format for its messages (see Figure 19.13).

An XML variant of FIX was created, named FIXML, primarily by John Goeller in the late 1990s. In 2003, FIXML was optimized to greatly reduce message size to meet the requirements for listed derivatives clearing. FIXML, with its simple structure, that mirrors the underlying FIX protocol upon which it is derived, has become widely used for regulatory reporting of derivatives positions and trades in the USA.

```
<FIXML v="FIX.5.0SP2" xsi:schemaLocation="http://www.fixprotocol.org/FIXML-5-0-SP2
fixml-order-impl-5-0-SP2.xsd" xmlns="http://www.fixprotocol.org/FIXML-5-0-SP2"
xmlns:xsi="http://www.w3.org/2001/XMLSchema-instance">
   <Order Typ="2" Side="1" TmInForce="7" Acct="String" TxnTm="2001-12-17T09:30:47Z"
ID="ABC-1234" Px="40.0">
      <Hdr SeqNum="2" TID="Exchange" SID="Broker" Snt="2001-12-17T09:30:47Z"/>
      <Pty ID="BROKER1" Src="F" R="34"/>
      <Instrmt Sym="IBM" SecTyp="CS"/>
      <OrdQty Qty="0.0"/>
   </Order>
</FIXML>
```

19.15.5 The eXtensible Business Reporting Language (XBRL)

XBRL, eXtensible Business Reporting Language, is an XML based data technology standard that makes it possible to "tag" business information to make it machine readable. As the business information is made machine readable through the tagging process, this information does not need to be manually entered again and can be transported and understood by computers and software applications. The tagging process includes certain metadata (information about information) for each tagged element. For example, a tagged number will include its meaning and, if the number has been the result of a calculation, can include the numbers that were used in its calculation.

Since its introduction, XBRL primarily has been used for the tagging and filing of financial statements, as the original idea for XBRL was developed by accountants to support financial reporting. Most recently the use of XBRL has expanded into the financial transaction processing area.

The following is an excerpt of an XBRL disclosure document required by International Financial Reporting Standards (IFRS 7) Financial Instruments: Disclosures.


```
<?xml version="1.0" encoding="UTF-8"?><!--Generated by Fujitsu XWand B0161C --
><xbrli:xbrl xmlns:xbrli="http://www.xbrl.org/2003/instance"
xmlns:ifrs="http://xbrl.ifrs.org/taxonomy/2012-03-29/ifrs"
xmlns:negated="http://www.xbrl.org/2009/role/negated"
xmlns:xl="http://www.xbrl.org/2003/XLink" xmlns:xlink="http://www.w3.org/1999/xlink"
xmlns:xs="http://www.w3.org/2001/XMLSchema" xmlns:ref="http://www.xbrl.org/2006/ref"
xmlns:xsi="http://www.w3.org/2001/XMLSchema-instance"
xmlns:xbrldi="http://xbrl.org/2006/xbrldi"
xmlns:iso4217="http://www.xbrl.org/2003/iso4217"
xmlns:example4="http://www.ifrs.org/xbrl/illustrativeExamples/Example4"
xmlns:link="http://www.xbrl.org/2003/linkbase"
xmlns:num="http://www.xbrl.org/dtr/type/numeric"
xmlns:nonnum="http://www.xbrl.org/dtr/type/non-numeric"
xmlns:xbrldt="http://xbrl.org/2005/xbrldt">
  <link:schemaRef xlink:type="simple" xlink:href="example4_2012-03-29.xsd"/>
  <xbrli:context id="y2012">
     <xbrli:entity>
       <xbrli:identifier
scheme="http://www.ifrs.org/xbrl/illustrativeExamples">Example4</xbrli:identifier>
     </xbrli:entity>
     <xbrli:period>
       <xbrli:startDate>2012-01-01</xbrli:startDate>
       <xbrli:endDate>2012-12-31</xbrli:endDate>
     </xbrli:period>
  </xbrli:context>
  <xbrli:context id="e2012">
    <xbrli:entity>
      <xbrli:identifier
scheme="http://www.ifrs.org/xbrl/illustrativeExamples">Example4</xbrli:identifier>
     </xbrli:entity>
     <xbrli:period>
       <xbrli:instant>2012-12-31</xbrli:instant>
     </xbrli:period>
  </xbrli:context>
  <xbrli:context id="y2011">
    <xbrli:entity>
      <xbrli:identifier
scheme="http://www.ifrs.org/xbrl/illustrativeExamples">Example4</xbrli:identifier>
    </xbrli:entity>
    <xbrli:period>
      <xbrli:startDate>2011-01-01</xbrli:startDate>
      <xbrli:endDate>2011-12-31</xbrli:endDate>
```

```
    </xbrli:period>
  </xbrli:context>
  <xbrli:context id="e2011">
   <xbrli:entity>
   <xbrli:identifier
scheme="http://www.ifrs.org/xbrl/illustrativeExamples">Example4</xbrli:identifier>
   </xbrli:entity>
   <xbrli:period>
       <xbrli:instant>2011-12-31</xbrli:instant>
     </xbrli:period>
   </xbrli:context>
   <xbrli:context id="e2010">
     <xbrli:entity>
        <xbrli:identifier
scheme="http://www.ifrs.org/xbrl/illustrativeExamples">Example4</xbrli:identifier>
     </xbrli:entity>
     <xbrli:period>
       <xbrli:instant>2010-12-31</xbrli:instant>
     </xbrli:period>
   </xbrli:context>
   <xbrli:context id="e2012_TradingSecuritiesMember">
     <xbrli:entity>
       <xbrli:identifier
scheme="http://www.ifrs.org/xbrl/illustrativeExamples">Example4</xbrli:identifier>
     </xbrli:entity>
    <xbrli:period>
       <xbrli:instant>2012-12-31</xbrli:instant>
    </xbrli:period>
    <xbrli:scenario>
      <xbrldi:explicitMember
dimension="ifrs:ClassesOfFinancialAssetsAxis">ifrs:TradingSecuritiesMember</xbrldi:exp
licitMember>
    </xbrli:scenario>
   </xbrli:context>
   <xbrli:context id="e2012_DerivativesMember">
    <xbrli:entity>
     <xbrli:identifier
scheme="http://www.ifrs.org/xbrl/illustrativeExamples">Example4</xbrli:identifier>
    </xbrli:entity>
    <xbrli:period>
       <xbrli:instant>2012-12-31</xbrli:instant>
    </xbrli:period>
    <xbrli:scenario>
      <xbrldi:explicitMember
```

•
•
•
•

...excerpted...

```
          </ifrs:DisclosureOfFinancialAssetsExplanatory>

<ifrs:DescriptionWhereGainsLossesOfFinancialInstrumentsMeasuredInLevel3OfFairValueHie
rarchyPresentedInStatementOfComprehensiveIncomeOrIncomeStatement id="XWAND00009999"
contextRef="y2012">Gains or losses included in profit or loss for the period (above)
are presented in trading income and in other
income</ifrs:DescriptionWhereGainsLossesOfFinancialInstrumentsMeasuredInLevel3OfFairVa
lueHierarchyPresentedInStatementOfComprehensiveIncomeOrIncomeStatement>
</xbrli:xbrl>
```

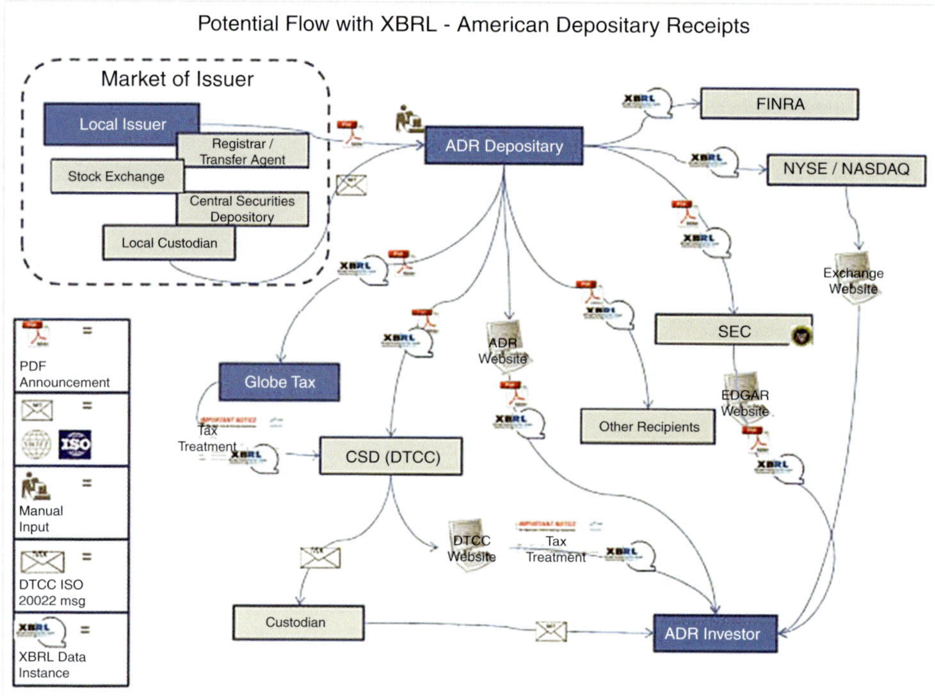

Figure 19.14 Illustrative corporate actions flow for ADR Corporate Actions.

19.15.6 ISO 20022 alignment with XBRL

In June 2009, SWIFT, DTCC (Depository Trust and Clearing Corporation) and XBRL US commenced an initiative to develop a corporate actions taxonomy using XBRL. Each of the elements in the corporate actions taxonomy corresponds to a message element in the ISO 20022 Corporate Actions Notification message. This initiative attempts to bring together a standard used by companies (XBRL) with the standard (ISO 20022) used by financial intermediaries to announce and process corporate actions events. (see Figure 19.14)

19.15.7 The Financial Product Markup Language (FpML)

The Financial Products Markup Language was created in response to the increased use and rapid innovation in over-the-counter financial derivatives. FpML was developed by JP Morgan and released in 1999. Meeting the need of the industry, the use of FpML spread quickly amongst the OTC derivatives dealers. The first usage of FpML was inside an organization, not between counterparties. Dealers used FpML documents to communicate internally with risk management and compliances areas of their organizations.

FpML documents are used to describe in electronic document format an OTC contract based upon master agreements created by the International Swap and Derivatives Association (ISDA). So close was the relationship between FpML and the ISDA Master Agreements that in 2001, ISDA took over management and operation of FpML.

There are two major versions of FpML in use today: the 4.x versions are the widest adopted. Given its document oriented focus, FpML in 4.x did not include business process definition or traditional messaging. With FpML 5.0, the re-architected version has a business process component integrated similar to FIX and ISO 20022.

FpML is highly coupled to the underlying XML Schema definitions; this can make FpML difficult to extend without detailed knowledge of the W3C XML Schema standard. However, knowledge of FpML and its relationship to the ISDA master agreements is indispensable for risk managers at organizations that have exposure to OTC derivatives.

Example trade request for information update message.

```
<?xml version="1.0" encoding="utf-8"?>
<!--View is confirmation>
<!--Version is 5-3-->
<!--NS is http://www.fpml.org/FpML-5/confirmation-->
<!--
  == Copyright (c) 2002-2012 All rights reserved.
  == Financial Products Markup Language is subject to the FpML public license.
  == A copy of this license is available at http://www.fpml.org/license/license.html
-->
<requestTradeReferenceInformationUpdate xmlns="http://www.fpml.org/FpML-5/
confirmation" xmlns:xsi="http://www.w3.org/2001/XMLSchema-instance" fpmlVersion="5-3"
xsi:schemaLocation="http://www.fpml.org/FpML-5/confirmation ../../fpml-main-5-3.xsd
http://www.w3.org/2000/09/xmldsig# ../../xmldsig-core-schema.xsd">
  <header>
    <messageId messageIdScheme="http://www.abc.com/messageId">111</messageId>
    <sentBy messageAddressScheme="http://www.clearingsvc.com/partyId">CLEARCO</sentBy>
    <sendTo messageAddressScheme="http://www.clearingsvc.com/partyId">ABC</sendTo>
    <creationTimestamp>2006-01-01T09:07:00+05:00</creationTimestamp>
  </header>
  <isCorrection>false</isCorrection>
  <correlationId correlationIdScheme="http://www.abc.com/conversationId/OTC">222</
correlationId>
  <sequenceNumber>1</sequenceNumber>
  <tradeReferenceInformation>
    <partyTradeIdentifier>
      <partyReference href="clearing-svc" />
      <tradeId
tradeIdScheme="http://www.clearingsvc.com/tradeId/OTC">SVC_001</tradeId>
    </partyTradeIdentifier>
    <partyTradeIdentifier>
      <partyReference href="member-firm" />
      <tradeId tradeIdScheme="http://www.abc.com/tradeId/OTC">ABC999</tradeId>
    </partyTradeIdentifier>
```

```
  </tradeReferenceInformation>
  <party id="clearing-svc">
    <partyId>CLEARCO</partyId>
  </party>
  <party id="member-firm">
    <partyId>ABCBIC123</partyId>
  </party>
</requestTradeReferenceInformationUpdate>
```


19.16 Reference data standards

The 2008 credit crisis has propelled one of the most neglected areas in the financial services industry to the forefront. Reference data, also referred to as securities master file data, define the financial instruments that are used in the financial markets. With the advent of more complicated financial instruments and rules regarding know your customer (KYC) reference data, following the issues surrounding the Enron collapse, data now are generally considered to include counterparty (legal entity) information.

The sources for reference data are data consolidators, such as Bloomberg and Thomson-Reuters for traditional equity and fixed income products. Exchanges are also a source for reference data, especially for listed derivatives and commodities. A relatively new player, Markit, provides information on OTC derivatives and credit products.

The current trend in reference data is to use vended solutions for reference data. A number of platforms have been created in response to this demand. We are now seeing a move to the cloud and SAS models for reference data.

The role of standards in the area of reference data is marginal at best. The existing standards for reference data are more or less themselves used as references for proprietary systems.

19.16.1 Market Data Definition Language (MDDL)

MDDL was a very ambitious initiative to create a comprehensive reference data model as an XML based interchange format and data dictionary for financial instruments, corporate events, and market related, economic, and industrial indicators, which started in 2001 by the Software and Information Industry Association's Financial Information Services Division (FISD).[48] Although the direct adoption of MDDL is very limited, the legacy of MDDL and its value as a reference model continue. The influence of MDDL can be seen in standards that evolved out of this initial pioneering effort.

[48] www.openmddb.org/

19.16.2 Financial Instrument Business Information Model (FIBIM)

FIBIM was initiated as a work item under ISO TC68/SC4/WG11. The goal was to create a reference data model for the financial services industry, focusing primarily on securities processing. The standard was to be ISO 19312. At the 2007 ISO TC68/SG4 Plenary meeting, the decision was made to integrate FIBIM into the ISO 20022 model, in order to provide an integrated business domain model (reference data) with the messaging model.

19.16.3 OpenMDDB

FIX Protocol Ltd. and FISD jointly developed a relational database model for reference data derived from MDDL in 2009. OpenMDDB also provides support for maintenance and distribution of reference data using FIX messages. The initiative was completed and a database model is available for download and use. The EDM Council, FISD, and FIX have entered into an agreement for EDM Council to facilitate and support evolution of the OpenMDDB by the community.

19.17 Business model standards

19.17.1 ISO 20022

As a result of the integration of what was to be ISO 19312 into ISO 20022, the financial messaging standard was expanded to be a business model of reference data for the financial markets in addition to the messages. The ISO 20022 business model was extensively refactored by SWIFT in 2010. The full impact of the ISO 20022 business model has yet to be realized.

19.17.2 Financial Information Business Ontology (FIBO)

The EDM Council and the OMG created a joint working group, the Financial Data Task Force, "to accelerate the development of 'sustainable, data standards and model driven' approach to regulatory compliance."[49] The initiative focused on semantics instead of traditional modeling techniques. The work continues on FIBO as of this writing (see Figure 19.15).

19.18 What is next for financial services standards?

It has been evident over time that an integrated approach to financial services standards is both possible and desirable. Users are faced with dealing with an unprecedented amount of data. Requirements to contain cost and the pressure to

[49] www.omgwiki.org/OMG-FDTF/doku.php; www.edmcouncil.org

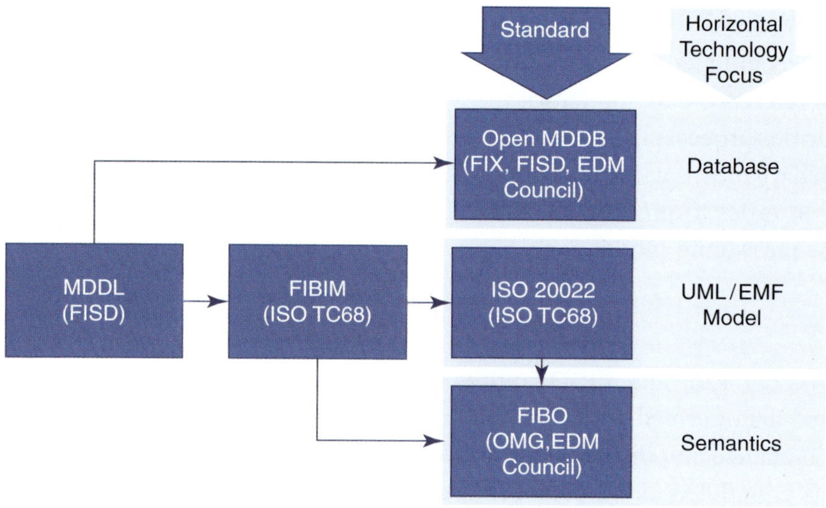

Figure 19.15 Standards evolution.

do more with less requires better solutions and places more emphasis than ever on delivering STP. Global regulators also are now developing and stating their requirements for data and reporting for systemic risk analysis. It appears that regulators have come to realize that to prevent drowning in information, and given the much increased volume of data they need to process, that they need standards. Also, some regulators have been getting involved on a more detailed level in the development and use of standards. For example, the ECB now requires the use of reporting templates for structured products for the posting of collateral. Also, in some cases regulators have been given mandates to set technical standards, for example ESMA in their oversight of ratings agencies.[50]

The focus on delivering solutions to address these needs has ebbed and flowed over the years and some ideas have been put forward such as, in the EU, the ECB work promoting a utility for registry of financial instruments. Most recently, the discussions with the US regulators especially have evolved past discussing legal entity identifiers (LEIs), progressing to discussing how to link the legal entities that the regulators will track to the financial instruments that they issue, trade and hold through an integrated approach. This combined with the continuing rapid improvements in the technology to model data, process, transport and store information presents an opportunity for realizing such improvements.

Working within the ISO framework, vertical industries and their technical committees can leverage the work of other TCs and adapt this work to their needs. In case of financial services, there is an opportunity for ISO TC68 to

[50] See www.esma.europa.eu/index.php?page=consultation

leverage work that has been successful for other service industries. In 2009, the chair of TC68 formed a Study Group to look at other ISO standards that addressed similar needs to those in financial services, such as identification of classification. One example is to leverage an ISO standard from ISO TC46, which focuses on standards for Information and Documentation. One of their SCs (ISO TC46/SC9/WG7) manages a standard called Digital Object Identifier (DOI), ISO 26324. This standard has been successfully deployed to support the identification, classification and relationship needs of several service-related industries particularly publishing of books, music and audio-visual media. In 2012, some of the members of the Study Group recommended further research into the technologies which ISO 26324 is built on, especially the handle system[51] and the related Digital Object Architecture developed at the Corporation for National Research Initiatives.[52]

The future of financial services standards will concentrate on collaboration, convergence and coexistence. The integration described above using the DOI standard is one example. Another opportunity for collaboration, convergence and coexistence is the ISO 20022 standard. One of the most interesting and useful characteristics of the ISO 20022 standard is that it provides a way to achieve the long-term convergence objective and at the same time provide a way to facilitate the short-term coexistence of several standards. This syntax-independent business modeling is a key strength of the ISO 20022 standard.

Another area in which to expand the relevance of the ISO 20022 standard is to add a semantics layer to the standard. This effort will concentrate on work to agree the definitions/meaning, and, converging on these, the ability to automatically map and correlate concepts and data across messaging platforms and dynamically integrate the resulting information should yield a new level of information integration within and across industry players.

19.19 Trends in the evolution of horizontal technologies: the Internet, the Web, and "consumerization"

Few organizations do not suffer from the "legacy overhang" created by years or decades of accumulated technology solutions. This provides a significant set of obstacles to taking advantage of the seemingly never-ending increases in computing power evident in today's society. Obviously, not all of these advances are relevant or useful in every enterprise, and assessment requires care and a clear-

[51] See `www.handle.net`

[52] `www.cnri.reston.va.us/papers/OverviewDigitalObjectArchitecture.pdf`

eyed view of the relevant specialties and their application to the issues inherent in the mission.

The evolution of microprocessor based computing technologies was largely driven by hardware factors for the period between 1960 and 1995. Advances in calculation, storage, and connectivity capacity have mirrored developments in the underlying physical engineering technologies. Two critical interdependent "social technologies" – Open Source software in the 1980s and the invention of the Web in 1993, both directly related to capabilities derived from early stages of the Internet – have been the "jet fuel" for the succeeding exponential increases in systems technologies.[53] The resulting exponential increases in data volumes and the distributed nature of the systems creating the data have led to new tools and practices that in some cases have created entirely new technologies or fields of study. Correlated developments in interdisciplinary research approaches bridging social and physical sciences have led to new observations and theories about the nature of complexity and complex systems. As an example, research in network analysis has been driven by measuring telecom traffic as the Internet has grown; approaches to analyzing social and biological networks have been heavily influenced by the results, current research into applying these approaches and techniques to financial market supply chains are immature but progressing rapidly.

In assessing the issues involved in planning the technology platform and toolsets for an enterprise, there are five interrelated "threads" that can be identified as having direct impact on the technologies and methods used to implement mission requirements. The table below does not seek to be a complete description – or one that experts in each field would necessarily find complete – but instead tries to classify the main thrusts in terms familiar to a lay audience. A "buzzword description" of some of the related technical approaches illustrates some of the areas named.

Theme or "thread"	Description	Technology approaches
Big Data	Exponential increases in data volumes that have overwhelmed traditional technical approaches and provided researchers with new abilities to generate observations and analyses not previously considered possible.	"Data Science" CEP (Complex Event Processing) Column Stores/Triple Stores/"No SQL" databases "Schema on Read" versus "Schema on Write"

[53] The important point here is that advances in hardware are now being driven by usage by well over a billion people – instead of ever increasing hardware capacity in memory, compute speeds, or network speeds defining the upper boundaries – they are now not the gating factor.

Theme or "thread"	Description	Technology approaches
Semantic technologies	First described in the seminal 1997 paper by Tim Berners-Lee and Jim Hendler, these approaches involve new methods of representing data such that they are encoded with a full context description. This means that other programs can use the information without having to understand the specifics of the application(s) that created it.	Ontologies RDF Schema Triple Stores "Data finds Data" – new information triggers inference engine(s) Rule engines
Crowds	Methods for involving large online groups in specific goal-directed applications. These can range from cleaning data to solving complex genetics questions by playing an online "game."	"Big Bang cleanup" – the possibility of utilizing crowd-based services to attack data cleansing and integration tasks previously regarded as too expensive.
Clouds	Massive capacity at a moment's notice. Security concerns may preclude the organization from utilizing these approaches for confidential data, but publicly available information can be managed this way, and the technologies and tools developed to support the growth of cloud systems should inform planning for surge capacity and other requirements.	Offers the potential ability to quickly deploy and extend online services and communities. A private site in the cloud could serve as a test room for a large variety of development for both tooling and UIX. Related technologies are crucial to the evolution of shared authoritative reference data systems.
BYOD – "Bring your own device"	"Consumerization" of IT devices, and the impact of tablets and other form factors – you will use your own hardware to interact with your employer's or organization's systems.	Profound implications for all aspects of systems and society, especially around personal privacy, organizational security, and information sharing.
Internet of Things	Everything, including the kitchen sink, is becoming network enabled – everyday objects can be separately identified and messaged.	

While by no means exhaustive, the list above should provide some guidance as to where advances in horizontal technologies will have an impact on the vertical technologies used in multiple industries. The rate and depth of impact will vary across vertical markets, but we anticipate that financial services will face significant challenges and opportunities in information management on an unprecedented scale in the next five years. We will use the online extension of this Handbook to assess these changes during that time, and invite your contributions to the dialog.

Bibliography

A Business Case to Improve Corporate Actions Communications, 2010, June, `http://xbrl.us/i2i/Pages/default.aspx`

Ferguson, Niall, 2008, *The Ascent of Money: A Financial History of the World*, Penguin Books.

Gleick, James, 2011, *The Information: A History, a Theory, a Flood*, Harper Collins.

Gridwise Architecture Council Interoperability Framework Team, 2007, Interoperability Context-Setting Framework, July, retrieved August 27, 2012, `www.gridwiseac.org/pdfs/interopframework_v1.pdf`

Groot, Martijn, 2008, *Managing Financial Information in the Trade Life Cycle: A Concise Atlas of Financial Instrument and Process*, Academic Press, Elsevier.

Hoffman, Charles and Liv Watson, 2009, *XBRL for Dummies*, Wiley Publishing.

Kay Review of UK Equity Markets and Long-Term Decision Making – Final Report, July 2012, `www.bis.gov.uk/assets/biscore/business-law/docs/k/12-917-kay-review-of-equity-markets-final-report%20downloaded%208/20/2012h`

Murphy, C. and J. Yates, 2009, *The International Organization for Standardization (ISO): Global Governance through Voluntary Consensus*, Routledge.

O'Brien, James and George Marakas, 2000, *Introduction to Information Systems*, Thirteenth edition, McGraw-Hill/Irwin.

Puffert, Douglas J., 2009, *Tracks across Continents, Paths through History: The Economic Dynamics of Standardization in Railway Gauge*, University of Chicago Press.

Sexton, Martin Christopher, 2008, *MDDL and the Quest for a Market Data Standard: Explanation, Rationale and Implementation*, Elsevier.

Sommerville, I. et al., 2012, Large scale complex IT systems, *Communications of the ACM*, **55**(7), 71.

Spewak, Steven M. and Steven C. Hill, 1993, *Enterprise Architecture Planning: Developing a Blueprint for Data, Applications, and Technology*, Wiley.

Spivak, Steven M. and F. Cecil Brenner, 2001, *Standardization Essentials: Principles and Practice*, CRC Press.

SWIFT Standards Team, 2010, *ISO 20022 for Dummies*, John Wiley and Sons.

Unicode Consortium, *The Unicode Standard*, Version 5.0.

Weinberg, G., 1986, *The Secrets of Consulting: A Guide to Giving and Getting Advice Successfully*, Dorset House Publishing.

20

Data storage and processing

Dilip Krishna

20.1 Introduction

Risk management depends heavily on analytics to generate information for decision making. It is no surprise, therefore, that computing has played a major role in risk management from the very beginning. In the past, risk measurement and reporting systems have largely been segregated by risk type – with separate systems for measuring credit, market and liquidity risk – though in some cases integrated market and credit engines have been developed. This approach is appealing because the systems are usually easier to implement when separate. It was also workable in an earlier age when it was tolerable for risk to be managed in silos. Recent developments have changed this dynamic significantly – not only do risk management systems need to integrate measurements across risk types due to needs like stress testing, but they also need to more readily integrate with transaction and finance systems to enable better Risk-Based Pricing (RBP) and Risk-Adjusted Performance Management (RAPM).

One unintended but favorable side-effect of segregated systems was that each computer system was not overly burdened by data processing requirements. The need for more integration has therefore posed a new and significant challenge to data processing technologies. Fortunately, we are in a period of tremendous innovation in data processing – data storage and processing technology have progressed to a point where these advanced needs are becoming eminently feasible. This chapter will first review the information needs of risk management, followed by a broad overview of data processing technology.

Handbook of Financial Data and Risk Information, ed. Margarita S. Brose, Mark D. Flood, Dilip Krishna and Bill Nichols. Published by Cambridge University Press.

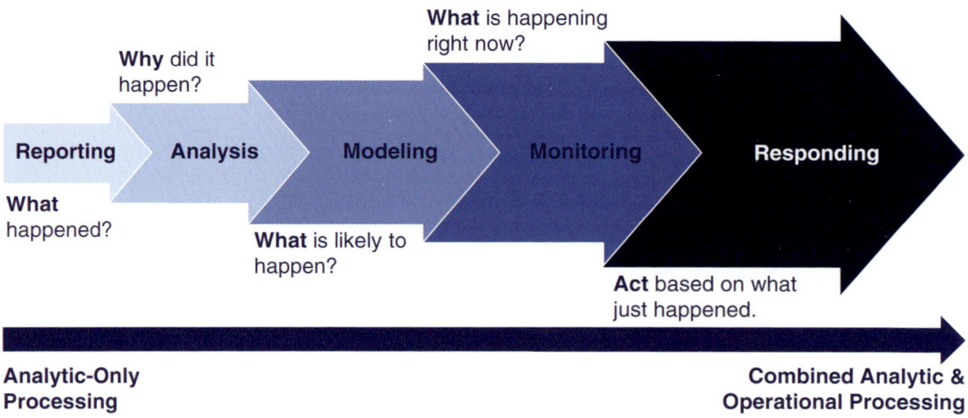

Figure 20.1 Risk management information access processes.

20.2 Analytic data processing and the needs of risk management

There are two broad patterns of information access – *operational* and *analytic*. Operational information access requires small amounts of specific kinds of information with a quick response time to support transactional decision making. Analytic information access, on the other hand, refers to processing large amounts of data, with number-crunching taking precedence over speed of response. Risk management uses both of these types of information access patterns.

Risk information needs can be classified into the five categories that use data in an increasingly sophisticated manner, as illustrated in Figure 20.1.

- *Reporting*: Risk reporting involves generating specific views of information for management as well as to regulatory authorities. Standard period-over-period reports usually involve data aggregation using functions like summation and averaging. High quality data are a key requirement. Since reports are simply snapshots of information, they can be created in batches for subsequent distribution to interested users. As such, they are usually not challenging to process in a timely manner; most of the latency in a reporting cycle is due to the time taken to create high quality data.
- *Risk analysis*: There is usually a need to "drill down" on specific aspects of a report to answer questions such as – "Why did the auto-loan portfolio consume more economic capital on a smaller outstanding than last year?" This is an "ad hoc" activity, where one question usually prompts another, more detailed and incisive question. Risk analysis activity requires access to detailed data. Since this is an interactive activity, the response time of an information system is an important criterion.

- *Modeling*: Risk modeling is a research activity. Analysts create models that are designed to predict specific risk events such as identifying potential delinquency. They theorize the predictive capability of a certain kind of mathematical calculation and then test the model against large historical datasets. Information systems must be capable of rapidly – but not necessarily interactively – processing new and unanticipated models.
- *Risk monitoring*: Risk models are embedded in systems (manual or automatic) to continuously monitor risk. Models are usually standard with short running times against small datasets, but quick – often real-time – response times are important.
- *Response*: Response to breaking a risk limit – for example, a VaR limit requiring to hedge trades – is still largely via manual intervention. However, for low-impact/high-frequency events it is becoming increasingly attractive to automate risk response. A classic case is the response to fraudulent credit-card transactions which is largely automated in most banks.

Each data operation shown in Figure 20.1 has distinct impacts on data processing and storage. While individual operations themselves do not push present-day technology to its limits, it is the need for simultaneity and scale that poses interesting technological challenges as one moves from the left side of the figure (purely analytic operations) to the right-hand side which combines analytic and operational requirements. It is worth noting that the purely operational challenges in risk management are usually trivial in contrast to those posed by other systems (say algorithmic trading applications which have to calculate and react to rapidly changing markets). Therefore, the rest of the chapter will focus largely on analytical (or combined analytical-operational) data processing.

Analytic processes, both manual and automated, are nothing more than a series of questions posed to the data. Often broad questions ("Which maturity bucket had the biggest liquidity mismatch?") lead to specific questions ("Which funding position in high-quality asset-backed securities is problematic?") that can indicate decisions that need to be made. Therefore, the ability to efficiently answer a priori unknown questions across large and often heterogeneous datasets can be a critical problem for risk analytics. It therefore can be instructive to study the simplest atomic component of an analytic process – known in computer science terms as "executing a query" – of asking a single question of the data. Analytic data processing can be thought of as sifting through a large amount of data stored in one or more related datasets. For example, one set may contain counterparty data, the second positions containing exposure estimates and the third transactions, with identifiers relating positions to customers and transactions are related to positions. Let us consider the following analytic query:

Return the average notional volume of CDS transactions for the counterparty with the largest overall exposure.

	Atomic Operation		Processing Component	Result
1	Read source data-set	Scan	**Disk IO**	Counter-parties and Positions
2	Relate data-set via redistribution	Join	**Bandwidth**	Related Counter-parties & Positions
3	Group positional exposure	Aggregate (Sum)	**Computation**	Counter-Parties and Aggregate Positions
4	Sort by exposure	Sort	**Computation**	Sorted Counter-Parties by Position
5	Filter CDS transactions	Filter	**Computation**	CDS Transactions for Top Counter-party
6	Calculate Average	Aggregate (Average)	**Computation**	Average Notional

Figure 20.2 Anatomy of an analytical process.

Many risk management analytic operations use a combination of such queries. Predictive risk models are nothing more than a combination of such questions – the questions being "trained" against a predetermined dataset with the objective being to create a combination that returns the desired result. The same questions, when applied to current datasets, become a risk monitoring application. Risk reporting on the other hand poses similar questions against periodic historical datasets.

Each question or query can be further decomposed into a series of atomic operations, as shown in Figure 20.2. The operations are as follows.[1]

- *Scanning* data involves reading individual data elements one at a time, which can be likened to finding a number in a phone-book by reading each name from the beginning. Needless to say, it can be time consuming for large datasets.
- Data can be *sorted* using one or more categories (exposure in the example). Since sorting requires data to be rearranged in a particular order, it usually involves intensive movement of data.
- Datasets frequently need to be *joined* using a common identifier. For example, positions with customer identifiers can be joined to customer records, creating

[1] Note that this discussion is more illustrative than definitive. For example, the data processing activities that come into play for each operation will depend on specifics of the query and the data set.

records with combined position and customer information. Joining datasets requires sorting and usually also involves data movement.

- Data can be *filtered* using one or more criteria. Filtering requires scanning the dataset, as well as computation to determine whether a particular record fits the required criteria.
- New data can be *derived* from existing data elements and involves computation of varying complexity.
- Finally, data can be *aggregated* from existing datasets. Summation, counting and averaging are some examples. Aggregation requires scanning a dataset followed by (sometimes complex) computation.

All queries are combinations of one or more of these atomic operations chained together to generate a resultant dataset. The derived data may be processed again via a further series of steps many times before final results are generated. A Monte Carlo simulation, for example, would issue such queries against data to collect the required dataset for an iteration of the simulation, create the results of the iteration and then write the data back to disk for further statistical processing of the results. Some of these steps may involve very complex calculations. In the Monte Carlo simulation example, a complex path-dependent calculation may be involved in the data derivation step. The aggregation steps in default models for Credit Default Swaps (CDS) or Collateralized Debt Obligations (CDO) would involve far more sophistication than just simple summation or averaging.

The figure also illustrates the three data processing components that are necessary for each atomic operation. The first is the *computation* component which is responsible for performing logical or mathematical operations on data. *Disk IO* components are required for storing and retrieving data (IO stands for Input-Output). *Bandwidth* refers to networks that move data between computational and storage components. These three components form the cornerstone of data processing, whether it is done within a single computer or at the scale of a massive data center housing thousands of machines. Therefore, the data processing challenge is to make as efficient use of available computational, disk and network resources as possible.

Data processing can then be seen as a complex combination of the atomic operations being executed against a set of data using some configuration of processing components. The efficiency and speed of data processing will depend on how well a processing configuration matches a given problem. While individual steps are conceptually easy to envision, the challenges arise due to the need to simultaneously satisfy a large number and wide variety of risk management queries. These challenges have been exacerbated by factors such as increasing regulation and sophistication in analytic techniques stretching risk systems in many ways.

- Extremely large datasets are increasingly being required. For example, to analyze a portfolio of CDOs along with their underlying assets like Asset-Backed Securities (ABS), large datasets of mortgages or credit cards must be analyzed in conjunction with CDO and ABS data.
- Large numbers of users concurrently accessing the same datasets.
- Queries are becoming more complex, requiring joining disparate datasets with extensive filtering and sorting criteria.
- Timeliness of data is becoming increasingly important. Though regulatory needs still can largely be satisfied by snapshots of information, the frequency and quality of such reports is increasing.
- Risk management "use-tests" put further demands on information systems, since monitoring and response requirements must be shown to utilize the same risk metrics as those used in reporting.
- Risk information systems must be as flexible as possible, open to quickly introducing new types of data for additional analysis.
- Risk processes are increasingly becoming critical to running the business both in a tactical (i.e., to execute transactions) as well as strategic (for regulatory reporting, capital management etc.) sense. This puts a premium on these systems being highly available, with little tolerance for planned or unplanned system downtime.

These trends have put pressure on risk technology designers to offer extreme capabilities across multiple dimensions such as the amount of data stored, the speed of reading, writing and processing data, and the flexibility to ask previously unanticipated questions of data. Hitherto banks addressed these requirements by decomposing the problems and implementing them in different environments. For example, market risk computations may be done on a separate system and the results aggregated into systems to calculate economic and regulatory capital as well as systems to perform pre-trade checks. These techniques, which can be less than optimal due to excessive overhead and problems of inconsistent data, were necessary because of the historical limitations of data processing technology, and were tolerable in a static risk management environment. Modern needs, however, demand a much more sophisticated approach that requires a data processing environment with the following qualities:

- *Firm-wide, integrated* and *timely* data available to any user that needs it,
- A data environment with *integrity* that inspires confidence, and *traceability* back to the source,

- Users have unfettered *access* to data and *flexibility* to analyze it in any manner they choose,
- A data environment that is *extensible* and able to accommodate new data.

These are formidable requirements that argue for a substantially integrated data environment. The process of combining hitherto separate risk environments can place a significant burden of performance on data processing systems. The rest of the chapter will address the technical challenges to extracting the kind of performance demanded, and the modern techniques available to solve them.

20.3 Risk information architecture

The previous section described requirements of risk information. To understand how they are implemented in practice, it is useful to think of five layers of functionality as shown in Figure 20.3.

While actual architectures within financial institutions will inevitably deviate from this idealized form, and are infinitely more complex, analytic systems can be classified into one or more of the following categories.

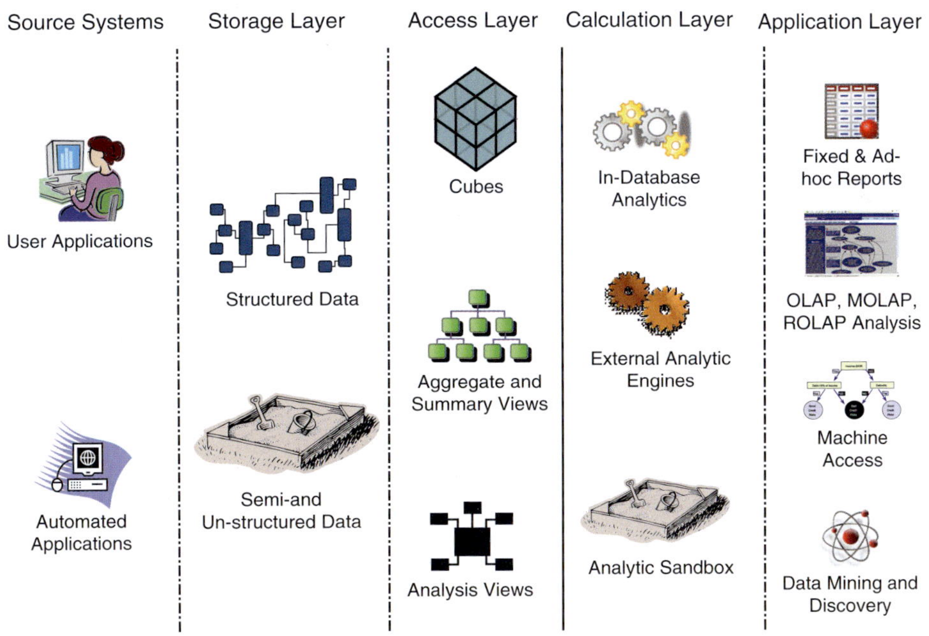

Figure 20.3 A framework for risk information architecture.

1. *Source Systems*: Data are generated as a byproduct of business activity, and are captured using source systems as encompassing and complex as firm-wide trading platforms and ERP systems, or as simplistic as Excel spreadsheets. Sources not only create new data, but also append to or modify existing data as in the case of financial posting or risk calculation engines. Sources typically create business transactions of interest, such as trades, but other sources may create reference data such as client and counterparty data or securities master data. In addition, there may be external sources of data from vendors such as Bloomberg, Reuters and CoreLogic.
2. *Storage layer*: Data from source systems must be stored for analytic processing. Rather than have a single storage area, financial institutions of size use multiple data storage systems. This multiplicity of systems is driven by many factors. Business needs, such as regulatory requirements to hold many years of data and the associated cost constraints, can argue for a separate storage system for cheap, plentiful storage. Separate systems can satisfy organizational needs for data ownership. System performance limitations are also often drivers for enterprise data to be broken up into manageable sizes.

   Since source system data cannot often be consumed as-is, some sort of transformation is usually required before data can be used. For example, data from source systems may not be clean enough for consumption (due to transcription errors in trades, for example). Another common challenge is the need to standardize data from different systems for enterprise reporting. This need for data transformation can be fulfilled in either the storage or the access layer. One approach is to store data in the storage layer in close to source system form, leaving most of the transformation to the access layer. The benefit of this approach can be reduced time and effort involved in storing data. Of course it means that the access layer requires more effort and attention, among other challenges. An alternative approach is to simplify the access layer and put more effort on standardization into the storage layer. This approach usually has the effect of increasing time and effort required to store data, but can reduce not only the complexity of the access layer but also, for technical reasons, usually reduces the storage space required in the storage layer.
3. *Access layer*: The data in the storage layer are often not in a format that can be easily used for analytics. Frequently they may simply take the shape of direct periodic data extracts from the source systems. In order to make the data useful for analytic purposes, it has to be restructured to make it suitable for data access. The access layer can be seen in many forms in a typical firm's architecture. Frequently a particular slice of data (say only positions and counterparty information) is taken from the storage layer and restructured to make it relevant to

credit risk reporting, for example. Other access layers may be represented as "cubes" – data structures that are optimized for "drill-down" or "slice-and-dice" analysis that is familiar to users of spreadsheet pivot-tables. Statistical analysis software requires data presented in analytic datasets, which can be created as extracts from the source layer. Lastly, an increasing trend is the need for data access by other systems via a so-called Service-Oriented Architecture (SOA), where systems programmatically express a desire for a particular kind of data to a software service, without specifying where it comes from. This service then determines the appropriate location to satisfy the request, translates the data stored in that location into an accessible form and returns it for consumption by the requesting system.

The storage and access layers are where raw *data gets converted to information* and therefore the heart of the analytic platform. This transformation process has led to an entire category of technologies and architectures. The plethora of solutions reflects the tensions common across analytic platforms. Different functions across an organization often share the need for the same data, albeit transformed in particular ways for their own purposes. Therefore, the need to standardize data to an enterprise view and share information competes with the other individual business functions' need for data to be viewed through their own particular lenses. A credit risk manager who cares about credit exposures may want to see positional information by counterparty risk hierarchies, while a market risk function will typically want to see the information derived from the same underlying data, but segmented by asset-class hierarchy. Decisions on where transformation will take place – the storage layer or the access layer – are influenced by a variety of factors including performance of technology components, speed of creating access layers, ability to preserve data consistency and cost.

4. *Calculation layer*: Stored data are consumed in two forms. First, information generated by the access layer can be directly consumed in the form of reports, dashboards and other applications in the application layer described below. However, for most purposes (especially risk management), data must be transformed into metrics via complex calculations. A VaR calculation, for example, is not a simple aggregation of underlying data, but could involve complex valuation and simulation processes. It is useful to think of these processes as belonging in their own calculation layer. In financial firms, manifestations of this layer are found in plenty – large banks often have hundreds of thousands of components which act as calculation engines. These range from Excel spreadsheets to stand-alone computation engines to "in-database" implementations where the storage, access and calculation layers are combined for performance reasons.

Calculations usually address a specific business need – regulatory and economic capital is calculated to address regulatory reporting and risk management needs respectively. However, there can be important ancillary uses of the resultant derived data – economic and regulatory capital are also useful inputs into relationship profitability systems. It is therefore not uncommon to see the output of the calculation layer being fed back into the storage or access layers for reuse by other business processes.

5. *Application layer*: The application layer is where information generated through the analytic process is used for business purposes. This is the layer where limit management systems, for example, make decisions on exposure reductions. There is a profusion of channels where information is used. The most basic is human consumption of information in the form of periodic reports depicted by the left-most circle in Figure 20.1. But business users also consume information in more complex ways such as On-Line Analytic Processing (OLAP) techniques where aggregate information can be quickly "drilled-down" to reveal detailed patterns. For example, this can explain which particular geography was responsible for an overall increase in consumed capital. Sophisticated consumption patterns include systems using an SOA interface to access data for further processing before delivering them to business users such as for on-line pre-trade approval. The most advanced use of information in the application takes humans out of the equation entirely (the right-most circle in Figure 20.1), letting computers make split-second decisions on rapidly changing data – this is the domain of algorithmic trading systems.

Figure 20.3 can be likened to an assembly line that converts raw data from source systems to a finished product – information – for consumption by users. Understanding the trade-offs involved in each layer is critical if one is to take advantage of the latest technological advances in designing the next-generation risk analytics information infrastructure. Such an infrastructure should be designed with the following considerations in mind.

- *Latency*: The process of converting data into information results in delays caused by data movement from one layer to another and by processing. In this context, data movement also includes all the time taken to cleanse data and transform it from the source format to one or more desired target formats. Indeed, data movement time is by far the largest contributor of latency in the processing lifecycle. While heavy processing requirements in financial institutions – like complex path-dependant Monte Carlo simulations – can sometimes take several hours, this time is usually overshadowed by data movement processes which, especially if they involve manual reconciliation processes, can easily take days.

- *Speed of implementation*: Modern business situations call for an agile response. A new risk can become critical within a matter of days and require fresh perspectives of information for analysis. The time taken to implement changes is critical and is influenced not only by technology but also by organizational aspects such as ownership of data definitions.
- *Ease of use*: Business users want instant access to the latest information, with an aggregate view that can then be drilled-down to visualize the detail to give them new and interesting perspectives. This implies data that are organized in ways that are familiar to business users so that they can directly manipulate data and have their complex questions answered easily without having to depend on a technology support to do so.
- *High quality*: Business users expect high-quality information which involves a complex trade-off between business process and technology. High-quality data require effective management of data characteristics like master data and metadata (Chapter 23). Though software tools are available (and effective) in these processes, they must ultimately be supported by a data governance framework consisting of people and processes to maintain and use the technology.
- *Cost*: This is normally a driving factor – the above must be accomplished within a (usually highly constrained) budget.

These considerations impact the choice of technology used to build the infrastructure. As discussed earlier, the technology fabric of an organization can be thought of as a combination of three primary resources: processing power ("computation"), data storage ("disk IO") and data transmission capability or "bandwidth." While it is possible with an unlimited budget to increase all three resources almost infinitely with current technology, doing so in a cost-effective manner takes careful design, which requires a deeper understanding of the current technological capabilities.

20.4 Data processing and storage

Data storage has been a feature in computing from its very early beginnings. The earliest "computers" like the Jacquard loom used punched cards to store patterns from which to create complex woven cloth. Today's complex risk management systems are direct descendants of this legacy, but with vastly larger storage capacities and faster, more powerful methods for data retrieval. The evolution of data management systems is not slowing down either; rather, the past few years have seen heightened innovation around data storage, retrieval and processing systems.

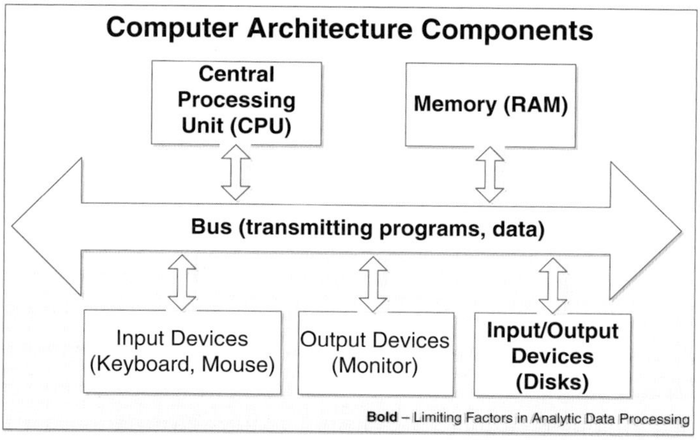

Figure 20.4 Computing architecture: a simple view.

Today's computers, fully electronic machines capable of processing vast amounts of data with high precision and speed, have a crucial flaw – their key input is electrical power. While computers lose their ability to perform calculations with the loss of electrical power it is imperative that they do not also lose their data, making the need to persist data across outages of power a critical requirement. It turns out that this is also an ultimately limiting issue in performance. To better understand these limiting factors and the techniques used to get around them, it is first important to get a bird's eye view into how a computer works. There are five basic elements[2] (see Figure 20.4):

1. A computational unit called the *Central Processing Unit (CPU)* whose function is to perform logical and mathematical operations on data that is fed to it;
2. *Random-Access Memory* (RAM) which is used to store instructions and data which are inputs to the CPU;
3. *Input Devices* that can accept data and commands;
4. *Output Devices* that distribute the results of computations; and
5. A *Data Transfer* mechanism between these components called the "bus."

While this paradigm of computing has not changed since its earliest days, the exponential growth in performance in most of the components[3] as well as the vast increase in sophistication in the input and output devices has made modern computers unrecognizable from their early counterparts. The ENIAC, the forerunner of modern

[2] This architecture is known as the Von Neumann architecture after computer scientist pioneer John Von Neumann.

[3] This growth has been long been predicted in what is known as Moore's "Law," after the co-founder of Intel, Gordon Moore, who first articulated the principle that the number of transistors on a chip can be doubled every two years. A number of digital electronic capabilities can be tied to Moore's Law.

mainframe computers, had a clock-speed of 100,000 cycles per second. In contrast, the Intel Westmere chip has a speed of 3 billion cycles per second – a 30,000 times increase over 65 years. Memory speeds have increased 600 times over 30 years. Even more dramatic are the improvements in input and output devices from the old punch-card and printer systems to today's haptic[4] feedback technology on modern phones.

Figure 20.4 models a modern data processing environment as well. While input and output mechanisms are critical components of any computer system they are by-and-large not limiting factors in the performance of risk data processing systems. Therefore the following sections focus on CPU, the storage mechanism and the bus in greater detail. They will focus particularly on their constraints with a view to giving the reader a broad understanding of the modern trends in data processing environments.

20.4.1 Computing technology

Data processing occurs in the CPU (see Figure 20.4) by executing a series of instructions of a computer *program*. Each instruction is executed in sequence to perform the basic arithmetical, logical, and input/output operations of the system. *Programmability* is a critical concept in computing – computers can execute completely different tasks simply by changing their program.[5]

CPU design has seen a tremendous amount of innovation to continuously and dramatically improve reliability and speed and reduce power consumption. The first CPUs were built using electric relays or vacuum tubes which performed the various logical steps. These progressively gave way to the transistor and Integrated Circuits (IC), culminating with the introduction of the microprocessor. Microprocessors integrate all the functions of a CPU onto a single IC, creating a multipurpose, programmable, device that accepts data as input, processes it according to instructions stored in its memory, and provides results as output.

The first microprocessors were developed in the early 1970s and have seen rapid performance improvements since then. An obvious means to increasing performance is to increase the rate of executing instructions (the *clock-speed*). Another is to cram as many transistors into a single chip as possible – today's CPUs have 100s of millions of transistors in them. A fundamental parameter of the microprocessor is the *bus size* which describes the width of the bus via which programs and data are

[4] Haptic technology is a tactile feedback technology that takes advantage of a user's sense of touch by applying forces, vibrations, and/or motions to the user (Inmon, 2005).

[5] This concept was first described by John von Neumann (1945). This attribution to von Neumann is not without controversy, however.

read and written. Bus size affects both the speed at which data can be moved and the total amount of data that can be processed. Computational requirements have accordingly driven a growth in bus size from 4- and 8-bits[6] in early microprocessors to 64-bits in modern CPUs.[7]

While these approaches to increasing performance have led to tremendous advances, natural boundaries of physics are beginning to be encountered. Therefore designers have tried to use other innovative techniques to boost the power of processors. A popular technique in today's processors is to use *multi-core* designs. A multi-core processor is simply a chip that contains more than one multiprocessor. This means that the processor's performance can potentially (but not always) be multiplied by the number of cores in the chip. Modern microprocessors such as the Intel Westmere chip use up to ten cores to deliver exceptional performance along with other techniques like multi-threading, pipelining and super-scalar execution to boost performance. Even with these improvements, a single microprocessor alone cannot deliver the performance demanded by modern analytics. Many of today's computers use parallel processing techniques to boost performance to effectively remove CPU power as a constraint in data processing. These are covered in detail in Section 20.5.

A noteworthy aspect of the development of the CPU is that, in contrast with earlier times, these techniques can be found in CPUs used in common desktops and even laptops.[8] The democratization of processing power has profound implications for data processing architecture. When CPU power is expensive, data have to be moved to the few locations where there is enough processing power to perform complex tasks. With today's powerful processors effectively eliminating this constraint, computer architects can create more natural designs for data processing.

20.4.2 Data storage technology

Data storage devices have also experienced the benefits of innovation. Computers use storage for three purposes. First, the computer needs to store the program that it will execute. Next, it needs a "scratch-pad" for intermediate results. Finally, a storage area

[6] A bit represents a single digital numeral and can exist in one of two states, 0 or 1. 8 bits can therefore exist in 2^8 states, yielding 256 possible values.

[7] There have been corresponding changes in bus length as well as width. Modern CPUs are usually too fast to be satisfied by main memory performance, so the challenge is to provide data to CPUs as fast as possible. It has become necessary to shorten the length of the bus length as well so that electrons have physically shorter distances to travel by locating data in *caches*. In multiple-CPU system configuration known as Symmetric Multiprocessing (SMP, see Section 20.5), caches are located closer to CPUs than other memory giving rise to Non-Uniform Memory Access (NUMA) architectures which are commonplace among modern microprocessors.

[8] This chapter, for example, is being written on a laptop with a dual-core, multi-threaded processor running at 2.4 GHz.

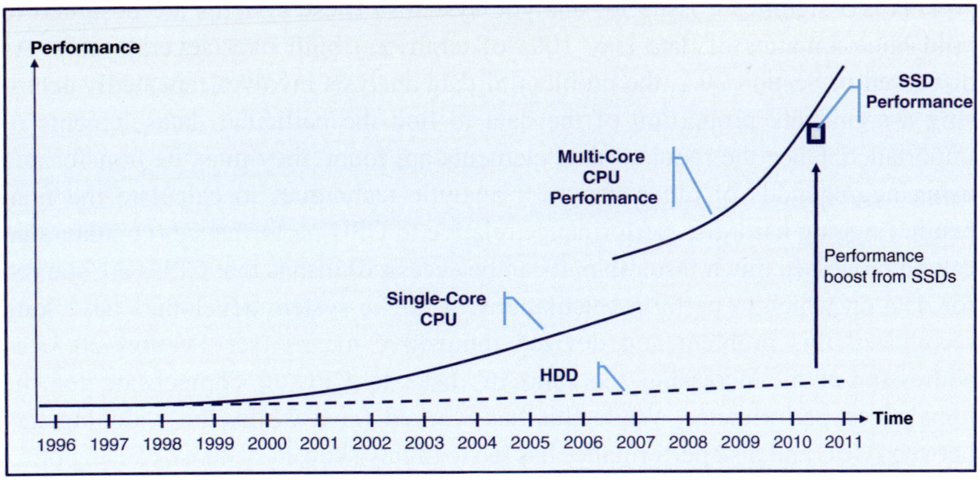

Figure 20.5 Relative performance of CPU and disk technologies.

is needed where data can be fed in from outside the computer so they can be processed and the results stored. Historically there were two key storage technologies: Random Access Memory (RAM – also known simply as "memory") and Hard-Disk Drives (HDDs or simply, disks). RAM, which is a purely electronic device, is volatile (i.e., will not retain memory through loss of power), while HDD is a non-volatile electromechanical device containing spinning disks and movable read/write heads.

Improvements in storage have come primarily in two forms – quantity of data stored and the speed at which it can be written and read (known collectively as "data access speed"). The quantity of data that can be stored cost effectively has grown tremendously. The first PC disk drives had storage capacities of a few megabytes of storage space[9] while today's smallest laptop comes standard with tens of gigabytes of disk. Unfortunately, data access speed has not been commensurate with this growth, nor has it kept pace with performance improvements in the other components in the system. This is illustrated in Figure 20.5 which compares CPU performance to that of storage technology. It is clear that CPUs have grown exponentially more powerful while the performance of hard disks has lagged significantly behind (though solid-state drives or SSDs discussed later have redressed the balance somewhat).

[9] Storage space is measured in terms of "bytes" – a megabyte is 1 million bytes, a gigabyte is 1 billion bytes while a terabyte is 1 trillion (10^9) bytes. Storage space has grown so much that the terms petabyte (10^{12} bytes) and exabyte (10^{15} bytes) are commonly used today. These terms are also commonly shortened to MB, GB, TB, PB and EB respectively.

This is a significant issue for analytic systems. These systems are designed to hold huge amounts of data (say 100s of terabytes) built over several years. As discussed in Section 20.2, the problem of data analysis involves repeatedly accessing a significant proportion of the data to find the particular data elements of importance. Once the required data elements are found they must be transformed using aggregations or other statistical analytic techniques to calculate the final result. Lagging hard-disk performance relative to CPU performance (i.e., data that can be processed much faster than it can be accessed) means that CPUs are starved for data on which to perform calculations. Analytic system developers have long recognized this problem and devised innovative means (see Section 20.5) of addressing it by increasing the ratio of disks to CPU to compensate for the unbalanced performance. While this has worked for a while, the widening gap between CPU and disk performance has led to highly skewed systems, causing other issues such as vastly increasing power and space requirements.[10]

Fortunately, a new technology – solid-state drives or SSDs – that promises to transform the landscape of analytic technology is sweeping the industry. An SSD uses solid-state memory (similar to RAM), but has the ability to store data in a non-volatile manner like HDDs. This ability to combine the best of both worlds makes SSDs an effective technology for use in analytic environments. SSDs are also many times faster than HDDs in reading and writing data[11] and generate much less heat and have lower power requirements because they contain no moving parts. The huge drawback of SSDs currently is significantly higher cost, but given trends in technology this disadvantage should disappear in the not-too-distant future.

The performance improvements afforded by SSDs mean that they have provided analytic system designers a way to better track CPU performance improvements as illustrated in Figure 20.5. While there is great interest in utilizing them, current cost limitations are forcing vendors to become innovative in introducing hybrid configurations with mixed HDD, SSD and RAM technologies. These types of technologies should be expected to become commonplace over the next 2–3 years and may revolutionize the data-storage landscape of risk analytic environments.

20.4.3 Networking technology

The third critical component of a computer system is the system bus. The function of a system bus is to move data between system components. In early computers, the various

[10] As the number of hard disks grows, so do power and space requirements. At the industrial strength scale of today's analytic environments, these factors can weigh heavily in the overall cost of operating an analytic environment.

[11] While the way in which data are read matters (for example, reading a huge block of sequential data is faster on a HDD than reading bits of data from different locations), SSDs are consistently 3–22 times faster given current technology.

components – CPU, memory, disk and other input/output devices – were connected by physical wires that were used to transfer data between them. While the advent of modern microprocessor technology has obviated the need for physical wiring, the modern notion of a computer network has transformed communication and data processing.

A computer network (or simply *network*) can be defined as a collection of computational devices sharing resources and information via communications channels. Various kinds of communications media such as copper wire cable, optical fiber or wireless technologies can be used to connect computers and other special purpose devices together. Starting with very small scale (such as connecting two computers together) in the 1960s, networks have now evolved to the global Internet which allows virtually any device to communicate with any other device via a standardized language or *network protocol*. A distinguishing property of a network – its ability to connect two or more devices together – can only be accomplished if all devices on the network adhere to a common set of standards. This can be complicated by the variety of communications media that can be used to connect to the network, a problem that led to the development of the Open Systems Interconnection model (OSI model) which defined a seven-layered framework.[12] In this idealized framework, each layer was modularized with a well-defined contract (or *protocol*) for the services that it would provide to layers above it. For example, Layer 1 – the physical layer – defined the basic electrical and physical specifications of devices and is mostly concerned with how to connect with a communications medium which could range from the common-place copper cable to something exotic like wireless media. Layer 2 – the data link layer – is then able to rely on the contract with Layer 1 and focus on the problems of transferring data. This approach has led to an explosion of network technology which has become the heart of modern data processing systems.

Network speed is important for data processing because the network is used to connect processing and data storage units to each other. A variety of network technologies have been developed to facilitate high-speed data transfer. Most current systems use different networking technologies for connecting internal components of a single computer (the common PCI interconnects in a PC, for example) and for connecting different computers together. New technologies like InfiniBand are attempting to provide one high-performance standard for both internal and external buses. The evolution is obsoleting the concept of a computer as an atomic entity separated from others by a network – effectively transforming a network of computers

[12] The framework consists of physical, data link, network, transport, session, presentation and application layers. Networking terms that may be familiar to readers can be identified in one or more of these layers. TCP, which forms the backbone of the Internet, is a transport layer protocol. Ethernet, the common wiring network used by desktops, spans the physical and data link layers. HTTP, which is the basis of the Web, is an application layer protocol.

into a potentially massive network of individual computational components that can easily and flexibly become one large supercomputing environment.

20.5 Parallel computing

Running several stress tests on a weekly basis with quarterly reporting cycles requires a thousand-fold or more increase in analytic performance over that required to run one manually consolidated, parametric test. Advancement of disk, CPU and network technologies alone – when applied to existing computing architectures – cannot deliver the massive increases in performance required for these sorts of modern analytic applications. This is especially true of disk performance. Even SSDs, which are potentially hundreds of times faster than HDDs, cannot match performance requirements of modern analytic applications. Simply put, it has not been possible for Moore's law alone to satisfy the demanding needs of modern analytic applications.

As briefly alluded to in Section 20.4.1, the answer is generally considered to be parallel computing configurations. Computer science has long depended on running tasks in parallel to speed up performance. The basic idea is to break up a task into multiple parts which can run simultaneously so that the overall time taken to complete a task is reduced by the number of computational units (so-called "units of parallelism") that are used. For example, if a task takes 1 hour to do, and it can be broken up perfectly into 60 parts that can be run on 60 parallel processors, the whole task should complete in 1 minute.

The simplest parallel configuration is one found in many modern desktops and is called Symmetric Multiprocessing (SMP). In this setup many processors access common memory and disk stores. This can potentially have a significant improvement in performance due to the ability to run multiple programs simultaneously, especially since each modern processor actually has many cores.[13] However, the actual performance can be limited when programs compete for access to shared memory and disk resources. Situations where CPUs can become starved of data are common in analytic applications that involve combing through lots of data. The result is that increases in performance do not linearly track an increase in the number of CPUs in a parallel SMP analytic system. This is shown in Figure 20.6 which illustrates processing bottlenecks in an SMP system.

In the partial parallelism case shown in the figure, bottlenecks in the SMP system can force some steps (such as aggregation and consolidation shown in the example) to run without parallelism. The implication to overall processing time is that it may

[13] A two-processor SMP machine with six cores on each processor actually has 12 computing engines running simultaneously.

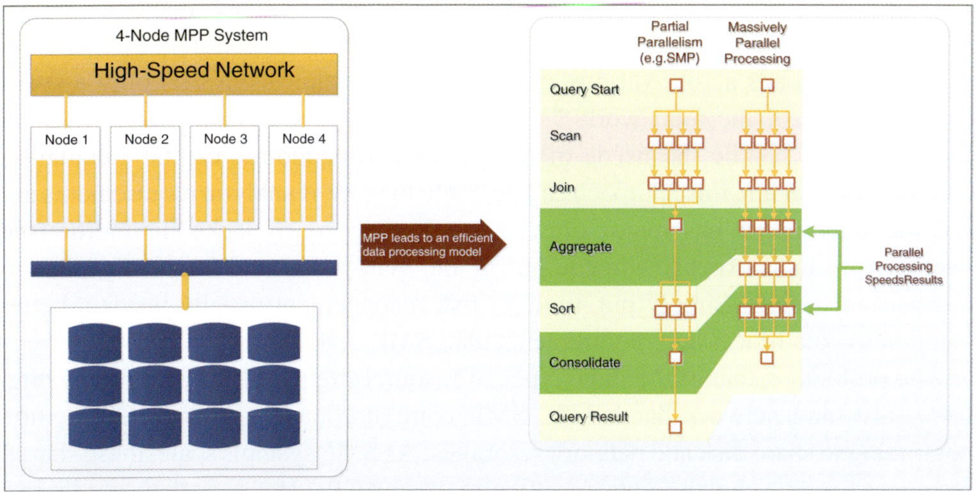

Figure 20.6 Massively-parallel processing (MPP).

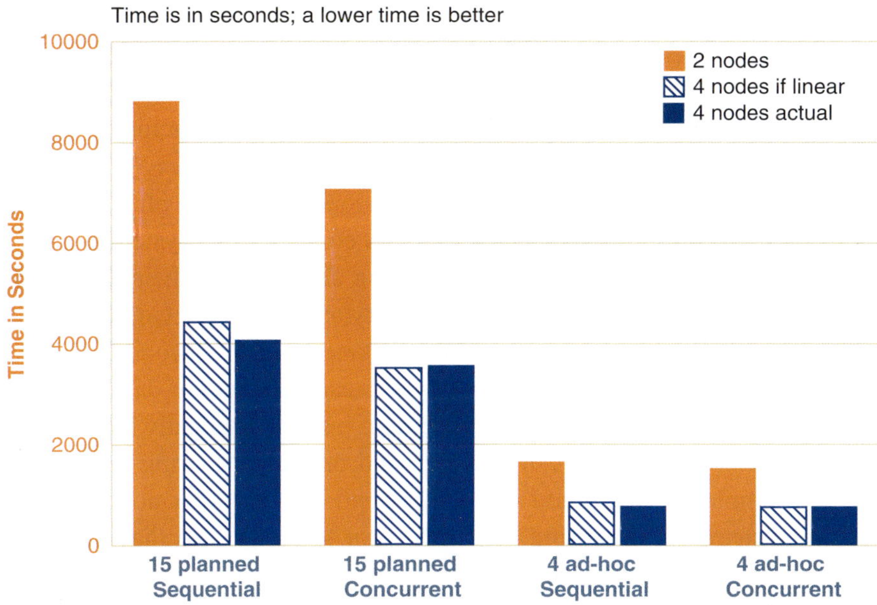

Figure 20.7 Illustration of linear scalability (illustration taken from Ballinger, 2009, Teradata Corporation).

not scale *linearly* with the amount of processing power being applied. The ideal situation may be illustrated by examination of Figure 20.7 which contrasts the time to compute a specific workload when the number of units of parallelism ("nodes") is doubled. The cross-hatched bars illustrate the target performance which is that processing must take exactly half as much time with four nodes as with two nodes.

Despite these problems, SMP solutions are commonly used in firms for analytical processing requirements. Most solutions get around SMP constraints by manually splitting workloads across various systems so that each system gets only a small enough subset of the total work.[14] For example, separate "data marts" may be provisioned to handle the needs of regulatory and economic capital calculation and reporting. While this approach can be made to work if analytic requirements are not too onerous, it has major drawbacks. The approach lacks flexibility (see Section 20.6.1), is expensive in equipment and man-power and frequently leads to inconsistent results. Simply put, modern risk management usually needs a better solution to data processing requirements than SMP systems.

The problem is that of the three types of resources required for data processing, CPU, disk and network bandwidth, SMP configurations only scale CPU while continuing to share disk and network resources. As SMP systems scale, these shared resources become choke-points, resulting in wasted CPU cycles as they sit idle and starved of data. Building a scalable data processing system needs a balanced approach which expands the scale of all the required resources simultaneously and avoids sharing any components. This is accomplished in so-called *Massively Parallel Processing* (MPP) systems. MPP systems (sometimes also referred to as "shared-nothing" architectures) string together a number of SMP "nodes," each containing its own storage subsystem and network bandwidth across a high-speed network as shown in Figure 20.6, and use specialized software to automate the splitting of data and workloads across different loads. Since the physical partitioning of data is not visible to users, they can simply connect to the system as a whole to perform data processing tasks without worrying about where their particular data slice resides. In the previous example, regulatory and economic capital users – and indeed, users who want both types of data simultaneously – will connect to the same system which would automatically know where each element of data was placed.

But smoothing workload involves more than just automatically allocating data. Compute and network power can also be used unevenly. During the month-end close, for example, financial and regulatory reporting users monopolize the data processing systems, while during the week prior to month-end they are very low intensity users. Therefore an efficient MPP data processing system will not only distribute data across its available resources, but will also distribute computing power across all its CPU resources. The difference in processing is shown in Figure 20.6. If the software performs well, each disk in the disk array will have 1/16 of the data, with each node doing a fourth of the computation. The high-speed network connecting the nodes is necessary to bring results of the computation together and marshal it to the user that has requested it.

[14] See Menninger (2011).

A properly balanced MPP system like this will not only make optimal use of its available system resources (CPU, disk, network) but can also scale reliably and predictably. Consider for example the case of a successful bank that doubles its size via acquisition. Clearly, doubling the size of data and processing workload without changing system resources will slow things down to at least half the original speed. As shown on the right-hand side diagram in Figure 20.6, it is indeed possible to double the size of an MPP system which does not have any bottlenecks and mitigate the performance impact. Figure 20.7 shows that an MPP system can closely approximate the goal of linear scalability. Note however that in most cases this will still be only an approximation; as illustrated in the figure the system behaves in a couple of cases somewhat better than predicted by linear scalability, and slightly worse in the others.

This makes the system a powerful tool to reliably predict performance and the investment required of a new application. The first data processing systems to use this technology on a wide scale were relational database vendors (see Section 20.6.1), a field in which this technology has become widely available.[15] In fact, modern MPP relational database systems can scale to extremely large sizes – the world's largest analytic systems boast 10s or 100s of SMP "nodes" strung together to form very large MPP systems that routinely analyze several petabytes of data daily.

MPP data processing has seen fresh impetus in this past decade as well. Web-based data analysis brings new issues that are not easily addressed by the limitations of relational databases (discussed in Section 20.6.2). A typical scenario for web-based companies (the so-called "dot-coms") is to rapidly analyze their customers' behaviors on the Internet via the trail of bread-crumbs they leave on their servers through the process of clicking through web-links. These "web-logs" are a treasure trove of useful information to web companies – the sooner they are processed, the faster companies can respond to rapidly changing customer needs. The challenge is that web-logs are usually extremely large – terabytes of new data per day are not uncommon. They are usually useful only for a short period of time, after which their utility drops precipitously. While modern relational databases can deal with large volumes of data essentially without limit, they do require time up-front for data preparation.

Therefore web companies have found it advantageous to develop bespoke massively parallel architectures to their particular problems. This is an area of tremendous innovation. Additionally much of this is open source[16] which allows for any interested company to use or extend the technology. Many dot-coms have

[15] See Gartner Research (2012) which discusses a number of MPP data warehouse platforms among others.

[16] Open-source software is computer software that is legally available in source code form allowing users to study, change, improve and at times re-distribute the software. Further details can be found at `http://en.wikipedia.org/wiki/Open-source_software` (downloaded March 25, 2012). For an interesting discourse on the open-source movement, see Raymond (2001).

taken advantage of the situation to refine these technologies for their particular situation, in the process rapidly advancing the whole field of data processing.

A well-known example of such a technology is MapReduce,[17] which since its inception has been rewritten via an open source project called Hadoop (see Section 20.6.2 for more detail). The Hadoop system can use 100s or 1,000s of nodes, each with its own memory and disk storage, which are connected to other nodes by a network. In this respect it is very similar to traditional MPP relational database systems as shown in Figure 20.6. MapReduce does not, however, require data to be structured in a systematic format (see Section 20.6 for a more detailed description), which means that web-logs can be analyzed without lengthy data processing. Data are processed via special-purpose programs called *Map* and *Reduce* – hence the name – which can be written in a variety of common computer languages such as Java, Python and Perl. Hadoop is a rapidly maturing product.

Each MPP approach has advantages and drawbacks. The long preparation times imposed by relational database systems are offset by the ease of retrieving and processing data from such systems for typical applications. While MapReduce systems virtually remove the need to prepare data, the specialized programming required to retrieve data from them make them unsuitable as general purpose data processing systems. The approaches are also converging. MPP relational database vendors are actively working to incorporate MapReduce techniques into their systems, while the open-source community is developing data processing and retrieval interfaces that obviate the need for bespoke programming.

It is clear that analytic data technology has come a long way in the past several years. Regardless of the direction that future refinements take, MPP processing systems are here to stay. The good news for developers of risk applications is that the rapid development of large-scale data processing systems means that complexity and data-size are no longer constraints to performing the kinds of data analysis that are needed.

20.6 Data structuring techniques

The physical structure of data on disk matters greatly both to technologists and data users, though they often have conflicting needs. Technologists care about the speed at which data can be ingested, processed and retrieved by the system. They also care about how quickly the system can be modified to adapt to new data requirements. Users, on the other hand, care about ease of use. They would like to state their request for data processing as simply and precisely as possible and have their requests satisfied promptly.

While storage technologies have improved in performance (see Section 20.4.2), the pattern of data on disk is still an important factor in how fast queries can be answered.

[17] See Dean and Ghemawat (2004).

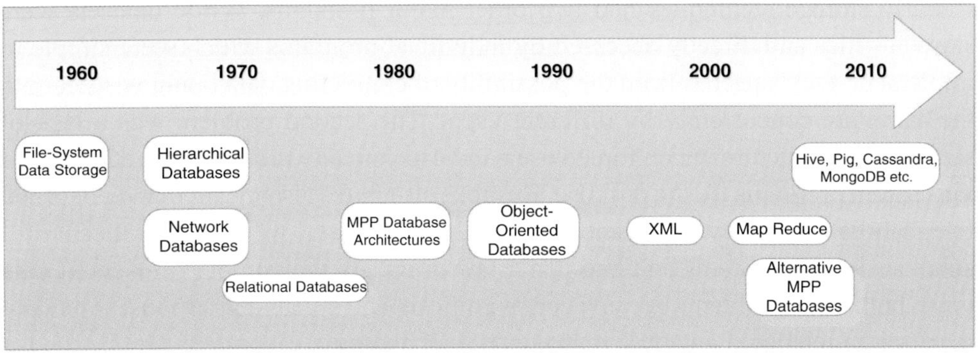

Figure 20.8 Chronology of database architectures (illustrative example).

The simplest way to improve response times is to reduce the amount of data read from disk, and to read it in fewer, bigger blocks.[18] However, reading few big chunks of data does nothing to address the "searching for a needle in a haystack" problem that is central to analytic needs. Users need to relate datasets to each other (e.g., counterparties to exposures), and need the flexibility to ask unanticipated questions without compromising response time. These competing requirements of users and technologists have led to the development of data structuring techniques. Figure 20.8 shows the profusion of innovation that such requirements have led to. It must be emphasized that this list is only illustrative, perhaps characterized as much by omission of many techniques as by inclusion of the ones that are discussed in this chapter.

Datasets were originally structured as simple lists of data on disk, much like writing information into a ledger which is extremely inefficient. Computer science has developed techniques like tree[19] and index structures to efficiently search for and access specific items of data. A tree structure assumes that data are organized much like the branches and leaves of a tree with each leaf representing a block of data. If one starts at the root of the tree, and knows which branch to take at each step, it is much faster to get to a particular leaf of a tree than examining each leaf. Index structures speed data retrieval analogous to the index at the back of this book. While these techniques work well in certain situations, they do not satisfy the flexibility requirements of analytic data processing (as many a frustrated reader has discovered when stumped by an index that does not have the particular item that she is interested in).[20]

[18] The necessity of reading big blocks is only a limitation of HDDs, not solid-state disks.
[19] See Bayer and McCreight (1970).
[20] These techniques are collectively known as IO reduction. This is important since performance of hard disks has not kept up with CPU. Many innovative solutions have been developed to address this problem over the years including indexing, materialization, partitioning, compression, column orientation and using field-programmable gate arrays to push IO filtering closer to the storage. These solutions can still be relevant in MPP systems to push the limits of performance.

Early storage techniques had two other major problems. While datasets were stored as files and directly accessed by individual programs which were simple to implement, they suffered from the possibility of conflicting data being written onto the same file concurrently by different users. The second problem was a lack of standardized structure which forced users to be technically proficient to retrieve data via custom programs for the purpose. Bespoke structures also meant that readers had to collaborate closely with writers of data. These problems were not significant on a small scale with systems that had just a few users. As the number of users of data grew, both these problems became very significant and demanded a robust solution – resulting in relational database management technology.

20.6.1 Relational database systems

A Database Management System or DBMS marshals requests for reading and writing from different users. In this way it can prevent conflicts between users performing various tasks and preserve the fidelity of data on disk. The problem of data access and standardization was solved by the application of *relational* technology to databases. The combination of these two technologies forms the basis for Relational Database Management Systems (RDBMS)[21] which are the dominant systems used for storing and retrieving data today.

Relational databases as based on four primary concepts, tables, columns, rows and relationships, illustrated in Figure 20.9 via a diagram known as a *data model*. Four *tables* for counterparty, position, security and trade are shown. Each table contains data organized in *columns* – for example the security table contains columns for security identifier (such as CUSIP or ISIN), security name, and numerous columns for details of the security. *Rows* of data are stored in each table, each row one data element for each column. *Relationships* between tables are established via identifiers (also known as *keys*). For example, the position identifier is the key by which trades are related to positions. In this way a large amount of data can be structured to simplify access. Other rules can be added – for example to enforce the concept that while a position can contain many trades, a single trade must belong to one and only one position. A formal, standardized technique for structuring data in this manner brings the enormous benefit of being able to define a standardized data access language. That language is known as *structured query language* or SQL (pronounced "sequel").

SQL makes data access simple and efficient. To understand how this works, consider answering the following question from the data structure in Figure 20.9:

[21] The seminal paper by E. F. Codd (1970) spawned the entire relational database revolution via the INGRES database product which is the forefather of most modern RDBMS systems.

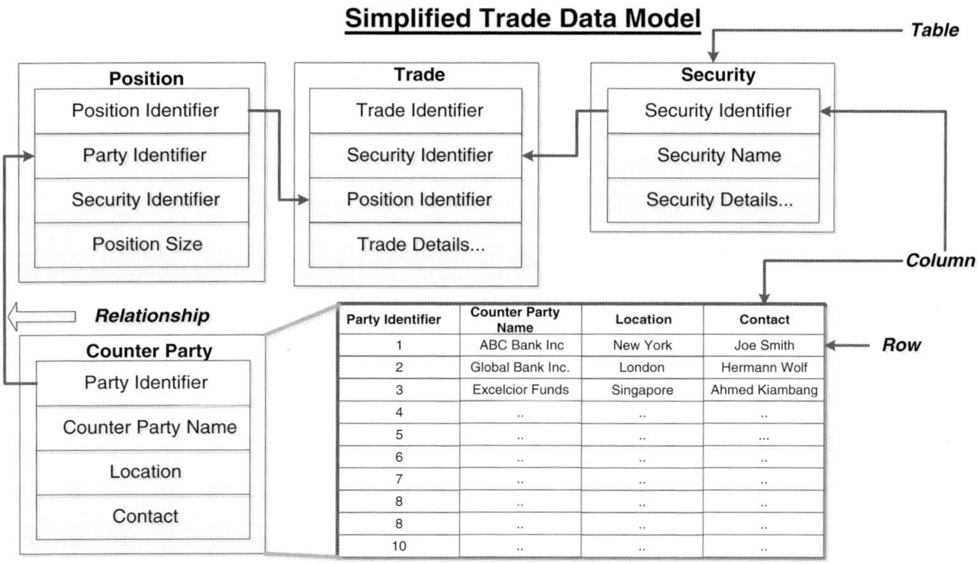

Figure 20.9 Example of a relational structure.

"Give me the aggregate position size for a specific counterparty." The procedure that may be used by a data processing system is as follows.

1. Scan the counterparty table until the specific record of interest is found. Store the party identifier for later use in a scratch-pad (e.g., in memory).
2. Read the position table in its entirety and sum exposures where the party identifier in the record matches the stored party identifier.
3. Return the sum to the user.

Improvements to speed up processing could of course be made in the way in which data are accessed, for example by using an index to speed up access to a particular counterparty.

Older non-standard data structures required users to write a customized program to execute the steps as described above. This *imperative* style of data access meant not only that users needed to be technically proficient but also that they needed to reinvent performance optimization techniques each time they accessed data, making queries expensive to write in terms of time and effort. In contrast, SQL uses a *declarative* style of access to simplify the user's experience to accessing data. In SQL the user simply states how the data should be returned. For example, in SQL a user only needs to state the above question in a formal syntax thus – "SUM (Position Size) FROM Counterparty and Position WHERE Counterparty->Party Identifier = Position->Party Identifier" – a form that even a casual reader can easily

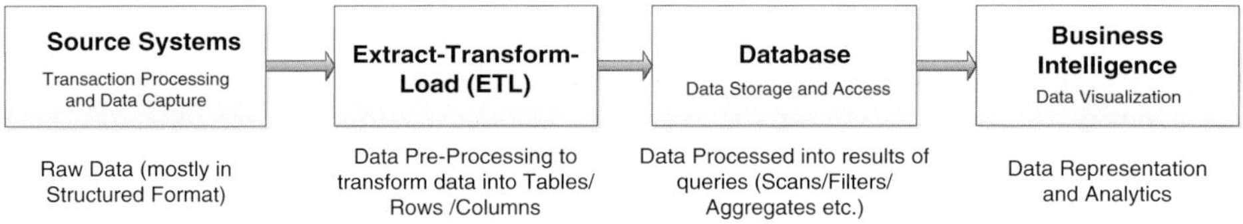

Figure 20.10 Analytic data processing assembly line.

understand.[22] Aside from simplifying user access, this approach also allows the database to figure out how to actually search and retrieve the data, giving it the freedom to innovate on ways to optimize data search and retrieval techniques. For example, an index created to quickly find counterparty identifiers in the position table (conceptually similar to an index in the back of a book) could be reused by other queries that list all trades for a given counterparty.

Relational technology has now enjoyed more than 40 years of maturity. Its near universal use has spawned an entire industry. Aside from database technology, which has itself had a tremendous run of development, specialized tools have been developed to pre- and post-process data. These are illustrated in Figure 20.10 as a "data processing assembly line." A whole sub-industry, called Extract-Transform-Load or ETL, is focused on perfecting solutions to transform data from transaction processing systems into a relational format. On the other end are Business Intelligence or BI tools, which transform data from the database into reports or other, more complex, visualizations. BI technology can also encompass complex statistical processing.

Over time, databases have specialized for various uses. Specialized On-Line Transaction Processing (OLTP) databases are built for processing a large number of transactions in real time. OLTP systems are usually seen in front office and trading systems. These optimizations go counter to the needs of analytical processing with its emphasis on aggregation, filtering and data derivation (see Section 20.2). To overcome the shortcomings of using OLTP systems for analytic techniques, computer scientists developed so-called "dimension data modeling" techniques.[23] Such databases built specifically for analytical purposes are referred to as Decision Support Systems (DSS) and address the needs of a particular business function. For example, credit exposure systems can be built to answer a range of questions around a bank's counterparty credits. Functional specialization allows technologists to increase performance from the database by specifying a particular view of data – in this case that of the credit risk manager – at the expense of other potential uses. A liquidity risk manager would find it very

[22] Note that actual SQL syntax is a bit more formal than that shown here, but not by much.
[23] Kimball et al. (2008).

difficult (and slow) to extract the information required from this sort of data structure. Of course one can create a new dimensional model that is focused on liquidity risk analysis. It is customary to refer to such specialized systems as Data marts.

Data marts are used in most large organizations. They are simple to develop and can be made to perform adequately. However, they are not flexible enough for situations with rapidly changing requirements. For example, combined credit and liquidity stress testing will likely require developing yet another data mart that combines data between two previously developed data marts. Over time, this can lead to an expensive data environment in software, equipment and manpower. Even more importantly, it is very difficult to keep data synchronized across various data marts, resulting in inconsistencies in analytics across data marts.

The simple solution is to create a single, reusable repository of data that can be used for multiple applications using "normalized" data modeling techniques. To contrast this larger scale effort from a data mart, the resulting system is called a *Data warehouse*.[24] Data warehouses integrate data from source systems into a single repository usable by many applications through a technique called *normalized* data modeling to distinguish it from dimensional modeling used in data marts. Normalized models have traditionally been difficult to use because they perform poorly on OLTP database technology. Initial implementations of normalized data warehouses also take longer and are more expensive than data marts, though they begin to offer unbeatable economies of scale after a certain point. For these reasons, companies have traditionally shied away from building data warehouses. Recent years have, however, seen practitioners combine normalized data models with MPP relational database technology (see Section 20.5) to develop an exciting class of new data warehouses that contain much of the data within an enterprise and are used by a large number of users across numerous business functions.

20.6.2 The age of Big Data

Though relational technology has revolutionized databases, it suffers from some significant drawbacks. The restriction that data are structured in rows and columns leads to unnatural data processing paradigms for some important applications where data have a naturally hierarchical (such as legal entity hierarchy) or graph structure which is integral to social media applications.

Another problem with traditional databases is that each row of data in a table is typically stored sequentially on disk. If the number of columns in a table is large

[24] See Inmon (2005) and numerous articles on data-warehousing at `www.inmoncif.com/home/cif-resources.php`

relative to number of columns of data that are actually required to be read, then a large amount of data will need to be read only to be discarded. This has led to the development of "column-oriented" databases where data are stored as columns rather than rows. Column-oriented database structures are particularly useful in applications such as retail credit analysis which might analyze a set of accounts with many of its attendant attributes, leading to very wide records.

Perhaps the biggest drawback is that naturally occurring data are often not structured in rows and columns, but occur in less formally organized forms. Data that have no formal, established structure (such as emails which may be important to analyze in an operational risk assessment) are called *unstructured data*. More common are *semi-structured* (also known as *machine-structured*) data that are often produced by computer applications in web-logs (see Section 20.5). Study of this sort of data is critical to many applications, ranging from web-marketing to online fraud. With relational databases, data in semi-structured or unstructured form must be converted through a time-consuming[25] process into fully structured tables and columns. Driven by the explosion of semi-structured data in Internet companies such as social networking sites to rapidly analyze constantly changing Web-usage patterns, new technologies with exotic names like HBase, Cassandra, MongoDB and CouchDB are being developed.

These technologies parallel the development of data access technologies like Hadoop discussed in Section 20.5. That section referred to the fact that Hadoop programs usually do not require any preprocessing. The reason for this is that basic querying in Hadoop involves writing programs to read and understand the underlying data structure before processing, allowing data to have *any* structure. In other words, where data processing prior to analysis in relational databases is divided into two phases (ETL and Database in Figure 20.10), in Hadoop it is combined into a single program. Data warehouse preprocessing results in data that can be queried very quickly, but the price for this is paid in ETL development cost and time.[25] This cost of developing ETL can usually only be made worthwhile if it is amortized over many (typically millions) of queries. The early Hadoop use-cases used data that had a limited half-life, so combining these processing steps and paying a smaller price in terms of specialized programs for each analytic was worthwhile.

As Hadoop and related technologies have gained popularity, they have been increasingly used in situations where the same dataset is queried repeatedly, making standardization important. Technologists have responded by developing databases that store standardized data but in formats that differ from the traditional table/row/column scheme and use languages other than SQL for querying them, resulting in

[25] Indeed, studies have shown that over 60–80% of the time taken in projects to create relational databases is devoted to this effort, see Strange (2002).

their being called "NoSQL" technologies. This is a fertile area of development with each of these technologies using distinct approaches. We will examine HBase[26] as an illustration of how these techniques differ from SQL, though it bears mentioning that by no means has HBase (or any other technology) achieved the kind of dominance that RDBMS have come to enjoy, at least not yet.

HBase, like Hadoop itself, is an open-source implementation of a concept originally developed at Google. As a remedy to the problems caused by Hadoop's lack of data structure, engineers at Google came up with a structure they called BigTable.[27] Unlike a relational structure (which the reader will recall is in table/row/column format), BigTable and HBase are described as a "sparse, distributed, persistent multidimensional sorted map." Deciphering this seeming gibberish requires understanding each word in the phrase. *Distributed* and *persistent* simply mean that HBase is expected to run on an MPP System (Hadoop) with persistent HDD technology.

The rest of the explanation refers to Figure 20.11, which shows a table in HBase storing counterparty information. In computer science, a *map* is a structure where a value can be retrieved using a key (e.g., the key for ABC Inc. is "Name"). HBase is a *sorted* map where keys are kept in alphabetical order, which results in important performance improvements in distributed systems (i.e., ABC Inc. is stored before DEF Inc. etc.). HBase distinguishes itself from relational databases in its *multi-dimensionality*; rather than each table strictly enforcing a two-dimensional row/column structure as in Figure 20.9, HBase can contain four dimensions with *rows, column families*, *columns* and *versions*. In the example in the figure, "Party Information" and "Exposure" are column families. Each column family can have an unspecified number and type of columns. For example, swap and FX exposure information can be stored for ABC Inc. and XYZ Inc. while DEF Inc. stores no information. This *sparse* data structure efficiently stores information only where needed. The final detail is that multiple *versions* can be stored for a single column. Accessing a particular value requires specifying the full set of information for table/row/family/column/version.

This structure stores this information in more natural format than relational databases. The information stored in one table in HBase will require three tables (counterparty information, party information and exposure) in a relational database for instance. However, it does come with drawbacks as well. A detailed comparison is outside the scope of this chapter, but more information can be found in the literature.[28]

While this section has presented relational database and NoSQL technology as distinct streams of development, there is a lot of effort underway to transplant the leading features of each technology into the other. For example, there are fairly

[26] This treatment borrows from Wilson. [27] See Chang et al. (2006). [28] For example see Gray (2008).

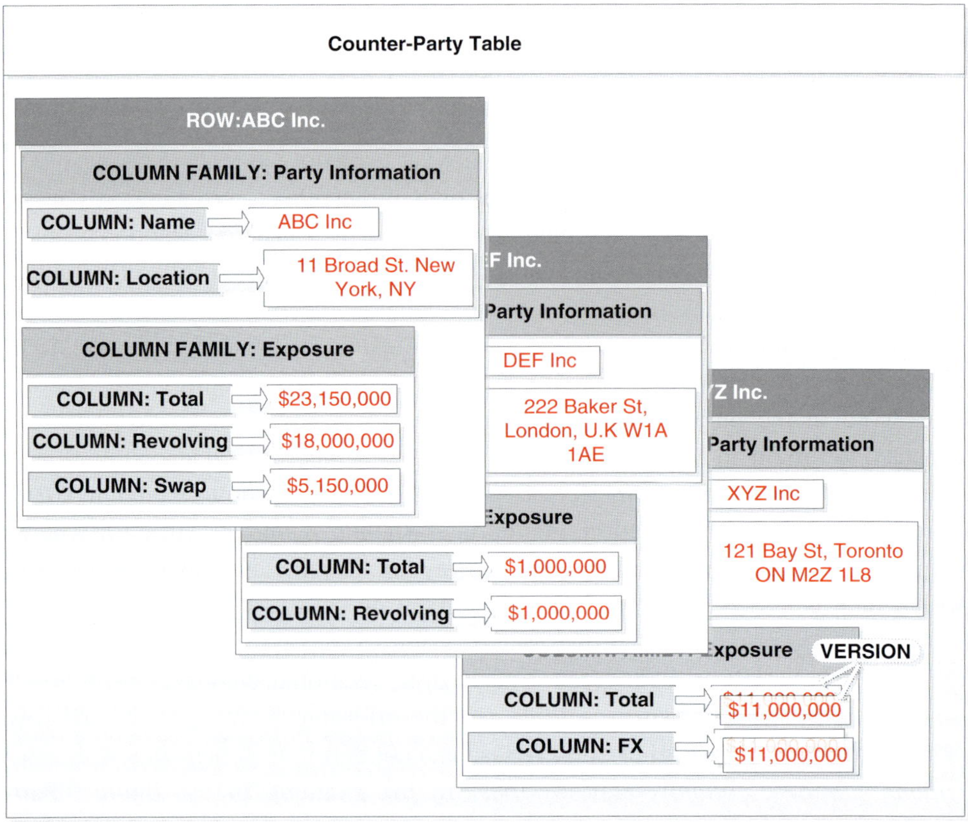

Figure 20.11 Storing counterparty information in HBase (illustrative example).

mature open-source products such as Pig[29] and Hive[30] that allow users to employ SQL or other simple languages to query Hadoop databases. Conversely, many relational data vendors are integrating MapReduce technology into their databases, enhancing the ability of relational technology to quickly analyze semi-structured or unstructured data.

20.7 Conclusion

As risk management increases in sophistication, increasingly stringent requirements are imposed on data processing systems. Fortunately, data processing technology finds itself in a moment of profound change. Processing power delivered by the

[29] See Olston et al. (2008). [30] See Thusoo et al. (2010).

various computational components is increasing significantly. Simultaneously, there is a revolution in parallel processing occurring that is bringing hitherto unimaginable amounts of data processing capability within the reach of almost any firm. These changes are happening both in the traditional relational database community as well as within the open-source movement. The numerous alternatives that are now available can confuse any professional interested in developing risk analytic systems. However, careful analysis of the available alternatives and their suitability to the requirement at hand can be very rewarding since given the present state of data processing technology there is almost no advanced risk management requirement that cannot be met.

Acknowledgements

The author would like to express thanks to the following people for their help in reviewing drafts of this chapter: Tom Manfredi and Bob Fintel. Any remaining errors are the responsibility of the author's alone.

Bibliography

Ballinger, C., 2009, The Teradata Scalability Story, Teradata Whitepaper,.

Bayer, R. and E. McCreight, 1970, Organization and Maintenance of Large Ordered Indices, Mathematical and Information Sciences Report No. 20, Boeing Scientific Research Laboratories, July.

Chang, F., J. Dean, S. Ghemawat, W. C. Hsieh, D. A. Wallach, M. Burrows, T. Chandra, A. Fikes and R. E. Gruber, 2006, Bigtable: A distributed storage system for structured data, OSDI'06: Seventh Symposium on Operating System Design and Implementation, Seattle, WA, November.

Codd, E. F., 1970, A relational model of data for large shared data banks, *Communications of the ACM*, **13**(6), 377–387, available at `www.seas.upenn.edu/~zives/03f/cis550/codd.pdf`

Dean, J and S. Ghemawat, 2004, MapReduce: simplified data processing on large clusters, OSDI'04: Sixth Symposium on Operating System Design and Implementation, San Francisco, CA, December.

Gartner Research, 2012, Magic Quadrant for Data Warehouse Database Management Systems, February 6.

Gray, J., 2008, Hadoop/HBase vs RDBMS, Presentation to Los Angeles CTO Forum on December 12, downloaded July 30, 2011 from `http://wiki.apache.org/hadoop/HBase/HBasePresentations`

Inmon, W. H., 2005, *Building the Data Warehouse*, Fourth edition, Wiley.

International Society for Haptics, Haptic Technology: An Animated Explanation, at `www.isfh.org/ch.html`

Kimball, R., M. Ross, W. Thornthwaite and J. Mundy, 2008, *The Data Warehouse Lifecycle Toolkit: Expert Methods for Designing, Developing, and Deploying Data Warehouses*, Second edition, Wiley.

Menninger, D., 2011, Secrets Revealed in Massively Parallel Processing and Database Technology, downloaded July 4, 2011 from http://davidmenninger.ventanaresearch.com/2011/01/19/secrets-revealed-in-massively-parallel-processing-and-database-technology/

Neumann, J., 1945, First Draft of a Report on the EDVAC, Moore School of Electrical Engineering, University of Pennsylvania, June 30, downloaded March 25, 2012 from http://qss.stanford.edu/~godfrey/vonNeumann/vnedvac.pdf

Olston, C., B. Reed, U. Srivastava, R. Kumar and A. Tomkins, 2008, Pig Latin: A Not-So-Foreign Language for Data Processing, downloaded July 30, 2011 from http://citeseerx.ist.psu.edu/viewdoc/summary?doi=10.1.1.124.5496&rank=1

Raymond, E., 2001, *The Cathedral and the Bazaar*, O'Reilly.

Strange, K., 2002, ETL Was the Key to this Data Warehouse's Success, Technical Report CS-15–3143, Gartner.

Thusoo, A., J. S. Sarma, N. Jain, Z. Shao, P. Chakka, N. Zhang, S. Antony, H. Liu and R. Murthy, 2010, Hive – A Petabyte Scale Data Warehouse Using Hadoop, downloaded July 30, 2011 from www.cse.ohio-state.edu/~zhang/Hive-ICDE10.pdf

Wilson, J. R., 2008, Understanding HBase and BigTable, downloaded July 30, 2011 from http://jimbojw.com/wiki/index.php?title=Understanding_Hbase_and_BigTable

21

The extracting, transforming and transmitting of data

Martijn Groot

21.1 Introduction

Accurate and actionable information is a necessary condition to any capital markets activity, whether it means doing a trade, informing an investor, selling a product or complying with regulation around the instrument and transaction lifecycles. The absence of easily accessible, consolidated information has caused enormous problems in reporting and insight into the overall state of the business.

Product proliferation and regulatory demands continuously foster the need for new vendor content and software products. These need to be integrated with existing infrastructures in turn. This is one of the reasons why, if we look at an institution's data architecture from a bird's eye perspective, it typically looks like a maze of applications, databases, reports, data feeds and point to point interfaces. New reports often get bolted onto existing processing, new data feeds are plugged in to feed existing applications and the processing in between more often than not is a mix of legacy systems interspersed with vertical silo applications.

Between the places of origin of financial information such as market makers, research agencies, exchanges and issuing parties and the ultimate users (whether end users or business systems), there typically is a convoluted information supply chain. This supply chain includes different intermediaries between the ultimate source and the point of consumption, including information aggregators, validation, data enrichment and distribution functions. Financial services firms realize they need to more actively manage this supply chain and devote as much attention to their data and the quality of their data as they do to other assets. The basis for all decisions is data and bad data means suboptimal decisions. Just like any other asset, information needs to be looked after.

Handbook of Financial Data and Risk Information, ed. Margarita S. Brose, Mark D. Flood, Dilip Krishna and Bill Nichols. Published by Cambridge University Press.

Any financial institution usually sources content from many different providers, including public sources, customers, trading counterparties, data companies and service providers such as fund administrators and custodians. The number of providers is correlated to the size and diversity in business activities, geographies, client types and financial products. Consequently, it has to perform substantial internal integration and quality assessment work. After purchasing, the information often has to be manipulated, relabeled, translated, cross-compared to other sources or internal information before it ultimately reaches users or applications where it can be acted on. After that, information is typically reformatted and collated again until it supplies information and statistics to external stakeholders that include shareholders, customers, trading counterparties and regulators.

The more intermediaries in the supply chain, the larger the time delay and the higher the potential for error. Unfortunately, all too often information arrives too late, incomplete or degraded at the place where it is needed. Due to frequent reformatting, information can get lost in translation. Effective content management is about getting the right content to the right place at the right time – for the lowest cost. Through this process it is important to keep track of an audit trail so it is always clear who touched or changed what.

Transparency as to *data lineage*, i.e., the origin of data elements used in pricing and reporting, is increasingly important for audit and regulatory reasons. The requirements in terms of operational due diligence are increasing, meaning the whole supply chain will be under greater scrutiny as a result of regulation and investor and client demands.

21.2 Different needs of various stakeholders

At the end of the information supply chain, there are different types of users. Data are needed at every point of the typical transaction lifecycle. However, the demands that different users have on information may differ. There are different quality and metadata aspects to information including speed, accuracy, completeness, retention and so on. A summarized supply chain is shown in Figure 21.1.

In a pre-trade environment, speed is of the essence. Typically the users will have a lot of context knowledge and are interested in getting up-to-date market prices with a minimum of delay.

In a research environment, completeness of data is important. By this we mean having, for example, an overview of all the terms and conditions of a security,

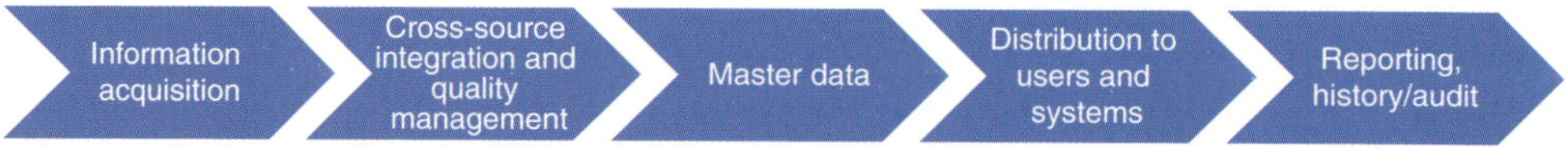

Figure 21.1 A summarized supply chain.

insight into the complete set of financials of a company or access to the complete universe of outstanding treasury bonds in fixed income research.

In a post-trade environment for clearing and settlement, it is critical to have unique and unambiguous identification of trading counterparties and instruments. Without accurate settlement instructions or without clarity as to which product has been traded, there can be failed trades and corresponding costs.

In a custody environment, a timely supply of corporate actions information is required. Whenever there are choices to be made (for instance between a stock or a cash dividend, to vote on shareholder meetings or to agree to a tender offer), the owners of the securities need to be informed well in advance and their responses need to be processed in time.

For regulatory reporting, the retention of information as well as the lineage is important. Firms need to be able to show where the information they used came from and what quality checks it has undergone. These criteria also hold for client reporting, for example performance attribution reporting.

The table summarizes some of the key characteristics of data quality at different points in the cycle.

Business area	Critical success factors on data	Potential ramifications
Pretrade	Speed	Missing out on business opportunities
Clearing and settlement	Unambiguous identification of counterparties and instruments	Failed trades, delays
Research	Completeness	Inaccurate research, wrong recommendations
Custody	Completeness, timeliness	Claims
Regulatory reporting	Accuracy, data lineage, retention	Fines, reputational damage

In the information supply chain, the recurring trade-off is that between flexibility and agility versus control. Because it is not always possible to have a single supply funnel that services everybody, there tend to be different information streams. Real-time data supply to sales and trading is separate from the supply of reference data to the back office. This in turn impacts data governance and the organization of data management functions.

21.3 A closer look at the information supply chain

The supply chain extends on both sides of the actual entry point of the information into the organization. Costs for the information from third parties often represent only a small part of the total costs of supplying users with the information they need.

The external part of the information supply chain includes the ultimate sources, aggregators, validators, enrichers and distributors and can also include, for example, position statements supplied by fund servicing companies and trade confirms by counterparties. As an institution often sources content from many different providers, it has to do substantial integration and quality assessment work. These steps of integrating, cross-comparing and enriching the original information purchased normally take place within the financial institution, although some have taken steps to outsource some of this.

In the section below we discuss the main stages in the supply chain including original creators, display, aggregators, utilities and the internal supply chain up to and including deployment to the point where data influences decision.

21.3.1 Ultimate sources of data

Financial information originates from new product issuance and trading and research activity. This information comprises the terms and conditions of new securities, the legal form of issuing entities, the prices given by brokers or from exchanges and the research opinions from rating agencies for example. Information can be created and come in from

- Portfolio statements
- Custodian information
- Trade confirmations from counterparties
- Legally required information, for investor disclosure, annual reports or prospectuses
- Government research mandate, e.g., national statistics agencies that produce macro-economic indicators
- To conduct or attract business, e.g., quotes and indication of interest through brokers
- Contract definitions from an exchange that launches new products
- To support doing business or to govern transactions, e.g., settlement instructions/ account information and master agreements that provide the legal framework around OTC trading
- Analyst and research firm data, e.g., ratings, macroeconomic or equity research, market analysis.

After creation, content can be sold directly, it can be posted on an order book or bulletin board in case of quotation data, and in many cases it is sold indirectly and bundled, aggregated and packaged with information created by other institutions to create a bundled content product.

The increasing use of standards such as XBRL could streamline the first part of the supply chain as issuers or their agents get more involved in supplying information about new financial products in a standard electronic format. On top of that, there are also increasingly solutions to help source and distribute the creators of for example structured product and fund information. With regard to information quality issues, it is always cheapest in the long run to get it right the first time and to fix issues at source.

21.3.2 Information aggregators: the financial content industry

Because financial information originates in many different places, it is very costly to pool it together. This function is done by data aggregators who directly source it from hundreds of exchanges, corporations, agencies, courts, brokers, newspapers, exchanges, chambers of commerce and so on.

The function of these aggregators is to bundle and pool information on a large amount of financial products and to standardize the format and presentation of the content. Typically the delivery methods range from standard "one size fits all" file products to sophisticated "slice and dice" options where clients select (sometimes interactively) what pieces of information about which financial instruments they want to receive and when they want to receive it. Increasingly, content providers offer more complete integration solutions such as software development kits (SDKs) and more flexible access to their content through web services to allow for example reports and business applications to tap into the required data as needed. This transition from bulk products to *information as a service* also has implications for the billing and metering of content usage.

Aggregators keep track of what they aggregate for who, i.e., the entitlements to information and attribution of revenues so that the original owners of the data such as exchanges can verify where their content has been distributed to and are compensated accordingly. The extent to which aggregators have to *interpret* information supplied to them in different formats and encoding conventions determines the level of potential value added, but also of subjectivity. It can be important to pass on the basis of these interpretations within the overall aggregated content product.

The ways in which commercial content is made available include the following.

- *File Based*. Here a vendor has split up its product over one or more files which can be retrieved by the user, for example through ftp downloads. Normally the files have clearly defined structures and sometimes files are specially cut for clients to reflect for example their investment mandate and portfolio coverage. In that case there is flexibility in selecting which information fields are included and which objects (instruments, companies etc.) are included. This selection can take place through for example a web browser or through supplying a shopping list file to the vendor. These are also called *portfolio type* products. A disadvantage of files for vendors is that they can be easily copied and redistributed without the vendor knowing.

- *Reports*. This is a special case of a portfolio type product in which information is requested and typically paid for on a piecemeal basis, for example a particular credit report, a particular analysis of a certain industry sector.
- *Streaming data*. Here a client retrieves a continuous price stream, either direct from exchanges via a direct market access (DMA) feed, through ticker plant software, through aggregators or also often directly integrated into software products such as trading or market making systems.
- *SDK/APIs*. In this case, applications at the user site can directly access vendor content through using the vendor's application programming interface (API). The API will outline a number of calls in which information can be retrieved. As data have to be requested and directly end up in applications, data usage can be more easily measured and controlled and the developer using the API gets a toolkit to quickly integrate the required data into business applications. SDKs and APIs help to address the increasing need for right data at the right place at the right time.
- The case of *Web services* is conceptually similar but makes use of a more standard architecture using XML and http. Client applications can invoke a vendor service to request data. For the vendor, this has the same advantage as far as keeping tabs on data usage is concerned.
- *Hardcopy*. The provision of paper reports or books, for example credit reports or handbooks containing listings of persons or companies.

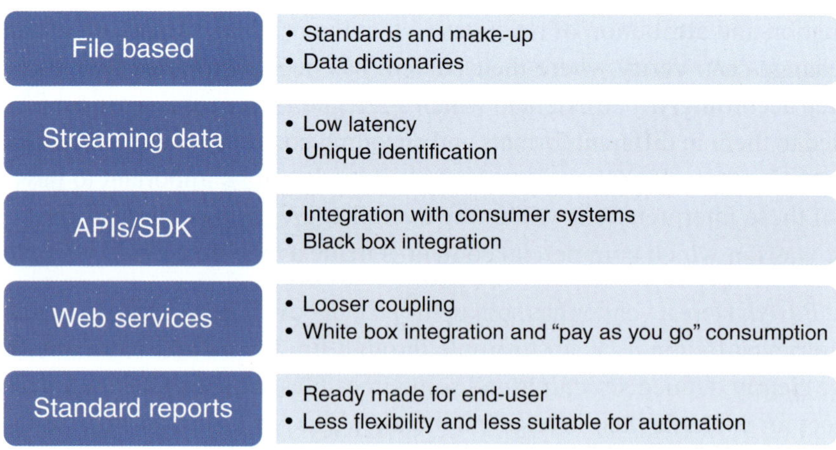

Figure 21.2 Different data sourcing characteristics.

There has been a long ongoing trend to tailor data consumption to data needs, i.e., to move from "bulk" data products to "portfolio based" data products where the information delivered corresponds, for example, to instruments held in portfolio or those in a wish-list or subscription list.

21.3.3 Industry utilities

Apart from companies specializing in the aggregation and bundling of capital markets data, other companies also provide data sources. This is typically as an extension of their core business so that these firms also put a lot of specific domain knowledge as add-on onto the basic data. These can include, for example, Central Securities Depositories, SWIFT and the ANNA Service Bureau. The distinction between aggregators discussed in the previous section and utilities is not black and white and is rather more of a continuum. Some of the larger data vendors are also the national numbering agencies (NNAs) for one or more markets and therefore responsible for issuing the security identifier (ISIN) in that market. Examples include WM Daten in Germany and Six Financial Information in Switzerland. Utilities would often provide additional services on validation, cleansing and integration around the data.

In some cases, data services firms act as "super" aggregators, combining the standardization into one format and the bundling of content with data validation services, typically in a specific field such as corporate actions or valuation prices. This standardization and validation of aggregated data may eliminate a lot of integration work and could be offered as a managed service as arguably many organizations do the same thing in-house, possibly together with other business process outsourcing services such as cleansing of data according to client defined business rules.

These cleansing rules could vary from client to client, making the scalability of these kinds of offerings somewhat difficult as we move closer to the area where the core business of a financial institution starts, the value-added proprietary business rules and models. Examples of input into these rules include:

- Which data source to use
- Which market makers to use as input sources to get for example a snapshot price
- How to fill gaps in time series (e.g., replace by fair value price based on bond cash flows and proxy spread, expectation maximization algorithm, interpolation, taking proxy return versus a benchmark time series etc.; in this case there can be a large diversity in rules).

The number, type and configuration of cleansing rules would depend on the data type and the downstream usage. The efficiency of the cleansing rules could be periodically reviewed by looking at the number of errors missed or, equally important, the number of false positives they have generated.

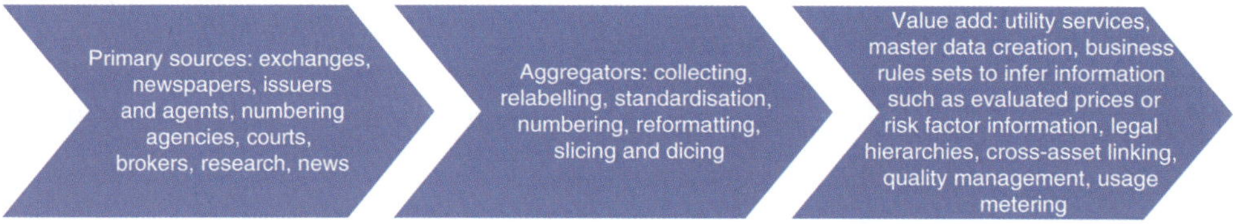

Figure 21.3 Primary sources, aggregators and information value-add.

Many aggregators have augmented their content products with additional inferred information. These "derived data" typically include evaluated prices for fixed income products and derivatives. In Figure 21.3 we provide some examples of and differences between primary sources, aggregators and information value-add.

21.3.4 The supply of content to processes within financial services firms

Most institutions have multiple content providers. There is typically not a single content provider – even the larger aggregators – that can address all information requirements. Besides, for business continuity purposes and in order not to depend on one vendor both operationally as well as commercially, institutions often opt to have multiple sources for the same information. Finally, by comparing and commingling content from multiple information providers, discrepancies and errors can be caught and corrected plus a fuller view of the market can be obtained by comparing, for example, prices from several contributors.

For transaction information, typically trade data are reconciled between internal records versus trade confirmations from counterparties using for example specific trade matching software.

Within the quality management process, we can distinguish between the following steps.

- *Sourcing* of the data from the provider. This can be in different ways with different levels of sophistication including direct on demand via web services. However, more mundane ways of sourcing still abound. For example, manual capture/entry of data through copying from faxes, emails, web pages and newspapers and the scrubbing, scraping or shredding of web or broker pages is still prevalent. Other ways of sourcing content include downloading it from an ftp server, or just reading a hardcopy report or a newspaper, making a phone call, scanning a prospectus, parsing an email and so on.
- *Reformatting*. Different providers of information use different standards, conventions or languages. Sometimes organizations have standardized on certain naming conventions and will "translate" external content to that representation. Reformatting

the data can involve ETL (Extract, Transform and Load) tools, a set of products that specialize in manipulating and reformatting data and XSLT tools that can transform one XML standard into another. Often a substantial amount of proprietary and local information needs to be part of the configuration of these tools so although they help solve the problem they are just part of the answer. This translation is not limited to data standards and also takes place at a lower level, for example the move between the ASCII and EBCIDC character sets with EBCDIC still common in mainframes.

- *Matching* different sources for the same instrument, transaction, corporate action and legal entity. For instruments, this can vary from exact matching on common identifiers such as ISIN code and place of trade for listed products to more "fuzzy" matching for legal entities where string comparisons are done on the name, or for OTC derivatives where matching has to take place by comparing a set of descriptive fields that collectively identify the product. For example, an interest rate swap can be identified by looking at the currency, tenor, reference interest rate and reset frequency.

  Comparing company names often includes text comparison and text manipulation including expanding abbreviations and stripping away legal form markers such as "Ltd.," "SA" or "Inc." In the case of corporate actions on the other hand, matching is more straightforward and typically takes place on a combination of fields that define the event type, effective date, identifier of underlying security. For transactions, it includes, for example, reconciling position statements received from a prime broker with those in the portfolio management system. Typically a hedge fund client would compare this with the information they have in their own portfolio management system or get from their custodian and do reconciliation. Big funds could use specific technology solutions for this. For instrument and legal entity matching, there are commercial sources that act as aggregators of identifiers and provide a cross-reference service that link different identification schemes.

  The notion of matching can be taken one step further to look for *similar* or *comparable* securities to check price movements for example. In this case, matching is generalized to put an instrument into its context, for example by finding other bonds from the same issuers, or finding other bonds with same rating, currency, issuer type and redemption characteristics.

- *Data integration and consolidation.* Once matched, the content from various providers can be compared or the internal view of transactions and positions can be compared to that from a service provider or counterparty. Blanks can be filled in as vendors provide complementary information, and at the same time discrepancies can be detected. Different providers may disagree about certain information or one may be sooner to report something than another.

- *Validation*. This can include formatting and syntax checks (e.g., a check on correct structure and check digit of the ISIN code), consistency checks (these can be within the terms and conditions of one product or can be by comparing different products, e.g., does the maturity date lie beyond the issue date, does a callable bond have a call date, are the market observed deposit rate and the deposit rate implied by FX forwards similar), and price validation checks (deviation against index), checking on the terms and conditions of a trade such as instrument, amount and settlement details. Validation often takes place with a four-eyes methodology employed which is required for certain data types in some jurisdictions. In this case, one pair of eyes does the actual correction or validation action and a second pair of eyes signs off on the work. For validation, workflow tooling is frequently used that groups quandaries about related issues together for easier routing and processing and distributes data issues over staff members, also for load balancing among staff. Issues can be organized for example by priority, by portfolio, by (internal) client, by owner and so on. This will depend on how an institution is organized. The approved data are often called *master data* or *golden copy data* and still need to be delivered to end-users and applications.

  When issues are not detected, the effects may not be immediately apparent. Bad data can lie undetected for a long time causing a minefield for unexpected reports. Typically, whenever something out of the ordinary is requested – for example an exposure report after the Lehman bankruptcy – issues come to light.
- The end of the information supply chain is new information. Financial institutions are essentially information processing businesses. Based on the execution of the processes fed by the content types described above, new information is put back to the client, investor, counterparty, marketplace and authorities. Since financial institutions are information generation enterprises, the costs (for commoditized products) and speed (for higher margin bespoke products) of pushing all the raw information through to produce new content is the critical measure. Complementing the comments on different information quality characteristics, we can evaluate the effectiveness with which an organization produces content itself. Similar to sources of information, aspects of quality of information produced strongly depend on the function of that piece of information, for example whether it is to be used as:
- Indication of interest
- Indicative quote
- Trade confirm
- Report to regulator
- Invoice to client
- List of transactions to be reported
- Portfolio statements.

Firms increasingly realize they need to actively manage this supply chain and that getting the most out of data is not a matter of creating more reports, adding more sources or databases. Instead, it requires an integral approach that cuts across business silos.

21.3.5 Data transformation and distribution

Firms can have a security master, a price master, a legal entity master, a settlement data master, a product master, a transaction log and so on. From the security master or price master store, information has to be brought to a place where it can be used, such as a report, business application or end user.

A key difference between client and proprietary information on the one hand, and vendor and public data on the other hand is at which point in the information supply chain it is stored. Data that are sourced from public and/or commercial sources are often stored in a separate data store decoupled from applications that may need it. They are often abstracted out of applications and are common to a variety of business functions. Proprietary and client data on the other hand are usually stored at the application level itself such as with a risk management, portfolio management or banking system which implies a far lower degree of standardization in terms of data model, taxonomy and semantics of for example client and counterparty data. Overlap and redundancies between master data stores and information kept at application level causes ambiguities.

A high level flow of information is shown in Figure 21.4.

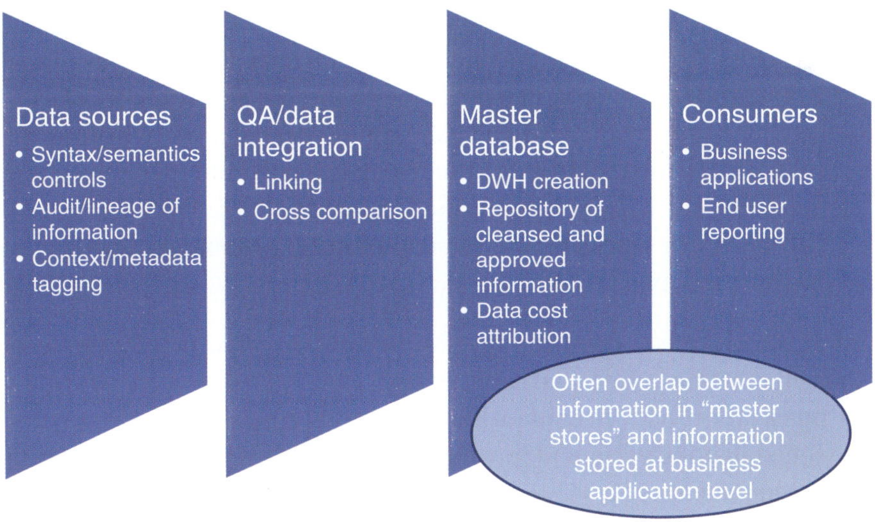

Figure 21.4 A high level flow of information.

Different ways in which data from a "master" source can be brought to the user include the following.

1. The approved data can be put into a distribution layer such as middleware which routs it to business applications such as a portfolio management, an order management, a trading or market making, a settlement or a risk system. This can be done on a publish subscribe model, where updates to a certain data universe such as for example a particular group of clients, a set of EUR corporate bonds, all cash dividends are pushed through to the using system, or it can be done on a request reply type basis. In a publish subscribe model there is normally a need to also store the golden copy data at the application level. In request reply setups the data consuming application requests data whenever they are needed and will go to a central store to obtain them.
2. The approved data can also be put into a downstream reporting database where it is available for on-line access. Typically, you would want to separate the data management and data processing part (loading, reformatting, validation) from data storage available for on-line querying. The reason for that is that you want to control what you make available (only approved data) and that you also do not want ad hoc queries to interfere with a scheduled process of loading, reformatting, cleansing to arrive at updated golden copy sets in time. At the place where the data has been processed prior to publication you want to keep audit information on who changed what when and you want to store the *lineage* of the golden copy data, i.e., which sources contributed what. Downstream data stores are called *data warehouses* (DWH) or *operational data stores*. To some extent, these terms are shades of gray and the meanings overlap. In general, data warehouses are larger data stores with broader coverage, they also often contain statistics on for example updates, which data are most used or control access in case there is some kind of (internal) cost allocation in place. Data warehouses are normally open to the whole organization. Operational data stores (ODS) contain more focused sets of data and would typically be used by a certain department, or certain business function. There would potentially be more of these, for example one for retail client information, one for equity derivatives prices and so on.

The information infrastructure and level of integration often reflects the organizational structure within a financial services firm.

Any financial services institution should carefully measure and control where information is distributed to and how heavily it is used. This is not just to satisfy business requirements and to conclude the appropriate content licensing agreements, but also to be able to attribute the cost of the information.

Within a business application, the requested data lead to actions and decisions (e.g., cash transfers, financial instrument orders, dividend processing, client portfolio revaluation, withholding tax calculations, regulatory reporting etc.). For whatever business function the application or end-user is using the data, at this point valuable new information is created which can in turn become the starting point of the next information supply chain as it is published externally to business relationships or regulators or internally to fuel other processes. The demands on flexibility and reporting speed have gone up drastically and the effectiveness in processing and generating new information and the extent to which a financial firm maintains the *overview* determines its success.

By and large, this supply chain holds for all commercially and publicly sourced data. Sometimes, some of the steps are skipped or combined into one. In the case of information retrieved from peers or clients and for proprietary data it can differ. Client or peer-to-peer retrieved data are data which are sourced on a piece by piece basis through a business relationship. Proprietary data are created as a result of internal business processes and include for example internal ratings, client profiles, pricing models and the institution's own trading portfolios. In this case the overall picture still holds but there is less automation and standardization. Some of these sources do not have the creation of information products as their core business but come in through emails, faxes and spreadsheets instead and can be more difficult to process automatically. Because this kind of information is also typically more often stored at the application level, it is also often entered directly through this application, in which case we arrive at a very much condensed supply chain.

The time delays that can occur as a result of a lengthy supply chain can be considerable. The delays incurred in each of the steps are becoming smaller as the process itself becomes more integrated, removing delays incurred by manual steps or process breaks.

The steps we see internally (cross reference, identification, error correcting) also occur to some extent at each of the steps in the external part of the supply chain. The difference is that in the external part there are multiple companies involved. Time delays can also have a technological nature because of manually rekeying data, sourcing information from printed material (such as newspapers, faxes) instead of electronically. The trade-off between speed and quality will be different depending on the business function serviced. To take two extremes: for real-time data speed is of the essence. The trader will know the market level and not be too bothered or distracted by the odd rogue quote. For regulatory reporting, speed is relatively less important and quality of information is critical as penalties and reputational damage can be immense.

During data distribution, different transformations can take place. The transformation of information means restating it into something else and includes mappings that we discussed in the content of the information supply chain such as standardizing on naming conventions. Many mappings have to be frequently used to be able to automate information flows between applications with different (implicitly) embedded data structures. Below we list a few examples of data transformations.

- *Scaling* prices to express them in a common base. Examples include restating penny based share prices (e.g., for London Stock Exchange listed shares) to British pounds, expressing prices in different currencies in the same reporting currency using FX rates and restating commodity prices into a common unit of measurement for financial reporting (example conversion in the case of natural gas can be to change the price from US$ per mmBTU into CAD$ per gigajoule).
- *Filtering* information through criteria such as AND/OR conditions, being in or not in a certain subset, being equal or not equal, arithmetic conditions or scanning for certain key words or phrases in a text. For all information that takes place in a certain time interval we could note the earliest, latest and the average arrival time. This could for instance be used to measure response times for execution venues, but also to rank custodians in how fast they relay and correspond about corporate actions. Filtering information is often necessary to prevent flooding users and systems with irrelevant information.
- *Arithmetical operations* such as adding, subtracting, multiplication or division of information by a constant, or by a dynamic element (e.g., a currency rate). These are the basic operations for example to create a spread curve or to move from one currency to the other or change the measurement unit as mentioned above.
- *Implied prices from related instruments.* Economically identical instruments should have identical prices. Through insight into price determinants and risk factors we know that often pricing information is implied, for example through related products which should move in tandem. The folowing are examples.
  - Extracting implied foreign exchange forwards and deposits. If the EUR 3M deposit, the EURUSD spot price and the EURUSD price in three months are all known, we can infer the USD 3M deposit price. Similarly, if the USD 3M deposit price was known, we could have inferred the EURUSD 3M forward price.
  - Foreign exchange triangulation. If the EURUSD and the USDJPY spot price is known, the EURJPY spot price can be inferred. This triangulation can also be used for FX options, for example if the at the money EURUSD and USDJPY option volatilities are known, the implied volatility for the EURJPY pair can be inferred.
  - Implying prices can also be done when information is complete, using implied prices both in the post-trade process as a validation against the prices used in revaluing the portfolio as well as in pre-trade to find arbitrage opportunities.

- *Implied additional pricing fields*. Often different price expressions exist for the same instrument and conversions take place between the two or one is added to the other. For example, changes from price to yield for fixed income instruments or changes from price to (implied) volatility in the case of options. If different pricing measures are used for different instruments, these conversions provide for a way to compare apples to apples. Note that clean reference data (for options strike, expiry date, option type, risk free rate, underlying, for bonds coupon rate, maturity date, payment frequency, daycount convention, redemption price and possibly optionality elements including call schedule and conversion price) is a requirement to be able to do these calculations accurately.
- *Screening functions* such as comparing data versus an index or a benchmark. Validation through rule checks such as checking fund composition versus a benchmark or to determine how *actively* a fund manager manages his or her portfolio and how much he or she deviates from the benchmark returns. Examples also include checking for reporting obligations when crossing an ownership threshold and checking an instrument on eligibility for inclusion in for example a UCITS fund or checking clients and counterparties on eligibility for trading certain products.
- *Translating* content to a certain standard, for example restating financial statement data in XBRL or providing mutual fund information in FundsXML.

Manipulations can also include technical operations such as table joins, filters on equality/inequality, checking on null conditions of table values, creation of substrings (to find common elements in, e.g., legal and security names), concatenations, one to one mappings, cross references and so on.

21.3.6 Data quality and data quality metrics

Data quality is an inherently fuzzy concept since different users will have different requirements and different notions of quality.

In this section we list a number of examples of data quality issues and various types of validation functions. We first look into the case of correcting for a missing value and filling in a missing value in a set of values. In the case of a time series, this can be a missing value which can be filled in by various ways with varying degrees of sophistication and use of context knowledge, for example

- Propagating the previous value,
- Interpolating through time between the previous and the subsequent value,
- Using statistical estimation techniques such as the expectation maximization algorithm,
- Using a proxy time series, either directly or by scaling the reference time series. For example, when missing a value for an equity, we could take the previous

value of the equity and apply the return of the index times the beta to it. Alternatively, it could be proxied more specifically by taking the return of a peer group and apply that to the previous value

Other cases where a missing value has to be filled in can be found in completing a yield curve for example. Often, fresh data are available for part of the curve but data for some intermediate points, for example, or the long end of the curve are missing. Tenor points can then be filled in using interpolation or extrapolation techniques such as:

- A linear or loglinear interpolation or extrapolation,
- Interpolation via different methods such as cubic splines or other polynomial methods,
- Finding a certain functional form/expression that characterizes or approximates the curve and then using that form to derive values for the required tenor points, for example using a model such as Nelson–Siegel which is a four-parameter model to describe a government bond curve. The parameters of the Nelson–Siegel model would be estimated by the available data; this results in a function which can also be used to obtain values for the missing points.

The specific interpolation method needs to be chosen in an intelligent fashion, i.e., reflecting knowledge of the product underneath. Cubic splines, for example, should not be used to extrapolate very far out in time or maturity, since because the interpolation elements are polynomials the values can become very large. Also, the result should always be verified to see whether it makes intuitive sense: when a spread curve or a credit spread becomes negative something may be wrong with the benchmark.

To find errors in price data that need to be repaired or screened, many validation functions can be run. These could be functions in the following areas.

- *Source comparison*. Compare two or more carriers of what should be the same information.
- *Semantic validations*. In some content areas filters with domain knowledge built in can be used: for example for a retail CRM system logic that knows the format of addresses, that has knowledge of postal codes format and street name writing conventions in different countries. It knows spelling and capitalization rules in languages and could expand on abbreviations of company legal forms for example.
- *Reference data consistency* and presence of information. Check various fields in the terms and conditions for consistency, for example if a bond is callable, a call schedule must exist and vice versa. In the case of an option, there must be a strike.

When a maturity date and an issue date are present, the issue date must be smaller than the maturity date.

- *Market data consistency.* In case of a quote with a bid, a mid and an ask consistency can be checked. Bid must be smaller than mid, which must be smaller than ask, and mid must be the average of the bid and the ask (leaving volumes aside). For a quote at end of day of a trading session which contains the open, high, low and close fields, the low must be the smallest and the high the largest fields. These sound like trivial checks but they can already help to uncover errors or gaps. More involved market data consistency checks can be done for economically similar or identical instruments, for example through foreign exchange rate triangulation and the forwards/deposits examples mentioned above.
- *Plausibility checks.* In this case price movements can be checked against historical standard deviations or versus the change in a benchmark. If the deviation is abnormal versus the historical data it will need closer inspection.
- *Tolerance settings* on price behavior. Different data repair and data validation functions will be chosen with different parameters depending on the instrument type. A simple example is a threshold on return and marking everything that changes by more than a certain percentage or certain number of standard deviations. In the case of curves we could use different percentage change tolerances for a curve, for example a different tolerance for the ON point versus that of the 10Y point. Stricter checks will be imposed on OECD currencies, for example, and there will be a higher tolerance for instruments where bigger swings are expected such as small cap stocks, exotic currencies and illiquid corporate bonds.
- *Transaction reconciliation.* This includes checking whether the instrument and counterparty referenced in the trade details exist in the master databases, whether the settlement instructions are present and whether there is an ISDA master agreement for that account. A metric often used in this context is the STP rate – the portion of trades which is processed without breaks and manual intervention.

When quality is defined and condensed to certain characteristics it becomes possible to track it and to set Key Performance Indicators (KPIs) on it. When selecting KPIs there are various considerations to keep in mind and we list some best practices here.

- KPIs should be *compared* to those of your peers or other departments, if only to see where you can improve and where you can improve most for a minimum effort. Ideally KPI comparisons would pool data from multiple companies that share a common KPI methodology, make sensitive or proprietary data anonymous and send back comparison on how you do relative to your peers. This benchmarking can be done *internally* from department to department or externally with a peer group.

- Take great care in choosing the right indicators that reflect the operational efficiency of the *specifics of the business* rather than copying a set of indicators from the neighboring financial institution. KPIs are after all a classic statistics issue: you want to condense information as efficiently as possible without losing the big picture and the main trends. In data terms, if you measure for example parts of the information supply chain, you would expect the KPIs to reflect the important *quality* aspects discussed above, and put the emphasis on the right mix.
- When applying KPIs to information streams between two parties, the selected KPIs should reflect the *purpose* of the specific relationship (in case of outsourcing to third party) or the purpose of the measurement. This purpose could be customer satisfaction – through metrics such as the response time for ad hoc queries or the percentage of portfolios adequately serviced with pricing models, (e.g., 100% accurate NAV, nothing contested, extent of mispricing, STP rate for settlement), the purpose could be cost (e.g., in $ per transaction), it could be the support of higher volumes through metrics such as the uptime of systems (5*9 availability, 24/7 support, the number of new instruments set up, number of clients onboarded, number of corporate events processed, number of portfolios serviced, latency, throughput in messages per second and so on). Similarly, it is important to understand what the result of improving on the current values of the KPIs will be. Will this be happier clients? More revenue? Lower costs? All of the above?
- Do not collect *too many* KPIs. KPIs can be strongly correlated so too many do not necessarily provide additional information. Furthermore, KPIs should be easily collected, ideally in a fully automated fashion. If they are costly or time consuming to collect, collection will not (always) happen or not happen in time and you will either have an incomplete picture or you will have an imprecise picture because people will rush to deliver something and interpret the data to be delivered in different ways. Reports with hundreds of KPIs will not be acted upon in an appropriate and efficient way and will probably not even be read: people will not be able to see the forest for the trees.
- Make the definition of the KPIs *consistent* among the various processes that need to contribute them. Inconsistent KPIs can never be rolled up into aggregate information or be the basis for operational risk or other charges allocation to business divisions. Consistency can also help the operational risk reporting and will definitely benefit the regulators that need to interpret and compare these reports.
- The KPIs should be under the *control* of the responsible party or of the party to which a process has been outsourced. For example, if an external party that manages the data is measured on lead time in setting up new instruments or customers in the systems, it could be that that lead time heavily depends on the

financial institution's own risk control procedures. There is a trend of BPO/ outsourcing companies to include business driven outcomes in outsourcing contracts so that, ideally, the business objectives between financial firm and BPO provider are the same. This could also be tied to a BPO's cost model, for example charging on a per loan or mortgage processed basis instead of fully fixed cost.

- KPIs should *support root-cause analysis* and predict trends.
- KPIs should have *owners* that are responsible to reach a certain value with the KPIs. There should be a feedback loop back to responsible people who can *act* on the KPIs.

Dashboarding and corrective actions on KPIs are the logical conclusion of collecting quality aspects of the information supply chain. The process starts with identifying which content is required and what the required service levels are for the processes that depend on that content. Setting up KPIs is the next step. As we have seen, the relative weights of each of the quality dimensions will vary by business function. Also note that data quality elements can represent competitive advantage; it may not be the goal to reach a certain *absolute* level of quality but you may merely need to be *relatively* better than everybody else. If that is the case a firm will have a competitive advantage even with far from perfect quality data.

21.4 Information governance, different setups and automation levels

It is sometimes said that the industrial revolution has bypassed the processing of financial information and that it is still very much a cottage industry with home-grown tools, techniques and spreadsheets to do the data supply job rather than relying on industrial strength scale and robust processing.

In this section we look at some of the ways in which the information supply chain is organized and discuss some of the organizational reasons why it is not always as efficient as it could be. We start by classifying applications according to business purpose: for example risk, portfolio management, reporting. You have systems which are primarily about data gathering, where information streams up, is progressively analyzed and ends in a dashboard or report. We can call these information delivery systems. You also have decision platforms which **need to communicate back** the decisions and where data flows both ways and where actionable information and the results of those actions need to be communicated to other systems and users such as trading, market making and order management systems.

Governance of data is a difficult issue. Data typically flow horizontally from for example trade inception through to risk reporting, crossing a number of silos.

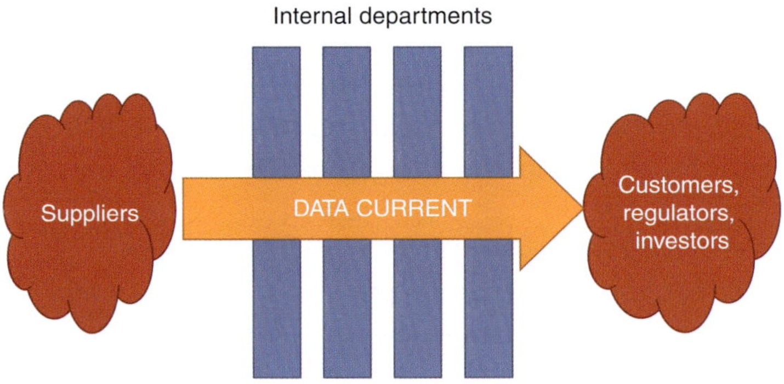

Figure 21.5 Data flows defy silos.

Flexibility in self-sourcing and manipulating information is very much a mixed blessing. The result of giving business users full flexibility over their data and models and the increased use of reporting and business intelligence embedded in business applications is the creation of independent data silos. On the other hand, the business knows where the data issues are because they have to work with them every day. Even though there are risks in having inaccurate or inconstant information, in general users will be reluctant to cede control over the data sitting in the business applications they rely on to do their jobs. Inevitably there is a delay between the speed with which new products have to be supported and brought to market and the infrastructure needed to properly settle, monitor and report on them post-trade. Typically automation has taken place back to front, so first the operational processes in the back office were automated (more prone to automation), then the middle office and ultimately the front office. Some organizations have recognized the need for a role that looks at information management separately and not within technology management. They have seen the need for a role that transcends departmental boundaries and have appointed chief data officers.

The use of spreadsheets and macros has put strong IT tools in the hands of business analysts and put increased pressure on the IT department, but without a management framework for version control, sharing and reusing, too much power and operational risk can be concentrated in whoever understands the 80MB spreadsheet...

An efficient yet robust information supply chain should allow for the following:

- Covering all instrument types and associated information on counterparties and corporate actions
- Ability to manage various frequencies of data
- Enterprise-wide consistent data standards

- Secure fitness for purpose of data via transparent validation rules
- Possibility of enriched data via application of various calculations
- Flexible and extensible data model to cater for new product and regulatory developments
- Reformatting of data to feed internal/external applications
- Easy reporting and disclosure of data to downstream users
- Robust security and audit functions
- Scalability to meet enterprise requirements
- Easy definition and collection of summary statistics and KPIs.

There are many reasons why this is easier said than done.

- Users have a narrow focus and want to keep control over the data in their silo.
- Finding a home for data in the agenda of the management team is difficult.
- For all data centralization projects as with all infrastructure issues, to make the business case you have to take a high level approach. In general, you need to show either to make money directly by allowing for example the trading of new products and markets or to save enough (or need a regulatory imperative).
- It is difficult to assess the impact of centralized solutions because of constant organizational reshuffles which make for difficult cost analysis and cost attribution. Besides, often data and data processing costs are not very clear in the first place and do not fully show up in management accounting.
- Stickiness of existing infrastructure. An information infrastructure is normally very sticky. It is hard to change it because it is costly, because of the vested interests of the people that look after it (not just vendors of content and software but definitely also the internal staff that has built up all the knowledge which will become worthless when it is gone).

Taking a closer look at how an information supply chain is set up allows investigation of the potential for improvement. Mitigating factors include knowing what you are doing, doing it in a consistent way, sharing information and avoiding redundancy. Setting up a security master twice does not *double* the work, it *triples* the work because you also need to reconcile between your two internal data stores. Mitigating factors that can reduce the data uncertainty and improve the efficiency of information lifecycle management include:

- Use of single data standards to solve the incompatibility and interpretation issues
- Number of staff doing reconciliations
- Separation of duties, clear permissioning and data owners per market or data category

- Controls and operational risk reporting framework
- Metrics and dashboards to monitor changes in quality and to spot trends
- Various tools and techniques listed in the section below.

How to go about implementing these mitigating factors depends on whether the business is organized on product lines, as a matrix organization whereby for example product lines, geographical areas and client segments all have functional owners and whether there are cross-vertical shared services which can drive standards and common services.

Even the best information architecture designed and set up from scratch will – to a greater or lesser extent – be a reflection of the business requirements *at that moment*. Because of changing needs it will be tweaked, abused and retrofitted over time. Even in the case of a sound underlying architecture, people will tend to take shortcuts. This is just one reason why *metrics* are important: to keep everyone on the right path.

Through monitoring trends on quality statistics, corrective actions can be taken to improve on a capital markets sourcing and distribution chain. In Figure 21.6 we give a schematic overview of the potentially virtuous circle of data, metadata, metrics and corrective actions.

The analogy of supply chain management can help to give a broader perspective here. What else is automation of processes and information flow but a conveyor belt of information? Also, it is useful both because the number of suppliers of content is very large and because certain parts of that supply chain or of the trade lifecycle (post-trade) will have been serviced by third parties. Less direct control over the various steps in the supply chain translates into a greater need for clear metrics to go

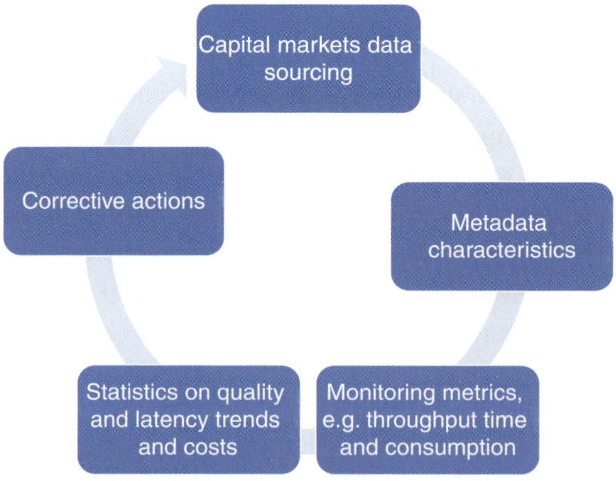

Figure 21.6 Continuous improvement cycle to improve sourcing and distribution of capital markets data.

into service level agreements. These metrics may help us judge the effectiveness of the information supply chain in fulfilling the needs of its customers that need to process/initiate the instrument and transaction lifecycle processes. It may help us define quality in terms that translate to business benefits for the users.

21.5 Tools and techniques

In this section we describe a number of tools and techniques that are used to manage, monitor and improve the process flow. Processes in the information supply chain often break because:

- They are too brittle (bad designs, misuse of formats, operationally complex, use of applications for new things without proper testing)
- The stress is too high (volume, tweaking of formats)
- "Glue" in the form of scripts and spreadsheets that was applied earlier does not work or is no longer relevant to the new situation.

Often, all three reasons are in place simultaneously. Some integration solutions focus on extracting, uploading and transporting sets of information; other solutions focus more on dictionaries, business rules, mappings and cross-references. Some system integration tools have come out of consulting companies who have done the same integration trick a number of times.

Different techniques and tools may be required in different organizational contexts. First of all, different departments have different lenses through which they view data. Below is a high level breakdown of typical information needs.

On top of that there is the organizational culture which can facilitate or block data integration efforts. For example, an investment bank can have a very centralized

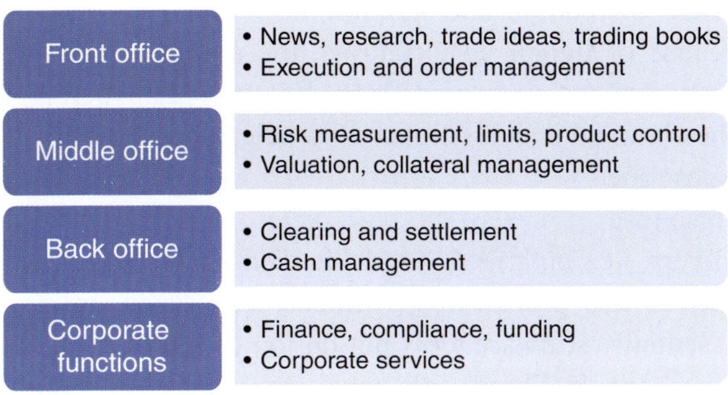

Figure 21.7 Typical high-level department split in a financial institution.

structure with one data department servicing everyone, whereas a heterogeneous universal bank can have a more decentral data management function. We call this the *central* data management versus *federated* data management setups. In the case of for example fund administrators or custodians, the data management department will have external clients.

Of the many tools and techniques offered we name the following.

- Methodological. Certain organizational setups such as clear ownership, governance and change control procedures of data assets. To what extent agreed procedures can help an efficient data supply chain depends very much on an organization's culture.
- Using third party data solutions to manage part of the supply chain. Through off-shoring, BPO or a Managed Services arrangement, organizations can put part of the data supply chain into the hands of third parties. These services are typically offered by larger system integrators or by larger software vendors. The pickup of these services may not be as large as expected.
- In the security reference data space often the most complex instruments are not fully covered by managed data services. Consequently, internal staff is still required to do the time-consuming work.
- Liability for data quality is still with the financial services firm, so they will still double-check, thereby negating a part of the potential cost savings.
- Not all data providers are willing to have their content channeled through a third party services firm.
- Data consumption tools. These are products that keep track of all products, contract terms and pricing policies of content providers so that a user has a Yellow Pages at his/her disposal to compare providers. On the other hand, these tools can keep track of how much data is used so that invoices from content providers can be predicted.
- Distribution solutions include many providers of middleware that also often keep the overview and keep track of where data have gone, verify whether they have arrived at their destination and monitor how much are used per consuming system. Because of tighter cost controls, the ability to allocate data costs to users and departments on a more granular level than before is increasingly required. The common data distribution paradigms include
  - Publish/subscribe
  - Request/response
  - Batch delivery, i.e., bulk feed via trickle feed via messages/event based.
- BPM tools (business process management, i.e., dashboarding and workflow tools) are essentially software tools put on top of a process to make every step in that process explicit. These tools could graphically show and summarize the data supply chain and show summary metrics through dashboards.

- Data tools, for example the use of standards, the use of markups to get more value and intelligence out of it. Ideally, you would want a form of "self-aware" and "self-describing" data. Data could turn into information nuggets if we add a descriptive layer with tagging and metadata info that answers on origination, destination and relations to other data elements.
- If data objects are also tagged with for example the roles they can play or what types of interactions they may have with other objects they can more easily be used as "Lego" bricks to be combined. Adding this kind of usage and lineage context helps to support more complex, flexible and more frequent ways to view, analyze, manipulate and adapt their business data.
- Data intelligence solutions. Some solutions provide sets of business rules specifically for certain data types, for example around transaction reconciliation in post-trade processing. Others focus on for example screening market data or corporate actions data.
- A separate set of products is focused on data manipulation: Extract, transform and load (ETL) tools. These are basically tools to help you move data from database A to database B, possibly reformatting and tweaking the data on the way. These tools can have a more technical focus or can be equipped with a data dictionary to add business domain knowledge.
- To counterbalance the prevalence of spreadsheets, solutions are offered that aim to put structure and version control onto spreadsheets.
  - Data aggregators/niche content providers and types of content products.
  - Data management tools. These are solutions that integrate with content suppliers and that include various sets of rule sets around sourcing, matching, validating, quality metrics, integrating and distributing content. Through this they can automate a substantial part of the information supply chain. Sometimes the label MDM for master data management is also used to refer to the management of all data that could conceivably be shared and are not user or business application specific. We should distinguish between generic MDM tools that offer general purpose, technical data storage and manipulation solutions and vertical MDM tools. In the latter case, there would be knowledge specific to a certain business domain (such as capital markets) packaged in the standard solution offering, for example, financial markets data models, financial product knowledge, validation rules, specific quality management and workflows.
  - Unstructured and semi-structured/machine-generated data transformation. Note that there is an increasingly blurry line between data and software products. Data providers are moving downstream through adding value added content derived through business rules (such as evaluated pricing) but

also through adding tracking, monitoring and data manipulation tooling. This category includes APIs and SDKs to help get their content more easily into reports and business applications.

21.6 Changing requirements

Technology and processes go hand in hand and you cannot have one without the other. Processes make sure that you actually get something out of the technologies. Technologies can enable and scale processes. Technology automates processes and information supply chains provide the required information based on the needs at a given point in time. Thus, requirements are frozen in a certain automated process while there will be continuous demands for additional information. The set of financial products for example is growing continuously, driven by investment, funding and risk transfer needs as well as by arbitrage opportunities. This causes continuous new demands on information supply.

The global financial crisis has led to traditional requirements being scrutinized as well as a number of new requirements.

- *Audit.* This poses demands on the retention as well as on the accessibility of historical information. It requires the keeping of records on who changed what at which point in time, not only on the information itself but also on for example any business rules that may have impacted it.

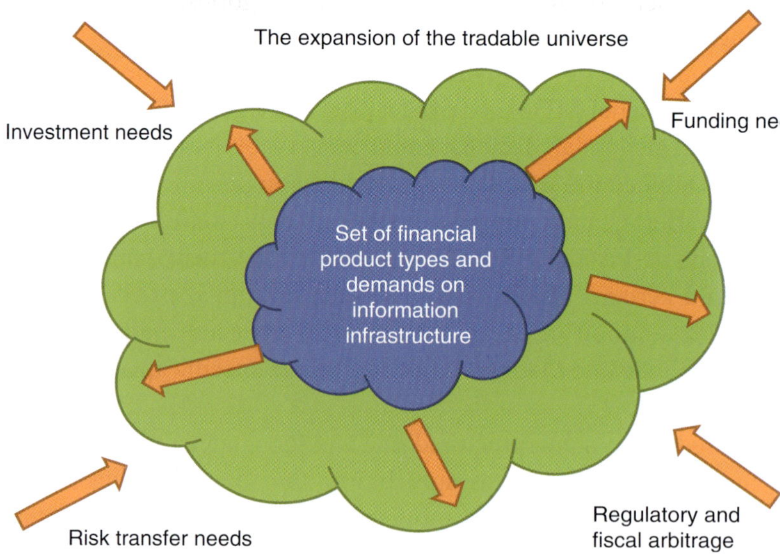

Figure 21.8 Growth of financial product types.

- *Provenance.* The ability to show where a certain data element came from. What was the (original) source and what signoff and/or quality checks has it undergone. To generalize this "data lineage" concept we can also speak of a "resume" or "curriculum vitae" of data, for example of a company or a security with a detailed history of all the changes, M&A, corporate actions etc.
- *Compliance.* Compliance is an overused term which in the context of data can both refer to abiding by content licensing agreements or collecting the required data to comply with various regulatory requirements.

Apart from the changing regulatory landscape for different financial services segments (banking, investment management, insurance, alternative investment), there are also significant cost pressures on the data supply chain. New projects initiated will therefore tend to focus on reusing existing components. Other than the requirements on how data are kept and the ability to show where they came from, the regulatory overhaul will also mean the emergence of new data sources. Swaps execution facilities (SEFs) and the move of clearinghouses to start clearing more product types can lead to additional data sources.

21.7 Conclusion

To summarize, we can note the following.

- The true value of an information supply chain is in the ability to store, process, use and reuse information of all types and from all sources quickly and to make it accessible anywhere for all capital market data categories.
- Technology can be the great equalizer in allowing access to information. It can also make people drown in information if not properly filtered, plus it can lead to decreased control, for example in decentral spreadsheet modeling.
- A weak data process/infrastructure has knock-on effects. It can slow down a business, have it miss out on opportunities and create a cause for client and regulatory concern.
- "Peeling the onion" in terms of drilling down into the various steps in a data quality process can be very cumbersome due to a spaghetti infrastructure. Often the streamlining of a data supply process is slowed down due to absence of global owner.

Through optimizing their supply chain, financial institutions can keep better track of what information sources they use, where that data ultimately end up and what happens on the way. An optimization process starts with knowing what the current situation is, and because of a typical spaghetti architecture where sources and systems have been layered on over time, this is often not a pretty sight. Yet, data standards and data management systems can help bring order to this picture, paving the way for a streamlined, transparent and more efficient way of delivering capital markets information to where it is needed. After all, clear and timely information is the foundation for any business, for normal operations as well as for supplying customers, investors and regulators with the information they require.

22

The visual analysis of financial data

Victoria L. Lemieux, Brian Fisher and Thomas Dang

22.1 Introduction

Financial risk analysis is cognitively complex and demands analysis and interpretation of diverse and often sizeable data sources. Whether the goal is to analyze systemic, market, liquidity, credit or operational types of risk, it is often the case that financial risk analysis and decision-making require rivers of data to be contextualized and analyzed across multiple dimensions on an almost real-time basis. The methods and technologies used to synthesize and analyze this flood of data, and to turn the resulting information into useful insights, are still few and far between. Traditional analytic methods and tools often force people to formalize their thoughts early in the process, and to adapt their way of working to the rigidity of mathematical formulas or computational methods. In contrast, risk analysis frequently requires open-ended exploration of large volumes of complex data, detection of previously unknown patterns, changes or anomalies, and seeking solutions to open-ended and even ill-defined questions. It is these challenges that visual analysis of financial data is designed to meet. Visual analysis technologies have the potential to reduce the time it takes to analyze complex financial data, to bring financial risk managers and decision makers to new understanding of their financial risks and to aid them in communicating about such risks. The goal of this chapter is to introduce the visual analytics approach, and to explore how it might support financial risk analysis and decision making.

In this chapter, we will:

1. Provide an introduction to information visualization (InfoVis)
2. Explain what visual analysis (VA) is, how it differs from InfoVis, and the reasons it is useful
3. Provide an overview of how VA has been applied in the domain of finance

Handbook of Financial Data and Risk Information, ed. Margarita S. Brose, Mark D. Flood, Dilip Krishna and Bill Nichols. Published by Cambridge University Press.

4. Discuss the components of VA and how these relate to its use in financial risk analysis
5. Provide two worked examples of the application of VA in the area of financial risk analysis.

The chapter is targeted at financial risk analysts and managers with little or no previous exposure to VA. Hence, the discussion will focus on the technical aspects of visual analytic system design only to the extent necessary to provide an understanding of the factors that determine the appropriate application of VA tools to particular analytic problems and datasets. Those with more of an interest in the design of financial VA tools will find the references provided at the end of this chapter a useful source of additional information.

22.2 Foundations

22.2.1 InfoVis

VA is defined as "the science of analytical reasoning facilitated by interactive visual interfaces" (Thomas and Cook, 2005). As the name implies, VA has to do with the visualization of information. Visualization is nothing new: pictures were used as a form of communication in early civilizations and cartographers have been making maps for centuries. Business information has been visualized in tables, outlines, pie charts, line graphs and bar charts for a long time (Tegarden, 1999). The history of computer-aided visualization is, of course, much shorter. Arising from the computer graphics community, the study of the use of graphics for visualizing data traces its origin to the 1987 report of the Workshop on Visualization in Scientific Computing held in Washington, DC (McCormick et al., 1987).

Visualization of data that lack spatial coordinates, such as financial records, is known as *InfoVis*. The first IEEE InfoVis workshop was held in Atlanta, Georgia in 1995 and included a talk on information animation applications in the capital markets (Wright, 1995). While often thought of as a subfield of visualization, InfoVis has its own unique challenges. From a design perspective, the abstract nature of the information that is visualized and the increased opportunities for interaction with that information place greater emphasis on graphics and interaction design.

It was the challenge of building an interdisciplinary approach to visualization in real-world applications that led to the creation of VA. Perhaps the most evocative call for this came from a short paper by a pioneer in computer graphics, Bill Lorensen. His 2004 paper "On the Death of Visualization" (Lorenson, 2004) cited a number of causes for concern, including the fact that the visualization community lacked application domain knowledge. He proposed that visualization researchers first "embrace our customers: find out the important problems they face." Second, they should "form alliances with other fields," and finally, they should "define some grand challenges."

In response to issues such as these, a panel of researchers drawn largely from the computer visualization community was assembled at Battelle's Pacific Northwest National Laboratory in the USA and asked to define a research agenda for a new interdisciplinary effort to design and evaluate technologies for strategic and operational decision-making for US homeland security. From the perspective of the decision-maker, homeland security concerns frequently pose "wicked problems" (Rittel and Webber, 1973) that resist computational solutions due to constraints on time and resources for computation and poorly defined criteria for solutions. Their efforts led to the publication of *Illuminating the Path: A National Research Agenda in Visual Analytics* (Thomas and Cook, 2005). VA is seen as complementary to visualization in its focus on analytic task performance per se, and in its choice of a scientific methodology as the mechanism for doing so. In keeping with Lorensen's recommendations, VA places a good deal of emphasis on development of technologies and analytical methods focused on the needs of a particular set of "customers," their data, problems, working methods, and organizational structures. This calls for a "translational research" approach (Fisher et al., 2011), a concept borrowed from the health science community and adapted for VA. The VA translational research cycle includes working with decision-makers in the context of their organizations to characterize data and solutions in the situations in which analysis takes place. This work defines research questions for laboratory investigation, the results of which guide the design of new interactive visualization technologies and analytical methods that are evaluated in partnership with the decision-makers and their organizations. This approach should enable organizations and individuals to more effectively utilize interactive visualization to solve problems that are ill constructed, where the characteristics of a successful solution are not well defined, and where data are massive, uncertain and changing over time. Thus, a common catchphrase in VA is "to detect the expected and discover the unexpected."

22.2.2 Human visual perception and cognition

Whether we are speaking of InfoVis or VA, visualization harnesses the power of human visual perception and cognition, lending truth to the old adage that "a picture is worth a thousand words." Visualization has advantages over other modes of communication because humans have evolved visual and spatial skills that include the ability to detect edges and discontinuities, things that stand out, variations in color and shape, and motion; to recognize patterns; and to retrieve information using visual cues (Kosslyn, 1980, 1994; Lurie and Mason, 2007). Each of these visual and spatial attributes can be transformed into a graphical image to provide a rich visual description of data. As these features can be observed with "pre-attentive processing," that is, they are perceived prior to conscious attention, they are understandable at a glance and much more rapidly than words (Roberts, 2003, Ward et al., 2010). Through encoding of data items and attributes into graphical images, visualizations can act as a repository of data which allows people to offload cognition to the perceptual system, using visuals as a form of virtual memory (Munzner, 2009). This can enlarge problem-solving capabilities by enabling the processing of more data without overloading the decision-maker (Tegarden, 1999). Visual images also have the advantage of being able to transcend the barriers of human culture since a map or a graph may be interpreted even by people who speak different languages (Ward et al., 2010). Finally, because visual cues stand out to human perception more than words, a picture sometimes "forces us to notice what we never expected to see" (Tukey, 1977).

22.2.3 Visual analysis

We can make a further distinction between types of InfoVis by the purpose of the visualization. Much InfoVis is created with the goal of presenting data to rapidly convey meaning or increase understanding. In the field of InfoVis for presentation, the two classic theorists are Jacques Bertin (1967) and Edward R. Tufte (1990, 1997, 2003, 2006).

In InfoVis for presentation, the objective is to design graphical representations that leverage human cognition to convey specific information. For example, many companies today utilize interactive data visualization to present accounting information to external users on their investor relations websites and to internal users in applications such as enterprise resource planning, balanced scorecard, network security, and fraud detection systems (Dilla et al., 2010).

In VA, on the other hand, visualizations are created for the express purpose of conducting analyses, such as recognizing patterns, discovering trends and identifying outliers, which can be difficult to do using traditional methods (Eick, 2000; Hanrahan et al., 2007). VA draws on human visual cognition and sense-making to generate inferences about the meaning of data. This is done through direct interaction with the visual representation to generate insights, such as by exploration, "information foraging," hypothesis generation and evaluation. This "human-information discourse" (Thomas and Cook, 2005) results in transformation of the visual representation in ways that reflect and reinforce the decision-maker's cognitive processes.

The same approach to design cannot be applied to both presentational visualization and visual analytics, because the expository, even persuasive, goals of presentation can directly conflict with the open-ended, exploratory goals of analysis (Roberts, 2003).

22.2.4 Visualization and VA in finance

We begin our overview by noting that there is little published scholarly research on the application of VA in finance, and related fields such as economics, compared to other domains. However, there is considerable research in the management information systems field on visualization in conceptual modeling (Bera et al., 2011).

The relative paucity of scholarly literature belies the fact that interactive computer-based InfoVis has been applied in the domain of finance in one form or another for at least two decades and its use is growing, both for purposes of presentation as well as for analysis and decision-making. Dilla et al. (2010) note that interactive visualization of data in financial reporting and auditing contexts is increasingly common. Several companies currently incorporate interactive data visualization features on their websites, including hyperlinked tables of contents, dynamic graphic images, and search engines (Kelton and Yang, 2008). Visualization tools such as dashboards, motion charts, multi-cube visualizations and a variety of visualizations too numerous to mention individually are in use in most financial services firms. Audit firms are using interactive data visualization to perform analytical procedures (Bay et al., 2006; Gunn, 2007). In 2008, the Securities and Exchange Commission (SEC) mandated that all publicly held companies furnish financial statement information in extensible business reporting language (XBRL), in part, to facilitate the development of interactive data viewers (SEC, 2008). The European Central Bank has also introduced an interactive visual data view for government debts, yield curves and CPIs (European Central Bank, 2011). Smartmoney's™ Map of the Market (2009) and FinViz™ (2010) provide examples of services that integrate financial market data with InfoVis for investors. In short, InfoVis has now penetrated just about every part of global financial markets.

Scholarly discussion on the application of InfoVis in finance can be traced back to the 1990s. One of the first examples, as previously noted, is Wright's (1995) paper on information animation applications in the capital markets. In this paper, the author observed that "Information animation applications are particularly suited to the securities industry because that is where we find huge amounts of data, the value of which declines rapidly with time, and where critical decisions are being made on this data in very short periods of time. Information animation technology is an important new tool for the securities industry, where people need to be in the decision-making loop without suffering from information overload" (Wright, 1995, p. 1).

Since this original paper, it is fair to say that the vast majority of research on the application of InfoVis and VA in finance has focused on its use to analyze securities market data, in particular stock tick data. Jungmeister and Truro's (1992) paper on adapting Tree maps to stock portfolio visualization, Dwyer and Eades' research (2002) on visualizing a fund manager flow graph with columns and worms, and Pasha Roberts' (2003) MSc thesis work on visualizing stock market tick data are three examples of this trend. This is probably so because of the relative public availability of such data. It is only recently – perhaps since InfoVis began to gain more of a commercial foothold in the financial industry – that a greater number of examples of research using internal, non-public data can be found. Among this research is Ieronutti and Pighin's work (2009) on approaches to visualizing multidimensional data in OLAP systems and Chang et al.'s (2008) work on applying VA to the analysis of wire transfer data, which involved collaboration with Bank of America. This research points to the benefit of financial sector partnerships with academia and granting researcher access to large sample datasets of internal financial data.

Most of the research on financial InfoVis arising from the InfoVis domain focuses on innovation in the graphical representation of information as opposed to the analytic problem space (what Amar and Stasko (2004) refer to as "representation primacy"). This focus has given rise to an incredible array of different visual metaphors for financial data, including tree maps, rings, worms, multi-cubes, spirals, network graphs and beads to name but a few. As Munzner (2009) notes, many of the designs reflect the rise of interactive 3D graphics as a new capability and incorporate 3D representations, though the vast majority remain 2D. Pasha Roberts' (2003) navigable 3D view of a single stock and Ieronutti and Pighin's (2009) 3D multi-cubes are two examples of 3D representations. In more recent literature on InfoVis, the task focus takes greater precedence, as in Savikhin and colleagues' development of PortfolioCompare, an interactive VA tool that helps consumers compare and choose among several portfolios consisting of different financial instruments.

The areas of financial risk analysis to which InfoVis and VA have been applied predominantly concern portfolio analysis and, in particular, extension of traditional statistical methods such as Value-at-Risk (VaR) and Average Time Value (ATR), where the issue of drawing analytic inferences from analyses based on limited known values and distributions can be misleading. Lei and Zhang's (2010) work with financial time series data to develop a visual volatility measure is illustrative of this branch of the research on InfoVis in finance as is Alsakran et al.'s (2010) work with tile-based parallel coordinates. In a similar vein, the need to understand overall market dynamics when assessing asset or portfolio performance has been the subject of some study, with Schneiderman and Wattenberg's (1992) research, Sawant's (2009) StockViz paper and Lei and Zhang's (2010) work all offering examples of studies motivated by the desire to contextualize asset or portfolio performance data to improve decision-making. Interest in other analytic problems has only recently emerged as areas of exploration; for example, the analysis of financial networks (Kubelec and Sa, 2010; Garratt et al., 2011) and compliance and fraud detection (Chang et al., 2008).

Although scholars of finance and economics have yet to fully embrace VA as an area of study or as a research methodology in the same way as it has been embraced in the commercial sector, greater collaboration between InfoVis scientists and researchers in finance and economics could generate innovative approaches in all of these disciplines.

22.3 The visual analysis triad

There are a number of frameworks that offer insights into the characteristics of VA (Munzner, 2009; Thomas and Cook, 2005; Ward et al., 2010). Each of these models approaches VA from the perspective of the VA tool designer. We introduce the Visual Analysis Triad (Figure 22.1), which combines elements of these models in a simplified framework to provide the end-user with a lens on VA. In this framework, three basic elements interact in VA: the analytic problem, the data and a VA tool.

22.3.1 The analytic problem

At the top of the pyramid is the analytic problem, which is the natural starting point for any consideration of VA since the science of analytical reasoning (the theoretical foundations of reasoning, sense-making, cognition, and perception) underpins it.

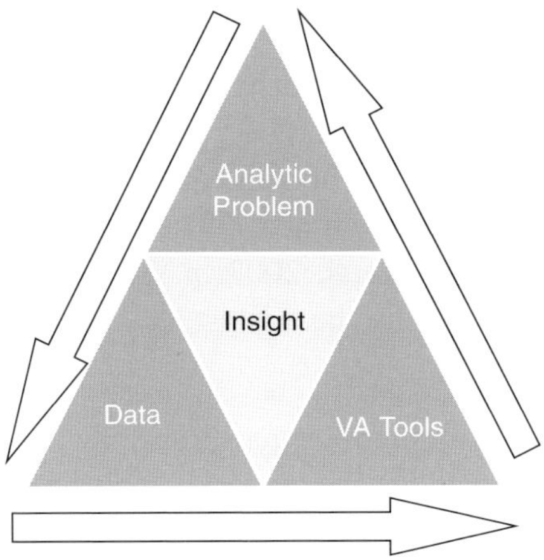

Figure 22.1 The visual analysis triad.

VA is said to extend human analytical reasoning capabilities in six basic ways (Thomas and Cook, 2005, pp. 3–33; Card et al., 1999):

- by increasing cognitive resources, such as by using a visual resource to expand human working memory,
- by reducing search, such as by representing a large amount of data in a small space,
- by enhancing the recognition of patterns, such as when information is organized in space by its time relationships,
- by supporting the easy perceptual inference of relationships that are otherwise more difficult to induce,
- by perceptual monitoring of a large number of potential events, and
- by providing a manipulable medium that, unlike static diagrams, enables the exploration of a space of parameter values.

Typically, VA is used when the analytic problem is not sufficiently well defined for a computer to handle it algorithmically. If the analytic problem can be solved completely computationally, then visual analysis is not necessary (Munzner, 2009). The purpose of VA is not to replace quantitative analysis, but instead to allow quantitative analysis to be more focused or to extend it (Grinstein and Ward, 1997). For example, computational methods are perfectly well suited to calculating portfolio losses and gains over time, whereas they would not work as well to detect patterns in portfolio managers' investment decisions over time.

VA is premised upon the human remaining a key part of the analytic process, blended with computational analysis. In conducting VA, people use dynamic interaction with visual representation of datasets to "... generate new hypotheses when exploring a completely unfamiliar dataset, to confirm existing hypotheses in a partially understood dataset, or to present information about a known dataset to another audience" (Munzner, 2009). Thus, researchers refer to VA as being about human intelligence augmentation (IA) instead of being about artificial intelligence (AI), where the goal is to replace human cognition with machine computation. For tasks suited to VA, to replace the human with artificial intelligence would be too rigid. Instead, the goal is to augment and boost human cognition through greater human/computer symbiosis (Roberts, 2003, p. 77). In general, then, VA is well suited to analytic tasks that require the analyst to:

- Process masses of complex data
- Answer an array of often ambiguous questions
- Keep humans in the loop and at the center of analysis
- Blend computational analysis with interactive visualization of the results of that analysis

- Provide quick answers with on demand improvement of analytical results
- Incorporate presentation linked with analysis
- Export easy to understand representations of results in real time (VisMaster, 2010).

There are many different types of risks in the financial domain and a multitude of ways in which to explore them (see for example Ramnath et al., 2008 or Brammertz et al., 2008). We suggest that, as diverse as financial risk analysis is, it is possible to understand the tasks associated with performing such analyses in terms of more generic perceptual, cognitive or knowledge-based tasks. One of the first steps in the application of VA, then, is to determine the analytic (also called insight) provenance of financial risk analysis tasks, that is, the perceptual and cognitive operations that analysts must perform in order to derive insight into the financial risk analysis problem space. Visual analysis researchers have used a variety of methods to gain insight into the analytic provenance of various tasks, including cognitive task analysis (Bailey, 1996), discourse analysis (Arias-Hernandez et al., 2011), and scenario-based design (Carroll, 1995).

Once the basic perceptual and cognitive operations have been identified, it is possible to determine the type of visualization that is likely to meet the task requirements. Wehrend and Lewis (1990) developed a domain agnostic scheme for classifying InfoVis techniques and their uses. Zhou and Feiner (1998) refined the Wehrend and Lewis taxonomy, organizing their taxonomy by visual accomplishments (e.g., inform or enable) and visual implications (i.e., what visual capabilities are called upon to achieve the visual accomplishments). Shneiderman (1996) has proposed a task-by-data-type taxonomy that crosses information-seeking visualization tasks (overview, zoom, filter, details-on-demand, relate, history, extract) with different types of data (1D, 2D, 3D, multidimensional, time series, network, tree).

To address what they see as a need to focus on the higher-order tasks that generate insight from visual analysis, Amar and Stasko (2004) have developed a knowledge task-based taxonomy aimed at addressing complex decision-making, especially under uncertainty, and learning a domain. The knowledge tasks they include in their taxonomy are:

- Expose uncertainty (i.e., in statistical models, data quality)
- Concretize relationships (i.e., between people and their work)
- Formulate cause and effect
- Determine domain parameters (i.e., determine the attributes of a dataset that shape our thinking about it)
- Multivariate explanation
- Confirm hypotheses.

In a subsequent paper Amar et al. (2005) proposed an empirically based model of low-level tasks. The tasks they identify are:

- Retrieve value
- Filter
- Compute derived value
- Find extremum
- Sort
- Determine range
- Characterize distribution
- Find anomalies
- Cluster
- Correlate.

Cognitive fit theory states that decision-making is improved when an information representation matches the problem solving task (Vessey, 1991; Tegarden, 1999). Tegarden (1999) has further elaborated that "It is possible that good visualizations are not only task dependent, but they may also be domain dependent. What is needed is a good taxonomy of domains, tasks, and visualizations." A comprehensive methodology still needs to be developed to map and subsequently validate the relationships between financial risk analysis, an abstract, high level activity, and more concrete, low level activities, such as perceptual, cognitive and knowledge-based tasks.

22.3.2 Data

The next component of the VA triad to consider is data. In any application of VA, the analyst is looking to derive insights from data, usually masses of it. VA therefore requires careful analysis of which data are required for the analytic task and of the attributes of the data in order to determine the best way to render them as graphical representations. Some important factors to consider in relation to data are outlined in the following sections.

Types of data

There are many types of data that may be used for financial risk analysis. Some examples from the literature of the type of data to which VA has been applied include:

- Stock portfolio data (Jungmeister and Truro, 1992);
- Fixed income and foreign exchange trading data exported from an Excel spreadsheet as a .CSV comma delimited file (Brodbeck et al., 1997);

- A high-dimensional, multivariate stock market registry dataset comprising approximately 3,000 portfolios made up of a selection of 2,000 different stocks in 54 market sectors in which the changing portfolio contents of all the registered fund managers is held at a granularity of approximately one month (Dwyer, 2003);
- Intraday stock tick data (Roberts, 2003);
- OLAP system data comprised of sales data by customer, broker and product (Ieronutti and Pighin, 2009);
- A dataset on prices and earnings by city (Girardin and Brodbeck, 2001);
- Wire transfers (Chang et al., 2008).

For the purposes of this discussion, we will deconstruct these types into data that are primarily numeric (or ordinal), data that are primarily textual, time series data, and data with spatial characteristics.

Data that are primarily numeric often are in the form of lists of records with one or more variables or attributes. Typically, these lists originate from data in tabular form (e.g., a spreadsheet or a relational database table). The rows in the tables usually represent categories of items (e.g., an asset, a client, a portfolio), while the columns represent attributes of items, or "dimensions," of the data (e.g., price, coupon rate, maturity date). Table cells typically contain numeric values for each dimension. Numeric values can be binary (assuming values of 0 or 1), discrete (e.g., taking on only integer values or values from a specific subset) or continuous (i.e., representing real values). Table cells may also contain nominal or non-numeric data. These nominal data may be either categorical (i.e., a value selected from a short list of possibilities such as person A, B or C), ranked (i.e., values that have an implied order, such as bond ratings of AAA, AA, or A), or arbitrary (i.e., a variable with a potentially indefinite range of values with no implied ordering such as addresses) (Ward et al., 2010). Financial risk analysis involves analyzing one or more dimensions of data for large numbers of items. As an example, we may want to conduct an analysis of a comma delimited .CSV file containing a list of records exported from a spreadsheet in which each line in the list represents a data point described by two types of attribute (a vector): a numeric continuous value representing the stock price and a value representing time.

In addition to predominantly numeric data, data may be predominantly nominal, taking the form of text in a structured form or in a relatively unstructured form, such as a document. Examples include emails, word processed documents, blogs, wikis, and tweets on a variety of topics. Such documents contain words that are organized into phrases, sentences, paragraphs, etc. according to rules of grammar (syntax) to create communications that have particular meanings within a given context (semantics). They may also contain other elements that are non-textual, for example, dates, which may also be analyzed. Textual documents also often have a format or internal structure, for example, a signature-block in an email communication that indicates the name of the sender of the email. Each word, phrase or other atomic element of a textual document or collection of documents may be analyzed. Attached to these documents may be a variety of metadata, such as a creator's name or the date of creation, which may also be analyzed, thereby adding to the dimensionality of the dataset. For this reason, it is the usual practice to reduce the dimensionality of textual datasets through the use of such techniques as Principal Component Analysis or Multidimensional Scaling. To visually encode textual data for visual analysis, it is necessary to parse the data and transform it into vectors. It is then possible to run algorithms suited to the analytic task in question in order to generate visual representations of the data (Ward et al., 2010). For instance, if the analyst were interested in detecting patterns within a textual dataset, (s)he could use a standard *k*-means clustering algorithm to identify groups of documents concerned with common themes.

Financial datasets often contain time series data. Typically, time is treated as just another attribute, or dimension, of data with discrete points in time matched with other attribute values to form a vector. These points in time may be distinguished by seconds, minutes, days, months, years or ranges of times. Time may also be represented in some datasets as dates combining numeric values or a combination of numeric values and nominal values, as in a textual document in which a date appears as day, the name of the month and year.

Some data are inherently spatial, such as a geographical location, or a field of measurements plotted in three-dimensional space. Financial data are less often inherently spatial in nature, though it is possible to find datasets which combine numerical values with categorical alpha-numeric data (e.g., city or country name) and geospatial data (GIS coordinates).

Size/volume

The number of objects in a dataset, or scale, is also an important consideration. Visualizations that work well with a small number of objects may not scale up well when the number of objects increases. For instance, a line chart that shows the prices for a single stock over a period of time is clear. Add five to ten others, and the visualization is still legible. Increase the volume of the dataset to thousands and the graph becomes impossible to read. One solution is to use a more scalable visualization to represent the data. Yet another approach is to use interaction techniques, such as the ability to filter the data, so only a subset is shown. Filtering can be combined with navigational zooming to focus in on only the data of interest. Another approach is to aggregate the data to reduce the amount to visually encode. This approach can be combined with providing overview visualizations that represent the aggregated data and another "drill down" view where the data are disaggregated and can be explored in detail. Finally, "Focus+Context" is an approach to data reduction wherein the user interactively chooses a subset of data to focus on whilst retaining a view that allows the user to see an overview of an entire dataset (Munzner, 2009).

Dimensionality

Dimensionality is another important term related to data and datasets that is essential to understand for the purposes of VA. A dimension is an element in a dataset that categorizes scalars, vectors and tensors into non-overlapping regions (e.g., a cell in a table). It is roughly equivalent to the value of a variable in statistics. Financial data are often of high dimensionality; that is, they consist of many different attributes or variables. As it can be difficult to transform high dimensional data into meaningful visualizations, it is sometimes necessary to employ techniques for reducing the dimensionality of data. A number of different techniques may be used.

Dimensional filtering

- *Slicing*: one dimension may be chosen for elimination and then shown in a lower level visualization (Munzner, 2009).
- *Projection*: no information about the eliminated dimension is retained (Munzner, 2009), retaining a small set of dimensions "that faithfully represent most of the structure or variance in the dataset," which may be among "the original ones, or synthesized new ones that are linear or nonlinear combinations of the originals" (Munzner, 2009).

Dimension reduction. This entails using statistical techniques to determine where there is a high level of correlation in the data so that highly correlated data may be treated as a group. Principal Component Analysis is a widely used linear method. Many non-linear approaches have been proposed, such as Multidimensional Scaling (MDS). These methods are usually used to determine whether there are large-scale clusters in a dataset indicating highly correlated data (Munzner, 2009).

Visual design. Another approach is to use a different visual design (e.g., 3D graphics) that permits encoding of a higher number of dimensions into a single visualization. Many of the innovations in InfoVis have come from the need to increase the dimensionality in representations and analysis of data. Using superimposed or juxtaposed visual displays, each of which contains some of the dimensions of the data, is another technique commonly used.

Number of parameters

An analyst may wish to subject a given dataset to univariate or multivariate analysis. Where multivariate analysis is the objective, the analyst may be interested in analyzing values associated with any or all of the dimensions of the dataset. These dimensions may represent dependent (i.e., variables that have a value that depends on other variables, such as price, which may be affected by time or type of asset) or independent variables (i.e., those that are not affected by other variables, such as city or country). To effectively map data to a graphical representation for the purposes of multivariate analysis, thought must be given as to how the mapping will represent the variables and facilitate the analysis.

Structure

Data may have relatively simply structures or semantics, such as is found in lists, uniform grids or tables, or they may have more complex structures. Structures may exist both in relation to the means of expression (syntax) or the types of interrelationships within given records in a dataset set or between records (semantics) (Ward et al., 2010). For example, portfolio data has hierarchical structure; that is, it comprises a top-tier of values representing portfolios (e.g., p_1, p_2, $p_3 \ldots p_n$) and a lower-tier of values representing the assets comprising the portfolio (e.g., a_1, a_2, $a_3 \ldots a_n$). Directory file structures offer another example of a hierarchical dataset. Datasets may also express relationships between items, such as the relationship between a portfolio and the assets within it. Thus, in mapping the structure of a dataset, a designer will want to consider what aspects of the structure are analytically significant to the task and map all of those to some element of the graphical representation. In datasets that have a clear structure (e.g., spreadsheets, tables) the process of identifying the data structure will be easier than when the data are comparatively unstructured (e.g., documents). In the case of unstructured documents, a designer may find it necessary to use some additional technique, such as natural language processing, to analyze the structure of the data.

Range and distribution

A problem of particular concern in the use of InfoVis for financial risk analysis is the issue of what to do about large and unevenly distributed datasets. Quite often, such datasets can lead to unappealing visualizations as some parts of the screen will be crowded while other parts will remain bare. Common techniques for addressing unevenness in datasets are scaling or filtering of the data on one or more dimensions to remove "uninteresting" data points. As these techniques can remove important characteristics of the dataset, such as a high number of outliers perhaps signifying volatility, they need to be used with extreme caution in a financial context. Tufte (1983, 1990) always advocates showing users the data rather than simply showing them aggregations of the data. Ward et al. (2010) also advise building visualizations that show both raw data and derived or cleansed data so that users are able to see the difference in the results. Whether working with raw data or aggregations, the analyst should carefully document every intervention involving the data, be it choice of dataset or the use of filtering or scaling techniques, so that anyone subsequently working with or interpreting the results of the visualization can take these interventions into consideration.

Dynamics

Financial data may consist of historic data, which are static and unchanging, or streaming data from live feeds that are constantly being updated (e.g., stock tick data). This is an important consideration in VA because some VA tools use algorithms for dimension reduction and for grouping or clustering items. Some of these algorithms work better than others when new records are added to datasets (Rooney et al., 2008).

Data quality

Another factor to be taken into consideration is the problem of "dirty data" as it is not uncommon to find that data have missing or incorrect values due to human error or technological malfunction. It may be necessary, therefore, to cleanse the data prior to VA. There are several techniques that can be used to address missing data values or to cleanse data. These are outlined in Table 22.1. Care should be taken to carefully note any pre-processing done on the data as this may affect the outcome of the analysis.

Table 22.1 *Data cleansing and noise reduction techniques (adapted from Ward et al., 2010, pp. 51–52)*

Technique	Advantages	Disadvantages
Discard bad record	Easy to implement	Not sustainable if number of bad records is high Information loss, e.g., bad records may provide interesting data
Assign sentinel value	Do not lose information (e.g., bad records easy to see when visualized) Easy to implement	Care must be taken not to perform statistical analysis on sentinel values
Assign average value	Minimally affects the overall statistics for the variable	May mask or obscure outliers or highly polarized data (e.g., average values may not work well if markets are volatile)
Assign value based on nearest neighbor	Can be a better approximation for a substitute value	Value of the neighbor may be affected by other variables and therefore not be a good substitute
Computer substitute value (imputation)	More time consuming and complex	High statistical confidence

Transforming data into visualizations

To create visualizations, data must be mapped to graphical marks in meaningful ways. There are a range of possibilities for data mappings, including the following.

- One-to-one mappings, where each data attribute maps to a distinct and different graphical attribute (e.g., color represents bond rating where green is AAA, yellow is AA, and red is A).
- One-to-many mappings, where redundant mappings are used to improve the accuracy and ease with which a user can interpret data values (e.g., where stock prices are mapped to color with green representing a high price and to size with bigger marks representing higher prices).
- Many-to-one mappings, where several or all data attributes map to a common graphical mark (e.g., where lines represent both the stock opening price and its closing price, but the positioning of the line indicates its meaning) (Ward et al., 2010, p. 259).

In the following sections, we present a brief compendium of types of InfoVis in finance, highlighting how information has been mapped to these graphical representations and providing examples of innovations in InfoVis that have built upon earlier visual "metaphors" to create more meaning-rich visualizations in support of financial risk analysis objectives.

Candlesticks

Possibly the oldest visualization in finance is the Candlestick Chart. The candlestick chart is used in stock trading and technical analysis in finance even today (see Figure 22.2).

Creation of a candlestick chart begins with a dataset containing open, high, low, and close values for each time period for one or more securities. The square at the center of the candlestick is the "body." The body may be either hollow, indicating that the security has a closing value higher than its opening price, or filled, which indicates that the security has closed down from its opening price. The thin lines above and below the body are called "shadows" (also sometimes referred to as "wicks" and "tails"). If the security closes above its opening price, the upper shadow (or wick) indicates the closing price and the bottom shadow (the tail) indicates its opening price. If the security closes below its opening price, then the top shadow represents the opening price and the bottom shadow represents the closing price. Using this system of graphical representation, it is possible to encode five dimensions of data for each security in a 2D spatial layout – the opening value, the closing value, the high value, the low value, and time periods. The use of "fill" in the body of the glyph reinforces the message about whether the security's value is up or down for the time period and also signals how the analyst should interpret the shadows.

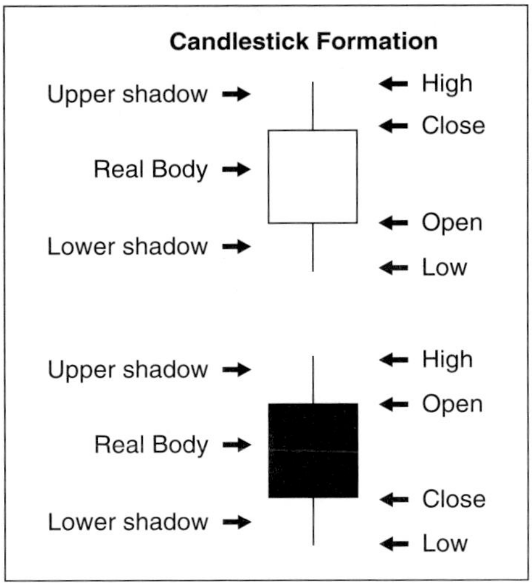

Figure 22.2 An example of the Candlestick Chart said to have been invented in Japan by Munehisa Homma in 1755.

Compared to bar charts, many traders consider candlesticks more visually appealing and easier to interpret (Sawant, 2009). An additional dimension of data – trading volume – can be encoded in a candlestick by altering the size of the glyph (see Figure 22.3). In this example, the longer the body of the candlestick, the higher the trading volume has been; the shorter the body, the lower the trading volume.

Line graphs, bar charts, pie charts and circle graphs

William Playfair is credited with the invention of the line chart, the bar chart, the pie chart and the circle graph (Friendly and Denis, 2005). Most financial graphs are a variation of the line chart, with time on the *x*-axis, price on the *y*-axis and volume as a separate bar chart below. To compare one security with other securities or indexes, some charting tool providers such as Google Finance and Yahoo Finance, allow the user to add data onto the observed chart (Lei and Zhang, 2010). Adding more stocks for comparison, however, makes these charts difficult to read (see Figure 22.4).

Ring maps

To overcome the limitations of traditional candlestick and line charts, Lei and Zhang (2010) devised a novel ring-based visualization (see Figure 22.5). These researchers observed that looking at the performance of one stock and assessing its riskiness on the basis of price change or a volatility calculation could result in a skewed impression – the stock may appear quite volatile and therefore risky, but an assessment of its volatility in relationship to the market as a whole is needed before a conclusion can be drawn. The ring view addresses this need by offering

Figure 22.3 Example of a candlestick diagram with elongated glyph to indicate trading volume.

"Context+Focus," which is the ability to see the detail without losing knowledge of the market as a whole.

To construct the rings, they first calculated each stock performance and placed it into a corresponding bucket. They created 40 buckets ranging from −10% and below to +10% and above. Each bucket was reviewed to ensure the number of data points was within the specified bucket limit. If it was over by 40% of the specified limit, the

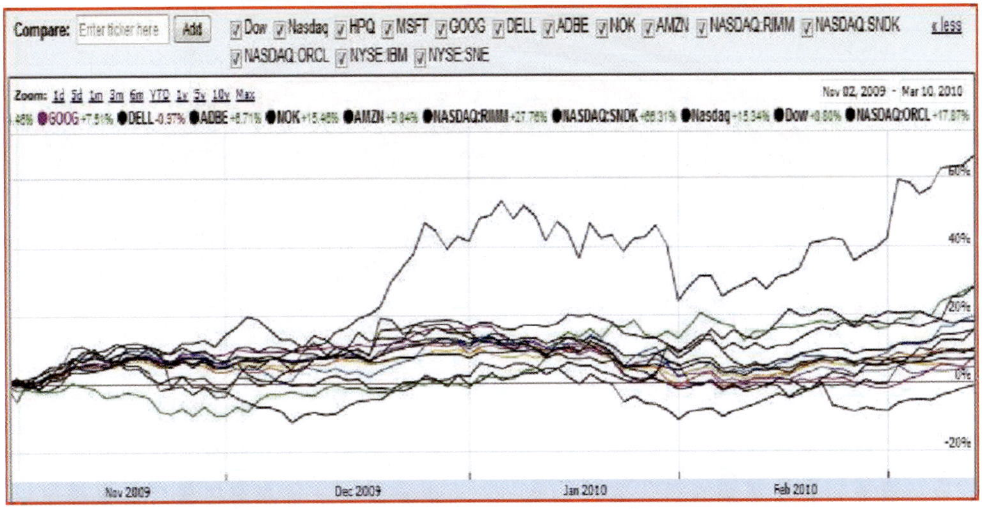

Figure 22.4 Stock market chart showing the values for several stocks. After several securities are added, the chart becomes difficult to read.

bucket was divided into two new sub-buckets. If it was less than 40%, data were distributed on the two extremes into the two neighbouring buckets. If the total of two adjacent buckets was less than the specified bucket limit, these two buckets were merged. They then drew the rings, with the data in each bucket forming a ring. They summed up the total capital size of all stocks in the same bucket and then calculated their proportions over the total. The arc length of each stock was then derived from its proportion of capitalization. They followed this process iteratively through each bucket and drew rings until all buckets were processed. Clustering was used to group stocks of the same sector together in a ring, and color variations were used to reinforce the allocation of stocks to sector clusters. Different ring sizes were used to indicate the scale of price change in percentage for all stock data, and different color background was used to show rising and declining stocks. Within a ring, the capitalization size of a stock was represented by its arc length. Special marks on the rings were used to show special stocks, while locate and identify functionality was achieved by allowing users to move the mouse pointer to any part of the rings to display details such as a stock codes.

Scatterplots

After Playfair's inventions, the most common type of InfoVis in finance is the scatterplot. It is also the first true 2D one (Friendly and Denis, 2005). A scatterplot is a plot of two variables, x and y, measured independently to produce bivariate pairs (x_i, y_i), and displayed as individual points on a coordinate grid typically defined by

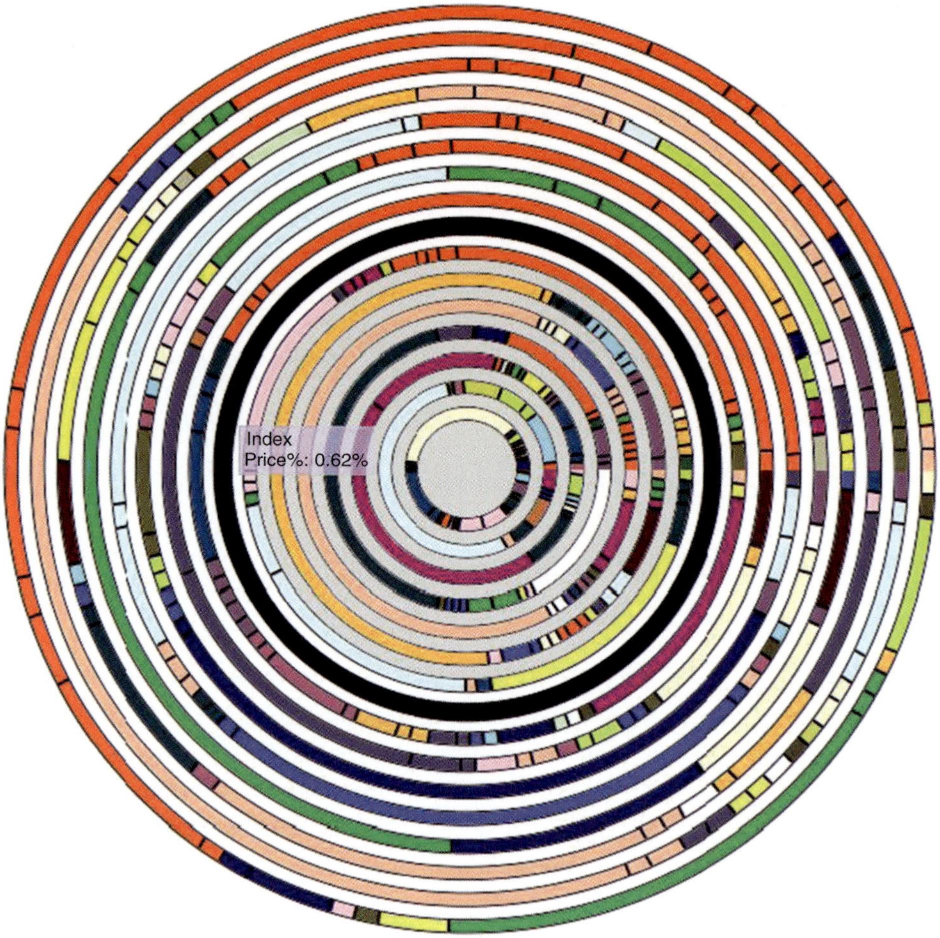

Figure 22.5 Lei and Zhang's ring diagram.

horizontal and vertical axes, where there is no functional relation between x and y (Friendly and Denis, 2005).

Modern enhancements of the basic scatterplot include a variety of methods designed to show more than a 2D view (e.g., glyph plots), to deal effectively with more complex data, and to allow the user to work interactively with collections of bivariate displays (Friendly and Denis, 2005).

Bead cluster diagrams

To extend the more traditional scatterplot visualization, Lei and Zhang (2010) used the price change and price value of a stock to create clusters of beads, wherein each bead represents the price of the stock at a given time (the horizontal axis) as well as the price change in percentage (the vertical axis) (see Figure 22.8). The color

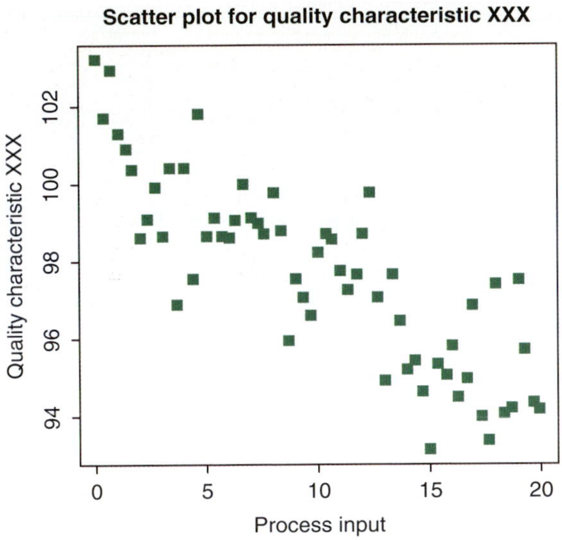

Figure 22.6 Traditional scatterplot.

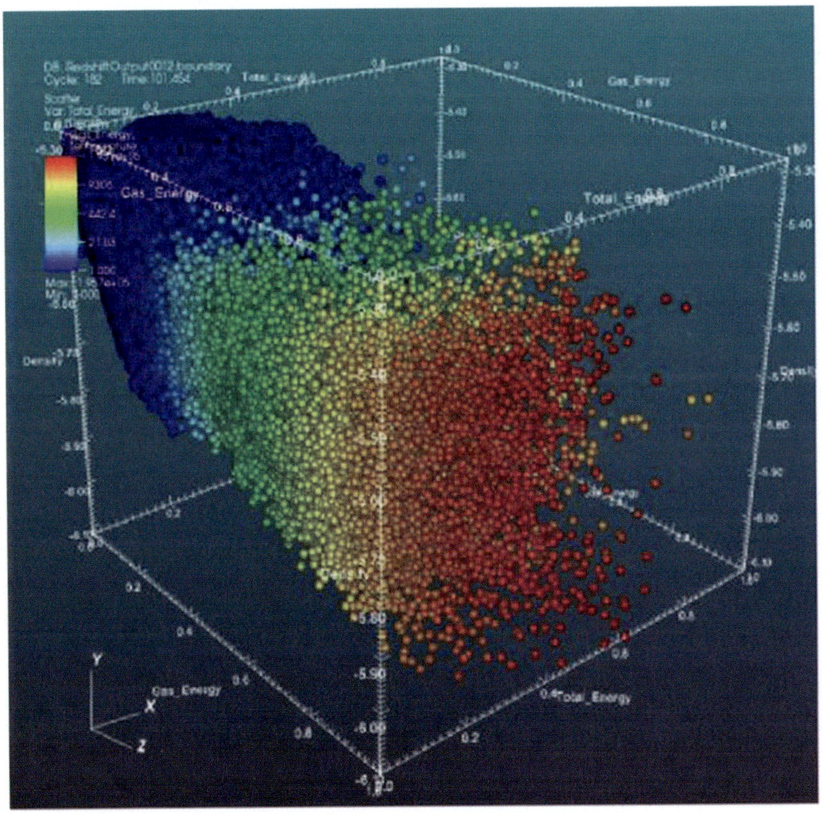

Figure 22.7 Example showing the extension of a scatterplot beyond 2D.

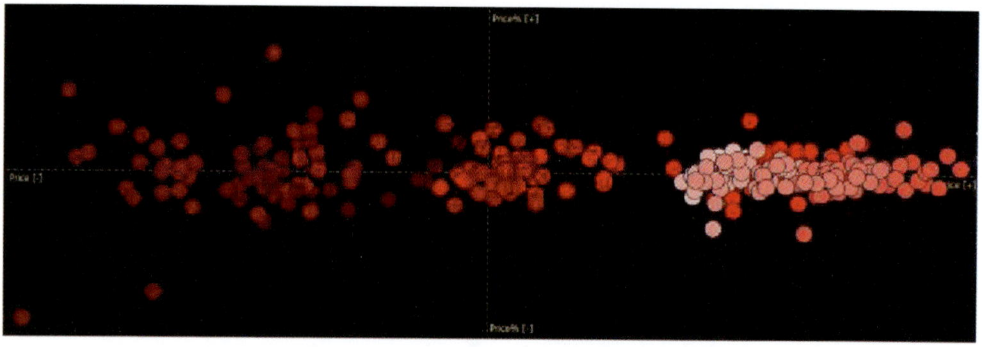

Figure 22.8 Lei and Zhang's (2010) bead clusters.

shading of each data point indicates the date. The darker the bead, the earlier the date it is recorded on. The advantage of this plot is that it condenses the data about price and price change over time within a finite display area and clearly shows the aggregated areas. Lei and Zhang used the plot to test the theory that there is a relation between price clusters and psychological barriers (Sonnemans, 2006). These psychological barriers can be interpreted as the resistance and support level of a stock. The implication of resistance and support is that the price of a stock should stay in between the prices at the support and resistance level. If the price drops below the support level, it will drop even further. Conversely, if the price breaks the resistance level, the chance of rising in the next few trading days is much higher. Reading the plot it is easy to see that the light pink bead in the lower left quadrant predicts the price drop evident in more recent times (the white beads).

Parallel coordinate plots

Parallel coordinate plots (PCP) were first introduced by Inselberg in 1985 as a mechanism for studying high-dimensional geometry (Inselberg, 1985; Ward et al., 2010). Since then, this visualization has become widely used for multivariate data analysis. In PCP, axes are parallel rather than orthogonal as in scatterplots, with evenly spaced vertical or horizontal lines representing a particular ordering of dimensions. A data point is plotted as a polyline that crosses each axis at a position proportional to its value for that dimension. To interpret the plot, the analyst looks for clusters of similar lines, which indicates partial correlation between pairs of dimensions. Similar crossing points indicate partial negative correlations. Lines that are either negative or have a slope that is significantly different from adjacent lines represent outliers. Like scatterplots, PCP is designed to show relationships between pairs of dimensions. As with line graphs and scatterplots, as the data size increases, a PCP diagram will become cluttered and difficult to read.

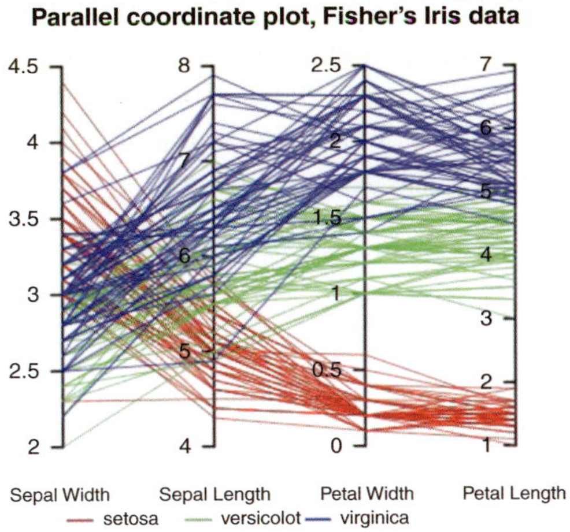

Figure 22.9 Parallel coordinate plot.

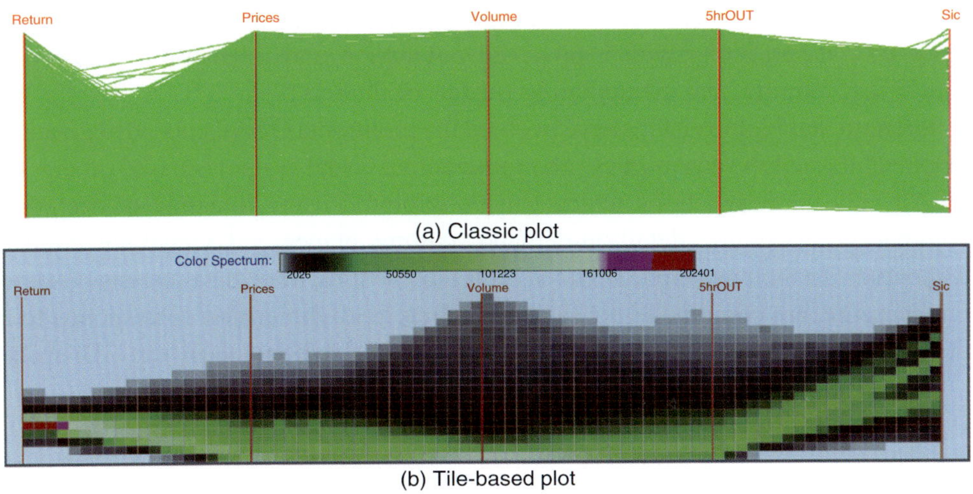

Figure 22.10 Tile-based PCP.

Alsakran et al. (2010) proposed an extension of PCP to overcome some of the limitations of the traditional diagram. They employed a tile-based parallel coordinates visualization, inspired by the art of mosaic, where the plotting area is divided into rectangular tiles. Each tile stores an intersection density that counts the total number of polylines intersecting with that tile. The intersection density is mapped to optical attributes, such as color and opacity, by interactive transfer functions. The method visualizes the polylines in accordance with the density distribution to reduce visual cluttering. They applied their approach in a case study on analysis of US stock mutual fund data for the year 2006.

OLAP visualization and visual OLAP

Data from databases often contain multidimensional data which can be difficult to query. OLAP techniques (Codd et al., 1993; Chaudhuri and Dayal, 1997; Gray et al., 1997) evolved as a means of overcoming the limitations of traditional approaches to querying high-dimensional database data and rely upon representation of multi-dimensional datasets as data cubes. A data cube is an *n*-dimensional matrix in which numeric facts called measures are categorized by dimensions. Typically, the result of an OLAP query is displayed in a 3D matrix format, where table rows and columns correspond to cube dimensions, while table cells correspond to measures (see Figure 22.11). An OLAP cube allows fast analysis of data and the capability of manipulating and analyzing data from multiple perspectives. This process is known in the literature under the terms OLAP Visualization and Visual OLAP (Ieronutti and Pighin, 2009). Cuzzocrea and Mansmann (2009, p. 8) describe visual OLAP as "An umbrella term encompassing a new generation of end-user tools for interactive ad-hoc exploration of large volumes of multidimensional data via providing a comprehensive framework of advanced visualization techniques for representing retrieved dataset, along with a powerful navigation and interaction scheme for specifying, refining, and manipulating subsets of interest."

Ieronutti and Pighin (2009) have evolved the traditional OLAP cube visualization to extend its analytic capabilities. Their interactive visual system is based on the 3D OLAP cube representations which can be explored using dynamic queries that combine color-coding, detail-on-demand, cutting planes and viewpoint control techniques. In their system, 3D cubes each representing three data dimensions and one measure, are combined into 3D hypercubes. Each dimension corresponds to a descriptive attribute (i.e., data dimension), while each point in the 3D space

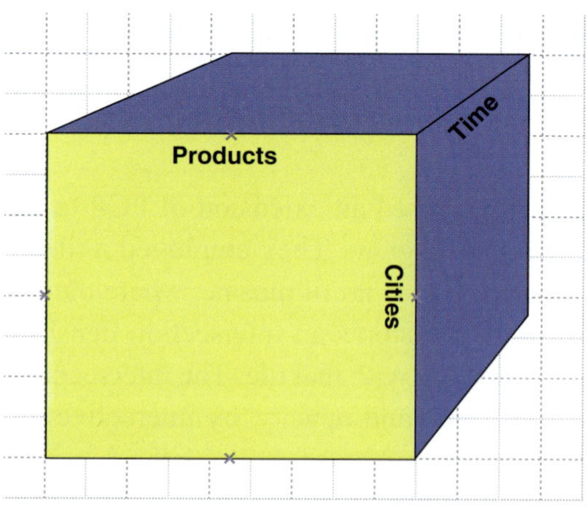

Figure 22.11 Example of an OLAP cube, showing three "dimensions" of data.

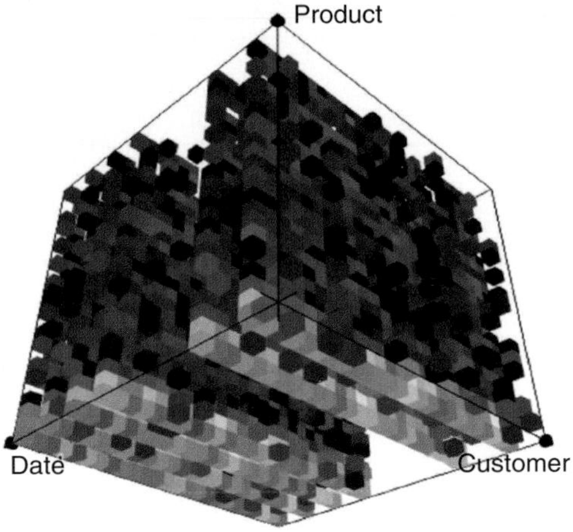

Figure 22.12 Ieronutti and Pighim's (2009) interactive InfoVis of multidimensional OLAP hypercubes.

corresponds to a numeric field (i.e., measure). Users are able to interactively explore each dimension of data at different levels of granularity by manipulating the hypercube (e.g., using point and click, details on demand, or rotate functionality).

Node-link diagrams

A node-link diagram, sometimes also referred to as a network graph, is a type of graph consisting of an ordered pair of nodes (or vertices) connected by lines, called edges (Bondy and Murty, 2008, p. 2; Caldwell, 1995) which represent a relationship between the two nodes. This type of visualization can be used to model many types of relations such as those in relational databases, ontologies, financial systems or social networks. Graphs draw upon graph theory, a body of knowledge that dates from the eighteenth century and that, in the 1950s, began to be applied in network analysis (Biggs et al., 1976; Barnes and Harary, 1983). There are a number of different types of such graphs: a simple graph has only one edge between two nodes, whereas a multigraph has more than one. If edges can only be followed in one specific direction this is called a directed graph. Graphs that have edges that can be followed in either direction are called undirected. A pseudograph has edges that begin and end at the same node. The placement of nodes and edges within the graph is usually governed by a layout algorithm, with different types of graphs using different layout algorithms.

One example of the use of node-link diagrams is Kubelec and Sa's (2010) analysis of the geographical composition of national external balance sheets from 1980 to 2005. In this study, the researchers viewed a financial system as a network, where nodes represent countries and edges represent bilateral financial assets. The researchers compiled a dataset and, using Pajek (Program for the Analysis and

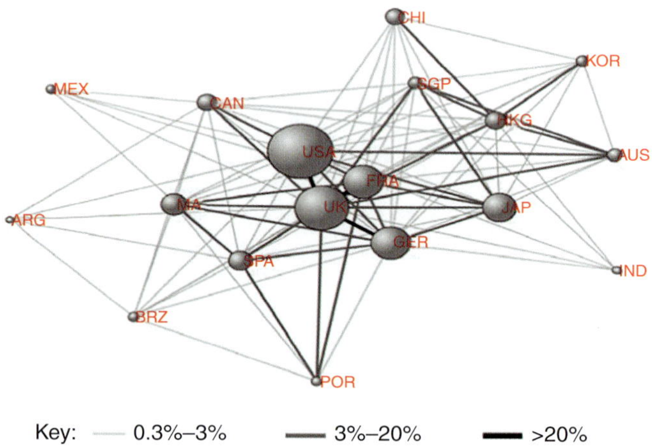

Figure 22.13 Kubelec and Sa's (2010) network graph.

Visualization of Large Networks) created both an undirected graph of the financial network that looks at bilateral financial assets and a directed graph that looks at the ratio of gross bilateral assets to GDP for particular countries. In designing the graph, the researchers also applied weightings. For example, links represent the strength of the connections between nodes, and not simply whether a connection exists or not, by the thickness of the lines. The size of the node is proportional to the country's financial openness, measured by the sum of its total external assets and liabilities. Pairs of countries with stronger links are placed closer to each other as well.

Chang et al. (2008) used a simple network graph in a VA tool which they call WireViz to show relationships between keywords in wire transfers in an application of VA aimed at detecting financial fraud.

Treemaps

Ben Schneiderman's (1992) and Martin Wattenberg's (1999) work applying Treemaps in the area of finance stands out as a significant contribution to innovation in the application of InfoVis in finance. Their work has been commercialized as Smart Money's ™ Map of the Market (Smartmoney™, 2009). The treemaps visualization is designed for hierarchical data structures, using juxtapositioning to show relationships as opposed to nodes and edges as in node-link diagrams. In the SmartMoney™ application of treemaps, the size of each company's market capitalization determines the size of its rectangle. The map has a hierarchy that is three levels deep: the entire market, market sectors, and industries. Within each level the stocks and their groupings are static, but moving the mouse over them gives information about each stock in a pop-up box. Clicking brings up a menu that allows drill-down into a variety of pages giving detailed information on a stock. Each stock is color coded to show percentage

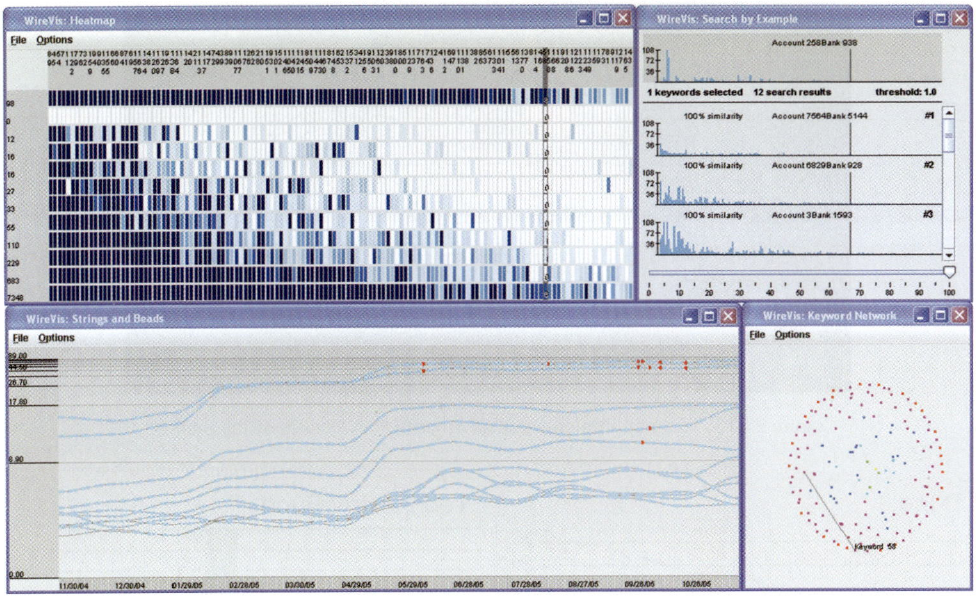

Figure 22.14 WireViz.

change, using a user-selectable red/green or blue/yellow spectrum to show loss/gain. Finviz.com™ uses a similar approach to provide an overview of the market, but has extended the approach by creating a 3D visualization to show the size of stock price increase or decrease. Many variations of this design have been developed since the introduction of treemaps, including squarified treemaps to reduce the appearance of elongated rectangles and nested treemaps to emphasize hierarchical structure (Ward et al., 2010).

Heatmaps

The heatmap displays a 2D table of items using color rather than text. All data values are mapped to the same normalized color-space, and each is rendered as a colored square or rectangle (Ward et al., 2010). Ziegler et al. (2008) employed a heatmap visualization that uses color and luminosity to indicate long-term performance of investment funds. The authors claim that their method has two advantages over traditional multivariate analytic approaches. A heatmap is more space efficient than a text-filled table, and the investor's areas of investment focus or preferences can be indicated through the use of weighting and a highlighted area of the visualization.

3D spatial layouts

Most visualizations are created as mappings of three dimensions or more of data represented in 2D spatial layouts and rendered in 2D (i.e., on a computer screen).

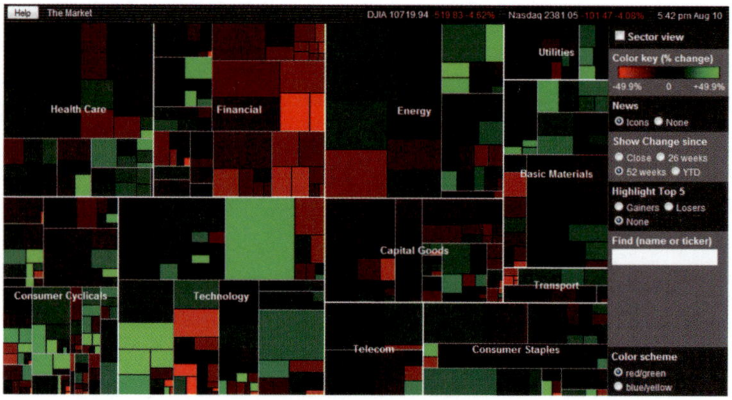

Figure 22.15 Treemap implemented as SmartMoney™ map of the market.

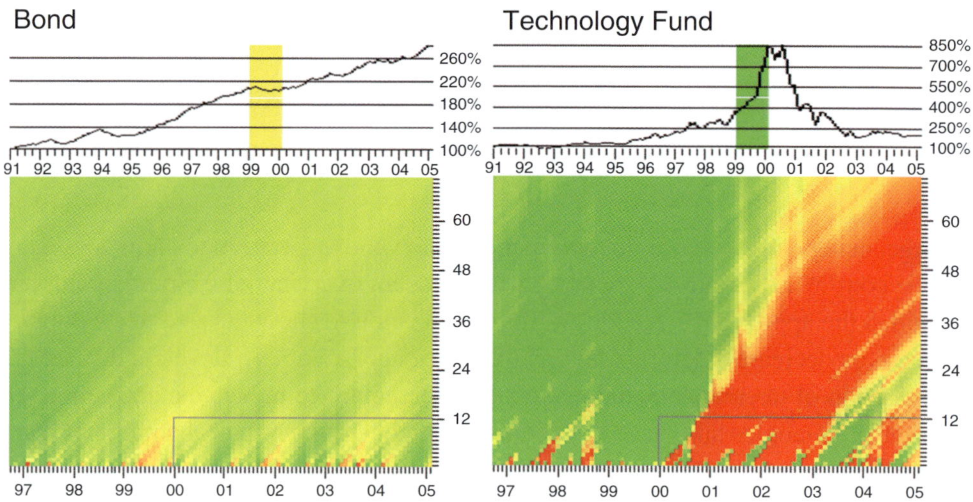

Figure 22.16 Zeigler et al. (2008) heatmap.

However, it is also possible to create representations of three dimensions or more of data shown as 3D spatial layouts but rendered in 2D. Munzner (2009, p. 687) notes that when computer-based visualization began in the late 1980s, and interactive 3D graphics was a new capability, there was a lot of enthusiasm for 3D representations. With 3D visualizations their increased surface area and potential for more visual features makes it possible to encode more information. There are costs to creating 3D graphical representations, however, so it is not always to be assumed that 3D spatial layouts are an improvement over 2D. One problem is occlusion, which occurs when part of the graphic is obscured by another part. To overcome this difficulty, navigational tools have been developed to help people rotate 3D graphical representations;

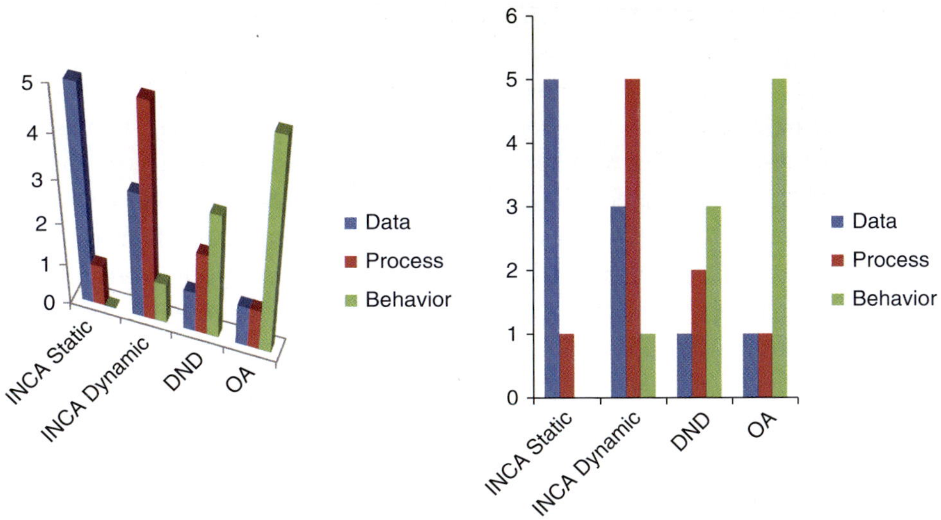

Figure 22.17 Figure showing a problem with visual perspective. It can be difficult to gauge the accuracy of the perception of height with visual representation of abstract objects, such as information.

nevertheless, it remains difficult for many people to use these navigational tools when navigation involves abstract visual dimensions (Plumlee and Ware, 2006; Munzner, 2009). Another potential pitfall of 3D visualizations over 2D is perspective distortion. Although people can often judge the heights of familiar objects in the real world, they cannot necessarily do so with completely abstract objects (e.g., a 3D bar chart) where the height conveys meaning (Munzner, 2009).

Nevertheless, the introduction of a third dimension does offer the possibility of encoding a greater number of dimensions of a dataset in a visualization.

22.3.3 Interaction techniques

Interaction with the InfoVis is a critical component of VA. It is by interacting with visualizations that the analyst is able to explore the data visually to gain new insights or explore hypotheses. There are a number of ways in which users of VA tools may be able to interact with those tools. Yi et al. (2007) have identified the following interaction techniques:

- Navigation in the visualization space by panning, rotating and zooming an image;
- Selection for further investigation or deletion of an object, a collection of objects or regions of interest in the visualization by highlighting, encircling, brushing or distorting a part of the visualization to draw attention to it;

- Filtering the number of items of data or data dimensions to reduce the amount of the data mapped to the visualization (e.g., identifying a range in the time axis by using a slider for further analysis);
- Encoding to change the graphical attributes of the visualization, such as point size, color, or line type, or even to change the type of visualization (e.g., changing from a scatterplot representation of multivariate data to a parallel coordinates representation). Many VA tools will support multiple types of visualizations because no single visualization is effective for all tasks. No best visualization for a particular type of financial risk analysis has been identified, so the analyst will benefit from using a variety of interaction techniques to re-encode the data to create different visualizations or visualizations of different subsets of data and to navigate through the visualizations in an exploratory fashion;
- Connecting to link different views of objectives to show related items;
- Abstracting/elaborating to modify the level of detail.

Ward et al. (2010) have studied a wide range of interaction techniques in VA tools and have developed a framework of interaction techniques consisting of:

- Type of operator (e.g., navigation, selection, manipulation, distortion, filtering);
- Space of interaction (e.g., screen, data value, data structure, attribute, object or visualization structure);
- Parameters of interaction operator (e.g., focus of the visualization, extent of the visualization space or underlying dataset, transformation of data or visualization, and blending of parts of space touched by more than one interaction).

Pair analytics (Arias-Hernández et al., 2011) is loosely based on "pair programming" from "extreme programming" software development methods (Beck, 2000; Gallis et al., 2003). In pair analytics a Subject Matter Expert (SME) and a VA Expert (VAE) perform an analytical task using an interactive visualization environment. As in pair programming, the VAE "drives" the software while the SME "navigates" the conceptual analysis of the problem. The VAE is trained in the operation of the visualization environment as well as in general strategies and methods for visualization analysis that make the most effective use of that environment. Their role is to act as the technical analyst, advising on analytic approaches, implementing them by controlling the software, and helping to interpret the resulting visualizations. The SME's deep understanding of the subject domain enables them to provide the conceptual interpretation of the visualization and its integration with concepts and approaches specific to the problem domain.

Immediate benefits for the SME and their organization are insight into analytic processes and how visual analytic tools might enhance them, guidance of effective tool selection and configuration, matching tools to characteristics of tasks and data, and training of SME analysts that might enable them to work on their own in many cases. During the analytic interaction, the SME may provide expert knowledge to suggest visual comparison of relevant variables, detect patterns and generate or test hypotheses. VAE analysts gain an understanding of real-world analytic problems and methods and how their analytic methods can adapt to them as well as the beginnings of SME understanding. The VA technology designers benefit from a better understanding of analytic practice, potential use of their tool, errors encountered, and so on.

The interaction of the partnership with the VA tool also generates a human-artifact dialog in which machine-models interact with human mental models. Visualizations created by this partnership may sometimes result in unexpected outcomes that do not fit into existing mental models due to the way the VA tool handles the data. In this way VA methods may support the SME to re-evaluate concepts and methods in the subject area, leading to new advances in their field of expertise.

All of these techniques may be used on their own or in combination with one another to explore datasets, though few VA tools support all of the interaction techniques described above, and Ward et al. (2010) note that much research remains to be done to identify which interaction techniques work best to support particular analytic tasks.

In the 2011 Records and Information Management for Financial Analysis and Risk Management Workshop, Dang and Lemieux (2013) proposed a functional framework for evaluating visualizations and interactions to support common analytics tasks in finance. An excerpt from this framework, showing a small list of visualizations and interactions techniques being evaluated against a set of well-defined elements (that analytic tasks can be decomposed into), is presented in Table 22.2.

In developing the framework, Dang and Lemieux acknowledge that the task of evaluating visualizations and interactions for analytics tasks is complicated by the fact that real life problems are complex, context-dependence, and very difficult to decompose into well-defined elements. The act of decomposing domain problems in a systematic manner involves three processes.

1. Narrowing down the infinite analytic problem space in the real world (even within the specific domain of finance or just a branch of finance) to a finite number of general problem classes that would represent a large number of common analytic tasks.
2. Defining, in precise detail, a finite list of elements that real-world problems can be broken down into. This list must satisfy the requirement of being both general enough that it can be used to describe a large set of analytics problems and yet concrete enough that scientific studies can be designed to measure suitability.

Table 22.2 *Evaluation of several visualizations and several supporting interaction features (adapted from Dang and Lemieux, 2013)*

Blank: Not suitable
"–": Possible but not preferable
"X": Suitable

Blank: No effect on Visualization
"X": Enhances or Enables

		Scatterplot	**Heatmap**	Treemap	**RSF (ringmap, sunburst)**	Times series	**Geocoding**	Similarity clustering (galaxy)	**Tables of values**	**Search and highlight**	Cursor highlight (picking)	**Seamantic zooming**	Graphical panning and zooming	**Overview and dtail**	Direct manipulation of data
Cognitive and perception tasks	Filter/locate	–	–	–	–	–	–	–	–	**X**	X	**X**	X	**X**	
	Find extrema	X	–	–	–				–						
	Determine range	X				–				**X**					
	Characterize distribution	X	–	–	–		**X**	X		**X**					
	Find anomalies	X	**X**	–	–	X	**X**					**X**	X	**X**	
	Cluster/associate	X	–	–	–			X		**X**					
	Correlate	X													X
	Compare through time					X			–						
	Compare between locations		–	–	–		**X**		–						
	Identify								–	**X**	X				
	Compare between entities	X	**X**	X	**X**	X	**X**		–						X
	Categorize			X	**X**										
Time-frame of analytic activity	Real-time transition	X	**X**			X	–								
	Rapid state comparison					X					X			**X**	
Visual scalability	1–100 data point	–	**X**	X	**X**	X	**X**	–	–	**X**	X				X
	Hundreds of data points	X	**X**	–	–	X	**X**	X	–	**X**		**X**	X	**X**	
	Thousands of data points	X					**X**	X		**X**		**X**	X	**X**	
Level of detail of value retrieval	Raw values shown			–	–				**X**						
	Quantitaitve estimates	X				X	**X**								
	Qualitative estimates		**X**	X	**X**	X	**X**	X							
Unit data type	Cardinal	X	**X**	X	**X**	X	**X**		**X**						
	Ordinal	–	**X**						**X**						
	Interval								**X**						
	Nominal			X	**X**				**X**						
	Structured text			X	**X**			X	**X**						
	Unstructured text							X	**X**						
Data model and organization	Flat (table-like)	X	**X**	–		X	X	X	**X**						
	Hierarchical (tree-like)		–	X	**X**				–						
	Network														
	Multi-variate								–						
	Unstructured collection (files)							X	–						

Columns 1–8: Blank, not suitable; –, possible but not preferable; X, suitable.
Columns 9–14: Blank, no effect on visualization; X, enhances or enables.

3. Making and acknowledging assumptions in the definition of analytic problem classes and their elements. At the point of this book's publication, there are many taxonomies of financial terms and financial analysis tasks, and little existing research regarding frameworks for breaking complex financial analytics tasks down into discrete elements. To create a more comprehensive list of elements and to settle upon a definition of, say, "portfolio monitoring," still require making many assumptions about the cognitive tasks involved for the sake of practicality.

22.4 VA tools

VA tools support the process of VA by facilitating pre-processing of data to enable its graphical representation, the actual encoding of the data into graphical form, and subsequent iterative interaction with the visualization(s) to arrive at analytic insight.

A preliminary assessment indicates that a number of commercial technology vendors and service providers offer VA (loosely defined) solutions. Many more VA tools have been developed in academic research labs. Without any intent to endorse particular products or services, we see the commercial VA solutions as being grouped into four broad categories: (1) those that are financial VA "pure play" tools (e.g., Quantum4D™), (2) those that offer more data analytics capabilities (e.g., Matlab™, Cognos™) and have incorporated visualization into their offerings, (3) pure play VA tools that can be used to conduct financial VA but also used for other purposes (e.g., Tableau™) and (4) those that are custom bespoke designs developed by InfoVis service providers or researchers. These solutions range from those that offer visualization as an "add on" mode of displaying data, analysis or search results, which in itself can be extremely helpful to users, to those that offer "true" visual analytic capabilities – the ability to have a human analyst interact with a visualization in order to delve deeply into data to structure, to organize, and, ultimately, understand them.

Generally speaking, the pure play VA tools offer more sophisticated visual interfaces and a greater number of options for interactivity than those that are add ons to financial analysis products. Financial analysis products are often superior to the pure play tools in terms of data pre-processing, however. A small number of financial analysis tools provide integrated advanced visualization as part of their offering, primarily for portfolio and risk management. In terms of configuration – pure play VA tools generally offer the greatest amount of configurability, while the Finance VA tools offer built-in domain-specific semantics and customized visualizations suited to financial tasks.

Though there are many proposed frameworks for comparing and evaluating VA tools, a comprehensive framework has yet to be developed. Nevertheless, all VA tools can be evaluated in terms of the degree to which they support the steps involved in undertaking VA; that is, by how they assist with the pre-processing of data, the graphical representation of the data, and how they facilitate visual interactivity. Moreover, as with other software tools, VA tools can be evaluated in relation to how much configuration is supported (configurability), as well as how much is required to achieve a successful implementation in a particular environment.

Though the VA tools are an important element of using VA for analysis and decision-making, the largest portion of the time and financial cost of developing an analytic support solution is usually in the requirement elicitation and design, not the technical development, and most currently available visualization products do not reduce this cost. Even powerful toolkits, such as Xcelsius and Panopticon, are (at the time of writing this chapter) not yet turnkey VA solutions, but only technological scaffolds for developing solutions. After licensing many VA products, an organization would still need to conceptualize the data, define the analytic problem, and develop, deploy, and maintain the VA solutions.

Last but not least, VA tools sometimes use algorithms for dimension reduction, for grouping or clustering items and for the layout of graphics in screen space. When these algorithms are part of commercially available VA tools, the type and mode of operation may not always be transparent. Each algorithm performs differently, has an underlying logic and may have weaknesses or limitations that it is important to understand. While it is key to understand how different algorithms function, it may not be easily achieved when a commercially available VA tool is used: such algorithms tend to be trade secrets and vendors may be reluctant to be transparent about their structure and how they function.

22.5 Applications of visual analysis in finance – case studies

Having explored the elements of the visual analysis triad, we illustrate the application of these concepts in two detailed case studies based on our original research on the use of VA in financial analysis and decision making and one case study contributed by the software company, Quantum4D™.

22.5.1 A case study on the application of visual analysis in behavioral economics

Analysis

The problems in this case study are focused on financial and economic decision-making tasks in which individuals tend to make sub-optimal choices. The sub-optimal benchmark is derived from economic theory that predicts that individuals will make utility-maximizing decisions. For example, the optimality of choice made in the financial portfolio selection task can be evaluated by Markowitz's Portfolio Theory (PT) (Markowitz, 1952), which predicts an efficient portfolio, one that has the greatest possible return for any level of risk, or the lowest possible risk for any level of return.

Certainly the notion of optimality and whether it represents reality is debated heavily in the field of economics, but this is not our concern because we are focused on the positive economics (Friedman, 1953) of PT. Positive economics, as a science, is concerned with the analysis of economic behavior and whether economics theory can describe and explain economic phenomena. This evidently leads to empirical studies and tests against reality on "how" people can reach optimality rather than "why" people are not optimal. Thus, the secondary problem of arguing for optimality is not within the scope of our work.

Economists have long struggled with a simple question: when an individual makes a sub-optimal choice relative to the prediction of the theoretical model, is this because he/she prefers to make the suboptimal choice, or because he/she is cognitively limited and has not evaluated all options?

Traditionally, economists have placed an emphasis on measuring the levels of cognitive biases or information overload, and as a result have overlooked the possibility of decreasing cognitive biases through new ways of exploring large datasets. Economists are divided in their response to this challenge, either ignoring bounded rationality and proceeding to provide more information to subjects, or acknowledging that sub-optimality of choice is due to inherent cognitive limitations exposed by the need to understand large datasets. When economists do attempt to improve understanding, it is usually through increasingly detailed instructions and examples to the subject, perhaps supported by multiple static graphs. Traditional methods cannot handle large datasets effectively because of the textual representation of these data and inability of subjects to interact with the data directly. The use of effective VA environments may allow us to focus only on preferences by removing the subject's cognitive burden.

VA can also provide experimental economists with much needed guidelines on the display of data in common tasks. To date, economists have not developed a standard for running an experiment – data are presented the way the particular researcher thinks will work best. This means that data are presented in different ways across various experiments. Often the way in which the data are presented goes unpublished, so that experiments are difficult to compare to one another.

Preliminary studies suggest that VA can help people make better financial decisions (Savikhin et al., 2008, 2011), but not all VA systems are equally beneficial. Laboratory experiments, with "real life" tasks and performance-based incentives, are required to determine their relative effectiveness (Savikhin et al., 2011).

The main challenge here is for economists to begin adapting VA in their field, and for VA experts to begin using laboratory experiments to evaluate the effectiveness of the VA produced for actual decision-making.

Data

The data used in our financial analytic studies are context dependent. For most experiments, quantitative values (e.g., monetary amount) are used for economic experiments. But qualitative data such as decision of users or description of choices are common as well. From our work, VA data can come from many secondary sources (i.e., databases), or be a primary source that is either artificially generated or user generated, and either simulated or extracted from a real source. The artificially generated data are useful for testing VA tools against an economic theory while the user generated data are mainly used for evaluating whether the VA tools help with reasoning. Generally the data used in laboratory experiments are artificial data. Data used in the analysis of decision-making include the actual choice made by the user, the intermediate choices made by the user, time spent on deliberation, post-surveys on usability and confidence, and sometimes mouse click/mouse movement or eye tracking data can enrich the dataset.

The data challenges in our current work are simply the storage and distribution structure for the data. For graphic depiction, other limitations may exist. For example, RGraph is a JavaScript library and JavaScript data structure is limited to arrays. Escaping this limitation would require prebuilt libraries that may place more restriction on data type and structure or in-house libraries, which would require a considerable amount of time and debugging.

VA tool: PortfolioCompare

PortfolioCompare was developed to help people understand diversification. It provides a visual display of how funds can be combined to reduce risk and improve return. The salient features are the scattergraph displaying the generated portfolios and sliders for selecting the weight of portfolios. The VA tool also connects to a distribution of risk plot (directly above the scattergraph) and has seamless link connection when the user hovers over a portfolio point. The data are encoded in a multidimensional array using processing (see Figure 22.18).

PortfolioCompare has three main interaction features. The first feature is the scattergraph hover feature where the information of each portfolio is summarized when a point is selected. The risk distribution graph above the scattergraph automatically displays risk distribution for the current portfolio and up to five previous portfolios can be selected for comparison. The slider interaction for adding funds to a portfolio lets users quickly see the outcome of different combinations. Users quickly realized which funds led to diversification. Finally, there are many click buttons that let the user delete, manipulate and create portfolio.

All three interactions are meant to let the users manipulate the portfolio and let them explore the concept of diversification. The VA tool clearly displays similar return and risk of each portfolio and feedback from users emphasized the ease of finding the best return-to-risk portfolio for a given return and ease of allowing them to realize how certain combinations of portfolio are uncorrelated and lead to diversification.

We found that PortfolioCompare helped the user select a portfolio that was closer to his or her risk preferences. The exploratory nature of PortfolioCompare was useful to subjects, as subjects continued to improve their selections during the decision-making process. The PortfolioCompare users reported that it was easy to

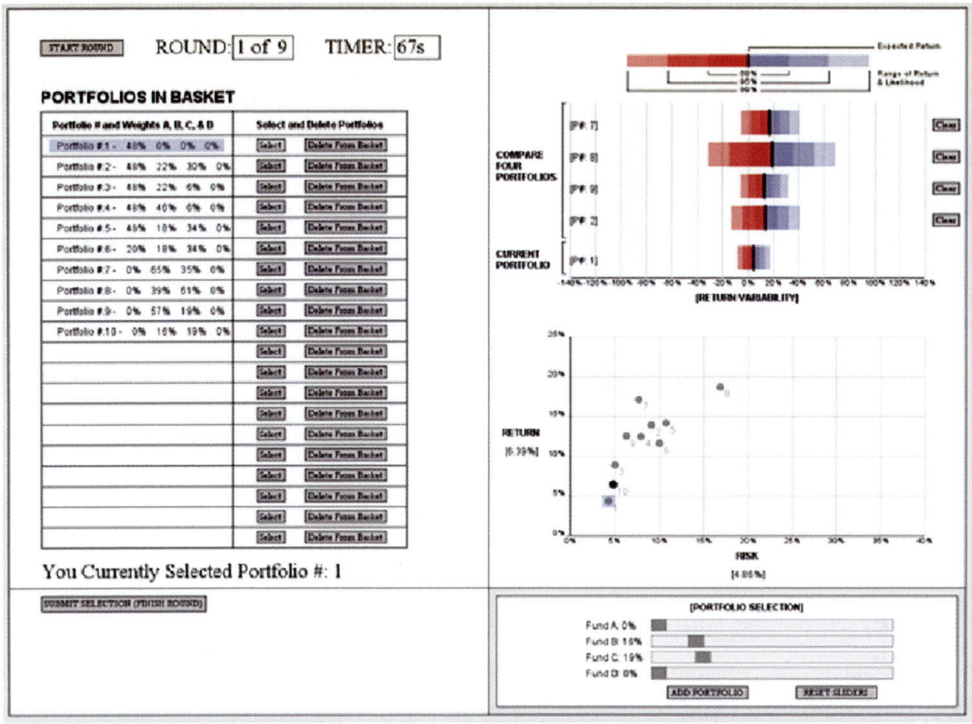

Figure 22.18 Screenshot of PortfolioCompare.

pick portfolios with specific return and the lowest risk. Users also found that combining certain portfolios would reduce risk. The word "diversification" was not used by all, but the narrative from the user survey is consistent with the concept of diversification.

This VA tool is better than the traditional method of diversification because it abstracted the tedious task of calculating funds' expected return and risk and presenting them on a scattergraph. This saves time for the user, and they can better focus on identification of optimal portfolios.

22.5.2 A case study on developing a custom VA solution for a boutique fixed-income asset management firm

Analytic problem

The target firm serves a specific niche, and treats each client separately, with separate plans and investment strategies that take into account more than just the technical analysis considerations of fixed income investment, but also the client's lifestyle, career plans, and risk tolerance. Since there are many different analytic actions in a firm that specializes in personalized asset management, we chose to address two common-denominator analytic problems that often arise for all clients.

1. *Portfolio monitoring*: An analyst will often want to understand at a glance which client portfolio is performing to expectation, and which is lagging, then zoom in to only a particularly troubled client and see accounts within, then the level of holdings within each account. This thought process corresponds with Schneiderman's mantra (1996) of visual information seeking: overview, filtering and zooming, and details on demand.
2. *Scenario analysis*: An analyst will often hypothetically change certain values in an account, or certain market or economic conditions, and observe the results of these changes on the performance of the account. The major performance measure being used at the firm for each client is Cost of Outstanding Income (COI), either in a given year or aggregated over a retirement plan. COI is defined as whether the income goal can be met exactly by all the fixed-income holdings that mature on that year. This is different from the common conception of portfolio performance: the profit and loss compared to the current market value of all the holdings.

In contrast to the optimal information seeking mantra, the existing processes of portfolio monitoring and scenario analysis at the firm involved very high cognitive load. Since there was no dashboard interface that showed the consolidated information of all the client workbooks at once, the analysts had to open and examine each client workbook manually, even when only an overview was required. For each client, there was a dashboard worksheet, but an analyst was forced to read through many numeric fields and mentally process them in order to understand the performance of just one client.

Data

The sample data consisted of a small series of Excel workbooks. Each of the Excel workbooks represented one anonymized client. Each client workbook had one or more accounts. The content of each account was not structured according to the instruments held, but according to the income streams generated by instruments maturing over the years. The workbooks contained many sheets, displaying both the data to be explained to clients during quarterly meetings as well as dense calculations meant for analysts only. We also encountered "bad data," such as naming and terminology inconsistencies. These issues in information management represented both an interface design problem and an opportunity for VA to make a difference.

Each workbook in turn contained four types of worksheets: dashboard worksheets, account sheets, formula sheets, and price data sheets. The account worksheet type is the core of each workbook. This is also the only type of worksheet with a mostly consistent semantic structure across all clients. Several notable fields in each account worksheet are the following.

1. Income Goal (IG): The sum of the income goals in a corresponding year within all accounts that belong to a person is that person's expected income for that retirement year.
2. Current Holding (CH): In our target firm, the value of this field in a given year does not mean the market value of the strip coupons held in that year, but rather, the aggregated value at maturity of all the coupons being held that mature in that year.
3. COI, as of a given year: This is defined as the actual dollar amount needed to balance a deficit year. COI = (IG − CH) × (Average) price of one or more coupons maturing in a given year/100 (coupons per bundle).

VA tool

The technology pipeline chosen to implement our solution consisted of the JavaScript language and the D3js visualization library. As with all JavaScript programs, a D3js visualization project can be developed, deployed and tested relatively quickly compared to using more comprehensive frameworks. Implementing the high-fidelity prototype via JavaScript also eases the process of creating a printout of the visualizations to include in a client report by leveraging the mature printing facility built into web browsers.

Our design involves a dashboard user interface (Figure 22.19) that displays at a glance the list of clients, the list of accounts in the currently highlighted clients, the information in the worksheet of the currently highlighted account, and price data. Aware of the cognitive load involved in switching rapidly between multiple visualizations (Klein et al., 2002), we used a minimal number of visualizations that would satisfy our data representation scheme: a bullet graph and a treemap. The analyst can edit the income goals/streams, and price tables to generate hypotheses for factor analysis, and the visualization would re-render instantly to reflect the results. Finally, the re-rendering of the visualizations is smoothly animated, which is both for aesthetics, and, more importantly, to draw the analyst's attention.

The treemap (Schneiderman, 1992) in our application is the major visualization for supporting the portfolio monitoring task. In the lowest level of detail, the default state, the treemap visualization depicts a bird's eye view of all the clients and their accounts. The underlying value of each node is the aggregated cost of outstanding income (COI) for all the retirement years within each account. We minimized the amount of information being displayed at any one time according to the principle of "overview, filter and zoom, details on demand," using semantic zooming (drill

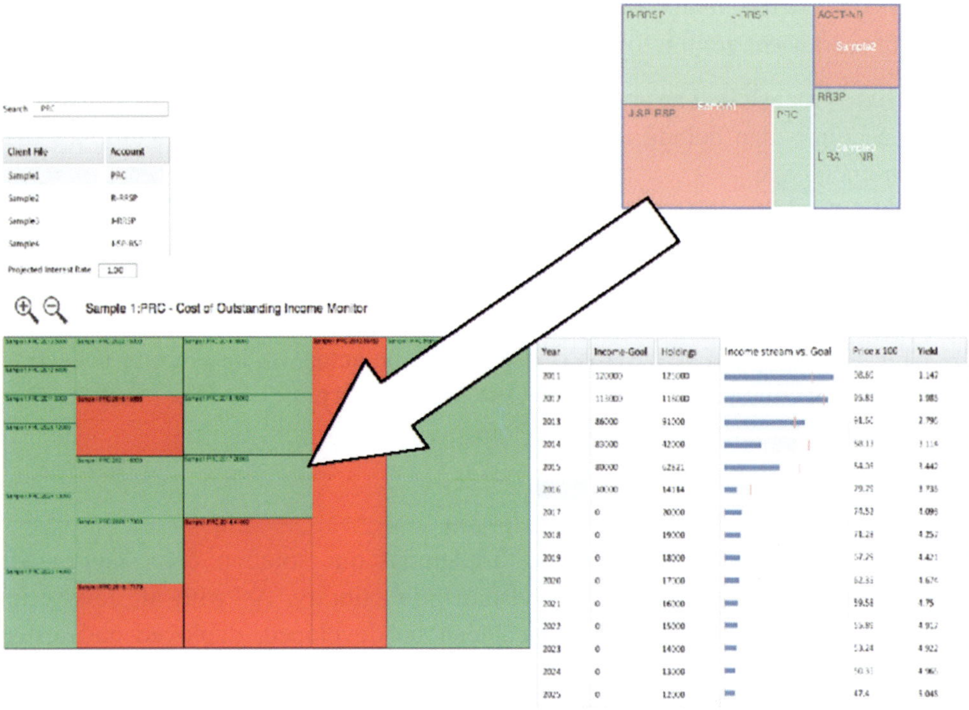

Figure 22.19 Visual dashboard with treemap showing two states: holdings in one account and list of client/account.

down) through different layers of the hierarchical data model (Weaver, 2004). Prominent clickable icons of magnifying loupes provide the zooming interactions. Another benefit of semantic zooming is to enhance the (already formidable) scalability of the treemap visualization to larger datasets.

Zooming in twice will drill down to the highest level of details, the list of holdings in an account. The underlying data also represent COI, but for each year instead of cumulatively. The color of the nodes encodes discrete states of account performance: whether the account is in surplus-green, neutral-white, or deficit-red. The red is darker than the green, so that the intensity of the cells provides redundant visual encoding to mitigate the effect of color-blindness or black-and-white printing. The relative size of the nodes represents the amount of the deficit/surplus, because relative sizes of rectangles can be compared very quickly and accurately via human visual perception (Cleveland and McGill, 1984).

The treemap was chosen to visualize the list of client, accounts, and holdings partly because of its efficient use of space. Compared to alternatives such as adjacency diagrams, treemaps do not have any whitespace, and do not need extra space for connectors between nodes. Treemap is also adaptable to similar data models. This treemap design can be readily modified to use conventional P/L measures instead of Cost of Outstanding Income.

The other visualization is a bullet graph (Few, 2006) that displays the income goals over time versus the cumulative value at maturity of current holdings over time (years). The three data types to be visually encoded include two numerical, cardinal values, which are very suitable for bar graphs, and discrete time series, which fit well into the *x*-axis of a graph. Income goals and current holdings are both denominated in the same currency, and are both in the same range, so they are ideal for bar graphs stacked side-by-side, or bullet graphs, where one bar graph is nested within the other to better utilize screen space. A bullet graph often also contains a marker floating on top of the inner bar, which is a red line in our design. This marker acts as an optional third "bar" to visually encode the "previous income goal" column in the data, which represents an ad hoc attempt by analysts to keep track of their own analytic history.

22.5.3 A case study on international debt as a pathway of systemic risk contagion using Quantum4D™

Analytic problem

Apart from its utility in assisting individuals and financial institutions with portfolio management and risk decision-making, VA has the potential to provide regulators and market participants with analytic capabilities essential to effectively monitoring risks to financial systems. A discussion of all of the different facets of systemic risk analysis is beyond the scope of this chapter; however, exposure to European sovereign debt has certainly proved to be a pathway of systemic risk contagion in recent years. This case study explores the use of VA to understand the pathways of systemic risk contagion in international debt.

Data

Data for this case study were drawn from statistics available from the Bank for International Settlements (BIS) such as Consolidated Banking Statistics (Table 9) and locational banking statistics such as BIS Reporting Banks – Summary of International Positions; External Positions of Banks in Individual Reporting Countries; External Loans and Deposits of Banks in Individual Reporting Countries; Local positions in Foreign Currency of Banks in Individual Reporting Countries; Currency Breakdown of Reporting Banks' International Positions; External Positions of Reporting Banks vis-à-vis Individual Countries; External Loans and Deposits of Reporting Banks vis-à-vis Individual Countries; International Positions by Nationality of Ownership of Reporting Banks. Typically, the dataset consists of numeric data in tabular form (.CSV file) representing the position (e.g., consolidated foreign claims of reporting banks on an immediate borrower basis on individual countries by nationality of reporting banks in millions of US Dollars) with the rows in the tables representing countries and the columns represent month and year (e.g., December, 2008, December, 2009, December, 2010, etc.).

VA tool

To achieve the analytic objective of understanding and tracking global capital flows to understand the pathways of contagion in international debt, the data from BIS Table 9B – Consolidated Foreign Claims of Reporting Banks on an Immediate Borrower Basis – were imported into Quantum4D™ and mapped to visual elements signifying entities (a 2D glyph representing banks by country), relations (a directional pipe) and data about entities and relations (e.g., percentage increase or decrease). In the visualization, the width of the relation was used to signify the amount of foreign claims of reporting banks on an immediate borrower basis: the wider the pipe, the higher the amount of foreign claims. Color was applied to relations and objects to represent the percentage quarterly change since December 2008 with intensity of color used to indicate how current behavior compares to the historical trends. The size of the 2D glyph representing country's banks was used to indicate aggregate (sum) of inflows or outflows, depending on country/bank. The example in Figure 22.20 focuses on Greece as at December, 2009. From this visualization, the analyst is able to see that Swedish, Swiss, Austrian, Belgian and Brazilian banks were all reducing their exposure to Greek debt at the time, with the Belgians apparently being the most bearish. Conversely, British banks increased their exposure to Greek debt by 22%. The analyst also is able to quickly see that German and French banks have the greatest exposure to Greek debt and a potential sovereign default. This visualization provides intelligence to provide financial regulators or market participants with a high-level "situational awareness" of the sovereign debt exposures for individual country's banks that can be rapidly perceived using the human visual system. The analyst is then able to use interaction features of the VA tool to further explore the data (e.g., zooming into subsets of the data, filtering subsets, adding aggregation to visualize derivative attributes not present in the original table, or to create new visualizations to compare Greece with other countries) or to build alerts into the system to provide an "early warning" system around key risk indicators.

22.6 Conclusion

The world of finance requires the processing of masses of data, and the effective analysis of such data is critically important to the management of risk within

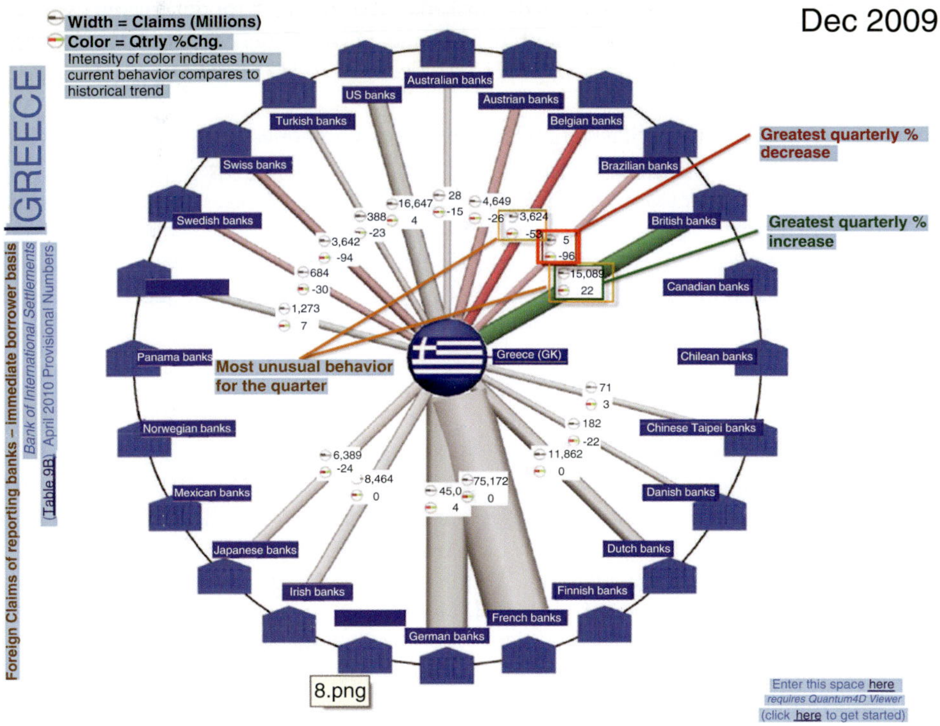

Figure 22.20 Foreign claims of reporting banks – immediate borrower basis: Greece using Quantum4D™.

individual firms as well as on a systemic level. Traditional approaches to financial risk analysis may not scale up well in the face of huge quantities of data: they may fail to reduce cognitive burden for the analyst, be too inflexible, or simply collapse when large-scale datasets must be processed. VA, which leverages human visual perception and cognition to overcome the limitations of traditional analytic approaches, offers a scalable solution to the analysis of large amounts of financial data. As a relatively new approach, however, there is still much research to be done on the general science of InfoVis and VA as well as on its specific application in the domain of financial risk analysis. It is our hope that this chapter has provided a foundation of knowledge on which to base future investigations.

Bibliography

Alsakran, J., Y. Zhao and X. Zhao, 2010, Tile-based parallel coordinates and its application in financial visualization, in: J. Park, M. C. Hao, P. C. Wong and C. Chen (eds.), *Proceedings of Visualization and Data Analysis 2010 (VDA 2010), January 18–19, 2010, San Jose, California*, SPIE.

Amar, R. and J. Stasko, 2004, A knowledge task-based framework for design and evaluation of visualizations, in: M. Ward and T. Munzner (eds.), *Proceedings of the 10th IEEE Symposium on InfoVis 2004 (INFOVIS '04), October 10–12, 2004, Austin, Texas*, IEEE Computer Society Press.

Amar, R., J. Eagan and J. Stasko, 2005, Low-level components of analytic activity in InfoVis, in J. Stasko (ed.), *Proceedings of the 11th IEEE Symposium on InfoVis 2005 (INFOVIS '05), October 23–25, 2005, Minneapolis, Minnesota*, IEEE Computer Society Press.

Arias-Hernández, R., L. T. Kaastra, T. M. Green and B. Fisher, 2011, Pair analytics: Capturing reasoning processes in collaborative VA, in: R. Sprague (ed.), *Proceedings of the 44th Annual Hawaii International Conference on System Sciences, January 4–7, 2011, Koloa, Kauai, Hawaii*, IEEE Computer Society Press.

Barnes, J. A. and F. Harary, 1983, Graph theory in network analysis, *Social Network*, **5**, 235–244.

Bay, S., K. Kumaraswamy, M. G. Anderle, R. Kumar and D. M. Steier, 2006, Large scale detection of irregularities in accounting data, in: C. W. Clifton (ed.), *Proceedings of the 6th Annual IEEE International Conference on Data Mining (ICDM 2006), December 18–21, 2006, Hong King, China*, IEEE Computer Society Press.

Beck, K., 2000, *Extreme Programming eXplained: Embrace Change*, Addison-Wesley.

Bera, P., A. Burton-Jones and Y. Wand, 2011, Guidelines for designing visual ontologies to support knowledge identification, *MIS Quarterly*, **35**, 883–908.

Bertin, J., 1967, *Sémiologie Graphique. Les Diagrammes, les Réseaux, les Cartes*, with Marc Barbut et al., Gauthier-Villars, Paris. (Translation by William J. Berg, 1983, Semiology of Graphics.)

Biggs, N., K. E. Lloyd and R. J. Wilson, 1976, *Graph Theory: 1736–1936*, Clarendon Press.

Bondy, J. A. and U. S. R. Murty, 2008, *Graph Theory*, Springer.

Brammertz, W., I. Akkizidis, W. Breymann, R. Entin and M. Rustmann, 2008, *Unified Financial Analysis*, Wiley Finance.

Brodbeck, D., M. Chalmers, A. Lunzer and P. Cotture, 1997, Domesticating Bead: adapting an InfoVis system to a financial institution, in: J. Dill (ed.), *Proceedings of the 3rdIEEE Symposium on InfoVis 1997 (INFOVIS '97), October 20–15, 1997, Phoenix, Arizona*, IEEE Computer Society Press.

Caldwell, C., 1995, *Graph Theory Glossary*, retrieved from `www.utm.edu/departments/math/graph/glossary.html`.

Card, S., J. D. Mackinlay and B. Schneiderman (eds.), 1999,. *Readings in InfoVis: Using Vision to Think*, Morgan Kaufmann Publishers.

Chang, R., A. Lee, M. Ghoniem, R. Kosara, W. Ribarsky, J. Yang, E. Suma, C. Ziemkiewicz, D. Kern and A. Sudjianto, 2008, Scalable and interactive visual analysis of financial wire transactions for fraud detection, *InfoVis*, **7**(1), 63–76.

Chaudhuri, S. and U. Dayal, 1997, An overview of data warehousing and OLAP technology, *ACM SIGMOD Record*, **26**(1), 65–74.

Cleveland, W.S. and R. McGill, 1984, Graphical perception: theory, experimentation, and application in the development of graphical methods, *Journal of the American Statistical Association*, **79**, 531–554.

Codd, E. F., S. B. Codd and C. T. Salley, 1993, *Providing OLAP to User-Analysts: An IT Mandate*, Codd & Associates.

Cuzzocrea, A. and S. Mansmann, 2009, OLAP visualization: models, issues and techniques, in: J. Wang (ed.), *Encyclopedia of Data Warehousing and Mining*, Second edition, Information Science Reference.

Dang, Thomas and Victoria L. Lemieux, 2013, Functional framework for evaluating visualization products: Development and application to the design of a custom visual analytics solution for a boutique asset management firm, in: V. L. Lemieux (ed.), *Financial Analysis and Risk Management: Data Governance, Analytics and Life Cycle Management*, Springer.

Dilla, W., D. Janvrin and R. Raschke, 2010, Interactive data visualization: new directions for accounting information systems research, *Journal of Information Systems*, **24**(2), 1–37.

Dwyer, T., 2003, A scalable method for visualising changes in portfolio data, in: T. Pattison and B. Thomas (eds.), *Proceedings of the 2nd Australian Symposium on InfoVis (Invis. au 2003), February 3–4, 2003, Adelaide, Australia*, Australian Computer Society.

Dwyer, T. and P. Eades, 2002, Visualizing a fund manager flow graph with columns and worms, in: *Proceedings of the 6th International Conference on InfoVis (IV' 02), July 10–12, 2002, London*, IEEE Computer Society Press.

Eick, S. G., 2000, Visualizing multi-dimensional data, *ACM SIGGRAPH Computer Graphics*, **34**(1), 61–67.

European Central Bank, 2011, Euro Area Yield Curve, retrieved from `www.ecb.europa.eu/stats/money/yc/html/index.en.html`.

European Central Bank, 2011, Government Finance, retrieved from `www.ecb.europa.eu/stats/gov/html/index.en.html`

European Central Bank, 2011, Inflation and the Euro, retrieved from `www.ecb.europa.eu/stats/prices/hicp/html/inflation.en.html`

Few, S., 2006, Bullet Graph Design Specification, retrieved from `www.perceptualedge.com/articles/misc/Bullet_Graph_Spec.pdf`

FinViz, 2010, Financial Visualizations, retrieved from `http://finviz.com/`

Fisher, B., T. M. Green and R. Arias-Hernández, 2011, VA as a translational cognitive science, *Topics in Cognitive Science*, **3**(3), 609–625.

Friedman, M., 1953, *Essays in Positive Economics*, Chicago Press.

Gallis, H., E. Arisholm and T. Dyba, 2003, An initial framework for research on pair programming, in: *Proceedings of the 2003 International Symposium on Empirical Software Engineering (ISESE'03), September 30-October 1, 2003, Rome, Italy*, IEEE Computer Society Press.

Garratt, R.J., L. Mahadea and K. Svirydzenka, 2011, Mapping System Risk in the International Banking Network: Bank of England Working Paper no. 143, Bank of England.

Girardin, L. and D. Brodbeck, 2001, Interactive visualization of prices and earnings around the globe, in: K. Andrews (ed.), *Proceedings of the 7th IEEE Symposium on InfoVis 2001 (INFOVIS '11), October 22–23, 2001, San Diego, California*, IEEE Computer Society Press.

Gray, J., S. Chaudhuri, A. Bosworth, A. Layman, D. Reichart and M. Venkatrao, 1997, Data cube: a relational aggregation operator generalizing group-by, cross-tab, and sub-totals, *Data Mining and Knowledge Discovery*, **1**(1), 29–53.

Grinstein, G. and M. Ward, 1997, Introduction to data visualization, IEEE Visualization Tutorial.

Gunn, J., 2007, XBRL: opportunities and challenges in enhancing financial reporting and assurance processes, *Current Issues in Auditing*, **1**, A36–A43.

Hanrahan, P., C. Stolte and J. Mackinlay, 2007, Visual Analysis for Everyone: Understanding Data Exploration and Visualization, retrieved from `www.tableausoftware.com/wp/visual-analysis-everyone`

Ieronutti, L. and M. Pighin, 2009, A novel visualisation and interaction technique for exploring multidimensional data, *International Journal of Business Intelligence and Data Mining*, **4**(3/4), 350–374.

Inselberg, A., 1985, The plane with parallel coordinates, *The Visual Computer*, **1**(2), 69–91.

Jungmeister, W. A. and D. Truro, 1992, Adapting Treemaps To Stock Portfolio Visualization, Center for Automation Research, University of Maryland, Baltimore.

Kaufman, A. (ed.), 1990, *Proceedings of the 1st Conference on Visualization '90, October 23–26, 1990, San Francisco, California*, IEEE Computer Society Press.

Keim, D. A., 2002, InfoVis and visual data mining, *IEEE Transactions on Visualization and Computer Graphics*, **8**(1), 1–8.

Kelton, A. S. and Y. W. Yang, 2008, The impact of corporate governance on Internet financial reporting, *Journal of Accounting and Public Policy*, **27**(1), 62–87.

Klein, P., F. Muller, H. Reiterer and M. Eibl, 2002, Visual information retrieval with SuperTable + Scatterplot, in: *Proceedings of the 6th International Conference on InfoVis (IV' 02), July 10–12, 2002, London*, IEEE Computer Society Press.

Kosslyn, S. M., 1980, *Image and Mind*, Harvard University Press.

Kubelec, C. and F. Sa, 2010, The Geographical Composition of National External Balance Sheets: 1980–2005: Bank of England Working Paper no. 384, Bank of England.

Lei, S. T. and K. Zhang, 2010, A VA system for financial time-series data, in: *Proceedings of the 3rd International Symposium on Visual Information Communication (VINCE' 10), September 28–29, 2010, Beijing, China*, Association for Computing Machinery.

Lorensen, W., 2004, On the death of visualization: can it survive without customers? in: *Position Papers of the NIH/NSF Proceedings of the Fall 2004 Workshop on Visualization Challenges, September 22–23, 2004, Bethesda, Maryland*, National Library of Medicine.

Lurie, N. H. and C. Mason, 2007, Visual representation: implications for decision making, *Journal of Marketing*, **71**, 160–177.

Markowitz, H. M., 1952, Portfolio selection, *Journal of Finance*, **7**(1), 77–91.

McCormick, B. H., T. A. DeFanti and M. D. Brown (eds.), 1987, Visualization in scientific computing, *ACM SIGGRAPH Computer Graphics*, **21**(6).

Munzner, T., 2009, Visualization, in: P. Shirley, M. Ashikhminand and S. Marshner (eds.), *Fundamentals of Graphics*, Third edition, A.K. Peters.

Ramnath, S., S. Rock, and P. Shane, 2008, The financial analyst forecasting literature: A taxonomy with suggestions for further research, in: *International Journal of Forecasting*, Elsevier.

Rittel, H. and M. Webber, 1973, Dilemmas in a general theory of planning, *Policy Sciences*, **4**(2), 155–169.

Roberts, P., 2003, InfoVis for Stock Market Ticks: Toward a New Trading Interface, Massachusetts Institute of Technology, retrieved from `http://dspace.mit.edu/bitstream/handle/1721.1/16668/56675083.pdf`

Rooney, N., D. Patterson, M. Galushka, V. Dobrynin and E. Smirnova, 2008, An investigation into the stability of statistical document clustering, *Journal of the American Society of Information Science and Technology*, **59**(2), 256–266.

Savikhin, A., R. Maciejewski and D. S. Ebert, 2008, Applied VA for economic decision-making, in: D. Ebert and T. Ertl (eds.), *Proceedings of the IEEE Symposium on VA Science and Technology 2008 (VAST 08), October 21–23, 2008, Columbus, Ohio*, IEEE Computer Society Press.

Savikhin, A., H. C. Lam, B. Fisher and D. S. Ebert, 2011, An experimental study of financial portfolio selection with VA for decision support, in: R. Sprague (ed.), *Proceedings of the 44th Annual Hawaii International Conference on System Sciences (HICSS-44), January 4–7, 2011, Koloa, Kauai, Hawaii*, IEEE Computer Society Press.

Sawant, A., 2009, StockViz: analyzing the trend of stocks in major auto, oil, consumer, and technology companies, in: H. Arabnia and L. Deligiannidis (eds.), *Proceedings of the 2009 International Conference on Modeling, Simulation & Visualization Methods (MSV 2009), July 13–16, 2009, Las Vegas, Nevada*, CSREA Press.

Securities and Exchange Commission (SEC), 2008, SEC Approves Interactive Data for Financial Reporting by Public Companies, Mutual Funds, retrieved from `www.sec.gov/news/press/2008/2008-300.htm`.

Shneiderman, B., 1992, Tree visualization with tree-maps: a 2-d space-filling approach, *ACM Transactions on Graphics*, **11**(1), 92–99.

Shneiderman, B., 1996, The eyes have it: a task by data type taxonomy for InfoVis, in: *Proceedings of the IEEE Symposium on Visual Languages 1996, September 3–6, 1996, Boulder, Colorado*, IEEE Computer Society Press.

SmartMoney™, 2009, Map of the Market, retrieved from `www.smartmoney.com/map-ofthe-market/`.

Sonnemans, J., 2006, Price clustering and natural resistance points in the Dutch stock market: a natural experiment, *European Economic Review*, **50**(8), 1937–1950.

Stasko, J. and E. Zhang, 2000, Focus+Context display and navigation techniques for enhancing radial, space-filling hierarchy visualizations, in: S. Roth (ed.), *Proceedings of the 6th IEEE Symposium on InfoVis (INFOVIS '00), October 9–10, 2000, Salt Lake City, Utah*, IEEE Computer Society Press.

Tegarden, D. P., 1999, Business InfoVis, *Communications of the Association for Information Systems*, **1**(4), 2–38.

Thomas, J. J. and K. A. Cook (eds.), 2005, *Illuminating the Path: The Research and Development Agenda for VA*, IEEE Computer Society Press.

Tufte, E. R., 1990, *Envisioning Information*, Graphics Press.

Tufte, E. R., 1997, *Visual Explanations*, Graphics Press.

Tufte, E. R., 2003, *The Cognitive Style of PowerPoint*, Graphics Press.

Tufte, E. R., 2006, *Beautiful Evidence*, Graphics Press.

Tukey, J. W., 1977, *Exploratory Data Analysis*, Addison-Wesley.

Vessey, I., 1991, Cognitive fit: a theory-based analysis of the graphs versus tables literature, *Decision Sciences*, **22**(2), 219–240.

VisMaster, 2010, Mastering the Information Age, retrieved from `www.youtube.com/watch?v=5i3xbitEVfs>`.

Ward, M., G. Grinstein and D. Keim, 2010, *Interactive Data Visualization: Foundations, Techniques, and Applications*, A.K. Peters.

Wattenberg, M., 1999, Visualizing the stock market, in: M. William (ed.), *Proceedings of the ACM SIGCHI Conference on Human Factors in Computing Systems (CHI '99), May 15–20, 1999, Pittsburgh, Pennsylvania*, Association for Computing Machinery.

Weaver, C., 2004, Building highly-coordinated visualisations in improvise, in: M. Ward and T. Munzner (eds.), *Proceedings of the 10th IEEE Symposium on InfoVis 2004 (INFOVIS '04), October 10–12, 2004, Austin, Texas*, IEEE Computer Society Press.

Wehrend, S. and C. Lewis, 1990, A problem-oriented classification of visualization techniques, in: A. Kaufman (ed.), *Proceedings of the 1st IEEE Conference on Visualization (Visualization' 90), October 23–26, 1990, San Francisco, California*, IEEE Computer Society Press.

Wright, W., 1995, Information animation applications in the capital markets, in: S. Eick and N. Gershon (eds.), *Proceedings of the 1st IEEE Symposium on InfoVis 1995 (INFOVIS '95), October 30–31, 1995, Atlanta, Georgia*, IEEE Computer Society Press.

Yang, J., M. Ward and E. A. Rundensteiner, 2002, An interactive tool for visually navigating and manipulating hierarchical structures, in: P. Wong (ed.), *Proceedings of the 8th IEEE Symposium on InfoVis 2002 (INFOVIS' 02), October 28–29, 2002, Boston, Massachusetts*, IEEE Computer Society Press.

Yi, J. S., Y. A. Kang, J. T. Staskoand and J. A. Jacko, 2007, Toward a deeper understanding of the role of interaction in InfoVis, *IEEE Transactions on Visualization and Computer Graphics*, **13**(6), 224–231.

Zhou, M. and S. Feiner, 1998, Visual task characterization for automated visual discourse synthesis, in: C. Karat and A. Lund (eds.), *Proceedings of the ACM SIGCHI Conference on Human Factors in Computing Systems (CHI '98), April 18–23, 1998, Los Angeles, California*, Association for Computing Machinery.

Ziegler, H., T. Nietzschmann and D. A. Keim, 2008, VA on the financial market: pixel-based analysis and comparison of long-term investments, in: E. Banissi et al. (eds.), *Proceedings of the 12th International Conference on InfoVis 2009 (IV' 08), July 9–11, 2008, London*, IEEE Computer Society Press.

23

Metadata management

Marc Alvarez

23.1 Metadata management technology

In one of the less precise vocabularies commonly used in the capital markets industry, the term "metadata" is typically taken to be described as "data about data."[1] Heads nod in unison and seemingly knowing glances are exchanged by participants, taking it as read that everybody shares a common understanding of the term.

Dig a little deeper, though – by comparing databases used by financial firms, say – and the imprecision of this definition rapidly makes itself known. To a computer scientist or engineer this presents a baffling situation – multiple firms working with the same data content treating the universe differently based on individual, proprietary implementations. As is common across many areas of the capital markets, few if any standards or even best practice conventions exist.

For a business, this lack of standardization almost always results in constraints on the ability to scale up operations to an enterprise level in a quick and efficient manner. For the fast moving capital markets this can easily result in missed opportunities to engage in new business opportunities, thereby having a direct impact on growth and performance. Truly, metadata sits at the intersection of the business execution and the underlying supporting technology infrastructure.

Ironically, though, metadata is ubiquitous to the acquisition, management and application of capital markets data. No firm can operate successfully without engaging in the definition and management of metadata. Indeed, those firms that invest in and maintain competence around metadata generate a dimension of competitive differentiation for their service offerings and operating efficiency. If there is one aspect of working with capital markets data content, mastery over the

[1] See Wikipedia: `http://en.wikipedia.org/wiki/Metadata`

Handbook of Financial Data and Risk Information, ed. Margarita S. Brose, Mark D. Flood, Dilip Krishna and Bill Nichols. Published by Cambridge University Press.

corresponding metadata content is key to developing profitable new lines of business while keeping associated risk visible and under control.

This chapter lays out a logical framework for working with metadata in the capital markets context. It aims to establish a foundation upon which to implement a metadata management capability that supports the firm's ability to leverage the power of metadata not only to take advantage of new business opportunities that would otherwise be inaccessible, but also to exploit existing operations to the fullest extent possible. Key concepts include:

- **Definition** – putting in place a precise understanding of the term "metadata" and applying it to the context of the business operations of a modern capital markets firm;
- **Sources** – identifying the sources of metadata within an organization and understanding the related behavior as it is applied to the business;
- **Applications** – identifying the manner in which metadata can be applied and the interfaces required to support it;
- **Governance and management** – understanding the need to apply governance to metadata and applying best practice to realize as many benefits as possible;
- **Emerging trends** – putting today's metadata experience into the context of the rapidly growing subject domain of metadata.

Above all, the most important lesson to take way is that metadata is organic and the product of the operating enterprise. There are no hard and fast rules to how best to apply metadata to any single firm. Rather, the business experience suggests that achieving the benefits from managing metadata as a core competence of the firm is a journey made up of many steps with incremental benefits realized from each step along the way.

23.2 Metadata defined

Metadata is a term that applies to any application for data content, and not just the capital markets industry. It is a foundation term that finds it roots in the generic technologies underpinning all IT applications. To be an effective resource to a capital markets firm, though, a precise definition that applies to the industry is an important starting point. Equally, the same is true in an effort to apply metadata to other industries – there is a fundamental need to understand the metadata concept in the context of the business to which it will be applied.

Start with the basic definition of metadata referenced above[2]:

[2] See Wikipedia: `http://en.wikipedia.org/wiki/metadata`

Metadata (metacontent) is defined as data providing information about one or more aspects of the data, such as:

- Means of creation of the data
- Purpose of the data
- Time and date of creation
- Creator or author of data
- Placement on a computer network where the data was created
- Standards used
- Rights and permissions associated with commercial licensing of content.

The basic information of a piece of music, for example, a digital image may include metadata that describes how large the picture is, the color depth, the image resolution, when the image was created, and other data. A text document's metadata may contain information about how long the document is, who the author is, when the document was written, and a short summary of the document.

Metadata is data. As such, metadata can be stored and managed in a database, often called a Metadata registry or Metadata repository. However, it is impossible to identify metadata just by looking at it because a user would not know when data is metadata or just data.

The above presents a good starting point to understanding metadata. The main weakness, though, is that it is generic – it can apply to pretty much any industry or business application. Therein lies a core characteristic of metadata: there is a close dependency between the underlying application of the business data content and the metadata required to manage it effectively. The metadata that is needed to make sense of a digital photograph, for example, can be expected to vary considerably from what is required to apply market data to the business needs of financial firms. While there may well be some common elements between the two, there is also a much larger collection of application specific metadata for each.

One of the big challenges in understanding the role of metadata in the capital markets is the diversity of applications that firms deploy. Each and every one requires significant amounts of business data content (such as prices or security master information), which by definition can consume different sets of market data content. It is this diversity that lies at the heart of a fundamental business goal for most financial firms – achieving economies of scale from the use (and reuse) of capital markets data content across the enterprise.

Here lies a second vital characteristic of metadata. The sources from which metadata is received can be very diverse. Indeed, the quality and usability of the metadata is directly affected by the manner in which it is generated and delivered to consuming applications. Since most applications consume data from multiple sources, this characteristic implies the need to integrate the corresponding metadata to some form of common standard within the application (a process called "normalization"). Ideally, all sources of metadata would adopt some form of standard and make this step unnecessary, however, this situation is not currently something that exists nor is it likely to in the foreseeable future.

This normalization and integration of content is not an easy task. It involves steps such as the following.

- Transforming content to a single format that can be consumed by the application. This requires rule sets to perform the transformations (which are metadata themselves) that need to be recorded and maintained.
- Storage needs to be considered along with an ability to produce an audit trail associated with rule sets applied to the data content.
- Methods and rules for applying the application of the content (such as a visual display application for example) need to be defined and managed.
- Records of linkages between metadata attributes as well as other data content need to be identified and recorded to support use of the content across multiple applications.

Therefore, the ability to combine various sources of metadata – both internal and external – is at the heart of leveraging metadata for the success of the business. Consequently, management capabilities over this content type are key to achieving business goals.

In order to better understand this goal, this chapter focuses on some very common and relatively simple use cases of loading and managing a relational database intended to support downstream business applications. To be sure there are many other applications that can be considered – financial firms are very creative and sophisticated places, often operating at the cutting edge of technology. However, the database use case is pretty much common across all data consumption models, so it follows that the principles associated with managing metadata should extend to a broader footprint across the organization.

For database applications, metadata typically fits rather nicely into three generally accepted categories.[3]

- **Technical metadata** – likely the most familiar form of metadata, this refers to the universe of attributes used to describe the physical implementation of the underlying business data content (the loading and management of content in a relational database in this example).
- **Business metadata** – refers to the information derived from the underlying business data content contained in the database that aids in interpreting and

[3] Wikipedia: `http://en.wikipedia.org/wiki/metadata`

understanding the content. A good example is the count of the number of records of interest to a particular user (e.g., "the NYSE contains over 4,000 listings").

- **Process metadata** – the universe of content required to actually make use of the database on a production basis. Typical examples include the number of concurrent logons, CPU usage per query or user, number of disk reads etc. An important application of capturing and managing process metadata is to put in place measurements such as *key performance indicators* (known in technology jargon as KPIs) which qualify as additional metadata content.

It should be clearly understood that all three categories of metadata need to be addressed in making an information service (such as a database) available to business applications and end users. In fact, for most applications it is a key success factor to have considered and planned for supporting this variety of metadata that leads to wide adoption and reliance upon the service by the end user community.

The industry is awash with examples of clever applications that can solve particular business problems that fail to scale up to support either increased usage or user loads. Consequently, in the user's eyes, the application is often deemed of limited value. In fact, however, attention paid to identifying and managing the metadata associated with the database underlying the application makes scaling out to support a wider audience (potentially the entire enterprise) very challenging.

A simple example of commonly used data content provides a good basis for examining the importance of these categories of metadata to supporting a successful information service. Figure 23.1 provides a very common example of a simple pricing matrix that can be found at the heart of any number of business applications.

To most professionals in the capital markets business, this example appears familiar and easy to understand. It is the basic layout for pricing a portfolio of shares. On the surface, it would appear to be self explanatory to most data users. However, there are some important assumptions that such a statement incorporates that are worth examining more closely.

Ticker Symbol	Market	Security Type	Currency	Date	Open	High	Low	Close	Bid	Ask	Volume
SCCO	NYSE	EQ	USD	31/01/2012	33.40	33.62	32.92	33.33	33.15	33.40	1,593,281
GE	NYSE	EQ	USD	31/01/2012	19.30	19.37	19.14	19.35	19.25	19.40	18,932,484
BP	NYSE	EQ	USD	31/01/2012	47.52	47.53	47.03	47.09	47.09	47.10	6,442,756
BCE	TMX	EQ	CAD	31/01/2012	39.95	40.16	39.92	40.03	40.03	40.04	1,300,320
RY	TMX	EQ	CAD	31/01/2012	54.39	54.82	54.16	54.63	54.63	54.64	2,552,365
BATS	LSE	EQ	GBp	31/01/2012	3,135.00	3,153.00	3,122.00	3,127.00	3,126.50	3,127.5	2,304,750
FXB	LSE	ETF	GBp	31/01/2012	157.00	157.59	157.00	157.50	157.49	157.53	30,706

Figure 23.1 Sample pricing portfolio.

1. First, the fundamental question is what technology is being used to render this table? While Figure 23.1 presents a simple table in a written document, this is an end product of the underlying technology used to record the values and generate the table. For the purposes of this example, a relational database is used. However, there are many different technologies that are just as valid to render this table including:
   - A spreadsheet application, such as Microsoft Excel
   - A non-relational database management system
   - An XML document
   - A structured file format of some kind, such as .CSV (*comma separated variable* – a commonly used file format in the capital markets).

   In fact, the choice of technology is the first example of metadata here – as the example will show, recording the underlying source (and by implication its technology) is an extremely important property of supplying the most appropriate business data content for a particular application
2. Second, are there any standards or conventions implied within the table? On the surface, each of the column headings would appear to be self explanatory. However, each can have a different interpretation, significantly impacting the larger table. Some examples include the following.
   - The term "ticker symbol" may mean something different between individual markets. For example, exchanges use ticker symbols as the primary means of identifying the securities they trade. They are typically short textual references that are unique to that exchange. However, there are no set rules between exchanges on the use of letters or textual strings so that ticker symbol "T" for the New York Stock Exchange (NYSE) identifies AT&T's common stock while the same ticker for the Toronto Stock Exchange (TSX) is used to identify a security associated with "Telus Corp."[4]
   - Notice the use of the currency field as it contains alpha-numeric coded values – is there a standard coding scheme in use and if so, which one is used?
   - What format for the date is actually used? Can this vary or is a single standard applied?
   - What actual values are used to populate the individual cells? Do all the markets in question have precisely the same definition for the values reported in this table? If not, where can the appropriate definitions be found?
3. Third, does this table have any logical grouping associated with it? Does it represent all prices for a single date, or can any combination of date and price be

[4] It is therefore impossible to rely on the use of ticker symbols in portfolios with securities sourced from more than one exchange. A unique identification of both the security and the market is required to ensure uniqueness.

included? Equally, are there any preset rules or preferences associated with the display of this content? And is there any implied logical validation of the values presented?

That is a fair amount of information to take note of in order to make sense of something as simple as a straightforward pricing table. In fact, it is answering these and a variety of other questions associated with the final display of the data content that lies at the heart of establishing and managing metadata.

The most obvious approach to exploring this in detail is to look at the metadata content based on the three general categories introduced above.

23.2.1 Technical metadata

Expressed as a conceptual database table design this appears as shown in Figure 23.2.

This is a pretty basic entity relationship diagram that directly represents the view of the content presented above. Some immediate observations are the following.

- A primary key is set by the combination of three fields (the ticker symbol, the market, and the date). As more than one value is used to identify a single row (or record) uniquely, this is an example of a composite primary key.
- By virtue of the three tables surrounding the pricing portfolio table (currencies, security types, and markets) it is pretty straightforward to work out that lookup lists are used to identify the specific values for each.

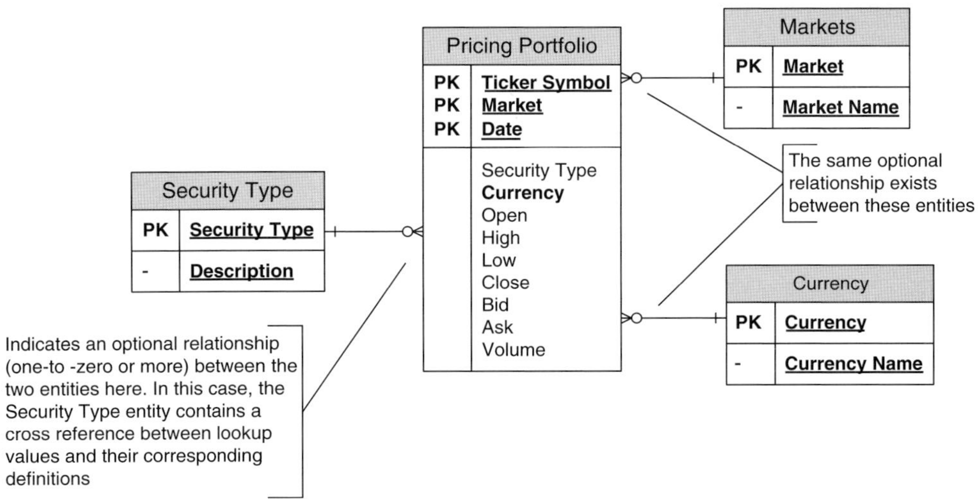

Figure 23.2 Database table structure.

- The currency value is bolded – this is database convention to indicate that a field is mandatory to the table. By implication then, the non-bolded fields are therefore optional – a value is not required in order for the whole record to be considered valid.

So that answers a few of the initial questions asked above, but not all. Creating and maintaining a record of these characteristics of the data table (or in data management terms their "database properties") represents the most fundamental technical meta-data associated with the content.

In fact, pretty much any relational database management technology requires that this information be set in order to create the database. Default settings are typically created when the table and its fields are set up, however this is simply an automation of the task of setting the technical metadata. As any good database manager will tell you, these values need to be set explicitly in order for the database to perform to acceptable levels – when there are multiple users of the same content spread across the broader enterprise, this act of defining and managing the metadata becomes mandatory. Nothing causes a database project to fail faster than inflicting poor performance on end users.

In order to construct this table in pretty much any relational database technology, the basic characteristics of the data need to be defined. These are typically referred to as "properties" and in this case the minimum set includes the characteristics shown in Figure 23.3.

The properties recorded here include the following.

- **Field name** – this is the unique alpha-numeric string used to identify the individual field in the table. Obviously it must be unique so as not to confuse users of the table.

Field Name	Data Type	Required	Primary Key?
Ticker Symbol	VarChar(60)	Yes	Yes
Market	Char(10)	Yes	Yes
Security Type	VarChar(60)	Yes	No
Currency	Char(3)	Yes	No
Date	Date	Yes	Yes
Open	Numeric(10,2)	No	No
High	Numeric(10,2)	No	No
Low	Numeric(10,2)	No	No
Close	Numeric(10,2)	No	No
Bid	Numeric(10,2)	No	No
Ask	Numeric(10,2)	No	No
Volume	Numeric(10,2)	No	No

Figure 23.3 Relational database properties.

- **Data type** – this sets the structure of the content contained within the field. It is a declaration by the owner of the database as to how the content should be treated by the database – for the user this setting lets them know the operations that can be performed on the data content (for example, only numeric values can be used for arithmetic operations).
- **Required** – informs the user whether a field must be populated if it is to be rendered as part of the table. Incomplete records are not allowed into the table unless the appropriate exceptions (more metadata) are made.
- **Primary key?** – identifies whether the field in question is used to identify unique records in the table. As the table implies, the primary key must consist of mandatory fields. However, not all mandatory fields need be part of the primary key – this is a control under the management of the database owner.

It is worth noting that all fields in a database need to have a type associated with them. Typically database systems will apply a default setting (such as variable length character string (or "Varchar" as used in the example). Setting the precise metadata of the content of the field is one of the key functions of a database – by defining the type, the manner in which the field is stored and can be used in querying is recorded. This allows the system to store and access the data in the most efficient manner possible (another example of normalization). There are many different data types that can be applied to a field which will affect the performance and efficiency of the database, so recording this specific metadata is a fundamental step to deploying any database application.

This is a pretty small snapshot of all the properties that can be associated with a database table. The key point to recognize is that the successful completion of this metadata content is required in order to make the business data content operational to the broader enterprise. While that is primarily a physical requirement in order to make the content available at scale to the enterprise (think in terms of a table with all the prices from all global equity markets for the past ten years, which is a pretty common requirement to drive charting and statistical applications), without this metadata content the authority and reliability of the content is completely open to question.

Here lies a vital characteristic of technical metadata: it is mandatory. No database or other data delivery can be successfully put into operation at scale without it. In addition, experienced database professionals and application developers know that there is a cumulative benefit in setting and managing technical metadata content. The more metadata defined, the better the richness and performance of the application. But the benefit does not come for free – as the volume and sophistication of the metadata grow, so too does the task of managing the content.

23.2.2 Business metadata

While technical metadata is required to make an information service available, to make the service meet (and preferably exceed) expectations of business users requires another layer of metadata altogether. As any experienced capital markets professional can verify, the world is replete with applications that may function and meet the minimum needs; there are precious few that are a joy to use.

In the world of metadata management, this class of content is referred to as "business metadata." This is the collection of metadata that serves to take something that works and turns it into something that becomes part and parcel of a successful business workflow. By definition this characteristic suggests that any lack of attention to business metadata is likely to place a constraint on the ability of the information service to meet operational requirements. A typical manifestation of this constraint is the inability to reuse the information service across more than one business function, leading to increased operating and other costs.

There are, therefore, many different kinds of business metadata. In fact, given a desire to reuse an information service across the enterprise, the only constraint is likely placed on the business requirements for each application of the data content. In order to set some boundaries here, the following commonly used business metadata examples are considered.

- **Viewing data content** – usually of top priority is the manner in which content is identified and displayed for visual purposes. These rules and conventions all fall into the category of business metadata.
- **Defining data content** – while the display of content may vary between applications, the underlying definition of individual items needs to be consistent. A single set of logical definitions is usually required to support multiple applications of the same content.
- **Other business metadata content** – attributes that record specific items appropriate to support the application of data content for individual business functions.

Database views

While valid, the layout for the data content in Figure 23.2 represents a relatively inefficient design for a database table. As the majority of fields are not mandatory, it is highly likely that the table could appear sparsely populated. This is a common occurrence as database size grows, as not all financial markets necessarily publish the same set of content.

The downside to this sort of table design is that performance tends to be compromised. Each unpopulated cell in the table represents a value NULL in the database. This increases the space required to store all the data content and by extension the time required to load and query this content. While not an issue for a

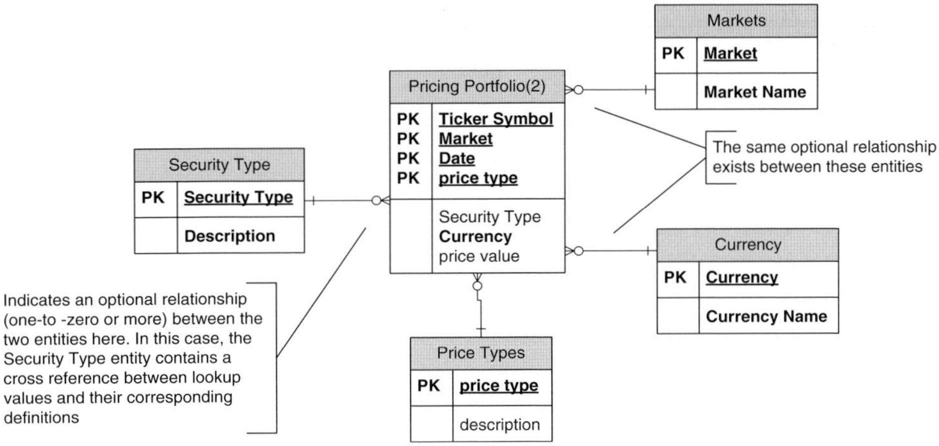

Figure 23.4 Normalized design of a database table structure.

small table as shown above, when expanded to the full scope of the global capital markets (over 11 million security records alone are supported by Interactive Data), the impact is rapidly felt.

In order to avoid this problem, database designers typically take a two step approach to the problem.

- The data content is written to the database in a manner that minimizes the number of NULL records and that minimizes redundancy (a process known as "normalization").
- The database creates what is called a "view" against which business users can issue queries and perform analysis. Since by definition the view, as illustrated by the table in Figure 23.1, returns the sparsely populated result set, it is referred to as "denormalized."[5]

Key points to recognize in this design are:

- Fewer fields are used to store the data content, resulting in a simpler structure to manage,
- The primary keys are different from the denormalized view,
- The structure eliminates the need to store NULL values in the database.

In this design, the layout of the pricing table will take a different form as illustrated by the example presented in Figure 23.5.

[5] Such sparsely populated result sets are also termed "stochastic" in the field of statistics.

Ticker Symbol	Market	Date	Security Type	Price Type	Price Value	Currency
SCCO	NYSE	31/01/2012	EQ	Open	33.40	USD
SCCO	NYSE	31/01/2012	EQ	High	33.62	USD
SCCO	NYSE	31/01/2012	EQ	Low	32.92	USD
SCCO	NYSE	31/01/2012	EQ	Close	33.40	USD
SCCO	NYSE	31/01/2012	EQ	Bid	33.15	USD
SCCO	NYSE	31/01/2012	EQ	Ask	33.40	USD
SCCO	NYSE	31/01/2012	EQ	Volume	1,593,281	USD
GE	NYSE	31/01/2012	EQ	Open	19.30	USD
GE	NYSE	31/01/2012	EQ	High	19.37	USD

Figure 23.5 Normalized data representation.

This representation of the content provides a more efficient manner in which to build the database table with an eye to improving performance for consuming applications when deployed at enterprise scale. However, it is also clear to see that there is a trade-off: improved performance is achieved at the cost of making the table more difficult to navigate.

It should be noted that normalization of data content is a key step in making data content accessible and usable by multiple applications. In fact, database normalization is a standard function of defining database design. Wikipedia defines the practice as:

> Database normalization is the process of organizing the fields and tables of a relational database to minimize redundancy and dependency. Normalization usually involves dividing large tables into smaller (and less redundant) tables and defining relationships between them. The objective is to isolate data so that additions, deletions, and modifications of a field can be made in just one table and then propagated through the rest of the database via the defined relationships.

The degree to which redundancy and dependency are removed from the structure results in the "normal form" of the database. There are currently six separate normal forms defined[6] as part of relational database design standards. As a rule of thumb, the higher the normal form of the database, the less likely it is to suffer from inconsistencies or other data related issues.

Equally, this behavior of data content suggests that the lower the normal form, the easier it is for a business user to interact with the database. This implies, then, that the higher the degree of normalization of the data structure, the more business metadata is needed in order to render a data display that is actually useful to a business user. This translation from something that is designed for storage and management of content to something that is recognizable to a business person is commonly referred to as a "VIEW" and is, by definition, business metadata.

[6] `http://en.wikipedia.org/wiki/Database_normalization#Normal_forms`

It should be clearly understood, however, that today's relational database platforms make it easy to combine the two so that the benefits of both improved performance and easy, consistent use across the enterprise can be achieved. The use of the VIEW concept allows the end user to visualize the data in the manner in which they prefer. In turn individual users can be permissioned to access (or prevented from accessing) an individual view.

Views are relatively easy to create using standard SQL techniques. At a very basic level, a VIEW is simply a derived image of the content of the database rendered in a way that meets the needs of the user of the data. It is entirely possible for multiple views to be created against the same data content stored in the database using a completely different representation. Characteristics such as field names and order, grouping and sorting can all be different for the same set of physical data content.

The creation of the VIEW, in other words, represents vital business metadata – it is the means by which the business user controls the data content that is actually selected for display or other use. This is an important point to reinforce. While technical metadata is required in order just to make the data content operational and available, the application of the corresponding business metadata is essential to taking the raw data content and making it into actual information for use in business applications. The business metadata in this example may not be mandatory, but it is an ingredient in the secret sauce that results in the business generating value from the underlying data input.

The data dictionary

The key lesson to learn from this discussion is that business metadata is what makes raw physical content actually useful to a business person. Successful management of this content is just as important as technical metadata. While technical metadata is the basic ingredient, business metadata provides the link between the physical, raw content and its application to the business.

An essential part of metadata in the capital markets context is the concept of a "data dictionary." Ironically, pretty much everybody in the industry seems to understand the need for and importance of such a resource, yet there are few common guidelines or standards in place. In reality, the data dictionary can be thought of as logical extension to the database properties introduced earlier and takes a form as shown in Figure 23.6.

Notice here how the combination of the technical and business metadata serve to lay out the navigation to the content of interest to a business user. In one place all the content of interest is defined to a level that a business analyst or other professional can understand. This metadata content is vital to making productive use of the underlying raw data asset.

Also notice in the table that some of the definitions are underlined. In fact, these represent the use of hyperlinks within the data dictionary content to

Field Name	Data Type	Required	Primary Key?	Field Definition
Ticker Symbol	VarChar(60)	Yes	Yes	An alphanumeric string used to identify individual securities within a given market. Ticker symbols are usually issued and managed by the exchange identified by the corresponding market identifier
Market	Char(10)	Yes	Yes	An alphanumeric coded value used to identify the individual market in which a security trades. Cross reference with the Market_Codes table for the textual definition of each individual value
Security Type	VarChar(60)	Yes	No	An alphanumeric coded value that identifies the asset classification to which the security belongs.
Currency	Char(10)	No	No	The three character coded value issued by the International Standard Organization (ISO)that identifies the currency units in which the prices associated with the security are quoted
Date	Date	Yes	Yes	The calendar date to which the pricing observations apply
Open	Numeric(10,2)	No	No	The first price reported in the market for the security on the date of trade.
High	Numeric(10,2)	No	No	The maximum price recorded for the security in the market
Low	Numeric(10,2)	No	No	The minimum price recorded for the security in the market
Close	Numeric(10,2)	No	No	The final price reported in the market for the security. Note that individual markets apply different rules in setting the closing price for the day. For more information see this reference.
Bid	Numeric(10,2)	No	No	The price a potential buyer of the security is willing to pay
Ask	Numeric(10,2)	No	No	The price a potential seller of the security is requesting
Volume	Numeric(10,2)	No	No	The total number of individual securities transacted on the selected data in the market

Figure 23.6 The data dictionary.

additional resources that can provide additional information about the field of interest. For example, the hyperlinks here are used to navigate to an external semantics repository that provides a baseline logical definition of the term. The hyperlink represents an additional metadata element deployed at the business level targeted at streamlining the application of the content to business problems.

Clearly, a data dictionary is a useful thing to have to hand – in fact most data vendor provided services provide some form of a dictionary in order to support integration of the content to business applications. Indeed, a good data dictionary can serve as the means by which to inform users and application developers to navigate the universe of content and determine appropriate selection of content for specific business needs. The complexity of making this selection – and costs when it is not done right – should not be underestimated. Precision in the use of data content and understanding the logical meaning of individual attributes can be critical to the success of an application.

As a result, a well planned and maintained data dictionary is of interest to the entire organization. Not only can it guide application developers in the use of individual content items (thereby reducing time and effort in development) but it can equally be of use to business users. Some of the key roles it can be applied to include:

- Establishing an inventory of content actually used by the organization,

- Populating help files and other documentation that support business applications and users,
- Providing business analysts with the background to design new applications on an enterprise level,
- Serving as the baseline for ensuring consistency of application of data content across the organization (a key success factor in meeting regulatory compliance requirements).

In fact, the data dictionary is a key resource that finds uses across the organization. However, it is only as good as its maintenance and the dedication put into keeping it up to date. When established as a corporate resource, a robust data dictionary is an important element in ensuring operational efficiency as it provides a focus on the content that the firm actually uses and promotes the reuse of that content for each subsequent application. It is therefore one of the key metadata investments that a firm can make that will yield tangible business benefit.

Additional business metadata

The key aspect that repeats throughout this chapter is that business metadata is not constrained whereas the technical metadata category is typically restricted to the minimum set required to make the data content operationally available to consuming applications. This lack of constraint opens up the possibility of business consumers of the content contributing to the process of scaling up accessibility to the content to an enterprise level where reuse of a single data subscription can yield significant economic benefits to the broader firm.

The universe of business metadata described here provides a basic introduction to the concept only. Business metadata is only limited to the uses to which the data content is put by business users and applications. Yet no matter the actual content selected, it all needs to be managed within a single framework and managed on an active basis. It is not a one time activity, but like all metadata requires ongoing support and management.

Some additional examples of business metadata include the following.

- **Quality indicators** – the act of defining the business data content within a data dictionary opens up the ability to record parameters that measure the perceived quality of the content. This is an emerging area that introduces a new dimension to metadata allowing for data quality checks (such as for timeliness, logical consistency, and accuracy) to be assessed and recorded with each insert or update to the database.
- **Derived and calculated information** – often database content undergoes further derivation or calculation and is stored alongside the original content. A good example is the calculation of statistical coefficients (such as volatility) in response

to the availability of a new pricing value. The rules for performing these calculations all fall into the metadata domain.

- **Database summary information** – many uses of data content can be satisfied by summarizing the contents of a database. For example, queries that produce record counts of individual tables within the database can be very helpful in sizing storage and performance requirements. In fact, most database platforms provide an array of tools to interrogate and summarize data content in this manner and have become essential in maintaining and managing the service provided to consuming applications.
- **Relationship metadata** – defining the relationship between different datasets within a database can be critical to consuming applications. If the connections are well understood, indexes to the related content can be defined (foreign keys are a common example) that make navigation to this content much more efficient and return to the user only the set of interest, despite being produced from different sources. The act of identifying and codifying these relationships is by definition the creation and recording of metadata.

The universe of business metadata and its applications is unbounded and far too rich a subject to be treated fully here. The point to recognize is that defining and managing the metadata associated with individual business processes is critical to achieving success in servicing the workflow. For more manual oriented tasks, the business metadata requirements can be quite small. However, for more automated workflows and functions, the need for metadata increases in parallel.

The opposite of this situation is also true. Failing to establish a sufficient set of metadata to support the workflow can put the business at risk. In highly automated environments, this level of risk can be severe, potentially causing service outage (think of a website failing), loss events, or worse. It is easy to see therefore, that metadata has a very close relationship with the operational risk a firm faces. In an industry of rapidly increasing data volumes this places a premium on the ability to define and manage the metadata that can be applied to service the business and its customers.

23.2.3 Process metadata

The last category of metadata to consider is what is referred to as "process metadata." Where the technical metadata reflects the content needed to put data into operation and business metadata provides the added value to making data content a business asset, process metadata is what keeps the service up and running. Without defining the appropriate process metadata, the level of service that can be provided rapidly becomes a likely point of failure.

Like the business metadata described above, process metadata should really be thought of in the context of the service to be deployed and the prospective consuming applications and use cases. It too, therefore, is an unconstrained collection of content that needs to be defined as part of the planning for the service. For the current purposes, two separate process metadata categories are considered here:

- **Operational metadata** – the set of content that focuses on maintaining the operation of the service in line with its availability and performance requirements,
- **Business process metadata** – the universe of metadata that is defined to make the service available to end users.

To be sure there are other categories of content that can be defined for process metadata. The two defined here represent some of the more common use cases encountered by today's capital markets firms. Each is explored in more detail below.

Operating parameters

Continuing with the relational database example above, the most obvious metadata to consider is the list of operating parameters required to ensure that the service meets the requirements of consuming applications and users. This may sound like common sense but all too often it is not taken into consideration when preparing to deploy a service to business groups. The downside of not taking this into consideration is likely very familiar to many professionals – an application containing data content of interest cannot be easily scaled up to support multiple users or broader applications. In the eyes of a business user, this amounts to a constraint on supporting the business and is typically considered to be a failure.

Operational parameters are therefore a critical set of metadata to consider. Examples of such metadata include the following.

- **Data volumes** – predicting the required disk space required to support a database and its projected growth is essential to ensuring that a data service can be made reliably available to the broader enterprise.
- **Data retrieval performance** – understanding the needs of the business application in terms of query resolution and time to render data content is vital to ensuring that the application gains acceptance.
- **Projected number of users** – a forecast of the projected user load along with their expected applications for the content. This content serves to aid in sizing the system, which has direct impact on the projected cost to the business.
- **Accounting and audit requirements** – identifying the usage reporting and other accounting metrics that will be required. Understanding perspective audit requirements can be vital in ensuring that visibility into actual data usage is supported for many lines of business. Anticipating audit support requirements can be vital to managing a firm's operational risk profile.

This is simply an indicative list of metadata that should be compiled as part of planning to deploy a service to the enterprise. For the most part the deployment technologies (such as a relational database platform) provide tools to capture this information and report it back to the database administrator (usually via some form of a command and control interface or console). However, the availability of the tools does not mean that the metadata requirements come nicely packaged and ready for deployment. The act of defining the metadata represents a key element in the intellectual property of developing the information service and related business applications and therefore should be viewed as a conscious decision that needs to be part of the scope of specifying the service.

In short, best practice for the deployment of a data service to enterprise makes considering these factors as part of the broader metadata framework pretty much a mandatory exercise. Failing to compile this content up front all too easily leads to obstacles to deploying a system that can cater to a successful rollout not only to support a single application, but also to planning for support across multiple business functions.

By now it should be clear that the metadata required to support any business application depends on the nature of the application itself. For some applications very little thought needs to be put into the metadata strategy while for others it can be a very demanding process. Recognizing the need to identify metadata requirements as part of the definition and specification process can yield significant benefits not only to rollout of an information service, but also to the ability to achieve reuse across the organization.

Business processes

Just as important as the operational parameters and requirements are the business processes related to the deployment of data service. This is where the rubber hits the road – the activities and business functions put into place to support the actual use of the information service within the firm. Two key functions immediately spring to mind:

- Implementing control over the access to the data content and, in turn, accounting for usage,
- Establishing and managing appropriate support and escalation for the service.

Both of these examples are rich in the potential metadata that can be defined as well as the uses to which it can be put in order to better manage a line of business. To be sure there are other examples that can be considered within the scope of this category, however, these two will serve to round out the definition.

Permissioning and entitlements Look back to the pricing portfolio table introduced earlier (Figure 23.1, above). It contains information that is likely of interest to multiple users of the organization. Can this information be made available to any users in the firm? Or are there any contractual or other requirements that need to taken into account? Equally, any business has a direct interest in identifying the total cost of ownership (TCO) of an application. Is this information easily calculated?

These are all examples of the universe of content making up the business process metadata category. They all focus on the use and management of the information service by individual business units. While the scope of this metadata category can vary quite considerably, in the current example it centers on the notion of applying permissioning control to the use of the content. This is by far the most commonly encountered form of business process metadata as it has direct economic impact on firms.

Many data suppliers link usage to the fee that a client firm is expected to pay. Consequently this requires that the consuming firm have the ability to control access to and report on data usage. This need lays out the requirement to support explicit permissioning for access to the content and support it with the appropriate administration processes.

By extension, the same reasoning applies to premium data content generated by a firm. In many cases such content is made available to customers and counterparties, sometimes on a commercial basis. Therefore, access control through formal permissioning is needed.

The act of defining what data content can be accessed is central to managing costs within today's firms and can take many forms. Take the portfolio pricing view introduced above. A very simple form of permissioning functionality would be to simply allow users to access the entire view or not. However, permissioning can rapidly become very complicated – perhaps some users can only see data content in the view from specific markets. Or perhaps only bid and ask values can be accessed.

The definition of these permissioning profiles is by definition an example of business process metadata. As part of the business process of accounting for usage, permission to access the content is explicitly assigned and administered as a business function. In turn, imposing this process on the users of the content results in the creation of additional metadata in the form of regularly scheduled usage reports which are used to substantiate billing with suppliers as well as to provide record of cost for accounting purposes.

Here lies a commonly overlooked task – accounting for the cost of services and calculating both the total cost of ownership (TCO) and the rate of return the information service generates for the firm. Without a commitment to defining the metadata requirements up front, it is very difficult for any firm to account for the expense of developing and maintaining the information service. Typically these costs are depreciated with the cost of deploying the first consuming application. This is by definition a constrained approach – if the metadata is laid down in advance (and the business process metadata in particular as this identifies the actual system usage), then a formal accounting framework can be put in place. Such visibility into the business aspects can then be applied to planning for extensions and maintenance to the service. More importantly it identifies the financial benefits to the firm arising from achieving reuse of the content across multiple applications.

The point to realize is that an additional sequence of activities is created as a result of putting the data service into production. Not only does this require additional administrative support, but it also creates a new group of stakeholders – the management and other business functions making use of the resultant metadata.

Application support There are other business processes or functions that can be included in the scope of process metadata. Equipping the enterprise with the appropriate support and escalation procedures is another good example. A sequence of steps related to the content of the information services and their intended users is defined in order to ensure expected levels of service are met. This process can be recorded as additional metadata intended to support the team assigned to perform the necessary product support.

The key point to note, however, is that all three categories of metadata are tightly linked. They are all essential to ensuring not only that the data content reaches the appropriate consumers, but also that it supports the application of that content and its management within the firm.

Here lies the true commercial value to identifying and managing metadata as a formal business function. By performing the appropriate metadata related tasks up front to define the application, its uses, and the manner in which it is managed, the broader enterprise positions itself to make use of the content in the most efficient manner possible and achieve reuse of data content in a consistent manner. For most firms this represents a departure from previous methods of application development, requiring investment and mastering new skills. However, the reward of achieving economies of scale over the acquisition, deployment and application of data content provides the return on investment for undertaking the effort.

23.2.4 Some final points on metadata

So far, this chapter has focused on the definition of metadata and identifying it as part of ongoing business workflow. This has centered on the relatively familiar ground of working with a simple relational database application. While understanding the general concept is an important first step, it should be recognized that metadata is a fundamental part of pretty much every business process within a capital markets firm today.

A key point to stress is that metadata is relatively organic in nature. It is generated from many different sources and has a variety of applications through the business workflow. This diversity means that different metadata subtypes (such as technical metadata) can be easier to identify and manage than others. However, when combined into a single framework, the value of metadata can easily exceed the sum of the constituent elements by improving the efficiency and performance of business applications.

Finally, it should be recognized that metadata is an emerging subject domain. The rapid and ceaseless growth of capital markets data volume requires better control and application of metadata as part of addressing business requirements. In order to make this data available in an efficient manner requires that it be integrated with an eye to reuse across multiple business functions in the organization. In turn data content can be expected to undergo further derivation and be applied as part of calculations in conjunction with data content sourced from different systems. Pulling all this data content into a single logical framework so that it can used by increasingly quantitative applications will place an even greater premium on developing management competence over metadata.

23.3 Metadata technology

So far the discussion has focused on the use of metadata within the readily recognizable context of relational database technology. To be sure this is the most common application in today's capital markets and is likely encountered by most market data professionals each and every day.

While a useful context, though, the use of the relational database technology implies a limit on the uses for metadata. In fact, the choice of the technology actually places a constraint on the extent to which metadata is used to support business applications.

The relational database use case highlights an important characteristic about the three different metadata content categories. There is a relationship between them that becomes obvious as part of the discussion of a common capital markets application. That raises the question as to whether the same sort of relationship applies to other applications in other subject domains. More subtly, the detection of such a deep interdependency between the constituent components of metadata suggests that perhaps there is a more generalized model that can be applied as a general discipline, independent of any one industry.

Like so much of the modern world, the answer can be found by looking to the Internet. More precisely, it can be found in looking at the technologies that are used across the Internet globally where metadata is emerging as a defined discipline in its own right.[7] More importantly, the International Standards Organization (ISO) has

[7] Wikipedia provides a very good primer on the current state of metadata management.

published a recognized standard for representing metadata for an organization in a metadata repository (see ISO/IEC 11179).

This is an important development for the definition and management of metadata to be used in a context free of any one technology selection, such as a relational database management system used in the previous example. While such technologies capture and store metadata, looking at metadata from a point of abstraction opens up a range of possibilities for managing metadata content in a professional manner.

The biggest benefit from taking this approach is that there are documented standards available to serve as a framework for implementing a metadata strategy. Adoption of the standards and the emerging technologies within the capital markets are at an early stage but the future direction is already clear. The ability to define and manage metadata as a business discipline without any dependency on any one technology opens up the opportunity to scale the use of data content in a cost effective manner across the enterprise. As major consumers of data content, this identifies significant economies of scale to be realized from streamlining the availability of consistent data content across all business users within an organization.

The remainder of this chapter takes the final leap in looking at metadata outside the context of any one technology. The availability of an ISO approved international standard provides the key starting point as it allows for the creation of methods and tools for managing metadata from a purely objective standpoint. In the world of the modern Internet, this has seen the arrival of two key items to drive the management of metadata.

- **The Resource Description Framework (RDF)** – a family of World Wide Web Consortium (W3C) specifications originally designed as a metadata data model. It has come to be used as a general method for conceptual description or modeling of information that is implemented in web resources, using a variety of syntax formats.[8]
- **The SPARQL query language -** is an RDF query language, that is, a query language for databases, able to retrieve and manipulate data stored in Resource Description Framework format. It was made a standard by the RDF Data Access Working Group (DAWG) of the World Wide Web Consortium, and considered as one of the key technologies of semantic web. On January 15, 2008, SPARQL 1.0 became an official W3C Recommendation.[9]

While there are a variety of related technologies and capabilities that have emerged, these two items in particular lay the foundation for managing metadata as a core discipline. Both are freely available as open source materials and both enjoy rapid uptake in the academic community world-wide (in common with many open source

[8] See Wikipedia: `http://en.wikipedia.org/wiki/Resource_Description_Framework`
[9] See Wikipedia: `http://en.wikipedia.org/wiki/SPARQL`

resources). Combined they are at the heart of a very active area of research and as they mature they are likely to have a major impact on how metadata is applied in the capital markets in the coming years.

23.3.1 Resource description framework

The Resource Description Framework (RDF) is a set of specifications published by the World Wide Web Consortium (W3C) that lay out the structure for a general method of describing, at a conceptual level, information that is intended for publication on the Internet. It can be thought of as a modeling method that serves to capture metadata as part of the process of publishing information content.

The most common syntax that applies RDF is that ubiquitous internet language, XML (eXtensible Markup Language). More precisely, XML is the serialization method (i.e., the actual publication form of the content) that can be applied to RDF so as to make the data content available for use by any RDF compliant application. In similar fashion to the Simple Object Access Protocol (SOAP – another W3C standard), RDF provides the basics for making content immediately accessible to any application that makes use of the standard.

For the capital markets industry this represents a significant step forward. In a nutshell, it means content that is published following RDF principles can be expected to contain not only the raw data content, but also a selection of metadata related to the application of that content. In effect this means that any payload of XML data content becomes what is called "self-describing" – it contains all the information needed to apply the content under business rules defined as metadata.

Sound familiar? Looking back to the relational database example provided earlier, this is precisely the same outcome from taking a well-structured approach to making an information service available to the organization. However, in this scenario both the metadata and the underlying business data content are physically linked with one another. This makes it very easy for a user of the data content to integrate that content into an application. From the business perspective this greatly simplifies the data integration process, offering significant improvements in the time and cost of deploying new content within a capital markets firm.

23.3.2 SPARQL

The actual XML implementation of RDF is called a "document." As the name suggests, it is literally a defined collection of text, usually rendered as a file. Given the scale of the data content produced by the capital markets on any given day, the size of such XML documents can be quite considerable – for example, a typical initialization file containing a current image of all municipal bonds produced by

Interactive Data totals over 20 gigabytes when uncompressed. While XML is noted for being easy to read and scroll through, documents of such size can be highly structured, requiring more sophisticated tools to analyze.

The most common manner of working with such files is to perform queries similar to the way in which relational databases are used. Querying is a powerful way to apply the power of the computer to analyzing the data content by interrogating the content using logical rules to render to the user the subset of content that is required at any one point in time.

The emerging popular tool for querying RDF/XML documents is called SPARQL (pronounced "sparkle"). The name stands for "SPARQL Protocol and RDF Query Language." It is a query language designed for interrogating RDF/XML documents and has now become a W3C standard.

SPARQL is implemented in a number of XML editing tools (the list seems to grow daily) and allows a user to interrogate the contents of the document without having to process it into a database. A similar example might be the use of the search and replace functionality in a word processing application – the data content is scanned and operated on in its native form (in this case that native form is a simple ASCII text file and not the proprietary document format).

Think about the implications of the SPARQL functionality for capital markets applications. Using tools like SPARQL, a business analyst can now interrogate raw data files (as long as they are RDF compliant of course) without having to parse the data into an analytical application (such as a spreadsheet) – there is no need for intermediate processing. Another example is the use of a machine interface like SPARQL in an application to identify and extract the content required as and when the document arrives, such as in rendering content for visual display to a browser screen. These are all direct benefits to business users of data content, leading to significantly improved productivity and reduced costs. Viewed at from the enterprise level, this benefit holds out the promise of significantly improved time to market for applying data content to emerging business problems.

More broadly, these capabilities introduce the power of Internet style data interaction to the modern organization. As more and more content becomes available as XML, these sorts of tools allow business users and applications to interact with not only the raw data content but also the accompanying metadata. This opens up entirely new opportunities for productivity growth and ultimately new business models for any firm, not just capital markets enterprises.

There are downsides to using XML in the capital markets. It is known to be very verbose and therefore requires a lot of bandwidth and/or disk space. This makes it inappropriate for high performance applications such as delivering real-time streaming data to trading and risk applications. It is ideal, however, for delivering data content with additional context and metadata on demand – a good example is the task of initializing a new instrument in a trading system. Using an RDF/XML message construct can deliver the combined raw data and associated metadata as a single payload to the consuming application. This allows for a lot of data to be communicated, improving the end-to-end performance of the setup function as it eliminates many of the problems inherent in the process when data are delivered in a piecemeal fashion or require additional manual intervention to complete the record.

Another drawback to XML is that it is known for slow performance. As the content is all provided as text, the operations that can be executed on this data type are limited. Secondly the nature of XML itself is quite verbose, a natural consequence of augmenting the underlying data content with metadata. Still, as price/performance ratios improve, so does the performance of XML technologies.

Fortunately, not all capital markets applications are as demanding as real-time trading systems. In fact, there are arguably more applications that can benefit from the use of XML structured data content than do not. Also, Moore's Law serves to ensure that the price to performance ratio of technology continues to improve over time. As more powerful processing becomes available at lower cost, many of these performance constraints can be expected to become less and less of a problem.

23.4 Managing metadata

The discussion has centered so far on putting metadata into the capital markets context and exploring the benefits of some of the modern tools and techniques. Together this forms the basic body of knowledge needed to embark on applying metadata as a discipline in its own right within capital markets firms. Given the significant impact that metadata can have on a firm's operations, though, it is essential that consideration be given to the manner in which metadata is managed. Failure to achieve consistent management over metadata content risks significantly compromising the goal of achieving enterprise scale operation of data services. Even at the operational level of supporting a stand-alone business application, poor metadata management can all too easily lead to service outages and ensuing loss of business.

One aspect of metadata that should be clear from the preceding sections is the diversity of the sources from which metadata can be generated. While the simple categorization applied here of technical, business, and process metadata provides a useful framework, it clearly provides only a generalized approach. Each business can be expected to generate its own universe of metadata and potentially classify it in a different way. While industry standards can help, there remains the need for a firm to define and embed the management of the metadata content within its own operations.

It is this diversity that presents a big challenge. No one management model from any single discipline within the metadata domain can necessarily be easily applied across the full universe of content and its application to business functions. It is for this reason that attempts to set up a single role within a firm to manage metadata typically meet with only limited success. There are simply too many moving parts from across the organization and the data content is so diverse as to make it a challenge to assign to any one group or team to manage.

On the surface this observation would appear to contradict the ongoing trend within the capital markets industry to establish data management as a distinct business function under the direction of a "Chief Data Officer" (also referred to as "Data Czar"). It is fair to say, however, that such roles focus on the acquisition and management of business data content within a firm and for the most part struggle with addition of metadata as a content category to the scope of the role. Besides, it is unlikely that any one person has the experience and knowledge to manage all the required metadata.

Consequently, no standard management framework or model has yet emerged as a best practice – all firms are forced to support their metadata requirements in parallel to their other data content. This is an ironic outcome as the primary benefit of investing effort into the definition and management of metadata within a firm is that it amplifies the ability to manage business data content and services. In a perfect world, management of metadata would be exploited to leverage as much business data content as possible with the least amount of direct effort. In the imperfect world of today's business landscape, however, the reverse is the usual case – applications consume data content on a stand-alone basis and reuse is the exception rather than the rule.

While this situation may appear to present a challenge, it does offer the opportunity to establish the metadata management function from a clean sheet. A good starting point is to look at the framework of applying best practices both within and across the organization as a whole. In turn, the outputs from applying such practices can then be grouped into a specific business function, such as a Chief Data Officer.

The decision on which practices to apply can be expected to vary by firm. This should not come as a surprise. If metadata management is to make a positive contribution to a firm's performance, then it conceivably forms part of the individual value proposition of each firm. Diversity in management over the key information asset of metadata can therefore be at the heart of establishing competitive differentiators between firms.

While diversity can be expected, there are some readily identifiable core areas that serve as logical starting points.

- **Inventory** – as with most data related projects, the place to start is to assemble an inventory of the specific data items of interest. In the case of metadata attributes, this can be viewed as an extension of the data dictionary used to define the individual business data attributes – for each data item, the corresponding technical, business, and process metadata attributes that are of interest can be identified as part of the original project. In turn, the source and management of each can be defined and tracked.

- **Communications** – where an inventory (such as a robust data dictionary) can be the basis for setting the scope of the metadata exercise, communicating the information to the full user audience is extremely important in achieving critical mass of interest.
- **Process** – defining, documenting, and enforcing the appropriate workflows to add, modify, and delete metadata content is vital. Disciplined process management serves to streamline the entire data service for end users by eliminating steps in the production chain
- **Quality** – ensuring the timeliness and accuracy of the metadata required by applications is a critical success factor. No other single factor can lead to slow take up than poor reliability of the data content.

If these parameters appear familiar, that is because these are all well-known aspects of data management. As such, many of the capabilities already in use today (such as data modeling tools, for example) can be turned to managing metadata content. Indeed, metadata management is often claimed as a key feature of such software packages. It follows, then, that these tools likely provide a good starting point for managing the metadata content needed to support applications consuming capital markets data.

Where this happy situation exists, it is therefore likely a good candidate for establishing formal management control over metadata management. Typically these tools will be deployed within data management functions. As a result, extending the data management function to include explicit support for metadata generated within the firm likely makes sense, although it will not come for free. Developing the skills and management workflows that meet the needs of the broader enterprise will require time, training and new skills. It is, therefore, a bottom-up exercise, not top-down.

Metadata management, therefore, should be viewed as a journey without a predefined destination. While there are costs to be considered up front, the benefits arising from achieving increased consistency and reuse across the organization are sure to result in a positive return on the investment.

To be sure this is only one of many options for managing metadata. But given how closely linked the management of metadata content is with the fabric of a capital markets firm's daily business operations, it does represent a good starting point. Ultimately, each firm needs to define the implementation that best meets their business needs. So while there may be some easy to spot starting points, each firm needs to set out on the journey independently.

23.5 Emerging trends and future impacts

A recurring theme of this chapter is that the metadata space is dynamic and evolving. New technologies and applications are coming to market with increasing frequency and the pace can be expected to accelerate over the coming years. It is worthwhile, therefore, to identify some of the emerging trends in the metadata space and attempt to ascertain their impact on the capital markets industry.

First, it is clear that metadata is a topic that cuts across industries. This means that a much wider pool of expertise is working on the technologies and solutions to business problems than if the issue was restricted to any one firm or industry. This suggests that best practices are very likely to make themselves known and support adoption by the capital markets industry. That means that a whole new set of technologies and methodologies will be readily accessible and easily applied to business problems.

Second, technology integration can be expected to accelerate. As the adoption of XML has shown, capabilities like SPARQL and RDF can all easily combine with existing business applications (such as a database or spreadsheet). These applications can be expected to continue to integrate these technologies, thereby making them available to business users and not just technologists. The tools for creating, managing, and applying metadata will increasingly become transparent to end users. Metadata will therefore increasingly become embedded in all aspects of business analysis associated with delivering market data content to the enterprise. Given the accelerating growth in the volume of this content, this availability should serve to support a firm's ability to take advantage of the opportunities that such growth will offer.

Management of metadata can be expected to become a major topic in the industry over the coming years. Various models will be proposed both from capital markets firms themselves and from the consulting and software communities. In all likelihood no one model will meet all the needs of any one firm. What works for one may not be easily transported to another – it will place a premium on the individual people who are comfortable and proficient at working with metadata. At the same time, case studies will be published that reinforce the benefits to a firm's bottom line from investing in improvements to metadata management.

Another trend to mark is that the ability to work with and manage metadata can be expected to become a highly sought after skill set. People who are proficient at working with the combination of technology, content and metadata will be at a premium in the marketplace. However, education and post graduate skills development can be expected to respond to the need for data experts who can support an organization's goal to achieve enterprise scale in their use of information services. Therefore, an increasing profile for the role of data management in business organizations can be anticipated, as well as within the academic community.

Finally, the pace of innovation and development of metadata technology can be expected to continue for the foreseeable future. Such progress will indeed be welcomed as firms scale up their expertise in order to meet the demands of the rapidly growing volumes and complexity of market data with which they need to contend. However, this pace of change and improvement can be expected to impact on business users as they grapple with successive iterations of capability over a short period of time. A good example might be to look at the rapid succession of cellular telephones over the past decade. New releases with new capabilities come to market so frequently that it can be hard to keep up.

Taken as whole, metadata management and technology can be expected to increasingly become part of the mainstream discussion about working with capital markets data content. While there will be effort required to take advantage of metadata, on the whole the ingredients are becoming available to make this a highly successful transition, not only improving the bottom line for capital markets firms, but also opening up whole new lines of business. Metadata will emerge as a standard tool for working with market data, familiar to everybody.

24

Data security and privacy

Michael J. McCabe

24.1 Introduction

Data security and privacy in the financial industry is a complex topic impacted by a number of state and federal laws. This chapter will review the landscape with a focus largely on the US regulatory environment. Where appropriate the non-US environment will be covered. A survey of state and federal laws is an ongoing project with evolution of the statutes and their interpretation a basic reality. Two concepts are addressed: data security and privacy. Data security is focused on the need to keep information secure and safeguarded. It addresses the question and methods of how to keep information protected from unauthorized access and misuse. Privacy, in contrast, is a discussion of the rights and limitations associated with the collection, use, maintenance and lifecycle of personal information. Any discussion of privacy needs to address what can be kept from individuals other than the owner and what cannot. Privacy has its roots in the concept of "the right of the individual to be let alone." Through a long history of cases the US courts have recognized the individual's constitutional right to privacy. While this does not necessarily extend to financial information, starting in the 1970s laws were adopted to protect the privacy of information. Interestingly, the USA takes the approach of specific laws governing specific data questions instead of broad encompassing rules. Typically the USA has addressed concerns through specific laws and regulations. This is in contrast to the EU where a broad directive provides the basis for protection.

This chapter looks at the rules and regulations providing the basis for data security and privacy and at the core concepts underlying compliance with them.

Handbook of Financial Data and Risk Information, ed. Margarita S. Brose, Mark D. Flood, Dilip Krishna and Bill Nichols. Published by Cambridge University Press.

24.2 Security and privacy implications for financial data

Customers of banks and other institutions assume that the non-public data that the institution holds about them is held securely. As has been demonstrated on numerous occasions, the release of financial information can result in significant harm. That harm is defined in terms of harm to the individual and harm to the organization. This can be in terms of financial loss, loss of reputation, loss of business and sanctions for failure to comply with existing rules and regulations. The costs associated with notification of customers and required remediation in the event of a data breach can be significant. A breach in data security that results in a violation of privacy can cost easily in excess of several million dollars per incident.[1] The implication for financial institutions is that data must be properly and effectively safeguarded.

24.2.1 Financial privacy regulations

The evolution of regulations regarding financial privacy has been an ongoing process. Constitutional privacy protection in the US is incomplete by its nature. Privacy is implied, not enumerated in the constitution. Over the years a patchwork of laws and regulations has been put into effect. The statutes, as enacted, represent the work of the political process in an attempt to address a particular concern at that point in time. As a result, US regulations tend to be overlapping and from time to time in conflict with each other. Congress, beginning with the Bank Secrecy Act (BSA) as well as the Fair Credit Reporting Act (FCRA), defined and limited the scope of financial privacy. There is a constant tension between an individual's desire for privacy and the expansive desire of law enforcement for unfettered access to data. The balancing act between these competing demands is defined in the current regulations.

24.2.2 Sensitivity considerations of financial data

Financial data are amongst the most sensitive information that will be held as data by a financial services firm. The data about your customer accounts contains information that must, by law, be kept confidential. The information held by M&A bankers is privileged inside information that must be segregated and protected. Your internal books and records describe the strengths and weaknesses of your organization. Trading positions define the exposure of your firm. Compliance information, such as SARS, is privileged, compartmentalized and governed by

[1] See PC World Business Center (2012).

regulation. Data must be kept secure from accidental loss or intentional misuse. In the modern technical environment, considerations must be given to the security of the systems that the data sit on, can be accessed from or can be viewed by. The drive towards distributed computing and the use of devices focused on the user experience (Blackberry, iPad, mobile tablets, cloud applications, etc.) may not provide the control over your data that is required. The risks associated with the storage of data within the systems of a firm, vendor systems, cloud providers, etc. all need to be considered. As noted above, there is an extensive range of laws and regulations currently in place, which must be complied with, to protect data and to control access to the data.

24.3 Government standards and financial security requirements

Information needs to be protected from misuse by internal and external actors. The basics of the design of systems need to consider the security requirements. To that end there are a number of standards that have been developed to guide the design of systems and to address these questions. While a single standard, for the design of information system security, does not exist, the following standards provide a useful framework that could be considered when developing new or updating existing systems. (The reader should note that these are standards developed for use with government systems but offer useful guidance to the private sector.) The standards reviewed are:

- NIST Special Report 800–53
- FIPS 199
- Federal Information Security Management Act
- ISO 27002.

24.3.1 NIST Special Report 800–53

This National Institute of Standards and Technology (NIST) publication is a part of the Special Publication 800-series that reports on research, guidelines, and outreach efforts in information system security.[2] The publication was developed by the Joint Task Force Transformation Initiative Interagency Working Group with representation from the Civil, Defense, and Intelligence Communities. It was developed as part of an ongoing effort to create a unified information security framework for the US federal government.

[2] See NIST (2009).

This publication provides guidelines for selecting and specifying security controls for non-national security information systems supporting executive agencies of the federal government. The publication provides guidance for meeting the requirements of FIPS 200,[3] Minimum Security Requirements for Federal Information and Information Systems. The guidelines apply to all components of information systems that process, store, or transmit federal information. The objective of the publication is to provide a set of security controls that can satisfy the breadth and depth of security requirements for information systems and organizations and that is consistent with and complementary to other established information security standards. Agencies subject to the NIST security standards and guidelines described in the publication should currently be in compliance with these guidelines. Information systems that are under development are expected to be compliant upon deployment.

The application of the security controls defined in NIST 800–53 represents the current leading practice safeguards and countermeasures for information systems. The security controls are annually reviewed by NIST, and updated as required. The minimum security controls defined in the low, moderate, and high security control baselines are also expected to change over time as well, as the level of security and due diligence for mitigating risks within federal agencies increases. Proposed additions, deletions, or modifications to the catalog of security controls and the proposed changes to the security control baselines in NIST 800–53 go through a public review process to obtain government and private sector feedback and to build consensus for the changes.

NIST 800–53 covers the steps, within a defined risk management framework, that address security control selection for federal information systems in accordance with the security requirements in Federal Information Processing Standard (FIPS) 200. This includes selecting an initial set of baseline security controls based on a FIPS 199 worst-case impact analysis, tailoring the baseline security controls, and supplementing the security controls based on an organizational assessment of risk. The security rules cover 17 areas including access control, incident response, business continuity, and disaster recoverability.

24.3.2 FIPS 199

Part of the certification and accreditation process for federal information systems[4] is the selection and implementation of a subset of the controls from the Security Control Catalog (see NIST 800–53, Appendix F). These controls are the management, operational, and technical safeguards defined for an information system to

[3] See FIPS Publication 200, NIST (2006). [4] See FIPS Publication 199, NIST (2004).

protect the confidentiality, integrity, and availability of the system and its information. The security categorization of the information system determines the baseline collection of controls that must be implemented and monitored.

24.3.3 Federal Information Security Management Act (FISMA)

The Federal Information Security Management Act of 2002 (FISMA)[5] is a US federal law enacted in 2002 as Title III of the E-Government Act of 2002. The E-Government Act (2002) addresses the importance of information security. Title III of the E-Government Act, entitled the Federal Information Security Management Act (FISMA), requires federal agencies to develop, document, and implement a program to provide information security for information and information systems that support the operations and assets of the agency, including those provided or managed by another agency, contractor, or other source.

The act defines the components of a security program. Such an information security program needs to include the following.

- Periodic assessments of risk, including the magnitude of harm that could result from the unauthorized access, use, disclosure, disruption, modification, or destruction of information and information systems that support the operations and assets of the organization.
- Policies and procedures that are based on risk assessments, cost effectively reduce information security risks to an acceptable level, and determine that information security is addressed throughout the lifecycle of each organizational information system.
- Subordinate plans for providing adequate information security for networks, facilities, information systems, or groups of information systems, as appropriate.
- Security awareness training to inform personnel (including contractors and other users of information systems that support the operations and assets of the organization) of the information security risks associated with their activities and their responsibilities in complying with organizational policies and procedures designed to reduce these risks.
- Periodic testing and evaluation of the effectiveness of information security policies, procedures, practices, and security controls to be performed with a frequency depending on risk, but no less than annually.
- A process for planning, implementing, evaluating, and documenting remedial actions to address any deficiencies in the information security policies, procedures, and practices of the organization.

[5] FISMA 44 U.S.C. § 3541, et seq.

- Procedures for detecting, reporting, and responding to security incidents.
- Plans and procedures to determine continuity of operations for information systems that support the operations and assets of the organization.

FISMA explicitly emphasizes a risk-based policy for cost-effective security. The Office of Management and Budget (OMB) requires executive agencies within the federal government to:

- Plan for security
- Ensure that appropriate officials are assigned security responsibility
- Periodically review the security controls in their information systems
- Authorize system processing prior to operations and, periodically, thereafter.

These management responsibilities presume that the parties responsible understand the risks and other factors that could adversely affect their ability to carry out the mission of the particular agency. As a key element of the FISMA Implementation Project, NIST also developed an integrated Risk Management Framework which brings together all of the FISMA-related security standards and guidance to promote the development of comprehensive and balanced information security programs.

24.3.4 ISO 27002

ISO/IEC 17799:2000, was revised in 2005, and renumbered in 2007 as ISO 27002 to align with the other ISO/IEC 27000-series standards. The standard covers generic information security management. It can be used as guidance by anyone that needs to establish a comprehensive information security management program or improve its current information security practices. The individual firm implementing an information security management program has the option to implement all or only a portion of the framework covered by the standard. Use of the standard by the private sector is not required but, as with the US government standards discussed earlier, it is a useful reference and source of guidance. The purpose of the standard is to provide a comprehensive framework. According to the title page, ISO IEC 27002 2005[6] is a "code of practice for information security management." ISO/IEC recommends that consideration be given to each of the practices when defining or updating an organization's information security management program. As such, it is an advisory standard. Organizations are free to interpret and apply the standard based on the particular information security risks faced by the organization. Application of the standard is flexible, allowing the practitioner to adopt the information security controls that make sense to their organization. As users have

[6] See International Organization for Standardization (ISO) (2005).

many options for implementation of the framework, testing for compliance is difficult to evaluate.

This standard contains 12 sections:

1. Risk assessment and treatment
2. Security policy
3. Organization of information security
4. Asset management
5. Human resources security
6. Physical security
7. Communications and ops management
8. Access control
9. Information systems acquisition, development, maintenance
10. Information security incident management
11. Business continuity
12. Compliance.

The standard is intended as a common basis and guideline for developing organizational security standards and effective security management practices. The first step defined by the standard is a risk assessment. The controls defined are intended to be implemented as a result of the risk assessment. If leveraged properly, the standard will provide an organization with a risk based solution appropriate to its needs.

24.4 Federal and other regulations impacting data security and privacy

24.4.1 Electronic Communications Privacy Act of 1986 (ECPA)

This law prohibits persons from tampering with computers or accessing certain computerized records without proper authorization. The Act also prohibits providers of electronic communications services from disclosing the contents of stored communications. In most situations it requires that the customer of the service provider be notified and given an opportunity to contest a government request for access to electronic mail or other stored communications. The ECPA protects wire, oral, and electronic communications while those communications are being made, are in transit, and when they are stored on computers. It applies to email, telephone conversations, and data stored electronically. ECPA includes three acts: (1) the Wiretap Act, which is an updated version of an earlier act commonly referred to as Title III; (2) the Stored Communications Act; and (3) the Pen Register Act.

The ECPA classifies all communications into three groups: (1) "wire communications," (2) "oral communications," and (3) "electronic communications."

The ECPA is divided into multiple titles.

- Title I of the ECPA, which is often referred to as the Wiretap Act, prohibits the intentional actual or attempted interception, use, disclosure, or "procurement of any other person to intercept or endeavor to intercept any wire, oral, or electronic communication." Title I provides exceptions for operators and service providers for uses "in the normal course of his employment while engaged in any activity which is a necessary incident to the rendition of his service." It provides procedures for Federal, State, and other government officers to obtain judicial authorization for intercepting communications, and regulates the use and disclosure of information obtained through authorized wiretapping.
- Title II of the ECPA, which is called the Stored Communications Act (SCA), protects the privacy of the contents of files stored by service providers and of records held about the subscriber by service providers, such as subscriber name, billing records, or IP addresses.
- Title III of the ECPA , which is called the Pen Register and Trap and Trace Statute, requires government entities to obtain a warrant before collecting real-time information, such as dialing, routing, and addressing information related to communications.

The ECPA had been significantly amended by the Communications Assistance to Law Enforcement Act (CALEA), the USA PATRIOT Act in 2001, the USA PATRIOT reauthorization acts in 2006, as well as the FISA Amendments Act of 2008. Because the ECPA was last significantly amended before widespread use of the Internet began, and because of the difficulty of applying it to new technologies (e.g., cloud computing), Congress and various interest groups have begun the process of reforming or rewriting of the act. As of 2011 a bill was introduced in Congress to update the act to reflect 25 years of evolving and changing technology.[7]

24.4.2 Electronic Funds Transfer Act (EFTA)

This act[8] requires banks to make extensive disclosures to customers about specific electronic funds transfer (EFT) transactions, both at the time the transactions are made and in the form of periodic statements. It is intended to protect individual consumers engaging in electronic fund transfers (EFTs). EFT services are defined to include transfers through automated teller machines, point-of-sale terminals, automated clearinghouse systems, telephone bill-payment plans, and remote banking programs. Customers are required to be notified, at the time they contract for EFT

[7] The Electronic Communications Privacy Act of 1986 (ECPA Pub. L. 99–508, October 21, 1986, 100 Stat. 1848, 18 USC § 2510–2522).

[8] Electronic Funds Transfer Act of 1978 (15 USC 1693 et seq.).

services, of their rights, liabilities, charges, procedures, etc., connected with the services, and who to contact if an unauthorized transfer is suspected. In the case of preauthorized periodic transfers, such as automatic bill paying, the bank must provide either positive or negative notice as to whether payments are being made on schedule. The Federal Reserve Board (Board) implements EFTA through Regulation E. Regulation E was amended to prohibit institutions from charging overdraft fees for ATM and point of sale (POS) transactions unless the consumer affirmatively consents. Additional updates addressed restrictions on fees, expiration dates on gift cards, and required that gift card terms be clearly stated. The Dodd–Frank Act granted rule-making authority under the EFTA to the Consumer Financial Protection Bureau (CFPB) and, with respect to entities under its jurisdiction, granted authority to the CFPB to supervise and enforce compliance with EFTA and its implementing regulations.[9]

24.4.3 Equal Credit Opportunity Act (ECOA)

The Equal Credit Opportunity Act, which became effective in 1975, first banned discrimination in credit access on the basis of sex or marital status. At the time it was common to deny creditworthy women credit or to require a co-signer on an application. Marital status, regardless of employment or personal wealth, was being used to drive credit decisions. Later the Act was amended in 1976 to include race, religion, national origin and age. The purpose of the legislation was to eliminate factors that were irrelevant to determining the creditworthiness of an individual.

This act restricts inquiries into a credit applicant's sex, race, color, religion, or marital status. It prohibits the retention and preservation of certain information by creditors and requires the preservation of specified records relating to credit transactions. The act regulates the manner in which information that has been collected by creditors may be used in making decisions regarding the extension of credit.

Notification of the applicant is required when credit is denied or revoked. The applicant must either be notified of the reasons for the decision or informed of his right to learn the reasons. In lawsuits brought for violations of the Equal Credit Opportunity Act, successful plaintiffs may recover actual damages, punitive damages, as well as attorneys' fees and court costs. Numerous Federal agencies have enforcement responsibility for the provisions of this Act.[10]

[9] Dodd–Frank Act §§ 1002(12)(C), 1024(b)-(c), and 1025(b)-(c); 12 USC §§ 5481(12)(C), 5514(c), and 5515(c).

[10] The Equal Credit Opportunity Act, 15 USC 1691.

24.4.4 Federal Deposit Insurance Corporation Rules and Regulations Part 332 – Privacy of Consumer Financial Information

This regulation[11] governs the treatment of non-public personal information about consumers by financial institutions regulated by the FDIC. These rules require: (1) a financial institution to provide notice to customers about its privacy policies and practices; (2) a description of the conditions under which a financial institution may disclose non-public personal information about consumers to non-affiliated third parties; and (3) the provision of a method for consumers to prevent a financial institution from disclosing that information to most non-affiliated third parties by "opting out" of that disclosure, subject to the exceptions as defined in the regulations.

This applies only to non-public personal information[12] about individuals who obtain financial products or services for personal, family, or household purposes from the institutions regulated. This does not apply to information about companies or about individuals who obtain financial products or services for business, commercial, or agricultural purposes. Additionally, this applies to the US offices of entities that are insured by the FDIC (other than members of the Federal Reserve System), insured state branches of foreign banks, and certain subsidiaries of such entities. The regulations specify:

- Initial privacy notice to consumers
- Annual privacy notice to customers
- Information to be included in privacy notices
- Form of opt out notice to consumers as well as opt out methods
- Revision of privacy notices
- Delivering privacy and opt out notices to consumers.

24.4.5 Fair Credit Reporting Act (FCRA)

Credit reporting was, for a long time, a problematic area. The resulting legislative response to the need for fairness, impartiality, and respect for the consumer's right to privacy was the Fair Credit Reporting Act of 1970. The Fair Credit Reporting Act (FCRA)[13] governs consumer reporting agencies that publish reports on consumers. A consumer report is any communication by a consumer reporting agency regarding a consumer's creditworthiness. It restricts those reports to specific uses and it

[11] 12 C.F.R. PART 332 – Privacy of Consumer Financial Information

[12] Non-public personal information (NPI) is defined as personally identifiable financial information that a consumer provides to obtain a financial product or service; information about a consumer resulting from any transaction involving a financial product or service; or information otherwise obtained about a consumer in connection with providing a financial product or service.

[13] The Fair Credit Reporting Act (FCRA) is a US federal law (codified at 15 USC § 1681 et seq.).

regulates the collection and use of personal data by consumer reporting agencies. The producers of the reports are required to adopt reasonable procedures so as to be responsive to consumer complaints and to respect the consumer's right to privacy. Information must be handled in a manner which is fair and equitable to the consumer, with regard to the confidentiality, accuracy, relevancy, and proper utilization of such information.

The act requires that when a service provider is engaged to prepare an "investigative consumer report" that the request for information must be disclosed to the subject of the report. The subject is then entitled to learn the nature and scope of the enquiry requested. An "investigative consumer report" is defined as an investigation into the consumer's "character, general reputation, personal characteristics or mode of living" by means of interviews with friends, neighbors and associates. Additionally, if a consumer report is used in any decision to deny credit, insurance or employment, the report user must tell the consumer the name and address of the reporting agency. Disclosure of consumer reports maintained by consumer reporting agencies is prohibited without consent unless such disclosure is made for a legitimate business purpose or pursuant to a court order.

The Act requires reporting agencies to use procedures that will avoid reporting specified categories of obsolete information and to verify information in investigative consumer reports that are used more than once. Providers of data are required to maintain security procedures, including procedures to verify the identity and stated purposes of recipients of consumer reports.

Individuals may sue consumer reporting agencies or parties who obtain consumer reports for violations of the Act. Individuals may recover for actual damages suffered, as well as attorneys' fees and court costs. Punitive damages or criminal penalties may also be imposed for willful violations of the Act. The Federal Trade Commission and other Federal agencies responsible for enforcing the provisions of this Act are also empowered to declare actions to be in violation of the applicable statute, issue cease and desist orders, and impose statutory penalties for noncompliance with agency orders.

24.4.6 Consumer Credit Reporting Reform Act of 1996 (Reform Act)

The Reform Act[14] requires more frequent and fuller notification to consumers, disclosure of all information (rather than only the substance of the information) in the consumer's file, and stricter reinvestigation when a consumer disputes the accuracy of information in his or her file. The Act imposes new restrictions on

[14] The Consumer Credit Reporting Reform Act of 1996 ("CCRRA"), 15 USCA § 1681s-2.

resellers of consumer credit reports and strengthens private enforcement rights for violations of the Fair Credit Reporting Act.

24.4.7 Freedom of Information Act (FOIA)

The Freedom of Information Act,[15] commonly known as the FOIA, was enacted by Congress in 1966 to give the American public greater access to the federal government's records. The Electronic Freedom of Information Act Amendments of 1996 expanded the scope of the FOIA to encompass electronic records and require the creation of "electronic reading rooms" to make records more easily and widely available to the public. Unlike the Privacy Act, FOIA procedures are available to non-resident foreign nationals. FOIA has been used from time to time to obtain information from regulatory and other government agencies. The typical financial services firm should always consider the impact of information provided becoming public via an FOIA request.

24.4.8 Gramm–Leach–Bliley Act

The Gramm–Leach–Bliley Act (GLB) is formally known as the Financial Services Modernization Act.[16] The Act addresses a number of areas and is focused on enhancing competition by dropping the barriers enshrined in the Glass–Steagall Act of 1933. The Glass–Steagall Act required the separation of banks, securities firms and insurances companies. From a security and privacy perspective, the Act requires financial institutions to explain their information-sharing practices to their customers and to safeguard sensitive data. Relevant provisions of the GLB Act include:

- The Financial Privacy Rule
- The Model Form Rule
- The Safeguards Rule
- Provisions that prohibit pretexting.

The Act:

- Requires clear disclosure by all financial institutions of their privacy policy regarding the sharing of non-public personal information with both affiliates and third parties,

[15] Freedom of Information Act 5 USC §552 et seq. (1966).
[16] The Gramm–Leach–Bliley Act (GLB), also known as the Financial Services Modernization Act of 1999, (Pub.L. 106–102, 113 Stat. 1338).

- Requires a notice to consumers and an opportunity to "opt-out" of sharing of non-public personal information with non-affiliated third parties subject to certain limited exceptions,
- Addresses a potential imbalance between the treatment of large financial services conglomerates and small banks by including an exception, subject to strict controls, for joint marketing arrangements between financial institutions,
- Clarifies that the disclosure of a financial institution's privacy policy is required to take place at the time of establishing a customer relationship with a consumer and not less than annually during the continuation of such relationship,
- Provides for a separate rather than joint rule-making to carry out the purposes of the Act; the relevant agencies are directed, however, to consult and coordinate with one another for purposes of assuring, to the maximum extent possible, that the regulations that each prescribes are consistent and comparable with those prescribed by the other agencies,
- Allows the functional regulators sufficient flexibility to prescribe necessary exceptions and clarifications to the prohibitions and requirements of section 502 of the GLB Act,
- Clarifies that the remedies described in section 505 are the exclusive remedies for violations of the subtitle,
- Clarifies that nothing in this title is intended to modify, limit, or supersede the operation of the Fair Credit Reporting Act,
- Assigns authority for enforcing the subtitle's provisions to the Federal Trade Commission and the Federal banking agencies, the National Credit Union Administration, the Securities and Exchange Commission, according to their respective jurisdictions, and provides for enforcement of the subtitle by the States.

Financial Privacy Rule

The Financial Privacy Rule requires financial institutions to provide each consumer with a privacy notice at the time the consumer relationship is established and annually thereafter. This privacy notice must explain the information collected about the consumer, where that information is shared, how that information is used, and how that information is protected. The notice must also identify the consumer's right to opt out of the information being shared with unaffiliated parties.

Model Form Rule

Under GLBA a financial institution is required to provide to customers notification of how the financial institution collects and shares information about its consumers. On November 17, 2009, eight federal regulatory agencies released the final version of a model privacy notice form to make it easier for consumers to understand how financial institutions collect and share information about them. The purpose of the model form is to provide a standard template that meets all of the regulatory requirements. The final rule provides that a financial institution that chooses to use the model form obtains a "safe harbor" and will satisfy the disclosure requirements for notices.

Safeguards Rule

The Safeguards Rule requires financial institutions to implement a written information security program that is appropriate to the company's size and complexity, the nature and scope of its activities, and the sensitivity of the customer information it handles. The Safeguards Rule applies to information on any consumers past or present of the financial institution's products or services. As part of the program, each financial institution must also:

- Assign one or more employees to oversee the program
- Conduct a risk assessment
- Put safeguards in place to control the risks identified in the assessment and regularly test and monitor them
- Require service providers, by written contract, to protect customers' personal information
- Periodically update its security program.

The Safeguards Rule requires financial institutions to examine in detail how they manage private data and to perform risk analyses on current processes. Such examination can be accomplished through standard controls audits. For example, it can be included in the scope of a common general controls SAS 70 audit.

Pretexting Provisions

This section of the Act prohibits obtaining customer information by false pretenses. The subsection defines as a violation obtaining or attempting to obtain, or cause to be disclosed or attempt to cause to be disclosed to any person, customer information of a financial institution relating to another person by making or providing:

- A false, fictitious, or fraudulent statement or representation to an officer, employee, or agent of a financial institution,
- A false, fictitious, or fraudulent statement or representation to a customer of a financial institution, or
- Any document to an officer, employee, or agent of a financial institution, knowing that the document is forged, counterfeit, lost, or stolen, was fraudulently obtained, or contains a false, fictitious, or fraudulent statement or representation.

The GLB encourages the organizations to implement safeguards against pretexting. Under US law, pretexting by individuals is punishable as a common law crime of False Pretenses.

24.4.9 Privacy Act (1974)

The Act establishes a code of fair information practices that governs the collection, maintenance, use, and dissemination of information about individuals that is maintained in systems of records by federal agencies. A system of records is a group of records under the control of an agency from which information is retrieved by the name of the individual or by some identifier assigned to the individual.

The Privacy Act[17] requires that agencies give the public notice of their systems of records by publication in the Federal Register. The Privacy Act prohibits the disclosure of a record about an individual from a system of records without thc written consent of the individual, unless the disclosure is allowed under one of the statutory exceptions. The Act also provides individuals with a means by which to seek access to and amendment of their records, and sets forth various agency record-keeping requirements.

24.4.10 Right to Financial Privacy Act (RFPA)

The Right to Financial Privacy Act (RFPA),[18] is a US Federal law enacted in 1978, that gives the customers of financial institutions the right to some level of privacy from government searches. Prior to the Act, the US government did not have to tell customers that it was accessing their records, and customers did not have the right to prevent such actions. The Act requires that Federal agencies seeking access to private financial records either notify the subject of the purpose for which the records are sought and provide the subject with an opportunity to challenge the disclosure in court or require the government to obtain a court order for direct access to the records if notice would allow the subject to flee or destroy the evidence. The Act also prohibits a Federal agency, that has obtained access to an individual's financial records, from disclosing the records to another agency without notifying the individual and obtaining certification from the receiving agency that the records are relevant to a legitimate law enforcement enquiry of the receiving agency.

The RFPA generally requires that the customer must receive:

- A written notice of the agency's intent to obtain financial records,
- An explanation of the purpose for which the records are sought, and

[17] The Privacy Act of 1974 – 5 USC § 552a.
[18] The Right to Financial Privacy Act (RFPA) (12 USC § 3401 et seq.).

- A statement describing procedures to use if the customer does not wish such records or information to be made available.

Certain exceptions apply which allow for delayed notice or no notice to be given to the customer. RFPA was amended by the USA PATRIOT Act of 2001. In October 2001, Section 358 of the USA PATRIOT Act amended the RFPA to permit the disclosure of financial information to any intelligence or counterintelligence agency in any investigation related to international terrorism.

24.4.11 BSA

The Bank Secrecy Act of 1970 (BSA, also known as the Currency and Foreign Transactions Reporting Act)[19] requires financial institutions to assist US government agencies to detect and prevent money laundering. The Act requires financial institutions to keep records of cash purchases of negotiable instruments, file reports of cash transactions exceeding $10,000 (daily aggregate amount), and to report suspicious activity that might indicate money laundering, tax evasion, or other criminal activities. As a result, most institutions today will no longer sell negotiable instruments when they are purchased by customers with cash, instead requiring the purchase to be withdrawn from an account at that institution. The BSA is sometimes referred to as an "anti-money laundering" (AML) law or BSA/AML. Several anti-money laundering acts, including provisions in title III of the USA PATRIOT Act, have been enacted to amend the BSA.

The BSA regulations require all financial institutions to submit five types of reports to the government. The following is not an exhaustive list of reports to be filed. The Report of Foreign Bank and Financial Accounts (FBAR) has an individual filing requirement, as detailed below.

- FinCEN Form 104 Currency Transaction Report (CTR): A CTR must be filed for each deposit, withdrawal, exchange of currency, or other payment or transfer which involves a transaction in currency of more than $10,000. Multiple currency transactions must be treated as a single transaction if the financial institution has knowledge that: (a) they are conducted by or on behalf of the same person; and (b) they result in cash received or disbursed by the financial institution of more than $10,000.
- FinCEN Form 105 Report of International Transportation of Currency or Monetary Instruments (CMIR): Each person (including a bank) who physically transports, mails or ships, or causes to be physically transported, mailed, shipped

[19] The Financial Recordkeeping and Reporting of Currency and Foreign Transactions Act of 1970 (31 USC 1051 et seq.).

or received, currency, traveler's checks, and certain other monetary instruments in an aggregate amount exceeding $10,000 into or out of the USA must file a CMIR.
- Department of the Treasury Form 90–22.1 Report of Foreign Bank and Financial Accounts (FBAR): Each person (including a bank) subject to the jurisdiction of the USA having an interest in, signature or other authority over, one or more bank, securities, or other financial accounts in a foreign country must file an FBAR if the aggregate value of such accounts at any point in a calendar year exceeds $10,000.
- Treasury Department Form 90–22.47 and OCC Form 8010–9, 8010–1 Suspicious Activity Report (SAR): Banks must file a SAR for any suspicious transaction relevant to a possible violation of law or regulation.
- "Designation of Exempt Person" FinCEN Form 110: Banks must file this form to designate an exempt customer for the purpose of CTR reporting under the BSA. In addition, banks use this form biennially (every two years) to renew exemptions for eligible non-listed business and payroll customers.

Additionally,

- It requires any business receiving one or more related cash payments totaling $10,000 or more to file form 8300.
- Monetary Instrument Log (MIL): The MIL must indicate cash purchases of monetary instruments, such as money orders, cashier's checks and traveler's checks, in value totaling $3,000 to $10,000, inclusive. This form is required to be kept on record at the financial institution, and produced at the request of examiners or audit to verify compliance. A financial institution must maintain a Monetary Instrument Log for five years.
- Suspicious Activity Report (SAR): The SAR must report any cash transaction where the customer seems to be trying to avoid BSA reporting requirements by not filing CTR or MIL, for example. A SAR must also be filed if the customer's actions suggest that he is laundering money or otherwise violating federal criminal laws and committing wire transfer fraud, check fraud or mysterious disappearances. The bank should not let the customer know that a SAR is being filed. These reports are filed with the Financial Crimes Enforcement Network (FinCEN).

24.4 12 EU Data Protection Directive

Up to this point this chapter has focused on data privacy issues in the US context. As discussed, the basis for US data privacy is a series of laws and regulations that have developed over time. There is no single unifying law or regulation underlying the

US framework. The basis for data privacy in the European Union is found in the Directive 95/46[20] also known as the "Data Protection Directive." The European Commission realized that diverging data protection requirements among EU member states impeded the free flow of data within the EU. The Directive is an effort to harmonize the various national regulatory schemes of the member states. Prior to the Directive there was a patchwork of privacy regulations across Europe. These were an outgrowth of individual member state regulations based on the Organization for Economic Cooperation and Development (OECD) issued "Recommendations of the Council Concerning Guidelines Governing the Protection of Privacy and Trans-Border Flows of Personal Data." Within the OECD's recommendations were seven principles governing the protection of personal data.

- Notice – data subjects should be given notice when their data is being collected;
- Purpose – data should only be used for the purpose stated and not for any other purposes;
- Consent – data should not be disclosed without the data subject's consent;
- Security – collected data should be kept secure from any potential abuses;
- Disclosure – data subjects should be informed as to who is collecting their data;
- Access – data subjects should be allowed to access their data and make corrections to any inaccurate data; and
- Accountability – data subjects should have a method available to them to hold data collectors accountable for following the above principles.

All seven principles were incorporated into the EU Directive.

Under the Directive data may be processed only under the following circumstances:

- When the data subject has given his consent
- When the processing is necessary for the performance of or the entering into a contract
- When processing is necessary for compliance with a legal obligation
- When processing is necessary in order to protect the vital interests of the data subject
- When processing is necessary for the performance of a task carried out in the public interest or in the exercise of official authority vested in the controller or in a third party to whom the data are disclosed.

Within the Directive, "personal data" are broadly defined as "any information relating to an identified or identifiable natural person ('data subject'); an identifiable

[20] See Council Directive (1995).

person is one who can be identified, directly or indirectly, in particular by reference to an identification number or to one or more factors specific to his physical, physiological, mental, economic, cultural or social identity."

This definition of data was specifically meant to be very broad. Data are "personal data" when someone is able to link the information to a person, even if the person holding the data cannot make this link. Some examples of "personal data" are address, credit card number, bank statements, criminal record, etc.

The responsibility for compliance rests on the shoulders of the "controller," defined in the Directive as meaning the natural or artificial person, public authority, agency or any other body which alone or jointly with others determines the purposes and means of the processing of personal data. Importantly, the data protection rules are applicable not only when the controller is established within the EU, but whenever the controller uses equipment situated within the EU in order to process data. Controllers from outside the EU, processing data in the EU, have to follow the data protection regulation.

Transfer of personal data from an EU Member State to a country outside the EU, with an adequate level of protection, is allowed under the Directive. Data transfers may not be made to a country outside of the EU which does not ensure the same levels of protection. Because of differences in the approach to regulating privacy the USA may not be considered to be in compliance with this requirement. To date, the USA has no single data protection law comparable to the EU's Data Protection Directive. To compensate for this difference in legal framework, the US Department of Commerce entered into negotiations with the European Commission. The result of the negotiations was the Safe Harbor Principles. These principles allow US firms to self-certify that internal procedures and controls are in place to provide protections comparable to those required by the Directive. The Safe Harbor Principles provide for firms operating to access, process and maintain data sourced from the EU.

24.5 Security and privacy technology

24.5.1 Authentication and authorization

Globally unique identity and identity management

Authentication requires that the system identify and confirm who is trying to access the system or data. A basic challenge in any technical environment is to uniquely identify and manage the identity of a particular component, application, file, database and/or user. There are a number of approaches to unique identifiers, often referred to as a Universally Unique Identifier (UUID). A universally unique identifier (UUID) is an identifier standard used in software construction, standardized by the Open Software Foundation (OSF) as part of the Distributed Computing Environment (DCE).

Common within the Microsoft and related technologies is the Globally Unique Identifier (GUID), a variation of the UUID. It is a unique 128-bit number that is generated by an application to identify a particular application, file, database entry, and/or user. Some systems identify user accounts by a username or a combination of username and computer domain. Based on this identification a GUID is assigned. Although GUIDs can be created in a number of ways, they usually are a combination of a few unique inputs based on a specific point in time (e.g., an IP address, network MAC address, clock date/time, etc.).

The use of UUIDs enables distributed systems to uniquely identify information without the involvement of a central system. It is important to note that the concept of a unique identifier should be taken to mean "practically unique" rather than "guaranteed unique." Identifiers are of finite size so it is theoretically possible for two differing items to share the same identifier. The identifier size and generation process need to be carefully defined so that collisions are improbable. If this is done properly, the UUID can be used to identify objects with reasonable confidence that there will not be a need to later resolve naming conflicts.

As noted above, one widespread use of this standard is in Microsoft's globally unique identifiers (GUIDs). Other significant uses include Linux's ext2/ext3 file system, LUKS encrypted partitions, GNOME, KDE, and Mac OS X.

Public key infrastructure (X.509)

Applications have requirements for security to protect against external threats to the communication of information. Secure communications can be accomplished by various means, one of which is an encryption key scheme. X.509 is an International Telecommunications Union (ITU) technical standard that addresses Public Key Infrastructure (PKI) and Privilege Management Infrastructure (PMI). X.509 specifies, among other things, standard formats for public key certificates, certificate revocation lists, attribute certificates, and a certification path validation algorithm.

Policies and entitlements

In recent years, there has been a push among large organizations to centralize the administration of user identities and their associated access privileges. One of the key drivers for identity and access centralization is security compliance. By providing a centralized solution the security environment can be applied consistently throughout the enterprise.

Security is a concern across the corporate entity and for the regulators of the entities as well. Costly data breaches occur regularly. The impact of such events can damage a firm's brand, investment return, and customer base. The time spent managing security incidents and managing related security risks is a drain on corporate resources that diverts assets that could be used to drive corporate growth. Organizations need to address security challenges by administering, securing, and monitoring identities, roles, and entitlements with appropriate lifecycle management, access controls, and related compliance auditing.

Despite current technology, security and privacy often remain too complex for individuals or enterprises to manage effectively or to use conveniently. The technical paradigm is constantly changing as the access to and use of data evolve. Within a technical environment the security and privacy technologies were often developed for a different, earlier purpose. As the use of systems evolves, care must be taken to ensure that the new configuration is not structured or operated in ways that are inadvertently insecure. The shift from system centric to cloud based is one such example. The user base may have grown significantly but the basic models for security and privacy may have remained unchanged. The net result can be that security features become clumsy and awkward to use. Features that are difficult to use are often seen as obstacles to getting things done and are prime candidates to be bypassed or ignored. Similarly, when security gets in the way of functionality, application designers and administrators are prone to de-emphasize it. The result is that end users often engage in actions, knowingly or unknowingly, that compromise the integrity of critical security features of the computer systems or that contribute to the unwanted release of personal or other confidential information.

By controlling and tracking what resources a user accesses, identity and access management schemes give organizations assurance that users are both authentic and restricted to those resources that they are entitled to access. The problem of centrally controlling and auditing an application or user's access to an application or resource remains. Identity and access management products are available from a number of vendors, including but not limited to IBM, Oracle, CA, HP, Microsoft, Novell, Entrust and RSA.

24.6 Digitally signed data

Digital signatures provide authentication and integrity protection. A digital signature or digital signature scheme is a mathematical scheme for demonstrating the authenticity of a digital message or document. A valid digital signature gives the recipient reason to believe that the message was created by a known sender, and that it was not altered in transit. Digital signatures are commonly used for software distribution, financial transactions, and in other cases where it is important to detect forgery or tampering. The digital signature itself is simply a small block of data that is attached to a document that has been signed. It is generated from the user's digital ID, which includes both a private and public key. The private key is used to apply the signature to the document, while the public key is sent with the file. The public key contains encrypted code, also called a "hash," that verifies identity.

Digital signatures can then be used to certify or approve documents. Certifying signatures verify the document's creator and show that the document has not been altered since it was signed. Only the original creator of a document can add a certifying signature. Approval signatures can be added by anyone with a digital ID and are used to approve documents, track changes, and accept terms stated with a document.

A digital signature can be used with any kind of message, whether it is encrypted or not, simply so that the receiver can be sure of the sender's identity and that the message arrived intact. A digital certificate contains the digital signature of the certificate-issuing authority so that anyone can verify that the certificate is real.

24.7 Secure transports and communications

Data can exist in one of four states. They can be at rest, in use, in motion or disposed. Each of these states has different security requirements. These states are defined as follows: "data in motion" includes data that are moving through a network, including wireless transmission, whether by email or structured electronic interchange, while "data at rest" includes data that reside in databases, file systems, flash drives, memory, and any other structured storage method. "Data in use" includes data in the process of being created, retrieved, updated, or deleted, and "data disposed" includes discarded paper records or recycled electronic media.

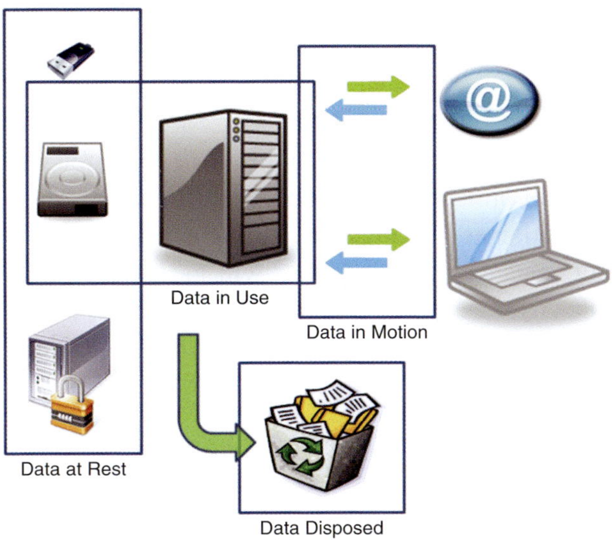

Figure 24.1 Potential data states.

Firewall and other security solutions should include data loss prevention capabilities to secure data in motion. Content scanning, blocking of communications containing sensitive data and encryption are basic techniques. Corporate policies and employee education are considered insufficient as a control unless appropriate technical controls are in place. Such controls should be tested and enforced.

24.7.1 Encrypted data at rest

The term "data at rest" is used to describe all data that are kept in computer storage of any kind. It generally excludes data that frequently move on computer networks or temporarily reside in computer memory. Data at rest includes but is not limited to archived data, infrequently accessed or rarely modified data. This typically takes the form of files stored on hard drives, thumb drives, backup tapes and portable disks. Data may be stored locally on a computer or on network storage devices. A vulnerability of this type of data is that unprotected data at rest are at risk to unauthorized access by users with system level access. In any system it is typically necessary to provide access to the system for maintenance and related operations. It is important to ensure that the administrative users are not provided with inappropriate access to the data stored on the system. On-disk encryption is typically used to protect data at rest. The separation of the data encryption from data access functionality can provide the appropriate segregation of security. Emerging law and regulations are beginning to recognize the importance of disk level encryption as a way of protecting data at rest. In recent years, in some states, it has been noted that it is not considered a data breach if encrypted data are subject to unauthorized access and the encryption key was not available to the unauthorized accessing party.

24.7.2 Encryption while in use

Protection of data while in use implies that the data being processed by an application are encrypted and protected from unauthorized access. As noted in the comments in the prior section addressing data at rest, access by users with system level access remains a constant issue. How data will be used within applications and protected during that use needs to be considered in both the application development and in the corporate policies addressing the use of the applications.

24.7.3 Over-the-wire encryption

Over-the-wire encryption protects data while in transit. One of the basic questions to ask, before encrypting and moving data, is do these data need to move? Do all of the data in a data record need to move? When data are in motion there is the largest number of possible loss scenarios. Structuring the movement of data, such that data can only move when appropriately authorized, can make a significant difference in the security of the data.

24.8 Conclusion

The regulations governing data security and privacy have been evolving and will continue to evolve. The drivers for this are the desire to protect privacy but also to provide access to certain information by a range of parties. The technologies that manage data will also continue to evolve and will often get out ahead of the regulations or expose the user in ways that have not been anticipated or initially understood. The result is that what is not regulated today may be regulated in the near future given changes in the perceptions of the regulators or the demands of the public. Predicting changes in the regulations remains difficult, if not impossible.

When considering data security and privacy one needs to consider the totality of the rules and regulations. Often it is helpful to look beyond the current area of interest to other regulatory areas for guidance and ideas on approaches to better address data security and privacy. Changes in technology can open holes in the most carefully thought through security strategy. The use of mobile devices, software as a service (SaaS), cloud computing, etc. can open your data to access in ways that may not have been anticipated. The desire to leverage the benefits of emerging technologies and maintain compliance with the various regulations may prove to be at odds with one another.

Being fully aware of the landscape of regulations and standards, as well as regularly confirming compliance with them, is an appropriate approach to providing security and privacy for your data.

Bibliography

Council Directive 95/46 of 24 October 1995 on the Protection of Individuals with Regard to the Processing of Personal Data and on the Free Movement of Such Data, 1995, Official Journal of the European Communities, (L281) 31.

Holtzman, D., 2006, *Privacy Lost: How Technology is Endangering Your Privacy*, Wiley.

International Organization for Standardization (ISO), 2005, ISO/IEC 27002:2005 Information Technology – Security Techniques – Code of Practice for Information Security Management.

National Institute of Standards and Technology (NIST), 2004, Standards for Security Categorization Federal Information and Information Systems, Federal Information Processing Standards Publication (FIPS), Publication 199.

National Institute of Standards and Technology (NIST), 2006, Minimum Security Requirements for Federal Information and Information Systems, Federal Information Processing Standards Publication (FIPS), Publication 200.

National Institute of Standards and Technology (NIST), 2009, Special Publication 800–53, Recommended Security Controls for Federal Information Systems and Organizations.

PC World Business Center, 2012, Cost of Data Breaches Falls for First Time in Seven Years, March 20.

Salomon, D., 2003, *Data Privacy and Security*, Springer-Verlag.

Schwartz, P. and D. Solove, Smith, E., 2012, Compilation of State and Federal Privacy Laws with 2012 Supplement, *Privacy Journal*.

Solove, D., 2008, *Understanding Privacy*, Harvard University Press.

Solove, D., 2011, *Nothing to Hide: The False Tradeoff between Privacy and Security*, Yale University Press.

Soma, J. and S. Rynerson, 2008, *Privacy Law in a Nutshell*, Thomson/West.

Trcek, D., 2006, *Managing Information Systems Security and Privacy*, Springer-Verlag.

PART VI

IMPLEMENTATION OF DATA AND ANALYTICS PROGRAMS

EDITORS' INTRODUCTION TO PART VI

Dilip Krishna

The previous sections of the Handbook have comprehensively covered the various aspects of consolidating data within a financial institution. These began with the regulatory imperative and advances in risk management. The Handbook also went into detail on various kinds of data found in these financial institutions. Where the first volume of the Handbook focused on the requirements for a strong risk information infrastructure, the second volume of the Handbook has focused on implementation issues. Data are intimately connected with the business processes that generate, modify and use them – this was covered in Part IV. Part V covered a plethora of technologies that may be leveraged to build a strong technology backbone for risk management. Yet for all the advances made in understanding risk data, and plentiful and relatively cost-effective technologies available for integrating information at this time, many risk information projects still struggle to accomplish their goals. Even when they do succeed, they often do this at the cost of large expense- and time-overruns. There are typically two reasons for this.

First, the scope and scale of risk information projects can be vast, usually covering a large part of a financial institution (not coincidentally, reflected in the scope and size of this Handbook). Project teams themselves may grow to 100 or more resources that may touch several hundred other people in a large financial institution. The challenge of putting together an effective project plan and executing it cannot be underestimated. This problem can be exacerbated by two factors common in such projects. First, requirements are often volatile and can change in the face of regulatory modifications, business process changes or in response to other initiatives within an organization. Additionally, systems in many institutions have been implemented and improved over time, with the resultant lack of institutional memory on the details of at least some of systems in the organization, which may leading to imperfectly defined requirements in many circumstances.

Second, managing data effectively and using them to provide effective support to risk management is a topic that transcends traditional organizational splits of such as "business," "risk management" and "information technology" groups. Data are largely originated in business groups, and manipulated by risk management users (among others) on systems that are usually, but not always, managed by an information technology group. This calls for an unusually high degree of cooperation that may be difficult to achieve among traditional organizational silos.

There have been significant advances in recent years that can be leveraged to address both these challenges, however. The discipline of project management has strengthened significantly through the development and increasing acceptance of strong, well-understood frameworks by such organizations such as the Project Management Institute (see `www.pmi.org`), advances that are matched by equivalent advances in managing IT projects at a tactical level through so-called agile techniques. There has also been a noteworthy convergence in thinking on data governance, which is the discipline that facilitates cooperation on data-related matters across organizational silos. Taking advantage of both these advances may help organizations set up their information infrastructure projects to meet their goals while keeping to time and budget.

Completing the project is only the first step, however. The resulting environments, especially those that are successful at meeting the needs of their users, tend to require heightened degrees of management focus. The reason is that information environments by design elicit insights that lead to further data needs. Therefore, such environments are constantly changing and growing. Management of such an environment, therefore, requires special attention to a continuous plan for incremental growth so that the user experience is not diminished as the scale of the environment expands. Of course, as the data in the environment are increasingly used in critical risk management activities, there is also a need to protect them from accidental loss or malicious access.

The section starts with Chapter 25 by Margarita Brose and Dilip Krishna, on project implementation. The chapter begins by describing the challenges of large risk information projects in some detail. From there, it describes how such large programs may be effectively managed by subdividing them into projects, and the important organizational factors for success such as stakeholder management, effective program communication and risk management. The second half of the chapter takes a break from the "macro" perspective to focus on "micro" issues of managing analytic projects to success. This section of the chapter discusses important aspects of analytic data projects, such as business requirements gathering, design, development and testing as well as some leading practices in on-going improvement. The chapter concludes with comments on topics that are important to these projects such as iterative development and global resourcing.

Chapter 26, by Jim Blair, addresses operations management, the discipline of administering and managing the infrastructure of a data and analytical environment.

The purpose of operations management is to provide a highly available, high performing and scalable environment to meet the demanding needs of risk management. The chapter begins by defining the unique characteristics of a Financial and Risk Analytics (FRA) ecosystem. It then lays out four pillars that assist with the successful management of such an environment: data delivery, systems availability, systems performance and integrating new projects into the ecosystem. Each of these pillars is then discussed in a significant amount of detail.

Finally, Chapter 27, by Deb Stockdale, is on data governance. Data are a valuable asset for financial institutions but are not often treated as such. The chapter, therefore, begins by a discussion on the importance of data governance and should help readers interested in articulating the importance of data to others in their organization. The chapter then progresses to discuss the various components of data governance such as data governance roles, identifying stakeholders and funding. Since data governance is intimately concerned with the intersection of business processes with data, successful solutions should be tailored to the particular culture of the organization to which it is being applied. The chapter, therefore, has an extensive coverage of various types of data governance models appropriate for a range of organizational structures and cultures. The chapter concludes with appendices that describe in detail some of its advocated techniques.

In summary, this section provides insight into the means to put into practice the concepts and technologies described in the previous sections of the Handbook. As such, it may be of interest to readers involved in managing any aspect of a risk infrastructure or data governance program, as well as those that are charged with the responsibility to "keep the lights on" the infrastructure once it has been established.

25

Project implementation

Margarita S. Brose and Dilip Krishna

25.1 Introduction

Project management is the bedrock for any successful system or process implementation. The practice and development of techniques for effective project management – especially in Information Technology (IT) projects – has come a long way in the past few decades. Best practices and IT management folklore popularized in seminal works like *The Mythical Man-Month* (Brooks, 1997) have been transformed into standardized, teachable techniques by large consulting firms as well as industry organizations like the Project Management Institute.[1] Many believe that implementation of these techniques has vastly improved the quality and timeliness of IT projects. Yet large IT projects still continue to offer up huge challenges in their implementations.

The challenges of project management are perhaps most evident in the implementations of large risk management, finance transformation and regulatory projects in financial institutions. For example, some estimates put the cost of Basel II compliance spending at as much as 10% of banks' overall IT costs (VanHoose, 2007). Certainly the largest banks in the USA, Europe and Canada have each consumed hundreds of millions of dollars on Basel II compliance. These have not all been entirely happy experiences. While typical large project management techniques have been employed in these firms, the often similar challenges experienced and worked through by project teams suggests that these projects are unique in nature, and different from traditional large application development and implementation projects. Due to the evolving nature of regulation, requirements definition typically poses a formidable barrier to their success. Their information intensive nature implies that they rely on a detailed understanding of the information

[1] www.pmi.org

Handbook of Financial Data and Risk Information, ed. Margarita S. Brose, Mark D. Flood, Dilip Krishna and Bill Nichols. Published by Cambridge University Press.

underpinnings of the organization – knowledge that is often incomplete. They are also massive undertakings cutting across multiple lines of business and functions, critically crossing the business–IT divide in multiple places in an organization.

Massive undertakings like large risk management and regulatory projects require the rigor and structure that come with time-tested project management methodologies. There are several different methodologies for approaching technical design and implementation, but for a successful project, a program or project manager should also employ techniques including managing expectations of stakeholders and project participants, as well as utilizing a proper communication strategy. Selecting the right people to execute in the project management role is also a key success factor. These concepts will be explored in this chapter.

25.2 Challenges

Financial and risk analytic transformation projects (collectively, "FRATs") are fundamentally different from regular IT projects for many reasons. Before discussing the methodologies for implementation, we present below some of the issues that present themselves to the team embarking on a large initiative of this type, and suggestions for meeting these challenges.

- Complexity – FRATs cut across many different functions, in what are typically massive global organizations employing tens or hundreds of thousands of people, and demand close coordination between many different constituencies. Business functions affected extend from the front office to the back office as well as numerous shared-services areas such as finance and risk management. On the technology side of the house, the project team should coordinate its own efforts with the numerous unrelated projects being run by business-focused IT teams as well as the IT infrastructure management functions that are critical to such a large effort.
- Managing scope – One of the tenets of modern IT project management is to tightly manage scope by properly defining requirements *before* implementation can begin. For regulatory projects, this is almost always unachievable. In many cases regulatory implementation is obliged to begin before all the regulations are even fully *interpreted*, let alone translated into IT requirements. Keeping agile with regard to requirements is a necessity in projects of this type.
- Long-term horizon – FRATs typically take 2–4 years to complete. This is long enough that there can be several external factors affecting the project – from internal reorganizations to mergers to budget constraints resulting from poor quarterly results. Each of these can have adverse impacts on the success of the project if they are not anticipated and planned for. Managing stakeholders and keeping the project visible in the corporation can help mitigate these impacts.

- Managing uncertainty – Even if the business problem did not impose uncertainty on requirements, the nature of such projects would do so anyway. FRATs are information and analytic intensive in comparison to typical IT projects such as implementing a front office trading system. Implementing information-intensive projects can be a very different challenge. Because most large financial institutions have mature systems built over several years (or decades even), it is impossible to completely define data and transformation requirements before implementing a solution. There is usually no one person in these organizations who has a complete understanding of the data in these systems, resulting in projects discovering, at cost to budget and timeline, unexpected issues and changes during implementation. Acknowledging these limitations and preparing for them can help manage the risk of encountering them in the project.

FRATs stress the business–IT interface to its limits due to the factors described above. With changing requirements, it can be easy for business teams to feel frustrated by the slow pace of delivery from IT teams. At the same time, changing requirements thwart the progress of IT teams. While changing requirements are usually to some extent a consequence of a moving regulatory target, it is often simply that the fast-paced nature of the projects leads to proper project management practices being ignored. Therefore, it is important to put the right people in place who have had experience running large and complex projects, and are able to identify these types of issues and manage them appropriately for the long-term success of the project.

25.3 Components of a successful risk information and analytics implementation

The implementation of a risk information and analytics system at an institution involves the complex orchestration of several components addressing portfolio modeling, trend analysis, stress testing, management information and other tools and outputs to support the risk manager. These components involve distinct activities and teams, but they should be integrated to help ensure that the overall goal of the institution is met. Therefore, such implementations are typically run as large-scale Programs consisting of several Projects, with the overall goal of making sure that all the Projects are executed in accordance with the design and directive of the Program. Projects are often composed of several workstreams, supporting each of the components of the project execution, for example, a system implementation segmented by business area or business process (see Figure 25.1).

Program managers are tasked with maintaining communication across all the project teams, identifying, escalating and managing the resolution of cross-project

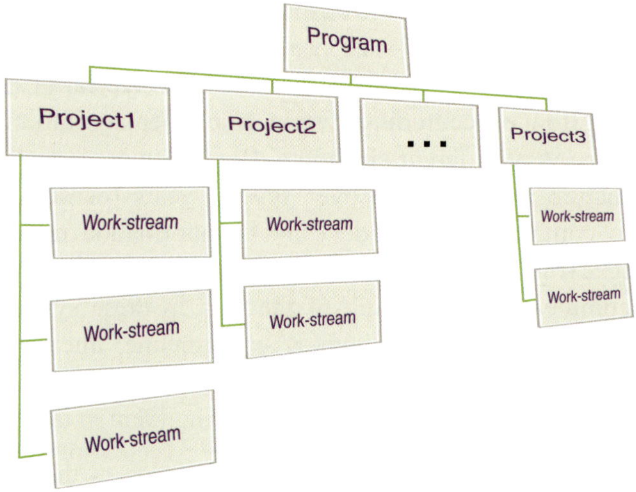

Figure 25.1 Illustrative program structure.

issues, and serving as "air traffic controller" for the various projects, making sure that that any cross-dependencies are managed appropriately and do not affect deadlines for implementation. A Program Management Office (PMO) can effectively make this possible under the leadership of a program manager. Tracking the progress of individual projects across the program and understanding their cross-dependencies allows the program manager to identify problems early and bring together teams to manage the issues timely. The PMO is also well positioned to identify cross-project synergies, and the opportunities for leveraging resources and tools across the program. Program managers also manage key stakeholders, and endeavor to maintain their continuing support and commitment to the program.

Project managers are responsible for making sure the outcome of their particular project is concordant with the expectations and specifications of the overall program. They are responsible for identifying and managing project issues, and escalating them when they rise to the level of needing executive attention or decision-making, especially if they may impact another project in the program. They are responsible for managing their staff and project resources, and their particular budget. Maintaining and tracking to the project plan is also a key and critical function of the project manager.

As shown in Figure 25.1, a typical large program will include many projects or workstreams (one author helped manage a program initiative with 42 different workstreams), with an underpinning infrastructure to support the work of the ongoing projects. Technology and Operations are often the key components of this infrastructure, along with Legal, Compliance and Risk Management. Early identification of the key members of these teams who will interact across the program can facilitate communication and the execution of project tasks.

The development of a project management infrastructure is just the beginning of a FRAT. Several other efforts should be undertaken to achieve success in the endeavor. Generally driven by the PMO, the development of a business case to support the initiative is often a critical first step, as is establishing a process to manage the resulting budget. Setting up a process to manage stakeholders throughout the course of the initiative, developing a broad communications program, planning for and managing change, and managing project risks are also key activities and key components to a successful project.[2]

25.3.1 Business case development

A FRAT may be driven by many factors, including regulatory requirements, by the results of an internal audit, an external audit, or a regulatory exam. The initiative may also be initiated in response to requests for better management information, or by a team in search of a more efficient process. Regardless of how it is initiated, one of the first steps is to build a business case for the required expenditure of time, energy and money. While regulatory initiatives usually do not result in monetary benefits required for usual business cases, the need to articulate the costs and indirect benefits is still important. A framework for the program, from expectations of what is to be delivered to estimated final costs, enables one to more effectively manage the effort and expectations of senior program stakeholders.

A Benchmarking Study is often a first step for a business that wants to examine what its competitors are doing and identify areas where it is not keeping pace with the industry (Wise, 2011). A benchmarking study is also an effective way to collect costing data which are needed to build the business case. A firm may conduct a voluntary survey of competitors to pull together the information desired, or engage an outside consultant to carry out the study anonymously. The results can then help set realistic expectations for a project, or encourage a firm to think more aggressively about its ambitions for its program.

Key elements of the business case include the benefits and costs of the project as well as any risks which can be identified and which should be considered before embarking on the project. Benefits of the project may include increased efficiencies in process or technology, or fixing an identified problem or gap in data use or availability. Costs of the project would include physical assets such as hardware, software and infrastructure management costs as well as an estimate of resources (usually comprising both firm staff and external consultants) that will be involved

[2] Many consulting firms have detailed approaches to project management developed in-house.

either as part-time subject matter experts or as dedicated project resources. Examples of risks that may be highlighted for consideration include possible regulatory changes in the course of the project, the possibility that staff needed may not be available, or the identification of other projects ongoing in the organization that may impact or reprioritize the proposed project (see Section 25.3.6 for a deeper discussion).

Often, the project planning team develops delivery scenarios. The base-case scenario will address the "must-have" mandatory regulatory requirement or executive directive. The team may build additional scenarios, which take into account "nice-to-have" components which may have longer time horizons for implementation and therefore higher costs but bring potentially significant benefits. The business case should also take into account unexpected changes to requirements (necessitating additional time and resources) or annual cost increases from vendors or internal partners.

25.3.2 Stakeholder management

FRATs are initiated for any variety of reasons in an organization: they can be driven by need or competition, by ego or inspiration, by regulation or regulator, or by executive edict. Perhaps more than most programs, risk and regulatory initiatives impact a wide range of business functions and units across the organization and have surprising, non-intuitive impacts on a variety of people. This makes it incumbent upon program leadership to identify the drivers and stakeholders of the program. Program leadership should constantly be alert to potential new impacts on the program, since both drivers and stakeholders can change with disturbing frequency.

Identifying the obvious and not so obvious influences on the program (also known as Stakeholder Mapping) is an often critical first step. Once identified, stakeholders can be canvassed for their expectations of the project, which should be incorporated into the fabric of the project plans, communication plan and deliverables. Frequent communication with stakeholders can help avoid political maneuvering behind the scenes and mitigate the risk of a project ambush. On a more positive note, transparency and communication of known issues and roadblocks, as well as progress and any assistance needed, can help sustain the support of key leaders in the organization.

Stakeholder Mapping is a practice that is often an effective and helpful tool used at the initiation of a program and then throughout the project to navigate minefields of personalities and agendas along the way. It can be used as a diagnostic tool to explain why the project is not moving forward, by helping to identify the individuals to recruit and educate regarding the project's goals and progress in order to bring it back on track. The program team begins by identifying the people who will be affected by the change. This includes people who are directly invested in the success of the program (i.e., are paying for it) or are indirectly impacted by the program in some way. Using a standard questionnaire, each of the stakeholders is interviewed. Data collected from the stakeholders include what they know about the project, what impact it will have on them, their experience with similar projects in the organization, and their expectations for success of the project. Once the data are collected, they can then be mapped on a 3 × 3 quadrant that shows the influence across one axis and commitment across another. Those who are placed in the box with high influence and high commitment are the strongest supporters for the project. Those people with high influence and low commitment are the ones who need to be engaged by the project team and other key influencers, to address their specific objections, so that they can be moved to support the project and the positive changes it will bring. In Figure 25.2, the

Support / Impact	**Low**	**Medium**	**High**
High	Low Support/High Impact **CTO**	Medium Support/High Impact **CIO**	High Support/High Impact **CRO**
Medium	Low Support/Medium Impact **Equity Head**	Medium Support/Medium Impact **COO**	High Support/Medium Impact
Low	Low Support/Low Impact	Medium Support/Low Impact **Fixed Income Head**	High Support/Low Impact

Red Box: Need to move this individual to the right
Amber Box: Would be helpful to build support from this person
Green Box: A critical person to keep in support of the project

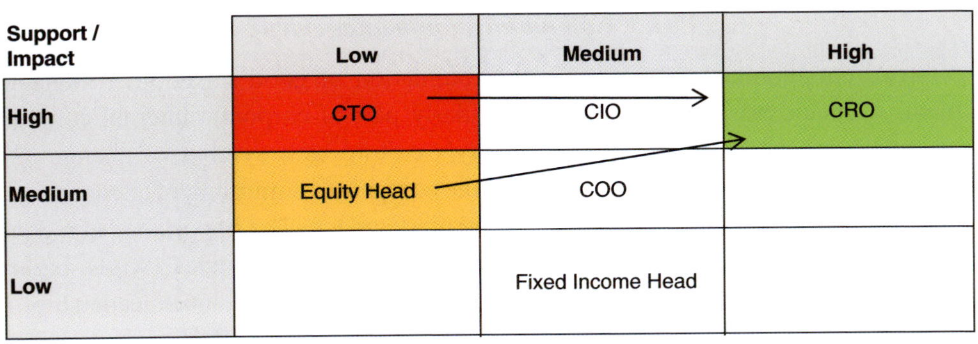

Figure 25.2 Illustrative stakeholder mapping.

CTO is clearly not well disposed toward the program. Her specific issues should be understood and addressed. For example, the issue may be that the program may be viewed as interfering with a data-center integration initiative intended to deliver major cost savings to the firm. Aligning the program's delivery approach with this initiative may not only remove the CTO's objections but in fact make her an enthusiastic supporter of the program. The point of Stakeholder Mapping is to help the management team understand where they need to spend their time in terms of managing change in the organization as a result of the project.

Once the interests of stakeholders are known, picking the appropriate stakeholders to sit on a program ownership committee (usually a Steering Committee) will allow the program manager to design appropriate ad hoc or regular meetings to inform the stakeholders/committee members of progress and look for executive decision-making regarding any program-level issues. A typical Steering Committee is chaired by the Project Sponsor and will include representation from numerous groups including Business Units involved in the project, Technology, Operations, Finance, Legal, Risk Management and Internal Audit. The meeting is generally managed by the PMO which is responsible for preparing and distributing the agenda and materials for the meetings. A steering committee agenda would typically include a report on established key indicators of success for the project (e.g., milestones reached, critical issues outstanding, and staff turnover on the project), as well as a discussion of issues escalated for that forum. While periodic Steering Committee meetings are a good source of information for senior management, program and project managers are well advised to make individual visits to the key stakeholders a periodic occurrence, to make sure that any questions related to a project can be addressed. One should not assume that views and opinions on the project will necessarily be shared by stakeholders in a large forum like a Steering Committee meeting. These touch-points also serve informally to gauge (and encourage) the continued commitment of the stakeholder to the overall program.

25.3.3 Communication management

A broad communication strategy is the foundation of good stakeholder management. A visible project, branded and regularly part of corporate internal communications, can become part of a company's culture as a symbol of change and forward movement. Its success can signal the company's commitment to continuous improvement or keeping competitive in the marketplace. The importance of a good communication strategy cannot be overstressed. Risk initiatives, while highly visible, usually do not bring obvious benefits. If the project loses leadership, it loses the opportunity to make its worth known to the organization in terms of improving efficiencies, improving controls, or managing risk. Risk initiatives should therefore be constantly nurtured, by properly communicating the benefits of the project to the organization.

Program communication is multifaceted, with different (but consistent) messages going up to management and down into and across teams. The failure of communication, either through its absence or by miscommunication, can compromise a program by wasting resources and affecting the working relationships of team members. Without a structured communication plan in place, a project can easily disintegrate into uncoordinated and wasted effort. If critical issues are not raised, acknowledged and addressed, resources that could be much better allocated are wasted.

An effective communication strategy will typically identify the modes of communication and the frequency and reach of the message. Daily team meetings, weekly status meetings, monthly steering committee meetings, and occasional "town halls" to celebrate milestones are common methods to promote communication and to build momentum in a project. There are a variety of tools available for communication, including intranet sites (e.g., SharePoint), Microsoft Project (for plan management and reporting), and more personal touch efforts like weekly newsletters, weekly team meetings, and social events.

The program manager usually owns the overall responsibility for communication across all project teams. The scheduling of regular meetings of the program and project teams (see Figure 25.3) helps ensure that all those involved hear the same information and updates regarding the program and project tasks and timelines. Publication of the program project plan and its milestones helps drive the message to the project teams regarding the need to meet deadlines and manage issues to meet those deadlines. Communication can range from simple tasks such as making sure that a list of all email addresses and telephone numbers is available to the team, to extensive efforts like a two-day offsite meeting to bring program stakeholders together to address urgent issues that require resolution.

A regular meeting schedule provides structure and reporting for a large program and its many projects. For example, a project team may find it useful to hold daily 30 minute meetings (sometimes called "stand up" or "check-in" meetings) to confirm

Monthly Communication Calendar

	Monday	Tuesday	Wednesday	Thursday	Friday
Week 1	Project Team Meeting			Status Updates Due to PMO	Program Progress Report Issued by PMO
Week 2	Project Team Meeting	PMO Meeting		Status Updates Due to PMO	Program Progress Report Issued by PMO
Week 3	Project Team Meeting			Status Updates Due to PMO	Program Progress Report Issued by PMO
Week 4	Project Team Meeting	PMO Meeting	Steering Committee Meeting	Status Updates Due to PMO	Program Progress Report Issued by PMO

Figure 25.3 Sample communication calendar.

work assignments, or to make sure resources are appropriately focused for the day. For smaller projects a weekly meeting may be enough to track progress and issues on the project. To maintain momentum and hold accountability in the project, periodic (e.g., weekly) "PMO" meetings with project managers can be held to review project status reports and discuss issues, risks and mitigation strategies. Early on in a project initiative, PMO meetings may be held more frequently as the project teams establish plans and resourcing, with their frequency falling off subsequently depending on the timing of deliverables. As a milestone approaches, their frequency may again increase to help ensure issues are raised and dealt with promptly in order to meet critical deadlines.

Meaningful management information on progress is most often achieved through the use of dashboard reporting: weekly reporting on the current status of the project(s) against timelines, including a summary of known issues and risks (see Figure 25.4). The so-called "RAG" rating (red, amber or green) is frequently used to highlight at a glance the current status of the project. If things are going well and as planned, everything is "green" or "Go." If timelines are slipping, or if there are concerns about the ability to resolve known issues, the status may be "amber" or in a cautionary state. "Red" usually refers to milestones that are past due, or highlights issues that need immediate resolution. These dashboards are typically used at weekly meetings and summarized for monthly Steering Committee meetings.

Communications should also be tailored to the culture of the organization. Some organizations are detailed oriented and so need specifics in detail, and numbers even, in contrast to others where aggregation, concept and strategy are preferred. The PMO should work with the program stakeholders to determine the exact type and form of management information to be provided throughout the program.

Just as critical as program level communication is regular and effective communication between the business and technology teams. This communication helps mitigate the danger that technologists will meet technological standards and leading-practices but fail to align with respect to the business needs. Starting with the development of business requirements, the technology team should participate with its business partners in conversations with respect to the vision of the system. Meetings should include representatives of both business (data) and technology, in addition to change management experts, who can help facilitate these conversations. Systems experts can help business people expand their vision of what a system can do, while also providing practical feedback on the limitations of those very systems. Data experts equally can be leveraged in the same conversations so that downstream users understand the types of data that they will be receiving, its origination and characteristics. These concepts are explored further in Section 25.4.

PROJECT MILESTONE STATUS

Programs and Projects	Owner	Suite & Application / Platform Selection			Requirements defined			Development & Testing			Conversion / Integration Plans			Integration Testing			Conversion & Post Conversion Operating Metrics			Readiness Reviews			Integration / Conversion / Implementation			Overall Status
		Status	Target Date	Revised Date	Status	Target Date	Revised Date	Status	Target Date	Revised Date	Status	Target Date	Revised Date	Status	Target Date	Revised Date	Status	Target Date	Revised Date	Status	Target Date	Revised Date	Status	Target Date	Revised Date	
BROKERAGE		GREEN			GREEN			GREEN			GREEN			GREEN			GREEN			GREEN			GREEN			G
System Roll-out (1)		C			C			Y			Y			Y			Y			Y			Y			Y
System Roll-out (2)		C			G			G			G			G			G			G			G			G
Clearing Platform		R			Y			Y			Y			Y			Y			Y			Y			
Mid-Office Integration		N			G			G			G			G			G			G			G			G
		C			G			G			G			G			G			G			G			G
		C			G			G			G			G			G			G			G			G
		C			G			G			G			G			G			G			G			G
		C			G			G			G			G			G			G			G			G

ISSUES AND FOLLOW UP

Programs & Projects	Owner	Key Accomplishments	Deliverables missed	Next Deliverable		Issues / Concerns / Dependencies
				Date	Deliverable	
BROKERAGE						
System Roll-out (1)						
System Roll-out (2)						
Clearing Platform						
Mid-Office Integration						

Figure 25.4 Example of a program dashboard.

25.3.4 Change management

Change management activities can be effective in driving the overall success of a project. Most projects require some element of change management. Implementation of a new system or new process is rarely, if ever, an easy transition – even for groups that are looking forward to the "new system" or the "new data." Having just one person on the staff who does not want to change can easily infect the rest of the team. A group that is not properly trained on the new system will not necessarily be inclined to embrace it, because "it's different."

Change management can be any type of activity coordinated to facilitate the project implementation and help ensure its success in the organization (Kotter, 1996). Change management activities can include team building events, cross-team learning sessions, "train the trainer" classes, visits to other companies or conferences to learn more about the technology or the process, or even weekly Friday morning coffee and donut sessions where the team can air its concerns more informally with the project management team.

As discussed below, by following a structured approach to change, the program leadership can help ensure that the components needed for success are in place. From developing a vision to training the team to be successful in the new environment, a well-designed change plan can provide the roadmap for managing change.

Change vision and strategy: building commitment

Change management begins with a vision for change, that is, the new environment, whether it is a regulatory requirement or a planned improvement in data collection and risk management. This vision for change drives the Change Plan, that is, the blueprint for effecting the change. Typically, the plan is the outcome of a change readiness analysis with stakeholders and others on whom the initiative will have an impact. For example, if a decision is made to implement a vendor-supplied solution for a risk management system, those involved with the decision should be canvassed for their opinions and suggestions for implementing the change. These interviews with key participants in the program, as well as with those on which the implementation will have downstream impact, is valuable input into the Change Plan. Participants are asked to articulate their own particular expectations or fears about the program and their sense of challenges and risks that the program will face. The Change Team can collate these challenges and risks and develop a plan that addresses them throughout the course of the program. The interviews help build support with stakeholders by seeking their input and validating their opinions. This process also results in a clear picture of the state of readiness of the organization for the change.

The stakeholder analysis (described above in Section 25.3.2), is also a contributing factor to the Change Plan. Having identified the key stakeholders and their support (or lack thereof) for the program, the team can develop an approach for managing and meeting stakeholder expectations, and articulate it in the plan. A communication plan (as described in Section 25.3.3 above) is also a strategic building block in building commitment for the overall program.

Preparing the organization for change: align culture and design new structures

Bringing a new process into the organization also requires the building of new structures to support that change. Job profiles, for example, should be written (or rewritten) to address the new environment. Technology requirements may need to be reassessed given the new process, and new employees hired to support the change. Another key ingredient that is often forgotten is the design and development of new behaviors to motivate adoption of the new end state, which should encompass new performance management measures. The new system or process may demand a cultural change in both types of people performing the job as well as the expectations for those who are doing the work. New job profiles and performance objectives would typically be deliverables described in the Change Plan.

Develop leadership

In addition to business users, business leaders need to be trained and coached in the new processes and expectations for data and risk management. The stakeholder analysis can provide guidance where challenges lie, or where champions can be found, to support the change. It is axiomatic that where leaders lead, others will follow. It is when the leadership embraces and reinforces the change that the organization is pushed to accept it. Providing regular updates to the business leadership, for example, enhances visibility for the program and provides the leaders with opportunities to influence and provide guidance. Having business leaders participate in communications to the project teams during the project, for example by presiding over Town Hall meetings or other types of milestone events, provides to employees validation of senior executive support and builds the momentum for acceptance of the change. The Change Plan should articulate approaches for developing business executives to champion and lead the change.

Training and knowledge management

Training and knowledge management (the transference of knowledge of the new system across the organization) are essential elements of a change management program. Involvement of the business users in the project from the development of requirements to conducting user acceptance testing (see Section 25.4) on the new system is an important way to build early adopters and encourages the business users to have some "skin in the game" with respect to the new processes and systems, and is an effective way to train new users. A well-developed training program, including training materials, a training schedule and objectives for the new process or system, should be part of the Change Plan.

25.3.5 Project financial management

Effective management of the finances of a project can often be the difference between project completion and the cancellation of a project. Project financial management includes all the planning, budgeting, accounting, and financial reporting for a project, with the aim of managing project resources properly and achieving the project's objectives. The ability to track and be able to provide timely financial information can provide the basis for better decision-making and prevent slowdowns or bottlenecks. Accounting for projects or programs independently from business as usual allows the company to better assess the return on investment of the individual projects and measure performance of those efforts.

The process begins with the development of a project budget, including all the costs in detail, including direct and indirect costs. Once the budget is finalized and the project begun, the program manager is generally responsible for overseeing the team that is tracking costs associated with the project. There are many different software products available to provide accounting support for project financial management. These products generally also have robust reporting capabilities that provide flexibility for creating different types of reports. For smaller projects, project management software or spreadsheet programs may suffice for tracking hours spent, software, hardware and other costs associated with the project.

The project budget will need to be continually assessed during the lifetime of the project. Any proposed design changes during the project need to be evaluated on a cost basis, as well as an implementation and time to delivery basis. Weekly reports of budget to actuals from the finance team should be utilized by the program management, and this management information provided to appropriate stakeholders. Trend analyses of monies spent and where they have been spent are also extremely valuable in anticipating changes and the need for additional budget.

There are many different ways to manage project finances. Earned Value Analysis (EVA) is one approach to tracking a program's financial performance. The methodology helps managers by looking beyond budget spent to actual work completed.[3] Other types of project financial management include basic budget tracking (budget to actual), resource tracking and time-to-completion tracking. Estimate to completion is another reporting device that gives the manager an

[3] See Nagrecha (2002). See also `www.pmtoolbox.com/project-management-templates/project-earned-value-analysis-template.html`

opportunity to track the burn rate of the project budget through the given timeline for the program. Costs and revenues that are allocated to projects may be further subdivided into a work breakdown structure (WBS), allowing for more detailed tracking and reporting.

25.3.6 Project risk management

Financial risks result in minor as well as major deviations of performance and can be measured by concepts like Value-At-Risk (see Part I, Chapter 2). In contrast, project risks are usually highly correlated and can result in major disruptions to cost and time. Numerous surveys have been done over the past 15 years that yield statistics[4] that are at least instructive if not truly shocking. For example, surveys indicate that more than 50% of projects fail in some respect, with a high percentage of projects never even being completed.[5] While the data have been challenged, the controversy is more about the definition of project success than any sense that the surveys are simply wrong. At the very least, the surveys indicate that projects can significantly miss initial estimates. Therefore project risk management is an important activity for a program manager.

Project risks come in many shapes and forms, and should be identified and managed throughout the project. These can range from operational risks, such as the turnover of key individuals on the project, to the failure of communication between teams. They can also be political risks, including the failure to identify a key stakeholder to the project who may become an opponent of the project and pose a serious threat to its completion. The risks can also be reputational; the failure to communicate progress within and outside the organization can adversely affect the standing of the project and impede its eventual success. A process to regularly identify key risks, track them in management forums and actively mitigate issues as they arise will facilitate the resolution of threats and blockers to the project and allow for a successful implementation.

Identifying project risks at the beginning of the project, as part of the project planning process, can allow the program manager to proactively mitigate possible issues even before they arise.[6] A Program Risk Assessment (as shown in Figure 25.5) is one way to initiate this analysis. The program leadership can involve the team and key stakeholders in identifying the various types of risks and then prioritizing them in terms of impact and severity. As mentioned above, the turnover of a key individual may be a high impact risk, but the probability may be low, and a plan can be put in place to monitor staff morale during the project, and also to train others

[4] IT-Cortex (`www.it-cortex.com/Stat_Failure_Rate.htm`) collates results of five surveys.
[5] For an amusing but all too real list of reasons projects fail, see Krigsman (2009).
[6] For a more detailed discussion of managing project risks, see Kendrick (2009).

Program Risk Assessment	People	Process	Systems	Regulatory	Organization
Project 1	Key person dependency (P-3, I-4)			Rules are still draft; not yet final (P-3, I-4)	
Project 2			New vendor relationship (P-3, I-4)		Governance structure needed (P-2, I-4)
Project 3	Need to hire resources (P-3, I-5)				
Project 4		New process needed (P-3, I-4)			Need to find new project sponsor (P-2, I-5)
Project 5				Rules are still draft; not final (P-5, I-5)	

Probability (1-5 High), Impact (1-5 High)
High Risk

Figure 25.5 Example of a project tracking matrix.

so that the possible impact might be softened. Failure to communicate effectively may be seen as a moderate risk, easily mitigated by a well-conceived communication plan, but the probability of missed or failed communications is high in any organization, and a program manager is well advised to monitor this risk regularly. In the analysis of risks to the project, all those identified as high risks should be addressed with a plan to mitigate and manage each risk appropriately. The plans should be part of the overall project plan, and monitored and tracked accordingly.

In Figure 25.5, for example, the identified risk regarding regulatory rules still in draft form in Project 5 could severely impact the project if the final rules are markedly different, as the probability assessment implies. A plan for addressing this risk might involve weekly check-ins with internal regulatory affairs groups or industry participants to monitor the progress on the rule-making or reviewing system planning to ensure that it is flexible to accommodate changes to the rules. The program leadership should devote time and effort to mitigate urgent high risks, and address lower impact risks as and when necessary

Another way to look at risks across the project is by scope, people and schedule.[7] There are various quantitative methods to approach an assessment of the probability and impact of each type of risk. For example, an increase in scope can have a negative impact on time and people, so that so-called "scope creep" needs to be managed effectively. The project leader can proactively build into the project plan additional

[7] Kendrick (2009).

time to address potential additional project demands, if the probability of this occurring is assessed at 50%. Rather than what might be expected as four weeks to complete system design, the prudent project manager might include a cushion of an additional week, if the probability that the design might change is considered a medium probability. Building in the additional week mitigates the risk that additional time needed for changes in design would affect the overall project timeline. This type of analysis applied across the whole project can lead to a much more realistic project plan and expected completion time frame.

Managing risks regularly through the project is critical so as to help mitigate unexpected impacts. Unidentified and unmitigated project risks can impact timelines, morale of the team, leadership support and program budgets. Tracking, monitoring and mitigating risks are the responsibility of the program and project managers, and while not guaranteeing project success, effective risk management can usually increase the probability of achieving the desired result.

25.4 Implementing an analytic data project

The past few decades of experience delivering large technology-heavy initiatives have resulted in many advances in project management, including the development of a standard Software Development Lifecycle (SDLC). The SDLC provides a framework of leading-practices for system designers and developers to adapt to their own projects, and consists of a set of steps or phases in which each phase of the SDLC uses the results of the previous one. The oldest and perhaps the best known of these is the so-called Waterfall Model which provides a sequence of stages of work where the output of one stage becomes the input for the next. While huge changes in technology and concomitant hyper-acceleration of project delivery schedules over the past two decades have seen the waterfall approach much maligned among the software development cognoscenti, it remains a primary framework of delivery for large programs. Much of the criticism of the model can be considered uninformed and invalid, however, as this model is almost never applied in a rigid and inflexible way. Rather, many projects embed the overall waterfall approach, while incorporating newer implementation techniques such as rapid prototyping techniques and extreme programming for implementing specific components within the overall program.

The Waterfall Model originated in manufacturing and construction industries, where changes during the project can be prohibitively expensive if not impossible,[8]

[8] Some of the material for this section has been adapted from the article on waterfall models found at `http://en.wikipedia.org/wiki/Waterfall_model#cite_note-0`

and was a good fit for early computer projects that were of necessity a combination of hardware as well as software components. While the model has been referenced as early as 1956[9] by Herbert Benington, the first formal description of the model was in a paper by Winston Royce in 1970 (Royce, 1970). Even in these early efforts the model was recognized as nothing more than a framework for software development with flaws that had to be understood and addressed to achieve a successful implementation. For instance, even Benington's 1956 effort used a prototype before actual engineering development began.

Royce's original model defined the following stages: requirements, design, development, integration, validation and implementation. The pure waterfall model requires that each stage is fully complete before moving onto the next one. For instance, requirements should be fully complete before design begins. This is usually impossible to achieve in practice. Requirements can change after they have been completed. Some aspects of design may be proved unworkable in practice, while in other cases new opportunities for improvement may be discovered during the process of implementation. This is especially true today, where rapid technology improvements sometimes render obsolete specific software choices during the course of a long project. Data intensive initiatives like FRATs add further uncertainty as described in Section 25.2, especially given that requirements changes should be expected through the course of the project. Notwithstanding these drawbacks, the basic premise underlying the waterfall model – that late-stage changes in the project can be prohibitively expensive – holds true in analytic projects, making the waterfall approach indispensible, albeit with significant modifications. An approach to implementing FRATs using a waterfall framework is shown in Figure 25.6.

It is apparent that there is much more complexity in this process than a simple waterfall with six phases. Nevertheless, it is possible to identify each workstream of the process with one of the phases identified earlier as indicated in the legend to the figure. The diagram should be read in two dimensions. The majority of FRATs are transformative and incremental, dealing with capability improvement (admittedly the scale of some improvements is large enough that the lines between evolution and revolution are sometimes blurred). The vertical axis generally deals with activities that transform the environment from current to future state. The horizontal axis deals with activities that progress through time in the typical fashion of a waterfall model. The waterfall phases used by this process model are presented in Table 25.1, along with the program workstreams that fall into each phase.

The next sections will describe these phases and workstreams in further detail.

[9] In a presentation on the SAGE system built for tracking Soviet bombers, Herbert Benington described the use of phases of development similar to the waterfall method (see Benington, 1956).

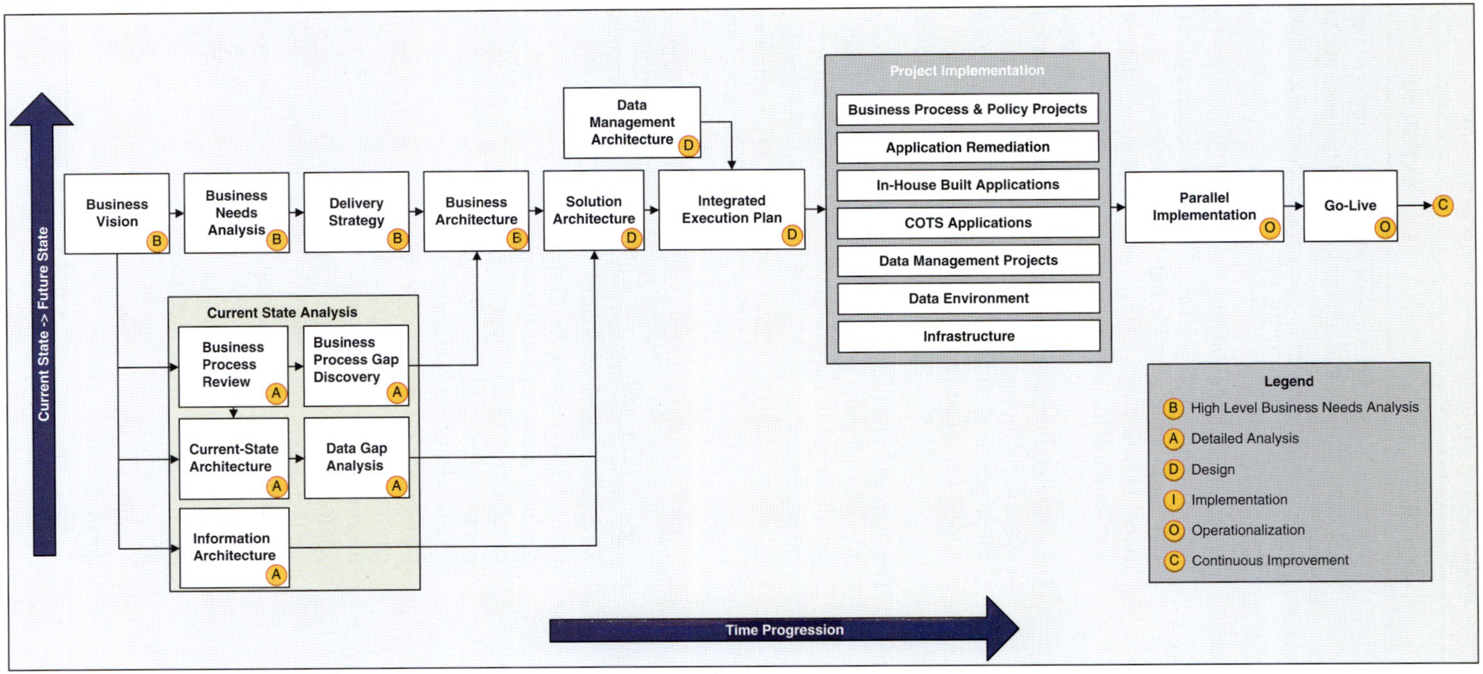

Figure 25.6 Illustrative program implementation methodology.

Table 25.1 *Example phases of a program*

	Phase	Description	Workstream
1	High level business needs analysis	Establish a high-level view of the intended project and determine its goals, map out an implementation strategy and identify costs.	• Business vision • Business needs analysis • Delivery strategy
2	Detailed analysis	Document current state and articulate the future-state environment to identify the gaps. Develop an implementation plan.	• Business process review • Business process gaps analysis • Current state architecture • Information architecture • Data gaps analysis
3	Design	Develop detailed design for the system, including business intelligence, data management and governance in detail.	• Business architecture • Solution architecture • Data management architecture • Integrated execution plan
4	Implementation	Develop and test the components for data integration, business intelligence, and operations management.	• Business process and policy projects • Application remediation • In-house built applications • COTS applications • Data management projects • Data warehouse build • Infrastructure implementation
5	Operationalization	Put the system into production. This phase requires careful orchestration across multiple business areas such as the user-community, IT and operations management. Implementation ideally occurs in incremental phases while keeping existing environments running, though in some cases a "big-bang" approach is required.	• Parallel implementation • Go-Live
6	Continuous improvement	Once the system is up and running, it is critical to loop back on a periodic basis to make sure that the business justification is being met, the design is still valid and the system is still producing as desired.	• Business value analysis • Long-term roadmap

25.4.1 High-level business needs analysis

The first step is typically to formulate a *business vision* for the firm which outlines the overall program goal. A program driven primarily by near-term regulatory drivers will likely be constrained primarily by timelines and likely not envision expansive process and system improvement, while one whose goals are finance or risk optimization will be more likely to consider such outcomes as integral goals for the program. Whatever the case, it is critical for senior management in the firm to clarify the program's business vision in a simple and easily communicable manner.

Business needs analysis examines the business demands of the program, as well as any other regulatory or competitive drivers. Given the data-intensive nature of FRATs, it is important to leverage business subject matter experts to take advantage of their knowledge of the internal environment, the firm's appetite for change and any applicable regulatory drivers. This analysis will usually result in a high-level business architecture that will determine the extent to which the program will rely on incremental change to meet project goals, and also identify those areas where radical changes will be required. The analysis should address the concerns of executive management, including the benefits and risks involved in the project.

25.4.2 Detailed analysis

Detailed analysis begins with a review via a *current state business process analysis* which decomposes relevant business processes into their elemental steps to understand the changes required to meet program goals. The outcome of this work will usually include such artifacts as

- Current State business organization including Lines of Business, International operations etc. and the mapping of products to business lines,
- Business processes that relate to the program at hand, for example, a risk management program in a bank may analyze processes for calculating collateral value, and
- Information flows between processes.

This work usually culminates in a *business process gap analysis* which describes the process gaps that should be closed to meet program needs (a typical process gap identified during Basel II – see Chapters 1 and 11 – implementations was the need to normalize risk ratings across lines of business). The outcome of business process analysis is a synthesized actionable list of process gaps for remediation that describes each process in detail and the actions needed to address them. In some cases, further detailed analysis will be needed to determine the actual action to be taken. In addition, at this stage a high-level list of data gaps (data that are not being collected) should be identified.

Current state architecture focuses on understanding the systems landscape present in the organization. Areas of interest typically include the range of existing systems in the organization, the function they play in the business process, the kinds of data that are generated or modified in each system and the vintage of the system (which is usually indicative of both the quality of data from the system and its potential appetite for change to effect quality improvement). Understanding data flows between different systems is important. In many large firms these are usually very complex and often contain carefully orchestrated bidirectional flows of information. It is critical to understand any changes in the landscape through new systems implementation or legacy systems retirement.

One aspect of current state architecture that requires particular attention in FRATs is the *information architecture*. This is a detailed description of data flows between the various systems that will be required to deliver the requisite information to achieve the program's requirements for calculation and reporting. Not all systems within the firm will be involved in creating, processing and delivering information required for a finance or risk information program. This architecture should include high-level diagrams detailing the subset of systems involved in the scope of the program and further elaborate relevant aspects such as those involving complex flows of data or user interactions, with the purpose of creating the documentation necessary for the implementation phase. In addition to information flow diagrams, details such as the business description of the data elements involved in processes should be documented. Information on data volumes and frequency is also very important. Since most information transfer between systems is on a periodic basis (such as daily) using batch files, information on these transfers such as the number of files, periodicity, etc., should be captured. In case of real-time transfer mechanisms, metrics such as transaction throughput, peak rates and times are essential. This is a daunting task in large organizations with systems spread across the globe. Much of this information is also likely to require the participation of system specialists who are busy with their own unrelated priorities, which makes the challenge even harder. Therefore, prioritization is required to meet the time constraints of a typical FRAT. The focus should be on broad and complete coverage rather than deep and precise documentation on any one system. It is unacceptable to have missed high-level details on an important system that sources, for example, equity derivatives, in favor of spending time detailing the data flows, throughput and volume of a cash-equity system.

The resulting current-state application and information architecture is an input to the next activity: *data gap analysis*. The purpose of this analysis is to assess the firm's readiness in collecting data required for calculations and reporting required by the program. During this activity, a list of required data elements is created followed by mapping them to data elements in source systems. This can be a laborious and manually intensive activity with a number of complexities. It is common for the same data elements to be found in multiple systems dealing with different products, so redundancies may need to be resolved. Systems that are in the process of being replaced may contain only a certain subset of data, with the rest of the data already having been transferred to its successor. A data gap analysis will therefore be imprecise but this does not detract from its importance. It should be viewed as a high-level overview of the data available within the firm, and an indicator of where gaps exist in content or data quality. Since achieving precision is impossible, data gap analysis should be done on a time-bound basis with successive iterations focusing on specific areas of interest. This not only saves cost but can also help ensure that the program does not get paralyzed by analysis. The result is a heat-map that indicates the readiness of each system in the firm to serve up particular kinds of data.

Table 25.2 shows an example of an aggregate view of data gap analysis for a single system. Data required for the entire program's calculation and reporting have been broken down into various data domains. As is typical, this particular system has data elements that match only a subset of these domains. The goodness-of-fit metric is calculated based on cut-offs on the match percentage – in this case match percentages of 75% and above are considered good matches. The shaded rows indicate domains where this particular system has ten or more data elements and so would be considered a viable source for data. The outcome of this analysis can be used for several purposes. It forms the basis for determining the kinds of systems that should be used to draw the data required for calculation and reporting. The gross-level information it provides about required remediation for each system can also be used as an important input into budgeting.

Table 25.2 *Data gap analysis example*

Domain	Matched	Gap	Match %	Goodness-of-fit
1	8	1	89	High
3	11	2	85	High
5	3	1	75	Med
6	7	4	64	Med
10	2	0	100	High
11	1	0	100	High
12	19	8	70	Med
13	4	3	57	Med
20	37	10	79	High
23	2	1	67	Med
25	0	1	0	None
26	1	1	50	Low

25.4.3 Design – developing the conceptual design for the system

The detailed analysis phase (see Section 25.4.2) is designed to produce the range of foundational information essential to designing a robust solution. To produce a proper design, business requirements should be synthesized with this information to produce an implementation plan. In a large, diverse program that is subject to changes through its course, it is impossible to fully document business requirements before beginning solution design and development. This "moving-target" problem, the downfall of many a project, can be circumvented by a broad understanding of the future state *business architecture*[10] which can act as a reasonable proxy for business requirements.

Business architecture defines the structure of the enterprise in terms of its governance structure, business processes, and business information and focuses on the business operations and analytics processes. Two specific views of business architecture are usually captured:

- a view that describes the primary business functions of importance in the program along with the responsible functional organization, and
- a view that describes the organizational units (both shared functions like finance and risk as well as client-facing business units) and their hierarchical relationship to each other.

This architecture view should reflect the future state of the firm after the program has been implemented. This should also necessarily reflect other changes that the firm may be undergoing at the same time, such as other transformational programs or the effect of mergers and acquisitions. The purpose of business architecture is to help ensure that the final solution is informed by a view of the business processes responsible for creating, modifying and using information generated by the program. This perspective can enable the program to detect incomplete or inconsistent business requirements early and avoid expensive re-work during the implementation phase.

Once the future-state structure of business processes has been defined, a *solution architecture* can be created to meet its needs. This framework ties together functional requirements with technology solutions. As with the other workstreams in the program, the solution architecture's scope should be constrained to address only its needs. The solution architecture usually starts with a high-level description of the overall system and information flow to address future-state requirements, and follows this with detailed descriptions of how particular aspects of the solution would work. Its purpose is to act as a communication mechanism between the various teams that will be involved in developing the final solutions. This will

[10] The following paragraphs have been drawn from the Object Management Group's definition of Business Architecture which can be found at `http://bawg.omg.org/business_architecture_overview.htm`

include teams that manage systems targeted for modification, those that may simply be required to feed new information to downstream recipients, as well as new systems that will need to be implemented during the course of the program.

Since the final solution architecture will have long-range impacts to the firm's ability to meet on-going business needs, careful thought should be given to trade-offs in design. There is often a choice between modifying existing systems and implementing new ones. Each situation should be examined on a case-by-case basis considering not only short-term implications (such as implementation cost or the impact to a particular on-going project) but also long-term ones such as the ability to enable new and potentially profitable business capabilities. It is also critical to ensure that there is deep and broad buy-in from the various stakeholders in the program to the options that are chosen in the final future-state solution architecture.

An example of solution architecture is shown in Figure 25.7. The overview is at an extremely high level, showing the systems involved in feeding information to the relevant processes in risk and finance, with a few brief annotations being made to qualify the flows. The detailed architecture comprises a series of snap-shots of individual processes (such as the process to calculate regulatory capital shown in the figure). Both these diagrams are uncluttered by detail on purpose;

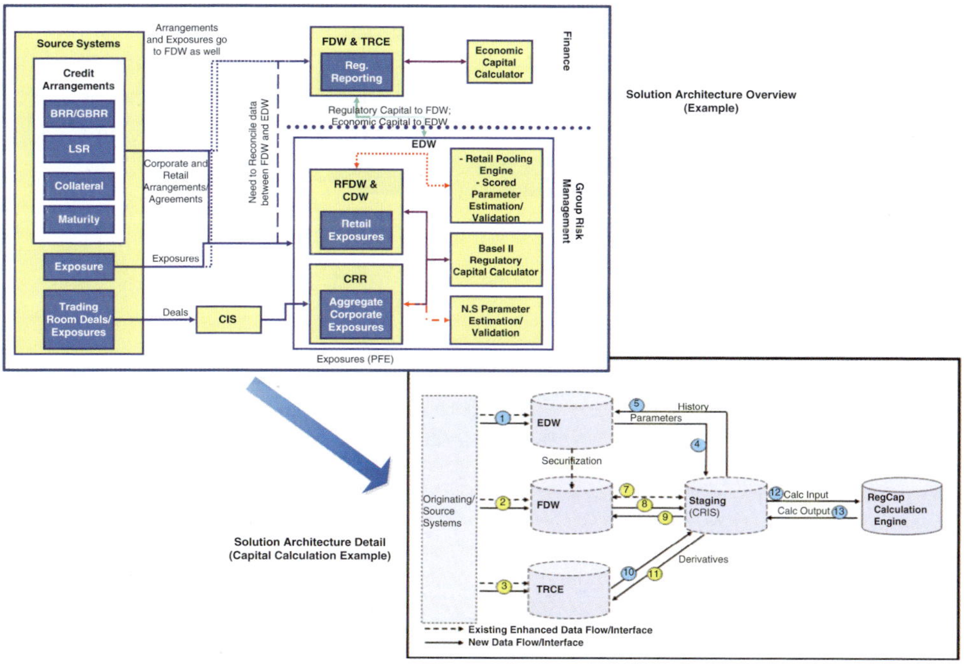

Figure 25.7 Example of solution architecture.

the typically low levels of communication between teams makes simplification necessary. If deeper discussion is required, further architecture diagrams can be created for the specific areas that require them. The diagrams do not stand by themselves, but are usually accompanied by a range of documentation such as detailed descriptions of system changes and tables describing data elements that are expected from each system. It is not uncommon for final system architecture artifacts to comprise tens of PowerPoint, Word and Excel documents. In keeping with the iterative nature of analytic projects, the solution architecture is a living body of documentation that will inevitably be modified as the program progresses. These changes may arise not only as a result of external factors discussed earlier, but also as a result of internal ones such as the discovery that the data from a particular system are inadequate for the intended purpose, necessitating further remediation.

In an analytic program *data management architecture* deserves special focus via its activity. Implementing proper data management procedures requires cooperation across business functions and especially between business and technology organizations. Data management is a complex interplay of the following components shown in Figure 25.8.

- *Data architecture* involves documenting the types and locations of data across the organization in one or more *data models*. These models are used to relate datasets together. Data architecture also describes the end-to-end process of collecting, organizing and distributing data.

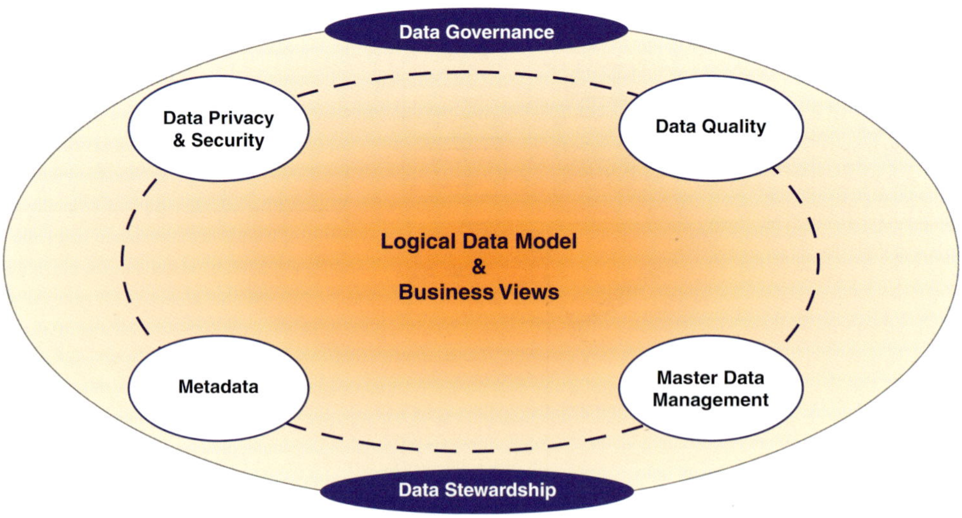

Figure 25.8 Data management components.

- *Data management disciplines* are involved in managing the data pipeline documented in the data architecture and involve a combination of technology as well as business processes. These disciplines include *master data management* (the process for achieving consistency in reference data – see Chapter 8), *metadata management* (keeping the meaning of data consistent – see Chapter 23), *data security* (ensuring that data are only available to authorized users – see Chapter 24) and *data quality* (enhancing the integrity of data).
- These disciplines should be supported and managed by operational business processes which come under the purview of *data governance and data stewardship* (see Chapter 27).

Focus on data management is essential to help ensure that the solution architecture goes beyond functional requirements to address the continued integrity of data necessary for financial and risk reporting and calculations. It is impossible to rebuild a full data management structure from scratch, at least in large and well-established firms, due to established business processes. Therefore data management architecture should necessarily take an incremental approach to addressing shortcomings in those areas of the firm that produce, modify or consume data relevant to the focus of the analytic program. It draws heavily from output of information architecture and data gap analysis activities (see Section 25.4.2) in outlining the future state of data management for the firm. Since the future state may include new tools for disciplines like data quality or metadata management, it may be necessary to conduct proofs-of-concept and vendor-tool evaluations prior to implementation. The outcome of the data management architecture work-stream is a series of documents that describe individual components in the future state, and how the business processes and technology in these components interact with each other.

The solution architecture and data management activities described so far synthesize the outcome of the two analysis phases (see Table 25.1) and are in themselves sizeable pieces of work, each of which can be expected to take weeks to months of time. A preliminary project plan is typically created at the start of the program to outline their timelines and dependencies. However, these two phases are only the foundation for the "real work" of program implementation, the bulk of which is conducted in the third phase. A significant replanning effort should be conducted at the close of the detailed analysis phase to integrate the various outputs of the preceding two phases into an actionable *integrated execution plan* that can act as the blue-print for the several projects that will need to be executed to meet the needs of the program.

Figure 25.9 shows an example of such a plan. This plan breaks down the entire program into three stages. The *analysis and design* stage roughly corresponds to the first three phases of the figure, the *construction* stage corresponds to the fourth phase

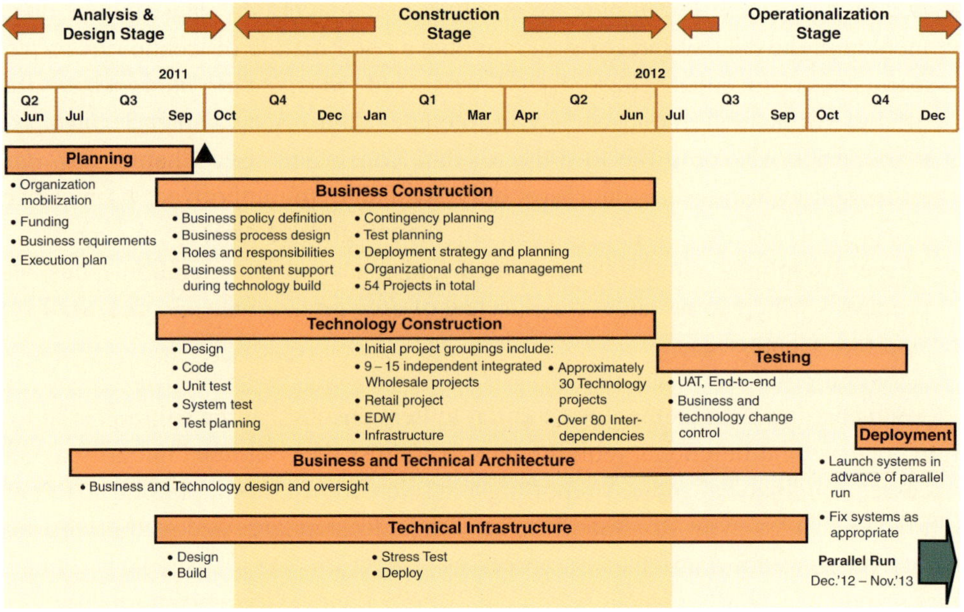

Figure 25.9 Example of an integrated execution plan.

in the table and the *operationalization* stage refers to the fifth phase.[11] There is an initial planning phase that launches the program and includes defining the business vision. Business and technical analysis and architecture activities start before the planning phase is complete to enable the final integrated execution plan (the artifact shown in Figure 25.9) to be completed, at which time the construction phase can commence. The plan uses overlaps of adjacent activities to avoid the pitfalls of the pure waterfall model discussed in the introduction to Section 25.4. To illustrate, consider the overlap of the construction and architecture phases. While architecture should initially be created before construction can begin, architects should work closely with the team designing and building business processes and technology to transfer knowledge as well as modify the architecture to reflect practical constraints (an example might be jurisdictional compliance constraints that were unknown at the time of development of the architecture). This overlap also allows the architecture to reflect new or modified business requirements as well. The plan also illustrates the fact that several major activities should be executed in parallel requiring careful firm-level coordination.

The integrated execution plan results in resource estimates for the projects in the construction stage. These resource and other cost estimates are required for

[11] Continuous improvement, the sixth phase, requires its own plans as described in Section 25.4.6.

individual projects' financial plans so that the overall program spend can be tracked bottom-up from each project. Along with tracking the financials of each construction project, proper project management discipline should be applied, as well including change control, project risk management and status reporting procedures (see Sections 25.3.3, 25.3.4, 25.3.5 and 25.3.6). These should be applied in a consistent manner across projects in the program so that aggregate tracking across the entire program is possible.

25.4.4 Development (SDLC) and testing

The construction stage (see Figure 25.9) can begin in earnest once the integrated plan is completed and communicated to all stakeholders. It is not uncommon for a firm to have 50–100 such projects running in parallel. Many of these projects will involve remediation of business processes or systems of some kind, while some will involve developing new applications. These are fairly standard projects amenable to traditional project management techniques that will not be covered further here.[12] The development of an analytic environment used for calculation and reporting, a critical aspect of FRATs, deserves special treatment and will be covered in greater depth.

Non-data warehouse projects

A FRAT in a large firm necessarily entails a number of different kinds of activities ranging from purely business process changes to implementing applications accompanied by process changes. The types of projects that are typically encountered in such programs include the following.

- *Business policy and process projects*: There will be many changes to business processes due either to regulatory requirements or internal risk and finance initiatives. Often these changes replace a manual, uncontrolled process with an automated one to provide more auditability and hopefully better efficiency as well.
- Many applications projects are also common in these programs, including the following.
  - *Application remediation*: Process improvement is often accompanied by changes to existing applications. Some changes may be simple modifications to existing configurations while others may be extensive development efforts.

[12] See, for example, Project Management Institute (2008).

- *COTS implementations*: The trend in most financial firms is oriented towards outsourcing their application development to vendors who can provide packaged solutions (Commercial-Off-The-Shelf or COTS applications). Vendor applications are usually, to a greater or lesser degree, customizable to the needs of a particular firm. Implementing vendor solutions typically involves compiling requirements and choosing the most appropriate of a range of potential vendor solutions. The firm's requirements should then be mapped against functionality offered by the vendor. An effective vendor selection process can help ensure that most of the firm's requirements will largely be available out-of-the-box or via easy configuration, which reduces implementation cost and time. A range of software modifications may still be required however. The requirements for such modifications deserve careful analysis and sometimes senior-management engagement to determine whether the firm can instead make simple process changes to leverage the application's existing functionality. Vendor applications will also need to be integrated into a firms existing system architecture, a non-trivial task that is often the cause of significant delays (usually due to data issues) if not approached with seriousness and rigor.
- *In-house application build*: In some cases the firm may decide that available vendor solutions do not sufficiently meet its needs. Other motivations can include the desire to in-source applications that are thought to provide competitive advantage. Such decisions should include analysis of not only project implementation costs but also the cost of on-going maintenance and improvement of the application. Such projects also call for exceptional project management discipline and vigilance against scope-creep.

- *Infrastructure implementation*: Large programs inevitably result in implementing new technology infrastructure. Such implementations are usually handled by specialized infrastructure teams with expertise in the process of procuring, installing and managing such infrastructure (see Chapter 26). Such projects can be very time consuming, often requiring many months of lead-time due to the complex orchestrations involved with managing data-centers. Therefore, program and project managers would be well advised to propel decisions on infrastructure as early as possible, and resist changes to these decisions unless they are absolutely unavoidable.

All except the most trivial of such changes should be run as an independent project, with the same project management rigor and focus accorded to the largest initiatives. This can help ensure that the whole program will not be brought to its knees by delays in smaller projects, since these projects can often become the weak link in the chain of a large program. A traditional waterfall approach (inasmuch as there is such a thing – see introduction to Section 25.4) can be taken to implement such projects. This includes the following groups of processes.[13]

1. *Initiating*: This step involves determining the nature and scope of the project. An important step is to develop the so-called project charter, in which these items are documented, so that changes to scope can be clearly monitored. During the initiation stage, the budget is developed for the project.
2. *Planning and design*: In this stage requirements are gathered and a solution designed. An end-to-end project plan is also developed for implementation, and resources identified.
3. *Executing*: These processes can help ensure that the work that has been planned in the previous stage is actually being accomplished. It includes implementing the steps required to achieve the final solution, such as developing code, writing policy manuals, developing control procedures and operationalization. It also includes quality-assurance activities such as software testing.
4. *Monitoring and controlling*: These processes can help ensure that potential problems can be identified and corrective action taken in a timely manner by regular measurement to identify variances from the project management plan.
5. *Closing*: This phase can help ensure that the project is formally accepted by stakeholders and includes administrative activities such as archiving documents and articulating lessons learned.

The analytic ecosystem

Several areas of a large firm are usually involved in data analytics. The analytic ecosystem refers to the entire motley collection of data-stores, some better managed than others, along with the analytic tools to access, manipulate and report on data. The data-stores in the ecosystem can be roughly categorized as shown in Figure 25.10. An analytic ecosystem contains the following components.

- *Data stores*: The many approaches used for data storage, modification and retrieval (see Part V, Chapter 20) include the following.
  - *Independent data-marts* which are usually special purpose data-stores designed to address a specific purpose.
  - *Data-warehouses* that are designed with a larger scope of data in mind. Their span of data coverage can vary considerably and range from serving the needs of a single business function to those of the entire firm.
  - *Dependent data-marts* are special purpose data-stores developed as subsidiaries to data warehouses. Data segregated are usually required to meet data

[13] The basics of the topics covered here can be found at `http://en.wikipedia.org/wiki/Project_management` with several useful references for further investigation.

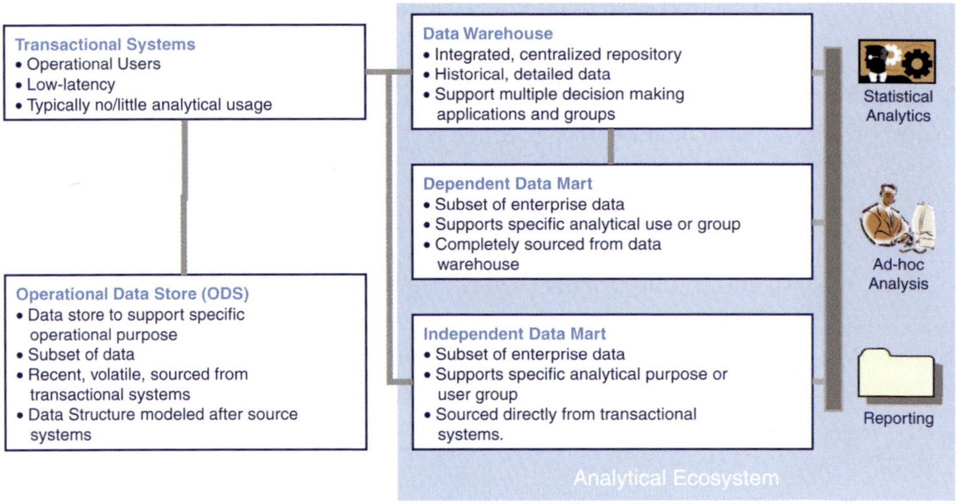

Figure 25.10 The analytical ecosystem.

access response time goals, but on occasion can also be required for security purposes (see Chapter 24).

- *Analytic tools*: The richness of the toolsets that enable users to efficiently analyze large amounts of data has increased over the past few years (see Chapter 22) and can be classified into the following types.
  - *Statistical analysis tools* are used to efficiently perform complex statistical calculations and are particularly popular for risk analysis.
  - *Ad hoc analytics tools* enable business users to interactively manipulate and analyze data to gain insights.
  - *Reporting tools* deliver data aggregated and formatted in pre-defined ways on-demand or on schedule.

Most FRATs inherit an analytic ecosystem at the inception of the program, and one of their main tasks is to determine the extent to which it can be leveraged to achieve the program's goals. Reusing components in an existing ecosystem can be a challenging for several reasons.

- Dependent and independent data-marts manipulate data for their own purposes. While they may contain data that are tantalizingly close to what is required to achieve the needs of the program (say, containing 90% of the required data elements) there may be great difficulty in closing the gap due to transformations that may be inconsistent with program needs.
- Data quality is a persistent problem. Data analytics for functions such as marketing can usually tolerate somewhat relaxed quality standards compared to those

for risk and finance. Improving the data quality of an existing data-store can be notoriously difficult (see Chapter 21).

- Data-stores within a firm can often differ in the interpretation of the meaning of a data element. The resulting unpredictability caused by metadata (see Chapter 23) differences can cause havoc to project timelines.

Given these challenges, it can pay to spend time up-front assessing the goodness-of-fit of candidate data-stores from these perspectives. These assessments might be best carried out in the information architecture work-stream of the detailed analysis phase (see Table 25.1). Depending on its goal and scope, the program may be able to use multiple sources of data for different domains of data or it may require a single consolidated data source. For example, a risk analytics program may be able to use independent data-stores for different kinds of retail products (say one data-store for credit cards, and another for mortgages), while a regulatory risk reporting initiative will need to use a single source of well-attested data. The work of upgrading the analytic ecosystem will therefore involve either upgrading an existing data-store (i.e., improving its data quality and extending the scope of its data elements), or creating a whole new data-store feeding in data from a range of existing systems like front office sources as well as other existing data-stores. While each of these activities is usually large enough to merit being treated as an independent project and can leverage the same six phases in the framework shown in Table 25.1, the work done in each project will differ considerably. Some examples of such workstreams include the following.

- *Data quality analysis* to assess the quality of data and identify data quality issues that could impact its use, such as integration with data for enterprise risk reporting purposes.
- *Data modeling* to ensure that the required data elements are documented and properly structured in physical data-stores to enable analytical usage.
- *Data integration* transforms and loads data into analytic ecosystem data-stores, often using software suited to this task, called ETL (Extract-Transform-Load) tools.
- The *business intelligence* workstream designs, builds, and unit tests software required to access and analyze information from data-stores in the analytical ecosystem.
- The *metadata management* workstream implements a metadata strategy which may include implementing a new metadata tool if the firm does not have one and feeding the tool of choice with metadata from various sources like data integration and business intelligence tools.
- The *data governance* workstream implements data governance and stewardship processes and policies across the organization (see Chapter 27).

Whether a data store is remediated or newly built from scratch will determine which workstreams will be part of the project. If an existing data-store already has all the data elements required with the right level of quality (a rare but not implausible event) only the business intelligence workstream will be needed. In contrast, when previously unavailable data elements are required to meet new regulatory requirements, a much more elaborate project is required which would include data integration, modeling, quality analysis and business intelligence.

Testing

There are many causes for errors in applications ranging from improperly specified user requirements and unexpected issues with data to developer error. Applications should therefore be thoroughly tested before deploying into production. The first step in testing is to create a predefined suite of tests that will determine whether the system is ready for deployment. Business use-cases defined as part of requirements are usually the most appropriate basis for testing. Test-suite creation should begin early in the project, usually after requirements are completed. In many systems, the scale of required testing is large enough to require automation. There are several automated testing tools available, though many projects have specialized requirements that require in-house developed testing tools. Test scripts should be developed for these tools based on the suite of tests that have been developed. The level of activity required for testing makes it a significant workstream in its own right so that is often run as a parallel track that starts early in the project and runs through its duration.

Testing analytic data-stores presents very different challenges from testing applications, mostly because the focus is on the data rather than on application functionality. There are typically two qualities of a system that need to be tested. Tests that confirm the system's ability to meet the functionality that is required by users are referred to as "functional" tests. Functional tests for an analytic data store usually address the following areas.

- *Data transformation*: Ensures that data have been properly transformed from source systems into the analytic data store.
- *Data quality*: Ensures that the data have integrity. These tests should also check that the data integration process properly detects and rejects poor quality data.
- *Data coverage*: For example, tests may check representative transactions in a range of user accounts to ensure that transactions for all products are in the data-store as expected.
- *Data access*: These tests verify that the routines to access data from the data-store produce the right data as expected.

The first three of these can be considered data storage tests. They should be completed prior to data access testing since in the absence of properly tested data it is impossible to say whether errors in that process are a result of improper access routines or underlying problems with the stored data itself.

The system should also be able to meet non-functional standards. It should consistently respond to queries within time required by users. It should also be able to protect data from loss in disaster situations. Finally, it should be able to scale with usage and the growth of the firm's business. Tests to confirm these qualities are called "non-functional" tests. Non-functional testing should be done only after all functionality testing is complete to help ensure that changes to the software to fix functional defects do not introduce changes to non-functional behavior as well.

The sequential nature of these tests causes test execution to be a significant component of an analytics project, often taking up to 20–30% of the time. This fact is sometimes overlooked when developing a project plan, resulting in rushed test-execution and missed timelines. The cost of repairing software defects can escalate dramatically through the project, so it is important to address issues as soon as possible. Testing is therefore organized into three successive phases as shown below, where defects can be detected and addressed at the earliest possible opportunity.

1. *Unit testing*: Large systems are developed in components that are small enough for individual developers to complete based on given detailed requirements. Individual developers use unit testing to ensure that the developed software meets these specifications. Unit testing can only fix component level defects and is typically done as an integral part of the development process itself.
2. *System Integration Testing (SIT)*: Once the individual components have been developed, they need to be integrated and tested as a whole. Since they have been unit tested by this point, SIT is designed to shake out issues relating to interaction between components. The goal is to ensure that the system works as intended from a technical perspective based on the project team's understanding of specifications.
3. *User-Acceptance Testing (UAT)*: Once the system meets specifications via SIT, users of the system will be invited to participate in a formal validation process to ensure that the system meets their requirements. This may seem redundant, but requirements are sometimes misinterpreted or incomplete. UAT therefore is an essential step before deploying the system to users. The leading practice is for users to develop their own set of tests in parallel so that the system is subjected to two congruent but different sets of tests.

25.4.5 Implementation

Once testing has been completed, the system is ready for deployment. Since the system will usually need to integrate into existing processes it is a challenge to do this for the entire user population in one fell swoop. Regardless of the rigor of UAT, some errors may appear once the system has been deployed. The new system may come with modified business processes that are unfamiliar to users. For these and other reasons, a phased rollout of a system is the preferred alternative, ideally starting with a small population of non-critical users and increasing the number over time until the entire user-population have been migrated. In this way, system errors can be shaken out in early releases when they will have the least impact. The impact can be further mitigated if the old "legacy" application can be kept running in parallel so that users can be comfortable that they could revert to the legacy application in case of some unforeseen error. These best-case situations are often encountered in simple reporting applications. The result of the low impact of these applications is that one can often accelerate production rollouts by combining the phased, parallel rollout with the UAT phase discussed in the previous section. A proper timetable can be important to help ensure full migration and shutdown of the legacy system to ensure that the parallel run does not continue for an unreasonable period.

Unfortunately systems often do not have the luxury of parallel rollout, or even the possibility of a phased implementation. Such "big-bang" deployments can be extremely high risk and require enormous preparation. The user testing phase should be extremely comprehensive and thorough, requiring innovation in testing that goes beyond the regular process described in Section 25.4.4. For example, data flows could be duplicated into new applications from the legacy application to help ensure as realistic a user-test as possible. It may also be necessary to stage mock go-live sessions to shake out any issues during the initial stages of deployment. Finally, it is imperative that the entire development and operational support team be at hand during actual go-live to rapidly respond to problems.

The first few days and weeks after go-live are critical regardless of the type of implementation. The system should be closely monitored during this period to help ensure that functional and non-functional issues are promptly detected, analyzed and remediated. It is common, at least for large system deployments, for a command-center to be set up during the first weeks after go-live. This room is usually staffed by a mixed staff including system developers, business analysts and operational staff who can rapidly come together to triage any reported problem. For globally deployed systems, it is not uncommon for these centers to be staffed for the full 24-hour day. The successful implementation should have as few errors as possible before going live, and rapidly respond to any that surface during implementation. This rigor can be critical in boosting user-confidence, resulting in stronger and faster adoption to help ensure the long-term success of the system.

25.4.6 Continuous deployment and improvement

Successfully implementing a system is a significant milestone. It usually requires frenetic effort as the project closes in on the targeted end-date. In fact, so much time and effort is usually expended during the last stages of a project that the organization often experiences a sense of collective fatigue at the end of the project. However, firms often miss lucrative opportunities by failing to capitalize on the implementation for additional benefits. This can be especially true of analytic systems, where the time, effort and investment made in integrating and cleansing data can be easily leveraged for new and interesting applications. Environments built for finance and risk analytics should integrate data from across the firm that are usually compatible to the needs of business applications. For example, customer data for risk analytics are largely reusable for revenue-generating customer-facing uses such as campaign management. Additionally, risk metrics can also be used to improve the quality of customer interactions through differentiated pricing that uses risk-based profitability. While there are many intuitively good reasons for leveraging the work already completed, large organizations operate on financial control principles such as budgeting and business cases, and any effort should be aligned to these processes. Therefore the concept of extending the analytic environment should be translated to specific projects against which return on investment can be quantified and budgets allocated.

To understand the possible return on investment, one should understand how businesses can leverage the analytic environment. Consider Figure 25.11[14] which describes the value chain for transforming data into decisions. In this construction, data are acquired (Step 1, denoted as Acquisition in the figure) from various sources such as trading systems and the branch network. Additional data can also be created in middle and back office activities. These data need to be refined and aggregated into usable information (Step 2, Refinement and Aggregation). However, this represents only the early stages of the value-creation process – once it has been refined information needs to be transformed into insights by analytic users and applications (Step 3, Hypothesis and Analysis). Finally, these insights need to be translated into action by decision-makers, releasing the latent value in data (Step 4, Action).

Many large financial firms collect sufficient quantities of data in a structured form. While there are exceptions (for example, data relating to specific terms and conditions on structured products may not yet be properly captured in a structured database format), data capture is usually not where many financial firms experience significant challenges with their analytic environments. Rather, the information creation stage is where many firms have difficulty, especially those that have not created strong factory-like environments. The result is that analytic users ("Smart People" in Figure 25.11) usually spend a significant percentage of their time (up to 80–90% by some estimates) in transforming data into a form that they can use. It is here that analytic projects can offer significant leverage, since they consolidate

[14] Adapted from Berry (2010).

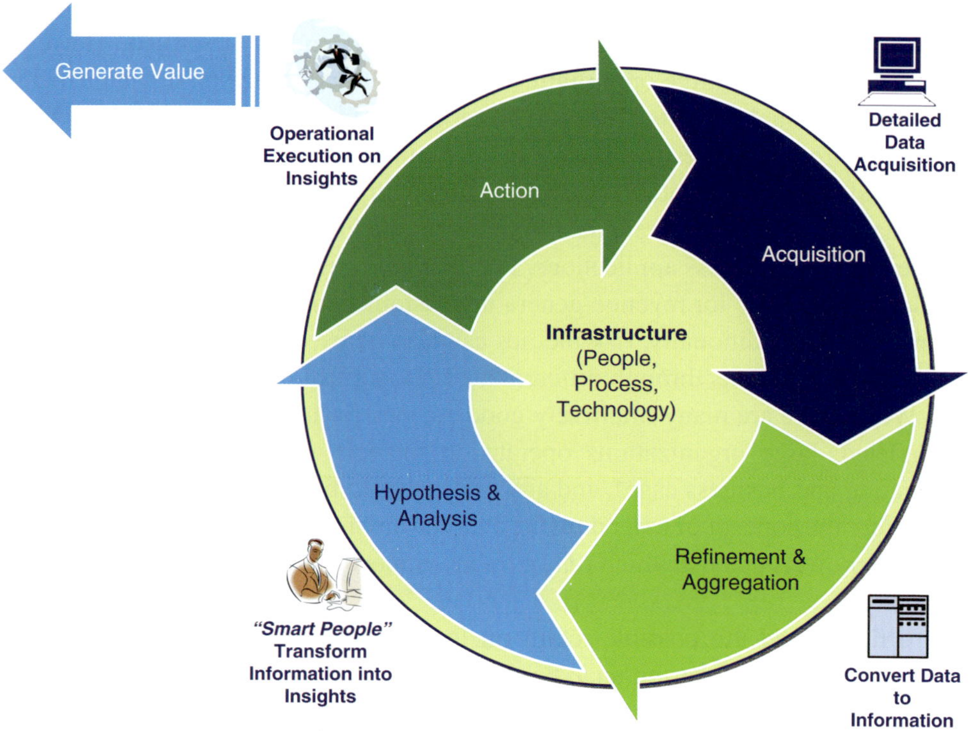

Figure 25.11 Illustrative: translating data to value.

firm-wide data into useful information. This consolidated data can then be reused by many analytics groups to generate new and potentially profitable insights.

There are usually so many projects that can leverage an analytic environment that it is necessary to prioritize among them. Several criteria can determine which projects take precedence, such as regulatory or business drivers, or existing budgets. An additional criterion that should be considered is the ability to generate data reuse as it can deliver cost and time savings to projects. The basic concept is to arrange projects in priority so that each project only needs to add an incremental amount of data to the existing analytic environment to achieve its goals. An example is shown in Figure 25.12.

The goal is to generate a list of projects that can be prioritized in such a way that each project needs to add only an incremental amount of data to the existing dataset to achieve its goals as shown in the upper left-hand diagram. If this can be achieved, the investment for each new project can be reduced as they progressively fill in any remaining gaps in data. This creates the "value-wedge" where the value added by each project increases while investment tapers off. The lower right-hand illustration in Figure 25.12 shows an example of the analysis required to achieve this. By laying

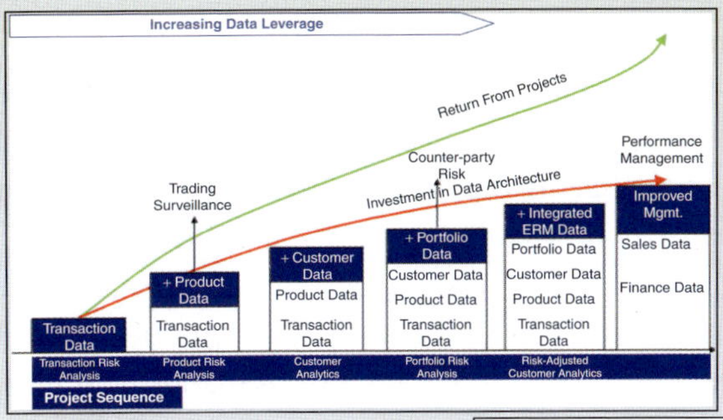

	Sales Profitability Analysis	Risk Management (Project # 1)	Risk Management (Project # 2)	Collateral Management	Pricing	Financial Management	Compliance (Project # 1)	Compliance (Project # 2)
Sales Profitability Analysis	100%	80%	69%	24%	41%	66%	5%	24%
Risk Management (Project # 1)	9%	100%	87%	28%	64%	58%	22%	32%
Risk Management (Project # 2)	6%	57%	100%	19%	83%	35%	17%	28%
Collateral Management	14%	100%	100%	100%	78%	100%	59%	100%
Pricing	4%	50%	100%	17%	100%	27%	17%	18%
Financial Management	14%	94%	90%	48%	58%	100%	35%	63%
Compliance (Project # 1)	0%	76%	88%	58%	66%	73%	100%	79%
Compliance (Project # 2)	7%	75%	93%	63%	80%	68%	50%	100%

Figure 25.12 Prioritizing projects by data reuse (illustrative example).

out projects in different functional areas in sequence, an overlap percentage can be calculated. In the figure, the columns represent projects which, if completed, create data that can be reused by other projects represented in the rows. For example, the Risk Management Project 1 creates data that covers 94% of the data required by the Financial Management project. Introducing this analysis into the project-prioritization process allows for significant cost and time savings. Using this criterion, and if other factors permit the ordering, cost savings can be achieved if Risk Management Project 1 is implemented ahead of Compliance Project 1.

25.5 Related topics

25.5.1 Iterative development

The waterfall model of project management introduced at the beginning of Section 25.4 described it as a flawed but useful model in software development. In fact, projects rarely if ever use the pure waterfall model as the exclusive management technique. Rather, large system projects use the waterfall model as a framework for organizing a number of smaller projects, each of which may in turn use other software development techniques. Many of these techniques seek to address the inflexibility of the waterfall model by promoting software development in rapid iterations. These approaches are collectively known as *incremental and iterative development (IID)*.[15] The short-comings of the waterfall model have been known since the very beginning of software development.[16] IID has consequently been employed from the very beginnings of software development.[17] However, the early instances of the use of IID were sporadic. In the 1990s, IID developed into an organized approach to software construction called *agile development*, which is today the preferred technique for software projects. Agile development includes such popular techniques as *scrum* and *extreme programming*. This section will only offer an overview but an interested reader can find details on each approach in the extensive literature on the subject.[18]

Many agile development methods promote development, teamwork, collaboration, and process adaptability throughout the lifecycle of the project. The overall project is executed over several iterations, each of which involves minimal planning and short time frames that typically last from one to four weeks. Rather than the time frame being determined by tasks as in a traditional waterfall, the tasks identified for a given iteration are determined by the fixed time (called a time-box) available. Each iteration is rigorous and structured and involves a team working through a full software development cycle, including planning, requirements analysis, design, coding, unit testing and acceptance testing. At the end of each iteration, progress is reviewed. Priorities are set for the next iteration based on business value perceived at the beginning of the iteration.

As such, each iteration can be considered a mini-waterfall that results in a working product demonstrated to stakeholders. The short project timelines minimize risk of not meeting stakeholder expectations and allow the project to adapt to changes quickly. Agile teams are usually small (less than 10 members), cross-functional and self-organizing. Team members take responsibility for individual tasks needed to deliver the functionality that each iteration requires. Teams usually include a representative of the stakeholder community whose responsibility is to help address user-requirement questions during each iteration.

Agile methods emphasize frequent, informal communication among team members. Written documents are discouraged in favor of working software unless the team is geographically dispersed. Most agile implementations use a routine, formal but brief daily face-to-face communication among team members such as a beginning-of-day standup[19] meeting that may last 15–30 minutes. Team members

[15] For a chronology of IID approaches, see Larman and Basili (2003).

[16] The following section has been adapted from "Agile Software Development" on Wikipedia, `http://en.wikipedia.org/wiki/Agile_software_development`

[17] The first reference to IID was made as far back as 1957 (Larman and Basili, 2003).

[18] See, for example, Larman (2004) and Highsmith (2009).

[19] The discomfort of having to stand up during the meeting ensures that these meetings are shorter, and more to the point, offers participants a chance to stretch their legs.

report to each other the previous day's achievements, that day's plans and any roadblocks, allowing problems to be exposed quickly. Most agile teams work in a single open office which facilitates communication and collaboration. Larger development efforts across multiple locations can be achieved by creating multiple teams working on different parts of an effort. Even when teams work in different locations, they maintain daily contact through communication media like videoconferencing, phone calls and emails.

Agile methods are not universally applicable. Rather, they usually work best in smaller projects with groups of experts where requirements are expected to change often. Agile techniques are also not well suited for critical projects. Therefore large FRATs are not well served by using purely agile methods of development any more than they can use pure waterfall techniques. Rather, they should use a blended approach with a structured, plan-driven process for the overall program based on the waterfall method complemented by subprojects that use agile techniques. This structure is suited to situations where the overall goal (e.g., the need to meet the needs of a specific regulation) is usually well specified but individual details are often less well defined and may change frequently during the course of the project.

Iterative development methods can have a significant impact on the productivity of projects and help mitigate project risks in large programs. However, they are not a silver bullet, nor should they be used to excuse lack of structure and project discipline. Rather, they require, if anything, more discipline on the part of the organization if they are to be successful. Also, all agile techniques are not equivalent. Each circumstance should be carefully analyzed to determine the agile approach which would be most appropriate for it. But used properly, agile development can significantly benefit an analytic program by reducing the risk of cost or time overruns.

25.5.2 Project resourcing

FRATs are by nature large and disruptive. They often involve large budgets, tight timelines and usually require a multitalented pool of resources to successfully achieve their goals. Firms can very rarely staff these programs purely with talent drawn from their existing resources. Firms have grown leaner and more productive over time by reducing redundancies in their staff, which gives business units less flexibility in allocating resources to lengthy projects of this nature. Firms increasingly distinguish between "Run-The-Bank" (RTB) operational functions and "Change-The-Bank" (CTB) projects which encompass regulatory and risk programs as well as other business initiatives such as improved customer marketing.

RTB resources require special skill-sets that make them valuable in their roles as revenue generators or deep knowledge of processes. They also typically favor

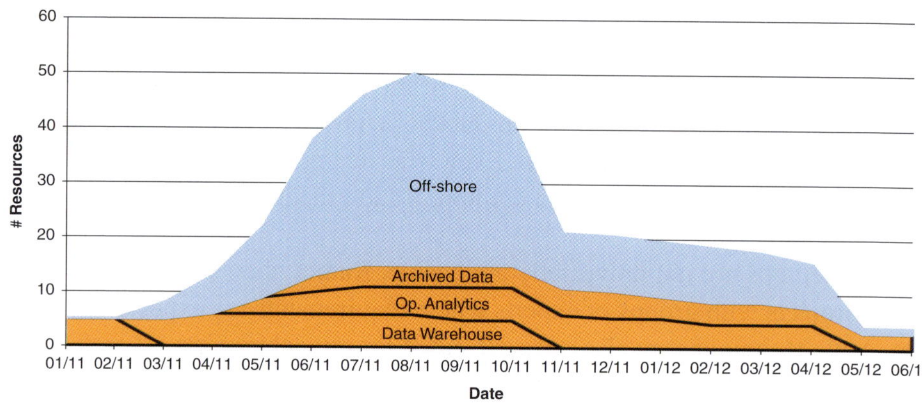

Figure 25.13 Example resourcing plan for an analytic project.

stability in their responsibilities over constant change. In contrast, CTB activities such as FRATs involve constant change and significant uncertainty. They are fast paced, typically with critical project deadlines. The most successful resources for such projects therefore are people with personalities that thrive on change. Aside from personality, these programs also require resources with different skill-sets from RTB functions. CTB projects need to be staffed with resources who bring a variety of skill-sets including knowledge of the business and technology as well as project management ability. Several of these skill-sets are unlikely to be found within RTB functions. For example, since the core business of banking requires extreme focus and care on technology operations, they are likely to staff these areas with highly competent professionals. In contrast, since software development is not a core business of banks, top developers are less likely to be found in RTB functions of banks. In addition, FRATs require a significant build-up of resources (see Figure 25.13) during the project followed by an equally dramatic reduction at the end of the program. Staffing these programs entirely with full-time employees of the bank is likely to be sub-optimal especially over the long term.

Therefore firms usually staff programs with a mix of internal and external resources. Internal RTB resources are critical to the program due to their knowledge of existing systems and processes. They are usually critical to the business needs and detailed analysis phases of a program as described in Table 25.1. However, they form only a small part of the project (often no more than 10–20%). The majority of the resources are drawn from a number of external organizations that play the dual role of flexibly augmenting the staff required for the project as well as bringing to bear specific knowledge and skill-sets that are required for successful completion of the program. The nature of the specific knowledge can range from specialized business information (e.g., detailed knowledge of requirements and interpretation of a regulatory rule) to a niche technical skill such as knowledge of how to configure a newly installed COTS application. Many such programs usually involve many different organizations working together. Firms therefore have to choose how to manage such a diverse set of resources – the most common models are hiring a single general contractor to manage the program or creating an office staffed by internal resources that would in effect act as the integration team bringing all resources to bear on the program.

The other aspect of resourcing is the physical location of resources. Driven by improvements in telecommunication technology, firms in many industries have aggressively been outsourcing work to off-shore locations such as India and China. While the initial impetus for sending work off-shore was lower costs, the practices have matured to a point where leading companies now use off-shoring as a way to leverage a global pool of talent.[20] Financial firms have also followed this trend and established large off-shore "captive" operations of their own. CTB programs such as risk and finance analytic operations can leverage these capabilities as well, usually in software development activities. The situation is further complicated in most large programs; due to the global nature of delivery and the involvement of several internal groups and external service provider firms there are usually a number of off-shore teams working together. The success of such initiatives therefore depends on the kinds of work off-shored, the experience of the teams executing the work and the project-management discipline that is applied to the process. The other challenge to be addressed in off-shoring analytic program work is the fluidity of change.

Typical analytic programs use an on-site/off-shore model for project delivery which depends on on-site leads who can direct work in off-shore centers. On-site leaders work closely with the various teams within the firm including users, system subject matter experts and project managers. These leaders also need to have a high degree of familiarity with the setup of off-shore teams, which is accomplished by either moving a few off-shore team members on-site to act as leaders or having on-site team members travel off-shore periodically to build the relationship. In addition to having on-site leaders, successful efforts also have team leads off-shore who can act as a single point of contact and understanding with on-site resources, and direct the activity of individual team members off-shore. On-site and off-shore team leads usually work together to decompose project tasks. Most of the involvement of off-shore teams is usually found in Phases 4 and 5 of the program implementation model described in Table 25.1.

Good results are obtained when teams can execute repetitive sets of tasks such as integrating data from source systems. Large projects may integrate more than 100

[20] Offshoring Research Network, Service Provider Survey Report (2007).

such systems, so both cost and performance efficiency can be gained from offshore delivery. The most successful efforts start small (say with 1–2 systems) and to develop a process that can then be scaled up to address an increasing number of systems simultaneously. The initial efforts also usually over-weight the on-site component of the team (with an on-site/off-shore ratio of, say, 50%) to help ensure that the process is properly defined and documented. Once this is done the ratio can go down to a more scalable and cost-effective number such as 30% or less. Off-shore teams can also be effectively leveraged for testing activities in the same manner.

In FRATs, one important concern when addressing off-shore delivery is the issue of security and privacy. There are many constraints in this regard, especially when dealing with personally identifiable data such as national identification numbers (e.g., Social Security Numbers in the USA). These considerations should be addressed ahead of time since addressing security policies in large financial firms is usually time consuming. They may involve detailed security reviews of off-shore premises, especially when external service providers are involved. Effective off-shoring may require creating obfuscated datasets where certain sensitive data elements (such as the SSN) are scrambled to make them less sensitive. Finally, there may be requirements for physical separation of data among systems which can have impacts on infrastructure setup time as well as project budgets.

For these reasons, global-sourced delivery should be an integral part of initial program design rather than considered an afterthought. Doing so will result in significant project acceleration and scale that can drive down both the cost and implementation risk of a large risk and finance analytic program.

25.6 Conclusion

This chapter has attempted to provide an introduction to project management methods that will assist the manager looking to implement a large risk management, finance transformation or regulatory project. Experience demonstrates again and again that without laying the groundwork for a large-scale project, failing to establish the framework for communication and management of stakeholders, or anticipating and managing project risks, a project has a greater chance of not meeting its objectives, in terms of scope, timing and acceptance. Similarly, understanding and following the fundamentals of project management methodology, as described in Section 25.4, goes a long way to enabling a successful design phase, system development, testing and, ultimately, implementation that meets business needs. Understanding the nuances of a data project can help the program or project manager not only plan accordingly, but be prepared to fully leverage the advantages of the data implementation for the benefit of the organization. While the basic tenets for managing a system implementation are straightforward, the adaptation of those guidelines are often challenges for the program manager and project managers and allowing for flexibility in application of those guidelines is often just as important as following the tenets.

Bibliography

Arrata, P., A. Despierre and G. Kumra, 2007, Building an effective change agent team, *McKinsey Quarterly*, November, www.mckinseyquarterly.com/How_change_agents_promote_operational_change_2049

Baekdal, T., K. L. Hansen, L. Todbjerg and H. Mikkelsen, 2006, *Change Management Handbook*, www.baekdal.com/downloads/changemanagement-en.pdf (downloaded April 9, 2012).

Beitler, M. A., 2006, *Strategic Organizational Change*, Second edition, Practitioner Press International.

Berry, A. B., 2010, *Information as an Asset: Transforming Data into Profit*, Teradata White Paper.

Brooks, F., 1995, *The Mythical Man-Month: Essays on Software Engineering,* Anniversary Edition, Addison-Wesley Professional.

Highsmith, J., 2009, *Agile Project Management: Creative Innovative Products*, Second edition, Addison-Wesley.

IT Cortex, Failure Rate, www.it-cortex.com/Stat_Failure_Rate.htm (downloaded April 9, 2012).

Kendrick, Tom, 2009, *Identifying and Managing Project Risk: Essential Tools for Failure-Proofing Your Project*, AMACOM.

Kotter, John P., 1996, *Leading Change*, Harvard Business Review Press.

Krigsman, M., 2009, *20* cynical project management tips, March 1, www.techrepublic.com/blog/tech-manager/20-cynical-project-management-tips/881

Larman, C., 2003, *Agile and Iterative Development: A Manager's Guide*, Addison-Wesley.

Larman, C. and V. R. Basili, 2003, *Iterative and Incremental Development: A Brief History*, IEEE Computer.

McKinsey & Company, 2006, Organizing for successful change management: A McKinsey Global Survey, *McKinsey Quarterly*, July, www.mckinseyquarterly.com/Organizing_for_successful_change_management_A_McKinsey_Global_Survey_1809

Nagrecha, S., 2002, An Introduction to Earned Value Analysis, March 16, www.pmiglc.org/COMM/Articles/0410_nagrecha_eva-3.pdf (downloaded April, 12, 2012).

Object Management Group, Business Architecture Overview, Business Architecture Special Interest Group, http://bawg.omg.org/business_architecture_overview.htm (downloaded November 19, 2011).

Offshoring Research Network, 2007, Offshoring 2.0: Contracting Knowledge and Innovation to Expand Global Capabilities, Service Provider Survey Report, https://offshoring.fuqua.duke.edu/pdfs/DukeServiceProviderReport_web.pdf (downloaded April 9, 2012).

PM Toolbox, Project Earned Value Analysis Template, www.pmtoolbox.com/project-management-templates/project-earned-value-analysis-template.html (downloaded April 9, 2012).

Project Management Institute, 2008, *A Guide to the Project Management Body of Knowledge: (PMBOK Guide)*, Fourth edition.

Rad, P. F., 2001, *Project Estimating and Cost Management* (Project Management Essential Library), Management Concepts Inc.

Royce, W., 1970, Managing the development of large software systems, *Proceedings of IEEE WESCON 26, August 1970*, 1–9, www.cs.umd.edu/class/spring2003/cmsc838p/Process/waterfall.pdf (downloaded November 12, 2011).

Shim, Jae K., 2011, *Project Management: A Financial Perspective*, Global Professional Publishing.

United States Navy Mathematical Computing Advisory Panel (29 June 1956), Symposium on advanced programming methods for digital computers, Washington, DC, Office of Naval Research, Dept. of the Navy, OCLC 10794738, http://en.wikipedia.org/wiki/Agile_software_development

VanHoose, D., 2007, Assessing Banks' Cost of Complying with Basel II, Networks Financial Institute, Indiana State University, September, www.networksfinancialinstitute.org/Lists/Publication%20Library/Attachments/83/2007-PB-10_VanHoose.pdf (downloaded October 9, 2011).

Wikipedia, Project management, http://en.wikipedia.org/wiki/Project_management (downloaded April 9, 2012).

Wikipedia, Waterfall model, http://en.wikipedia.org/wiki/Waterfall_model#cite_note-0 (downloaded November 12, 2011).

Wise, C., 2011, Achieving High Performance: The Value of Benchmarking, Accenture White Paper, www.accenture.com/us-en/landing-pages/management-consulting/benchmarking/Documents/pdf/Accenture_Achieving_High_Performance_The_Value_of_Benchmarking.pdf (downloaded April 14, 2012).

26

Operations management

Jim Blair

26.1 Introduction

Success or failure of a financial and risk analytics ecosystem not only depends on the derived business value of that system, but also relies heavily on how the financial and risk analytics ecosystem and data stores are managed behind the scenes. Support, stability and overall performance of the complete financial and risk analytics environment are key factors to success. A financial and risk analytics ecosystem, henceforth abbreviated to FRA ecosystem, includes the complete set of tools and technologies needed for financial analysis. This includes analysis tools on the desktop and in intermediate servers, as well as all the underlying data stores, data-marts and data-warehouses. Support of this FRA ecosystem is typically called operations management. The following sections will discuss in detail the best practices of operations management for an FRA ecosystem.

26.1.1 The FRA ecosystem

The FRA ecosystem comprises many different components, including source systems and business intelligence (BI) tools. In Figure 26.1, you will see "source systems" on the left. These are the Operational systems that run the business and are inputs to the FRA ecosystem. Operational systems are often called transaction systems because they process the transactions of the business. Examples of Operational or Transaction systems are Credit Card processing, General Ledger systems, and Accounts Payable systems.

Also in Figure 26.1, you will see on the right side of the diagram both "Desktop" and "BI Tools." The desktop is just that – either a laptop or computer on a desk where the analyst conducts his work. BI tools are software and hardware. BI tool

Handbook of Financial Data and Risk Information, ed. Margarita S. Brose, Mark D. Flood, Dilip Krishna and Bill Nichols. Published by Cambridge University Press.

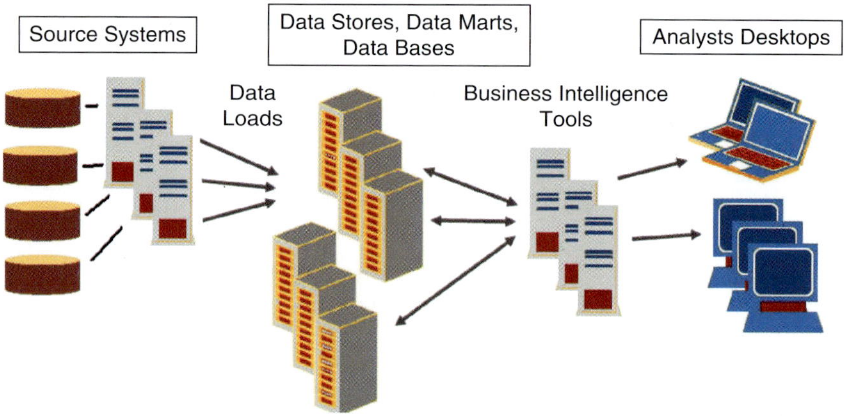

Figure 26.1 Risk analytics ecosystem.

components (like Excel) on the desktop allow the analyst to run reports. There are also components that perform the processing on another computer in the data center. Examples of BI tools are Business Objects and Crystal Reports from SAP, Cognos from IBM, Microstrategy and SAS. These component BI tools exist between the business analyst and the databases and are often transparent to the analyst other than the "screen" on their monitor.

The data-warehouse is one major component of the FRA ecosystem. Different people have different definitions for a data-warehouse. The most popular definition comes from Bill Inmon (Inmon, 1995) ("A data warehouse is a subject-oriented, integrated, time-variant and non-volatile collection of data in support of management's decision making process.") The various terms are explained below in further detail.

- *Subject-oriented*: A data-warehouse can be used to analyze a particular subject area. For example, "risk management" can be a particular subject.
- *Integrated*: A data-warehouse integrates data from multiple data sources. For example, source A and source B may have different ways of identifying a product, but in a data-warehouse, there will be only a single way of identifying a product.
- *Time-variant*: Historical data are kept in a data-warehouse. For example, one can retrieve data from 3 months, 6 months, 12 months, or even older data from a data-warehouse. This contrasts with a transactions system, where often only the most recent data are kept. For example, a transaction system may hold the most recent address of a customer, where a data-warehouse can hold all addresses associated with a customer.
- *Non-volatile*: Once data are in the data-warehouse, they will not change. So, historical data in a data warehouse should never be altered.

A data-warehouse deals with multiple subject areas and is typically implemented and controlled by a central organizational unit such as the corporate Information Technology (IT) group. Often, it is called a central or enterprise data warehouse. Typically, a data warehouse assembles data from multiple source systems. However, the data warehouse is only one major component of the FRA ecosystem. A data mart is a simple form of a data warehouse that is focused on a single subject (or functional area), such as sales, finance, or marketing. Data marts are often built and controlled by a single department within an organization. Given their single-subject focus, data marts usually draw data from only a few sources. The sources could be internal operational systems, a central data warehouse, or external data. Unlike a data warehouse, a data mart usually addresses only a small number of subject areas and is often controlled by the department whose information needs it serves (such as finance or market risk management).

The FRA ecosystem enables two functions.

1. *Decision support*: Supports business and organizational decision-making activities, for example, determining the sales volumes last month for each department. A decision support query helps decision makers compile useful information from raw data, documents, personal knowledge, and/or business models to identify and solve problems and make decisions.
2. *Deep analytics*: Supports deeper enquiries such as trend analysis and "what if" analysis, for example, using data to compare department 12 to all other departments over the last two years in terms of sales volume for the three major product groups.

The FRA ecosystem is all the necessary hardware, software, and processes that empower both decision support and deep analytics.There are three basic phases of every corporate initiative, no matter how large or how small: plan, build and run. Operations management is about the last of these – the "run" phase of the FRA ecosystem. The FRA ecosystem will last as long as the data are still in use. That could be a lifetime and could be used every day of every week through the year. Operations management is a $7 \times 24 \times 365$[1] effort. Operations management is the enduring force that keeps all components of the FRA ecosystem and data available and high performing. When you see a duck gliding across a pond, it appears effortless. However, we know there is a lot of paddling going on under the surface. Think of operations management as the "paddling" that goes on unseen in the back office that keeps the FRA ecosystem gliding along smoothly.

[1] This is a common term meaning that the system is required seven days a week, 24 hours a day, 365 days a year. This is a high bar to meet, but one that is not uncommon in large, global financial institutions.

26.1.2 Four major pillars on a solid foundation

In managing the FRA ecosystem there are four major "pillars" that sit on top of a general operations foundation to support financial analytics. In the following sections, best practices of these four major pillars are described, and how they apply to operation management of a FRA ecosystem.

1. *Data delivery and data access*: This pillar is about "Who" has access to "What" data and "When" – "What" data are loaded or added to the FRA ecosystem, such as sales, finance, marketing, and/or customer data, and "When" the data are loaded to the FRA ecosystem, i.e., hourly, daily, weekly or monthly. Once data are loaded, access permissions to read and or update the data must be granted to specific people.
2. *Systems availability*: Just having data and access to it is only part of the game. The FRA ecosystem must be up and available whenever needed by analysts. System availability includes every component from the desktop where the analyst is working to the actual data stores that contain the data. It also includes everything in between such as analytical tools used by the analyst to evaluate the data.
3. *Systems performance*: Once data are loaded and available, it is critical to provide the highest level of performance, measured in terms of response time. User response time is the time it takes from when the analyst hits the enter key on his computer to when he gets a response back. This time obviously includes all parts of the FRA ecosystem (see Figure 26.2). The response time of various reports and queries, applications and jobs running in the FRA ecosystem (some of them interfacing with users and others running in the background) must all be acceptable to the user community. If the response time of the FRA ecosystem is unsatisfactory, users will go elsewhere to satisfy their analytic needs.
4. *Integration of new projects*: Managing a stable environment with little change is comparatively easy. Managing and controlling change is critical to the smooth running of an FRA ecosystem. You must consider how and when to add new data sources and new projects to the FRA ecosystem. Smooth integration will protect the existing integrity of the FRA ecosystem. Standards and conformity insure the integrity of the existing components.

26.2 Pillar 1: Data delivery and data access

Data are at the core of the FRA ecosystem and are the reason for its very existence. The first pillar of operations management is about making data available to users. Data in a FRA ecosystem most often will reside in a variety of data stores such as data marts, databases, and data warehouses. Requirements for loading data, granting access to the data and providing data security are the major components of data delivery and data access.

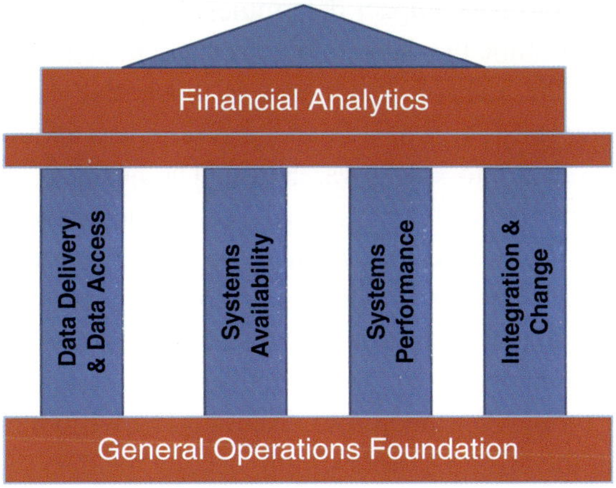

Figure 26.2 The four pillars of operations management.

26.2.1 Requirements for loading data

Loading data is like putting books in the library. What data to load and when to load them are determined jointly by the business unit who generates the data and the users who need to use it. It is critical that all data should have a defined owner. Data owners should be identified by the Governance committee or council (see Chapter 27). The data owner must be someone in the appropriate business area who knows and understands the data, since it is the data owner who determines the business value of the data and business impact of the data. Without clear data ownership, data requirements are hard to define.

The business units working in conjunction with IT must meet and agree on what data should be loaded and when they should be loaded. The agreement must be clear and documented via a Service Level Agreement (SLA), which describes the level of service that will be provided by IT and agreed to by the business. An SLA should exist for loading data, and as such must be clearly and precisely defined and agreed upon. It is not enough for data owners to ask that they "get data daily." The details must be documented such as the time for delivery (by 8am, by end of day, before start of the day, etc.). Additionally, is every day treated the same, such as weekends and holidays? The term "daily" means many different things to many people. Therefore an SLA must be very clear, specific and concise.

Importantly, every SLA must also be driven by business value and business impact. Thus the specifics of an SLA are determined by the business, i.e., the data owner. IT needs to reach agreement on what is technically possible as to what data are made available and when they will be available. In addition, the business must have business justification as to their needs, since SLAs are about satisfying business "needs" not "wants." In this way, business impact determines the parameters of a data delivery SLA such as frequency, time the data are required and the data's latency.

An example from the author's experience is illustrative of these terms and the process of negotiating an SLA. A new financial system was about to be implemented and required loading data into a data warehouse. The business owner indicated that they needed the new data available to the analysts by 6 am daily. The IT project team looked at the upstream processes that created the data. After analysis, it was determined that the upstream processes typically finished at 4 am. There would be only two hours available to load the data after the upstream processes completed. Since the data load processes would itself run for 3–4 hours, it was impossible to meet the 6 am target data delivery time. After thoughtful discussion between the business and IT, it was determined that the business users simply needed the data more frequently than weekly and an agreement was reached between data owners, business users, and IT to have the data available by 6 am, but be one day behind. This made it possible to load Monday's data into the data warehouse Tuesday evening, so it was available to analysts by 6 am Wednesday. Data were loaded daily, but the users were always one day behind. This was completely satisfactory to the business owner, met the business needs and was achievable by the IT staff. In the above example, the SLA was written with a frequency of "daily by 6am," with a latency of "one day behind."

Every SLA should have a test period to determine whether the assumptions are correct, and determine whether the SLA is realistic and achievable. Without a test or trial period, an SLA is simply a gamble. It is better instead to try it out for a short period of time to really see if it is realistic, modifying the SLA metrics as necessary at the end of the test period. The final, achievable SLA will then be published after the successful trial period.

In addition to the frequency and latency parameters for SLAs, an operational target is always included in the SLA. The target should consider that problems inevitably come up. For a daily load, one might consider a 98% target for meeting SLAs. That would allow for one error every two months (such as the load happening at 9 am for a 6 am SLA) with no impact to the service level agreement. When implementing an SLA, be sure to set a trial period. The trial period will be used as a test period to validate whether the SLA is realistic or not.

An SLA should be a simple document that contains the following sections.

1. Service description – what you are trying to achieve
   (a) Metrics (include % to be achieved)
   (b) Frequency and latency
2. Assumptions and constraints
   (a) Upstream dependencies
   (b) Special situations, such as during month-end processing

3. The time periods when the SLA is in force, i.e. which hours of the day days of the week and or month is the SLA in force
4. Relevant contacts within IT and the business
   (a) Who to call about specifics of the SLA
   (b) Who to contact when the SLA is in jeopardy
5. Escalation process when the SLA is in jeopardy or missed
   (a) Method of notification such as direct call, voice mail, email, or central communication point such as a website
6. Trial period
   (a) If a daily SLA then test for 30 days
   (b) If it is a monthly SLA then test for three iterations
7. Reporting requirements
   (a) Who receives reports on this SLA
   (b) How often are reports produced
8. Signatures to show agreement
   (a) IT and the business owner(s)
9. Versioning for changes.

A sample SLA document is shown in Figure 26.3.

SLAs are relatively simple to create and execute once negotiated. There are six simple steps as follows.

1. Develop a first draft of an SLA for discussion (often called a "straw-man" or starting point) (authored by IT)
2. Negotiate specifics between IT and all business areas involved

- Description: Finance data from the AR system will be loaded in to the data warehouse on a daily basis 98% of the time
- Times for this SLA
  > Daily by 6am except on Company Holidays
- No user access to financial data will be allowed during the load
- Contacts for status and escalation
  > IT Operations, Mr. Tom Dixon 1244-6632
  > Finance Coordinator, Petre Andrews 1299-8830
- Trial Period is Sep 1, 2010 through Nov 1, 2010
- Reporting
  > Monthly to Director levels in Finance, Sales, and IT
- Signatures (at director level only)
  > ....
  > ....
- Version 1a, effective August 18, 2010

Figure 26.3 Sample SLA.

3. Implement the SLA for a trial period
4. Assess the validity and success rate and adjust as needed
5. Report on the SLA
6. Review the SLA annually and change accordingly.

When considering the timing of the data load, upstream dependencies are always a definite constraint. If the upstream data feed is late, will there be sufficient time to load the data and meet the SLA? Upstream dependencies should be defined and documented during the analysis of the SLA. Since there are most often historical data on the upstream processes, with a little bit of analysis one can determine with high accuracy when they typically complete. That will provide a good basis for discussion on possible parameters for an SLA between IT and business owners.

While business owners of data usually make decisions on SLAs based on business need or impact, it is not atypical for there to be no identifiable data owner, i.e., no one has ever stepped up to take "ownership" of the data. It is critical to have a business owner of data to make decisions about "needs versus wants." SLAs for data that have no owner are meaningless, since there is no real need for an SLA if there is no owner. Data without an owner become unmanaged and unreliable. Therefore, as a last resort, IT can step in and become the data owner. Even in that case, all the above concepts still apply.

Another consideration when defining the requirements for a data load is data quality. Only the business owner can determine what level of quality is required. Numerous questions, such as what follows, need to be answered by the data owner.

- Should all data be loaded even if a known value is wrong?
- Should the erroneous data be NOT loaded and handled by some error process?
- What is the error handling process?
- Will there be duplicates allowed?
- How will the data quality be monitored?
- Will the load simply have validation routines that identify data errors?
- How complex are these routines and do they cover every possible error that could happen?
- Is the validation only at a summary level
- Records in = records loaded.
- Dollars in = Dollars loaded.
- Even if the source system has bad data, do you load it regardless of its quality?

While data loading processes will handle most errors, additional checks and balances must be in place to handle potential outliers. The best practices for data quality include three levels of "firewalls."

1. A moderate level of checks and balances run during the actual load;
2. A second set of checks run after the loads, called *validation routines;* and
3. Data quality processes to monitor, profile, analyze and correct data errors that have a significant business impact as determined by the data owner.

26.2.2 User access during data loading

The *load window* is the time that the load process is in actual execution. Load processes are often designed so that there can be no access during the load window, essentially keeping users "in the dark" during the load window. Even though the technical environment is up and running, the system is completely unavailable from the perspective of users. They cannot get to the data, run reports or do any analysis. This factor must be taken into account when considering and calculating system and data availability. It is a business decision whether data will require access or not during the actual load time. Compromises are possible. Analysts can often make do with previous period data (yesterday's or last month's data). In architecting the load process, data access requirements during the load window must be taken into account. If analysts require, it is not difficult to provide access to the previous period's data during the load of the current data. IT must work with the business on this very important topic early on in the development process to ensure that the business requirements for data access are met.

26.2.3 Communication to the business

Establishing a simple communication plan on the status of data loads (success or failure) is essential. This can be as simple as an email sent out to a large distribution group, or more sophisticated like an internal website that shows the finish date and time of the loads. Business users always want to know when new data are available. Being proactive in notifications via frequent communications every day, week or month is strongly recommended, rather than being reactive and only notifying the business users when data are late.

26.2.4 Data access and data security

The second aspect of Pillar 1 has to do with data access and related data security parameters. Having data in a FRA ecosystem is only useful if the right people have access at the right time. It is also critical to deny access to unauthorized people and keep the data secure. It is the business owner of the data is that determines all aspects of access to "their" data. IT will simply execute on those rules once authorization has been approved. Data owners must address five considerations regarding data access.

1. Who gets access This should be defined by specific name, which is most often an assigned ID. Names can be placed into groups, and these groups may be called *roles*. A sample role may the fraud detection group, or credit analysts group. Everyone assigned to a specific role will have the same access, with each role having different data needs. Once roles are set up, assigning access becomes easier and straightforward. Since the process of granting access to data must be auditable, creating and granting access only through defined roles greatly simplifies this process.

2. What data are accessed One must be very specific about what data are accessible by whom, since privacy concerns determine which parts of data will be accessible to whom (see also Chapter 24). This is particularly the case with Personally Identifiable (PI) information like a Social Security Number (SSN). As an example, accounts payable group may be allowed to have access to all accounts payable data except PI data. Data access should be defined at the appropriate level as follows:

(a) Database level, i.e., grant access to an entire database or databases
(b) Access to portions of a database such as tables or views
(c) Data groups such as specific rows for various departments
(d) Data elements such as SSN, Personal Health Information (PHI), corporate financial data, and personal identifier information, such as street address or email address.

3. When will data be accessed Specific times (hour, day, week, etc.) that data are accessible also must be clearly defined. Multiple times can be set up and/or excluded such as a rule that says that the fraud detection group only has access to financial data between the fourth and last day of the month. An additional consideration is whether data are accessible during the actual load window. There are several options, as follows.

(a) No system access. This means that when data are being loaded, users are not allowed to access any data. In effect the users are locked out of the system.
(b) Allow completely open access. In this case a user report or query could slow down the load process by using system resources such as CPU. It is very possible for a user to block a load when they are reading the same block of data that is being loaded. This could result in a deadlock and stop the load process dead in its tracks. This level of access will often result in reading some new data along with some old data. This can clearly lead to misinformation. These are often called "dirty reads," meaning the data are only partially valid (dirty) with mixed old data and new data.
(c) Allow access only to the previous version of the data. This would require preserving a copy of the previous data prior to the load, switching user access

to that data and then loading fresh data into a new area. When the load is complete, users are switched to the new version and the old data dropped.

For example, if the current database contains 32 weeks of data and the 33rd week of data is about to be loaded, the procedure will consist of the following steps.

1. Create a copy of weeks 1–32
2. Switch users to access this original version of data
3. Load week 33 into the new copy
4. When the load is complete switch users to the new version (which now has all 33 weeks)
5. Delete the old version with only weeks 1–32.

This scenario allows the most complete access to data for the user community. And it ensures no misinformation (i.e., dirty data). It is, however, costly in terms of additional space since effectively twice the space has been used. These trade-offs are usually acceptable when user access to data is absolutely required at the very times that data are being loaded. A monthly data load process may often take 10–20 hours of load time, and a business cost-benefit analysis may determine that it is useful to bear the additional cost of space in order to grant users access during the load window.

4. What type of access is granted When granting access it is critical to specify the exact operations that can be performed on the data. The acronym CRUD (which stands for Change, Read, Update and Delete operations) is often used to discuss the types of data access permitted. Data in an analytic environment are almost never deleted, so users are never given the permission to delete data. However, each user's ability to perform change, read or update operations must be considered individually (or as part of a role's permissions).

Another important operation is the ability to extract data from the analytic environment and copy it elsewhere. The target may be as simple as the desktop or as extreme as another database environment. When users are allowed to extract and copy data to another location, all control is lost. Extracted data can be changed, updated, and deleted at will. They can then be reported on, yet there is no audit trail. A common problem is caused by users extracting data and placing them in a spreadsheet or desktop database such as Microsoft Access. Such spreadsheets of data are often colloquially referred to as *spread-marts*. These data are then further modified and used to generate "production" reports. The lack of controls results in a high potential for misinformation and erroneous reporting. Spread-marts are the bane of auditors and can result in poor (sometimes even disastrous) financial decisions.

5. How is access granted The process of granting access must be simple and automated. If the process is cumbersome users will find creative ways to bypass it. The automated processes should be well documented and published. Thus the actual data owner simply needs to define access permissions and have IT execute on those procedures. A major consideration in automation must be full reporting capabilities and auditability, especially in the context of increased regulatory scrutiny.

26.2.5 Data security

The level of security that is placed on data is driven by business requirements. Different types of data require different levels of security. The business data owner defines the level of security needed while IT implements it. Some common security considerations follow.

- Encryption: Should the data be encrypted "at-rest" within the database? If not encrypted in the database, should they be encrypted when the data are moved to other data repositories inside the company? What about when data leave the company premises? The same questions are also pertinent in regard to data coming from sources outside the firm.
- Data obfuscation: There are times when a proxy to sensitive data is required for users who do not have permission to view the data itself. Teams of developers require all data elements in order to develop and test the relevant routines, but most often should not have permission to view PI data. Data can be obfuscated in two ways – *masking* and *randomizing*. Masking is simply covering up the contents, for example changing a SSN to all Xs. Randomizing changes the values, but leaves them as meaningful, for example, changing the SSN to a random number that has no direct connection to the source. The choice between randomizing and masking is a business question to be answered by the data owner in conjunction with the group requiring obfuscated data. The data owner will know which fields to mask and will know what an acceptable level of masking is. For example, instead of masking the entire number, it may be sufficient to simply mask the last four digits.

26.2.6 Backup and recovery

Over time the analytic data environment can hold large amounts of immensely valuable data which need to be protected from loss by backing up. There is a commonly used mantra – "If you load it, then back it up" – in other words data that are worth loading are also worth saving by way of a backup. To properly determine backup requirements, one must start with requirements for the restore process. A key question to be answered is how quickly data must be restored. Even for a single warehouse, there are usually several right answers to the question depending on recovery requirements for each type of data. The correct approach is therefore to have the business owner drive the backup and restore strategy.

In developing a backup-and-restore strategy, the following three scenarios should be considered.

1. A portion of the data becomes corrupt or inaccessible: A quick recovery (measured in minutes or hours) is typically required in this instance. This requirement would dictate that a near real-time copy (via fast media like a hard-disk or fast-tape) be available on site at all times.
2. The entire data warehouse is corrupt or destroyed (for example through hardware or software failure): In this case a phased recovery may be more appropriate with mission critical data recovered fist, followed by less critical data. In a recent data center mishap in 2010, over 80 terabytes of data was lost. All the data were available on backup tape. However, there was no properly documented recovery plan, nor had recovery ever been tested. In addition, the backup tapes were housed offsite 60 miles away and had to be retrieved. Even worse, attempts to restore data caused additional problems due to the poor lack of preparation. The end result was that major portions of the data were unavailable for up to a week. This situation could have been avoided by creating an agreed upon plan for restoring data and testing it for capability and accuracy.
3. A major disaster causing the majority (or all) of the data center to go out of commission: In this scenario, it is important to understand the short-list of processes critical to the company's operations. Restoring the FRA ecosystem often does not make the short-list and therefore is deprioritized in favor of more mission critical processes. Mission critical processes usually include everything from restoring email systems and call center systems to restoring other operation systems that are core to the business and drive revenue.

To develop an effective backup strategy, therefore, it is important to consider the outage scenarios described above and determine the requirements for restoring data.

26.2.7 Data availability and data access summary

Data access and data security are determined by the business and data owners. Loading data, granting access and implementing security are simply executed by IT. All SLAs for data delivery, level of data access and data security are agreed upon between the business owners and IT, then documented and published. Backup processes are driven by recovery requirements, and recovery requirements are driven by a business need. IT is simply the custodian of the data and ensures business requirements are met.

26.3 Pillar 2: System availability

Having data and providing access to the data is only part of the necessary support of the FRA ecosystem. It must be up and available to the user. System availability must be defined taking into account the requirements of all the business owners and users of data. The components of the FRA ecosystem that must be considered are shown in Figure 26.1 and repeated here for convenience:

1. User's desktop
2. Business analysis tools and software
3. Data marts, data stores and warehouse (where the data are stored)
4. Network.

Each of these components is critical to the operation the ecosystem, which is only as strong as its weakest link. If any of the components "goes down" (i.e., becomes non-operational for any reason), from the perspective of the user the complete system is down; it is immaterial to users which component has failed. Therefore, it is important that each of these components has a SLA defined for availability consistent with the overall availability goals defined by business users. Availability SLAs are defined in much the same way as a data delivery SLA and should have similar level of business justification. The SLAs also should be agreed upon by all associated business departments and by IT, documented and published. IT groups are then responsible for executing to SLA requirements. For instance, a high-availability solution must be implemented only if there is appropriate business justification, such as a requirement for the environment to be available 7 × 24 × 365. This is important since more stringent availability requirements come with exponentially higher costs.

A key time consideration for availability is prime-time versus off-hours, since the business impact of these times is usually quite different. Having the system unavailable for two hours in the middle of the day in the middle of the week usually has far more impact than being down at 2 am for two hours on a Sunday morning. For that reason, it will be necessary to have separate SLAs that address different time periods. Peak versus off-peak hours is a common way to segregate availability SLAs, as illustrated in the following example.

- *Peak hours (7 am–6 pm, Monday through Friday): target SLA of 99.9% calculated on a monthly basis.* 11 hours per day times 21 days per month times 60 minutes per hour = 13,860 total available "peak" minutes per month. 0.1% is 13.8 minutes of downtime per month during peak hours.

- *Off peak (6 pm–7 am Monday through Friday, plus weekends): target SLA of 99.5% calculated on a monthly basis*. 7 × 24 for 30 days = 302,400 minutes, minus the peak hours of 13,860 = 288,540 total available non-peak minutes per month. 0.5% equates to 1442 minutes or over 24 hours downtime per month outside of peak hours. Although this sounds like a lot of downtime, this allows for a dedicated maintenance window described below.
- *All SLAs can have exceptions and require business sign-off*. No SLA can provide for every possible situation and thus exceptions must be considered. An example may be that the company is exercising a disaster recovery test over a weekend. During the test there will be no power to the data center. Therefore, the FRA ecosystem is unavailable. There is no need to try to identify all exceptions in the SLA document. However, it is important to describe how exceptions will be handled, authorized, and signed-off.
- *Planned downtime*. From time to time it will be necessary to make systems unavailable for the purpose of performing housekeeping tasks described below. These downtimes are scheduled and planned and users are notified of the outage. This can be distinguished from an unplanned event (such as a server going down unexpectedly), or a data center being flooded by a break in a water supply – these are referred to as *unplanned downtimes*.

Additional concerns for availability are the need for *maintenance windows* for planned outages such as preventative maintenance and upgrades. It is good practice to perform regular preventive maintenance to avoid potential longer outages. Additionally, hardware and more specifically software will require upgrades to newer releases. During portions of these processes the system will be unavailable. If the hardware or software in any part of the system is unavailable to users during the maintenance window, then for all intents the system is down and users cannot run reports or execute queries. However, the existence of previously agreed upon maintenance windows does not mean that day-to-day business impacts can be ignored. Rather, when considering a hardware or software upgrade, IT should work with business stakeholders to agree upon a schedule so that unexpected software or hardware version dependencies are not encountered. Preventative maintenance may be performed on a more regular basis according to pre-determined schedules.

Availability statistics must also consider load-windows. The hours of unavailability (both planned, as in a maintenance window, or unplanned, such as a system failure) must be incorporated into the SLA. Any unavailability, planned or unplanned will count against the SLA. The majority of hardware and software comes from third parties with whom service levels with must be established to ensure the ability to meet agreed-upon SLAs. These typically are part of the purchase contract and should include the following considerations.

1. Hours of service: When is the vendor available to provide support?
2. Response time to phone requests: What is the guarantee maximum time for a reply?
3. "Phone home" capabilities: Does the hardware and/or software have the capability to self-diagnose and call the vendor without human intervention when an error is encountered?
4. On site response time: Will the vendor come on-site to actually fix the problem? How close are they to your site?
5. Spare parts: Are these kept on-site or not?

These vendor SLAs should be backed up by financial penalties for non-performance. Maintaining good relationships with your vendors will help greatly in solving technical problems with the hardware or software. Good vendor relationships are often partnerships and result in a "win-win."

An example from the author's experience will serve to illustrate the importance of a proper vendor management. A small software vendor only offered 8 am–6 pm support on weekdays, with no one available after hours to support any possible issue. This fact was omitted during the procurement negotiations and only uncovered once the software had been installed and running for several months. Subsequently, an error occurred when running the software after hours at 2 am on a Friday night. There was no vendor support until Monday, resulting in the entire system being unavailable for almost 32 hours.

26.3.1 System availability summary

Availability is an easy metric to understand when considered from the perspective of the business user. All components of the system must be considered when negotiating and documenting an availability SLA with business users. Separate SLAs must be defined for prime-time versus off-hours, times which must themselves be defined by the user community. SLAs must consider all possible reasons for any part of the system going down, including maintenance windows and time for upgrades as well as possible load-windows where users may be locked out. Finally, SLAs can only be met with users if a solid contract with definitive SLAs enforced by financial penalties exists with third party vendors for each of the system components.

As with all SLAs, first agree, then document and publish and allow for exceptions when necessary. Lastly, manage, track and report on the metrics of the availability SLAs and agree on necessary changes.

26.4 Pillar 3: System performance

Once data are loaded and system availability addressed, performance is the next critical point in managing an FRA ecosystem. Performance is of paramount importance. Performance to the end user is the time it takes from when he submits a report or query until he receives a response. In a FA analytical environment there are numerous components as described in Figure 26.1 above. To the end user, the "system" should be considered as the complete set of components in the FRA ecosystem that are in use when a report or query is submitted. If response is slow, users become frustrated, complain and possibly will go elsewhere to get their work completed.

Echoing the discussion on availability, users usually do not have visibility to underlying technologies and nor are they typically concerned with them. When a user submits a report to run, their only concern is how long it takes to get their report. Therefore it is important to address all four components of a system (see Figure 26.1) when considering performance. It is the end-to-end performance across all these components that determines the response time. Even though the database may be running fast, if BI tools or network are slow, then the system will be deemed slow. It is meaningless to guarantee performance at the database level while ignoring the BI tool. For that reason, best practices have ownership of all components of the FA system managed by the same group or department. If different groups manage different pieces of the system, it becomes difficult to get to the root cause of a performance problem.

Perceived user performance is determined by two factors: available system resources (see Chapter 20) and the level of workload on the system. Workload is determined by the number of tasks the system is performing simultaneously. This includes data loads, automated reports and user initiated queries as well as other system maintenance activities like data backup jobs. All these activities compete for resources from each component of the system. Assuming invariant system resources and applications,[2] the only variable that can be managed to deliver performance is prioritizing the various activities on the system. Prioritization, like most aspects of an FA system, should be decided based on business impact. Priorities may change depending on the day of the month (for example finance users may have priority during the monthly close cycle) or even during the time of day. Often load jobs get priority over user requests during off-hours so they can complete before prime-time begins.

[2] Performance often can be boosted significantly by modifying applications to use resources more efficiently. It can often, but not always, be significantly improved by adding system resources in the form of more (or faster) CPUs, network or disks (see Chapter 20). However, operations groups have relatively little control over either modifying code or adding to resources, owing to the significant time, effort and investment required to make these changes.

Each system component usually comes with tools available to capture the workload on the system.[3] It is critical to capture history for a period of time to get a clear picture of how the system operates. This history must be granular enough to yield a clear picture of user activity at different times of the day, week and month. Figure 26.4 depicts an 18 day period of performance and is color coded to show when the "system" is slow (red) or fast (green). Each column is a day of the week, and each row is an hour of the day. This graphic is colored so that white means 0–20% busy, green is 20–50% busy, yellow is 50–80% busy, and red is greater than 80% busy.

Another useful graphic that shows performance is the heartbeat graph (Figure 26.5). Two queries are made every ten minutes of the day. The same two queries run each time and should have predicable run time. This graphic shows elapsed time for each of the two queries over a 30 day period. As you can see, it looks much like a heartbeat or EKG. When there is a spike the query is running longer, implying that the system is working harder. This graphic will provide a quick look at the overall health of the system based on standard queries that run at regular intervals.

Once sufficient history has been compiled (say three months of information) one can now start to model the performance, identify bottlenecks, and see when the system is idle. Identification of "peaks" and "valleys" in workload[4] is important, since it is often possible to gain significant performance by moving non-essential activity from times of peak-usage to valleys in usage where system resources are comparatively plentiful. Based on this workload history, data, business and IT groups can determine workload priorities. A good procedure is to develop a "straw-man" of a possible priority scheme, usually drawn up by IT based on historic performance data. This scheme can be modified with the appropriate business departments and finalized into an implementable scheme. Priority rules should be phased in carefully by implementing a test period which allows for reevaluation and changes.

Here are some additional things to consider in managing workloads.

- Number of concurrent users allowed: Concurrency is the number of users that can be actually running reports and requests at the same time. A common option is to lock out users when a threshold is met.
- Limit the number of concurrent requests for each user: A single user may be able to submit many concurrent requests, effectively locking out other others. It is like trying to put too many people into a revolving door at the same time. If you

[3] Indeed, an important criterion to be considered when acquiring a particular piece of software or hardware is the tools it includes to monitor system performance.

[4] Workload peaks are times when many jobs (user requests, loads etc.) simultaneously impact the system, while valleys are times when the system is comparatively under-utilized.

	04/23	04/24	04/25	04/26	04/27	04/28	04/29	04/30	05/01	05/02	05/03	05/04	05/05	05/06	05/07	05/08	05/09	05/10	05/11	05/12	05/13	05/14	05/15	05/16	05/17	05/18	05/19	05/20	05/21
	Mon	Tue	Wed	Thu	Fri	Sat	Sun	Mon	Tue	Wed	Thu	Fri	Sat	Sun	Mon	Tue	Wed	Thu	Fri	Sat	Sun	Mon	Tue	Wed	Thu	Fri	Sat	Sun	Mon
0	36	17	18	20	11	11	22	34	30	24	99	41	19	19	18	51	19	19	19	45	20	19	19	24	54	37	34	12	35
1	32	28	38	34	27	14	26	20	29	44	63	30	26	25	30	88	34	36	33	38	16	29	38	36	29	37	46	2	23
2	34	35	35	27	25	1	25	34	37	92	96	46	36	25	36	91	38	37	44	36	13	38	38	37	43	38	47	1	26
3	45	55	48	43	44	1	5	45	46	100	95	44	5	5	45	87	47	45	45	17	1	44	45	50	56	48	29	11	46
4	7	8	8	7	9	4	5	19	8	100	68	10	5	8	8	46	7	8	6	20	2	7	8	7	7	7	17	42	14
5	16	6	22	17	18	24	6	18	26	100	50	16	15	15	18	84	16	16	8	26	2	11	13	13	6	12	13	13	11
6	8	9	19	14	14	35	15	14	28	99	80	16	15	15	12	12	25	18	17	24	4	15	14	30	25	29	15	13	67
7	47	59	54	13	44	44	26	32	50	99	43	60	44	22	63	27	98	51	18	20	6	42	50	61	20	57	19	8	72
8	53	49	64	33	74	13	6	7	35	99	26	99	30	14	62	85	69	41	76	16	21	25	86	99	80	46	21	51	
9	69	94	100	85	99	3	27	58	84	99	99	100	68	12	94	68	48	76	100	28	9	60	97	88	100	68	25	13	
10	95	100	97	72	100	29	22	98	68	94	99	100	90	51	73	47	45	84	100	8	20	99	93	87	75	78	43	8	
11	100	100	73	95	100	1	4	100	100	99	95	93	38	44	100	78	79	78	90	17	42	99	79	93	75	59	34	1	
12	52	98	66	86	65	1	1	100	97	100	85	98	57	96	55	58	59	50	34	33	18	99	92	65	67	21	50	1	
13	52	53	49	81	86	1	1	100	100	100	100	100	17	89	37	52	31	56	79	1	19	97	99	77	68	10	1	1	
14	80	59	72	68	76	1	1	75	100	92	100	99	12	69	62	66	61	56	97	1	6	97	100	94	52	23	1	1	
15	68	88	93	88	92	1	1	70	100	81	100	99	17	13	98	99	54	99	100	6	5	98	100	94	32	44	3	3	
16	100	65	67	77	52	1	2	62	100	100	99	92	5	51	98	99	70	99	58	3	4	99	98	58	63	40	1	1	
17	60	52	44	40	11	1	3	53	91	97	99	89	2	1	79	38	49	100	29	4	3	21	95	25	30	2	1	1	
18	7	2	23	2	3	1	1	10	45	96	96	58	21	1	27	2	57	39	4	1	11	4	46	14	2	2	4	1	
19	1	2	16	1	9	1	7	82	66	100	49	1	1	1	7	1	3	4	1	1	1	2	12	1	1	10	11	1	
20	1	3	3	2	13	1	5	96	85	100	45	1	1	5	2	1	5	9	1	1	17	1	5	2	33	9	1	5	
21	13	8	5	4	8	3	5	50	94	94	20	3	23	4	3	4	5	9	21	3	3	30	4	5	41	54	3	3	
22	7	9	9	7	9	1	3	9	14	54	86	6	17	1	6	10	9	8	27	1	1	42	7	7	7	8	1	1	
23	1	1	1	1	2	1	1	5	1	83	18	1	9	3	20	36	2	1	35	1	1	1	1	1	1	1	1	1	

Figure 26.4 System performance chart example.

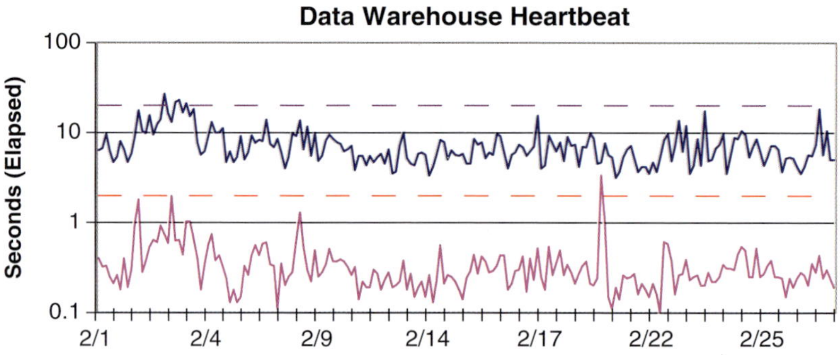

Figure 26.5 Heartbeat graph.

manage the flow, everyone is happy; if you clog the revolving door, it will eventually stop.

- Priority of "production work" versus ad hoc.
- Production requests are those that have been through a higher level of rigor, testing and certification, and thus are optimized. These may be prioritized at a higher than average level, similar to the express line in a grocery store.
- Other considerations.
  - Users running truly useful analytic reports requiring intensive processing, versus users testing reports (i.e., "try it and see" type of reports) in the production environment.
  - Thresholds on the types of activities that can be performed at various times of the day.
  - Limits to the amount of data that can be accessed or extracted.

26.4.1 Benchmarking

Benchmarks are an excellent tool when trying to determine performance metrics. A benchmark is nothing more than a comparison. For example, one may want to compare how things run today versus how they ran last month. Benchmarks are often also used for comparing vendors in their capabilities; this type of benchmark is often called a POC, or Proof of Concept.

IT is the department that should create and own a performance benchmark process for the FRA ecosystem components. Creating a performance benchmark that is repeatable is an excellent way to determine if you have the right level of performance on your system. Running a consistent performance benchmark periodically and comparing results is a great way to see if your system is still "working as planned." A performance benchmark should be a moderate sized body of work that simulates the real production workload. It is recommended that the run time for a benchmark be about 2–4 hours in length.

The following applies to the IT team that owns the benchmark. There are six critical points to remember about benchmarks.

1. Creating a “real life” benchmark. A good benchmark should be a body of approximately four hours of real workload. The benchmark must be of sufficient size to be of real benefit, yet not too large that you cannot run it often. Include all types of user queries and reports. Capture 25–40 different queries/reports and extrapolate those into about 400. Remember you want to mimic production. Do not use static data. The data used in a benchmark should grow as production grows, therefore, use production data. Your benchmark should have predictable results.
2. Run the benchmark with consistency. Run the same number of concurrent queries each time, i.e., start the first 25 queries and when one batch of queries finishes, start the next one. Thus you will keep 25 queries running for the duration of the benchmark. Run the same queries each time. Capture metrics on each query/report in the benchmark process.
3. Establish a baseline. Run the benchmark during the lowest period of system usage. Stop all other uses and processes from running queries or reports when the benchmark is running if at all possible. Capture the elapsed time of the benchmark. This becomes your baseline for all future comparisons. Remember, you are interested in performance only!
4. When to run the benchmark process. Run the benchmark a minimum of every three months. It is not always necessary to stop all other users and processes on the system, but you will need to run the benchmark at a time of low usage. Then document what else is running during the benchmark for future comparisons. Running the benchmark before and after every upgrade is a great metric to have; whether the upgrade is hardware related, software related or there is a major application implementation. Knowing the performance of the system before and after every major change will tell you the impact of the changes you have implemented.
5. Be able to run the benchmark on demand. Automate the process for running the benchmark with scripts and appropriate processes. The benchmark should be able to be run on demand. If you suspect you have a performance issue, your first point of analysis should be to run a benchmark and review the results. Automating the benchmark execution and reporting will greatly ease the running of the benchmark. If automated, you can run the benchmark at will.
6. Track, graph and analyze results. Document what has changed since the last run.

Benchmark example At one company, the benchmark was run monthly. After six months, the benchmark runtime had grown by 28%. What was running in three hours was now running in three hours and 50 minutes. After analysis it was determined that the cause was data growth. The databases used in the benchmark had also grown by just over 30%. The benchmark was running true to form.

26.4.2 System performance summary

Performance is in the eyes of the end user. The workload across all components of your FRA ecosystem will ultimately determine performance. It is critical to establish, document and publish solid workload rules that cross the complete risk analytics environment. These workload rules must be based on business value and business impact. IT and the business must define and agree on workload priorities. Base your performance SLAs on elapsed time across components. And lastly, yet most important, create and automate an internal performance benchmark and use it to measure performance.

26.5 Pillar 4: Integrating new projects, managing change

The one constant in an FRA ecosystem is change. The curse of a successful FRA ecosystem (at least from an operations perspective) is the "Law of More" – more users wanting more applications and reports with added complexity and variety of analysis against increasingly large datasets with frequent updates in a more timely manner. Therefore change management becomes an important and central aspect of managing a successful FRA ecosystem where constant change is the norm.

Changes typically fall into three categories.

1. Fixes: Correcting something previously implemented,
2. New applications or enhancements: Adding new functionality, features or new data,
3. System changes: Upgrades/new releases to the software or the hardware, usually driven by increasing workload.

All of these changes must be carefully managed to preserve the integrity and viability of the existing system. Changes are usually (but not always) known well ahead of time, and can be planned and managed to a controlled implementation. The first principle of change is to "do no harm," and ensure that the change does not cause a problem in an existing application. It is very important therefore that the FRA ecosystem creates its own robust change management policy. While the policy will need to interoperate with corporate change management processes for operational systems, decision support systems and business analytics have a different set of needs. The key difference is that these environments undergo far more frequent and subtle changes than operations systems, especially since a key criterion for success in these environments is the generation of new insights and uses for data. Indeed most successful FRA ecosystems undergo some level of change almost every week.

The goal of change management is to control and manage approved changes while mitigating risk as much as possible. It deals with how changes to the system are managed so they do not degrade current system performance and availability. Change management is especially critical in today's highly decentralized, network-based environment where users themselves may be applying many changes. A key cause of problems is the implementation of changes by those who do not fully understand their implications across the operating environment. Therefore for effective change management, all modifications must be identified, planned, thoroughly tested and evaluated prior to implementation. Even after this preparation, it may be necessary to reverse changes, so back-out procedures should be established in case unanticipated problems arise. Too often project managers neglect to engage the operational support staff early on in the project design and development. There is nothing better than to have a voice at the table early in the development process so that you can manage change and be proactive.

The change management policy can use the following six-step approach.

1. Document and communicate a change management process which includes the following.
   - Policies and procedures: The methods for introducing change into the environment.
   - Roles and responsibilities: Responsibility and accountability for each step od the change process.
   - Measurements: Measurement of success of changes.
   - Prioritization: The order in which changes go in. Are some changes delayed until another time?
   - Fall back procedures: If the changes are unsuccessful, how do you back out the changes to get back to where you were?
2. Establish a single change coordinator to receive change requests. This coordinator will also be responsible for notifying all affected parties.
3. For each change, create a plan for implementation of changes that includes the following components: resource requirements, analysis of risk, impact to FRA ecosystem, Back-out procedures and implementation schedule.
4. Implement and monitor the changes. If required, implement back-out procedures to reverse the change. Otherwise carefully monitor the system for at least the first few weeks to determine that the changes do not produce any adverse effect.

5. Evaluate and report on changes implemented and confirm that the planning process was sufficient, and determine when certain changes are more prone to problems.
6. Modify the change management process if necessary.

Of note in Step 3 is the mention of analysis of risk and impact to the system. This step is most often taken lightly and more significantly is not often reviewed by the FA operations management team. It is in this area where the ball is dropped and "runtime" problems occur. The project team might have done a great job of planning and executing the changes, but more often than not, there was minimal consideration for the "running" of the system.

Some common problems that are encountered in frequently changing environments include:

- Changes not being applied in a timely manner
- Increase in volume (users, data, reports, etc.) not being properly factored into the change
- Too many changes being implemented concurrently
- Changes not being properly integration tested via a robust quality assurance process
- Operations management not being integrated into the sign-off process
- Not considering system availability problems or challenges to other SLAs due to the change.

26.5.1 Change management and growth

When it comes to change, there is still one more major consideration. That is the natural growth of the existing system. Growth can be due to an increase in any of the following:

- Data due to growth of history, or addition of new datasets
- Number of users
- Number of reports and queries
- Complexity of reports, queries and dashboards.

Addressing natural growth requires a proactive approach by tracking the history of users, data, and reports over a minimum rolling six months history, and extrapolating this out over the following six months. The resultant metrics will enable proper capacity planning and forecast need for new system resources. Figure 26.6 displays a usage trend graph showing 12 months actual usage and 12 months extrapolated out over time

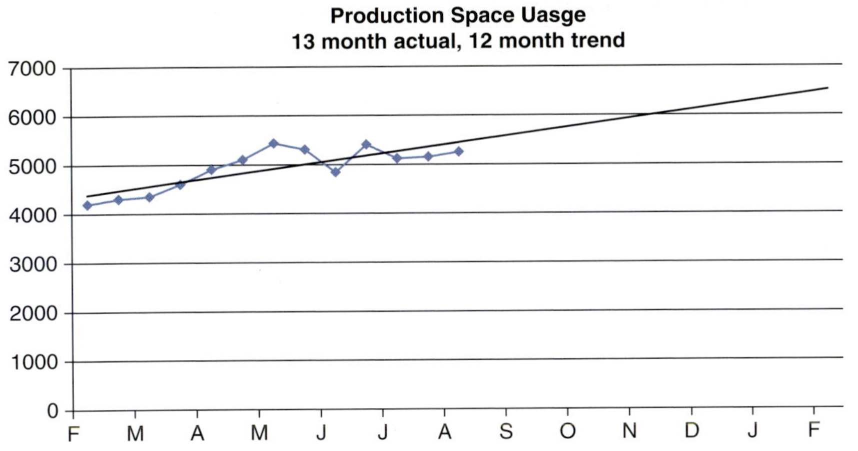

Figure 26.6 System usage trend example.

26.5.2 Change management and testing

A major part of managing change is testing. Unit testing is fairly straightforward and is done by the individual making the change. Integration testing must also be done to characterize the impact upon existing workload on the system. A common problem in integration tests is omitting to include other changes being implemented at the same time. A great method of integration testing is to run the benchmark process previously mentioned while running your integration testing. This method will allow you to run the new processes and simulate production processes, via the benchmark, at the same time and see if the new processes slow down the current production processes and adjust if needed.

Another part of testing that is often omitted is volume testing. Both the initial volume for when the change is implemented as well as potential for growth over time must be considered and tested for. It is also worth noting again that increases in volume must consider not only data but also growth in users and complexity.

26.5.3 Change management summary

Change is a constant in an FRA ecosystem, and therefore must be planned for. Every change must consider the impact to the existing system. It is highly recommended that operations management be involved early on in the new projects themselves to ensure the production FRA ecosystem is protected. The most important firewall against unexpected adverse effects of change is testing. All tests such as unit, integration, benchmark, and volume tests must be completed. It is best if operations management signs off on every change implementation early on in new projects. And lastly set test periods for each new SLA to determine if the SLA is realistic and achievable.

26.6 The foundation of operations management: general operations tasks

In addition to the four major pillars in operations management, there are other operations tasks that are necessary to keep things humming.

26.6.1 Help desk

The help desk is a call center designed to provide users with assistance, often with computer-related or other technical issues. A help desk is usually operated by individuals that have an extensive background of the appropriate system, so that they can walk users through step-by-step solutions to their problems. Rather than using a central IT or corporate help desk, it is far better for the FRA ecosystem help desk to be in the operations group supporting it, since the operations group consists of individuals that know the interrelationships of all the components of the ecosystem and can more quickly solve user problems.

26.6.2 Data archive and purge

Operations teams often need to manage the volume of data in the FRA ecosystem by running either archive or purge routines. An archive is a move of data to another, presumably lower cost, location. A purge is a permanent delete after which data cannot be recovered or recreated. These operations are usually executed daily, at month-ends, or annually depending on the need and type of data. With increased need for auditability (driven for example by government regulation), data purges are exceedingly rare. A critical requirement of archive data is that they must be "reloadable," i.e., one must be able to get the data back to an environment where they can be used. On the archive side of the equation, the business must decide how much data to keep on the active systems.

As with everything else, both archive and reload requirements must be driven by business need. The size of the active system, determined by the amount of data required for analysis, will determine archive cycles. Likewise, only business need can answer questions on reloading target databases, amount of space needed to be reserved for this process and time availability for reloads. Of course, IT must provide guidance and work with business users in this matter, especially since technology advances in performance and cost-effectiveness are rapidly enabling large amounts of data to be available in "near-line" archives that allow analysts considerably more flexibility in accessing archived data.

26.6.3 Hardware and software upgrades

The operations management team will need to stay abreast of the latest versions of the software and hardware that comprises the business analytics ecosystem. New functionality becomes available with each release, and new releases often promise better performance as well. To take advantage of newer releases, the operations team needs to work closely with vendors of hardware and software. It is very important that these upgrades be managed by the same group since there are critical dependencies between each component.

Software and hardware are typically certified by the vendors to be compliant with specific levels of other software and hardware. The vendor will support their software only if it is certified. As an example, ABC Software version 7 is compatible with XYZ Software version 5.3. Thus version 6 of ABC software will not be supported. Even if it does work, the vendor makes no representation of support for debugging any issues discovered with this combination. Each component of the entire FRA ecosystem must therefore be compliant with every other component. Therefore, when facing an upgrade one must first look at all the components across the entire FRA ecosystem for compatibility.

Often an upgrade to one piece of the ecosystem will be impossible unless one or two others are upgraded as well. This ripple effect of sequencing of other upgrades often causes upgrades to expand in scope. Many companies find that they cannot upgrade software to gain needed functionality because they are behind in versions of other software in the ecosystem. This is particularly problematic if different groups "own" different pieces of the FRA ecosystem, causing problems with system ownership, coordination and prioritization. Another common issue is determining the resources needed to perform the upgrade.

Testing for upgrades needs to encompass the entire ecosystem. The performance benchmark process discussed in Section 26.4.1 above becomes a major testing step to be executed before and after every upgrade. Specific regression testing will also be required to confirm continued functionality. This testing must be done across the complete FRA ecosystem, from desktop all the way to the database and back.

26.6.4 Communication to users

Communication to the business community is essential in a diverse FRA ecosystem. It is essential to crafting the appropriate message to the users about important events like ongoing issues and the status of an upgrade. This task must be given to a single group. Messages must be in language that the business can easily understand. A common forum for user communication is an intranet site like the company intranet or a Microsoft SharePoint site dedicated to operations management. While emails alone are not sufficient, they certainly have their place in providing information to specific targeted audiences.

Communications should be as complete as possible. The information to be communicated should include:

- Current system status
- This should cover all components
- If any component is unavailable, estimates for it to come back up should be provided if available
- Calendar of upcoming events
- Planned enhancements
- Planned maintenance efforts
- Planned outages
- Contact list
- Help desk
- Data owners
- Data governance team
- IT contacts
- Monthly usage reports
- SLAs for data delivery, system availability, and system access
- Data access process.

A user communication vehicle like a common website can also be used for more frequent messages in situations like an upgrade or major implementation. This can be an effective way to keep the implementation team as well as the business users informed as to the progress of the major changes. Information to the user should be as proactive as possible without sending out a flurry of emails.

26.6.5 Management reporting

Operations management is the keeper of all the metrics for the FRA ecosystem. A monthly report should be produced and posted on the communication website referenced above. A short list of suggested reporting is given below – good practice would suggest providing 13 months of trending information as part of the report.

- SLAs met/missed for the month
- Data delivery
- System availability
- System capacity graph

- Usage by department, number of users and volume of use
- New projects implemented
- Significant outstanding issues with status.

26.6.6 Audits

Source systems are usually the source of record and most information should come from the source systems. There is one prominent exception, and that is identifying who has access to what data in the FRA ecosystem. Since operations management is in charge of granting access to the data as defined by the business owners, who better to document who has access to what than the FA operations team and provide that information to the auditors. Although information security may determine who gets access to what data, it is often the operations team that executes to those directives and can generate a report of who has been given access to what, as well as who has in fact been acting on those permissions.

26.6.7 Disaster recovery and business continuity

Disasters can be caused by many situations. Some are related to natural phenomena, like earthquakes and tornadoes, but other conditions may also result in data centers being fully or partially closed.[5] Disasters really fall into three categories.

1. Single point of failure: One component is down, yet the rest of the business is up and running. A localized water-leakage situation could be an example.
2. Full disaster: The entire data center is down. This is the typical disaster situation which is planned for.
3. Business interruption: The data center is up and systems are working. However, staff are unable to get to work or get the work done. This is often called business continuity and could be related to a natural event (unexpectedly severe snow-storms, forest-fires etc.) or other events like a virus outbreak.

Recovery from each of these three scenarios is quite different and can only be defined by discussion between the business units involved and IT. Here are a few points to consider.

Disaster 1: Single point of failure

This should be considered the highest priority since analysts cannot get their work done even though the rest of the business is up and working, due to the business

[5] In one situation, a coolant leak in a data center caused the forced evacuation of all personnel, a "disaster" by any standard.

analytic environment being down. All hands should be on deck to recover the FRA ecosystem.

Disaster 2: Full disaster

In this situation the entire data center is down and business is completely unsupported. In this situation, most often operations are moved to the Disaster Recovery (DR) site where systems must methodically be brought up in an order prioritized by business need. Typically, the FRA ecosystem is not the first system that should come back online. Operational systems including communication systems (voice mail, call center, financial systems, etc.) must be back up first and then analytical systems can be addressed.

Disaster 3: Business interruption

In some cases the data center may physically be operational but workers may still not be able to get to work. Examples are often weather related, such as a blizzard, or even an epidemic such as a quarantine situation.

In this scenario, the business must decide on a definitive level of recovery, i.e., what to recover, and how quickly. In the FRA ecosystem, often all the data cannot be recovered without an extended period. Business owners have command of priorities for recovery based on business impact, while IT has knowledge of recovery processes and costs associated with providing defined levels of service in recovery. Therefore business owners and IT must work together to determine the kinds of recovery systems implemented after incorporating the risk to the business, the cost-benefit of quicker recovery times and the feasibility of recovering specific types of functionality or data in the analysis. For example, a single-system disaster (type 1 above) can be easily resolved by implementing a full dual redundant system, but it needs to be cost-justified by business value and/or risk management considerations.

Most companies draw up disaster recovery plans as well as a business continuity plan. FA operations management must play an integral role in the planning process to ensure that the FRA ecosystem is protected and recoverable in each of these scenarios.

26.7 Conclusion

Success or failure of a business analytics environment not only depends on the derived business value of the ecosystem, but also on how the systems and the actual data are managed behind the scenes. The business will set the priorities on data delivery, system availability, who has access to what, and what is an acceptable level of performance. It is up to operations management to execute on these business requirements and ensure the stability and performance of the entire business analytics environment.

Operations management needs to work with the business to design the environment based on business requirements. The agreements reached with business users is documented in Service-Level Agreements (SLAs) which describe in detail what the business can expect in terms of the system's functionality and performance at various times of the day, week and year.

It is up to operations management to deliver data according to the SLAs. All system components must be up and available according to the SLAs. System performance must be managed to meet the needs of the financial analysts. And lastly, as change occurs due to new projects, new data, or changes in business, the operations management team must protect the existing production processes, plan for the new changes, and ensure new and old work together.

Just like a duck gliding on the surface of a pond, things should look calm on the surface to analysts. Yet we know there is a great deal of activity beneath the surface being performed by the operations management team. Without this support from the operations team, havoc may ensue in an FRA ecosystem resulting in the data being much less accessible and usable than it should be.

Bibliography

Blazewicz, Jacek, Wieslaw Kubiak, Tadeusz Morzy and Marek Rusinkiewicz, 2003, *Handbook on Data Management in Information Systems*.

Inmon, W. H., 1995, What is a data warehouse? Prism Technical Topic, **1**(1).

Information Builders, Business Intelligence and Integration Without Barriers.

Jarke, Mattias, Maurizio Lenzerini, Yannis Vassiliou and Panos Vassiliadis, 2003, *Fundamentals of Data Warehouses*.

Kimball, Ralph, 2002, *The Data Warehouse Lifecycle Toolkit*.

Orand, Brady, 2011, *Foundations of IT Service Management with ITIL*.

Sarsfield, Steve, 2009, *The Data Governance Imperative*.

Singh, R. Jaiteg, 2011, *Understanding ETL and Data Warehousing: Issues, Challenges and Importance*.

27

Data governance and data stewardship

Deborah Stockdale

27.1 What is data governance?

27.1.1 Background

Data governance, the exercise of decision-making and authority for data related matters, has developed over the past years in response to organizations recognizing the need to have better control over continuously growing volumes of enterprise data. The term data governance has a myriad of definitions and this chapter will present a sample of those views and attempt to rationalize definitions into an actionable understanding.

Data governance broadly refers to the statues, regulations, policies, standards, guidelines, business rules, organizational bodies, decision rights (deciding how to decide), accountabilities and enforcement methods for people and information systems as data related processes are performed across the enterprise. Practices and frameworks have been developed to include multiple enterprise data performance activities including data quality, data protection, data visibility and security, data archival, data definitions, content management, data warehousing, metadata management, and master data management.

According to the Data Governance Institute (Definitions), a well-defined and complete data governance solution is a system of decision rights and accountabilities for information-related processes, executed according to agreed-upon models which describe who can take what actions with what information, and when, under what circumstances, and using what methods. There are other valuable views and definitions of data governance.

Handbook of Financial Data and Risk Information, ed. Margarita S. Brose, Mark D. Flood, Dilip Krishna and Bill Nichols. Published by Cambridge University Press.

27.1.2 Emergence of data as an asset

The planet is awash in data. Cash registers ring up transactions world-wide. Web browsers leave a trail of cookie crumbs nearly everywhere they go. RFID literally announces inventory's presence enabling firms to precisely journal every step their products make along the value chain: "I'm arriving in the warehouse," "I'm on the store shelf," "and I'm leaving out the front door" (Gallaugher). A study by Gartner Research claims that the amount of data on corporate hard drives doubles every six months, while IDC states that the collective number of those bits already exceeds the number of stars in the universe. Wal-Mart alone boasts a data volume nearly 30 times as large as the entire print collection of the US Library of Congress (Evans-Correia, 2012). This flood of data drives a tidal wave of opportunity.

Increasing corporate data volumes, and easy access to rich third party data, combine to enable a new age of data-drive, fact-based enterprise decision making. The ubiquitous availability of reporting, ad hoc queries, data exploration, data modeling and analysis, statistics and quantitative analysis, explanatory and predictive models, all offer management fact based results and insights to drive granular and targeted decisions and actions.

The benefits of timely, fact based and insightful data analysis all combine to drive competitive advantages at organizations such as Zara, Netflix, Google, Wal-Mart, Harrah's Casino Hotel, Caesar's, and Capital One (Ayers, 2007). These organizations used data to identify profitable and valuable customers and drive programs which rendered their performances a full ten times greater than the S&P 500. Data-driven decision making is even credited with helping the Red Sox win their first World Series in 83 years (Lewis, 2003), and with helping the New England Patriots win three Super Bowls in four years. To quote from the Bloomberg/BusinessWeek cover story on analytics,[1] "Math will Rock Your World!" Clearly, data represent corporate memory and therefore must be considered a corporate asset. Data can easily be defined as a collection of stored truths which may be viewed as a set of captured facts, measurements, and ideas used to make decisions.

What does it mean to treat data as an asset? An asset is any item of economic value owned by an individual or corporation. Can facts, measurements and ideas be items that are of economic value owned by an individual or corporation? The obvious answer is yes; data can be bought and sold like commodities or used to generate business revenue and make smarter decisions.

The economic value of data may be looked at in one of two ways. Firstly, data may be sold as a commodity such as ebooks, training, music, movies, and so on. Customers are willing to pay to gain access to this data for their consumption. This

[1] Steven Baker with Bremen Leak, "Math will Rock Your World," Bloomberg/BusinessWeek, January 26, 2006, accessed March 13, 2012.

directly implies that there is an economic value for data in the form of a commodity because customers see a value in obtaining it. Secondly data may be used in making smarter business decisions that allow for companies to become more profitable and/ or reduce their potential operational risks. Trends may be formulated based on existing data that allow companies to forecast data so that they can make strategic business decisions based on sound forecasted data.

27.1.3 Data asset inventory and control

Corporate assets are inventoried, controlled and used to generate revenue. The Information Asset Profiling (IAP) (Stevens, 2005) process is a Carnegie Mellon developed approach designed to provide organizations with a common, consistent, and unambiguous understanding of information. Specifically IAP defines asset boundaries, owner or owners, and information security requirements for each asset. Descriptions of where assets are stored, transported, and processed offer an opportunity to determine the asset's value to the enterprise. See the appendix, Section 27.3, for additional details.

A common, consistent and unambiguous understanding of data helps with the selection of controls for asset usage and protection. When this valuable information is introduced early in the system development lifecycle, controls can be designed to ensure that the access, quality, ownership and security requirements of an asset are enforced. Asset profiles can also provide context and meaning to compliance and audit activities enabling management to make informed decisions on how to respond to regulatory findings.

27.1.4 Data asset usage in generating revenue and value

Revenue generation is part art and part science. Historically revenue generation has been done based primarily on intuition, gut instinct, and "knowing" the market. While art and intuition bring insights and experience to bear, they should be backed up with data and facts, science. With the dynamic and global competitive landscape combined with rapidly changing customer behaviors, it is more important than ever to make data driven, fact based decisions to identify and generate revenue.

The first driver behind this shift is competitiveness. The value a company extracts from its data can be a significant strategic differentiator. Organizations make both operational and strategic decisions based on their data, and a better information environment will lead to better insight and hopefully to better decisions, thereby providing a definite competitive advantage.

The second driver for managing data as an asset is regulatory compliance and risk management. As regulatory scrutiny grows and executives are held responsible for the accuracy of the financial information their company communicates to the public, the importance of good data management becomes clear – from the highest levels of the organization on down. An ad hoc, unstructured approach to data management can lead to costly errors and open the enterprise up to potentially devastating regulatory risk.

Making data driven revenue decisions is not as simple as printing a stack of reports from the operational systems. It also goes beyond having a number of subscription based reports and tools as well as a mass of spreadsheets requiring constant updating. It is really about how data assets are efficiently utilized and enabling test and learn as well as being able to ask any question at any time. Utilizing data to make decisions will also aid in determining success factors. Whatever strategic or tactical choices are made, there should be an end result in mind. It is important to set expectations unambiguously to enable a clear assessment of decision results. By measuring results, better future decisions can be made by understanding what data and approaches worked well and which did not. Bringing the data assets to the forefront of decision making is vital. There are many systems and tools available to aid in the decision-making that are all data based. Embrace the use of data to make effective, revenue-positive decisions.

Data asset usage manifests value in many ways.

1. Enhanced information management capabilities driving new efficiency opportunities, and increased profits through increased effectiveness in revenue generation, and customer profiling.
2. Improved alignment of business initiatives to drive increased business value through data availability and enrichment, enabling insightful strategic planning and execution.
3. Increased profit through improved efficiencies enabled through centralized standards.
4. Compliance/regulatory adherence to satisfy auditors and regulators by developing data management environment that comply with specific requirements.

Effective data usage also establishes an improved platform for measuring, monitoring, and improving business performance by tying operational metrics to business performance measures and facilitating reporting and management of critical processes.

When organizations manage their data as a corporate asset, focus is turned away from the expense associated with acquiring and storing data and towards the business value that can be obtained from using the data and the full strategic lifecycle of the asset. Managing data as an asset requires accountability from the business side of the organization, not just from IT. If an enterprise is spending millions of dollars to manage its data assets, the business must be held accountable for generating some measurable return on those assets. Managing the value of data also requires commitment from the highest levels of the organization, direct ties to the strategic goals of the organization, and the involvement of both business and IT stakeholders.

Clearly, managing data as an asset is not easy, and it requires significant shifts in an organization's mindset as well as its business and technology processes. So why are enterprises more willing than ever before to make the changes necessary to treat data as a corporate asset?

Some organizations are finding that their existing methods for ensuring the accuracy of information just are not feasible anymore. In extreme cases, an organization might have to employ an entire group of people whose job is to reconcile financial data produced by multiple systems to make sure everything balances before presenting it to C-level executives. This approach to data management is costly, unsustainable in the long run and prone to delays and human error.

Finally, an organization that fails to treat its data as an asset will not reap the full benefit of investments in business intelligence, performance management, and other initiatives that are information based. Millions of dollars spent on enterprise resource planning (ERP), customer relationship management (CRM), and other transactional systems can go to waste if the information they capture, use, and produce is not managed to ensure accuracy and consistency.

27.1.5 Data management

Data management is the process of managing data assets and includes the development and execution of architectures, policies, practices and procedures in order to manage the information lifecycle needs of an enterprise in an effective manner. Figure 27.1 denotes typical data management failures and their direct business impacts.

How does data governance improve poor data management?

- Data governance creates an oversight structure surrounding data management activities to direct, control and ensure alignment with corporate objects and business value.
- Data governance assigns responsibility and accountability to individuals with quality measurements.
- Data governance policies, standards, processes, guiding principles and rules are key activities for successful data management execution.

27.1.6 Data asset risks[2]

All organizations face the challenge of protecting valuable corporate data assets against threats and risks. Many financial institutions have incurred data losses in recent years which have alerted us to the responsibility of managing data. If data assets are lost or

[2] See Chapter 24 for more detail on this subject.

Data Management Challenges	***Effect on Enterprise Business***
Data Quality	Inaccuracy or contradiction in financial reports increases non-compliance risk Marketing campaign sub-optimization Uncontrolled customer communications
Master Data Management	Inaccurate customer and product linkage Purchasing effectiveness reduced by poor price/item linkage Product to market delays Customer profitability unknown without linkage
Data Integration	System based data conflicts inhibit reconciliations Unnecessary data complexity and inconsistency Performance challenges and interface duplications
Data Privacy and Security	Regulatory risk increased Credibility reduced with customers, partners and regulators Authorization abuse Identify theft
Metadata Management	Poor data understanding results in poor and inaccurate decision making Development costs are increased and delivery times extended Data asset unused because it is not understood or reliable
Data Architecture	Data is inconsistent across business units and subject areas Redundant and unreliable data Data understanding, access and usage costly and time consuming Incorrect usage of data

Figure 27.1 Business impacts of poor data management (Stockdale, 2013).

leaked, otherwise spotless reputations can be damaged, financial losses may be incurred due to contractual indemnification, or commercial secrets leaked to competitors. Data losses may also result in identity theft for individuals and fraud. Enterprises may be subject to hefty fines from regulatory bodies for data asset losses.

Information assurance (IA) is the practice of managing data asset related risks. Data and information systems are protected by ensuring confidentiality, integrity, availability, authentication and non-repudiation which are defined below.

- *Data asset confidentiality* is about ensuring that only those authorized have access to data. Typically, information should have a confidentiality level denoted based upon organizational definitions and structures, i.e., protected, restricted, confidential, proprietary, etc. Data aggregation can increase the confidentiality level of data. For instance, name and address details may be one level of confidentiality, but when paired up with bank account details, the aggregation will become a higher level of data access because it is more sensitive and more valuable, and potentially more damaging if lost or compromised.
- *Data asset integrity* is about protecting the data and ensuring that it has not been tampered with. This means data cannot be created, changed, or deleted without

proper authorization. It also means data stored in one place agree with other related data stored elsewhere. The unintended modification or loss of data is the most prominent cause of loss of integrity. This could be through human error, hardware or software errors, or through physical causes such as extract files and physical hazards. Intentionally altering data is potentially the most dangerous risk for data integrity and requires defined processes and controls to mitigate.

- *Data asset authenticity* is necessary to ensure that the users, documents and data are who or what they say they are. Proliferation of data through extracts, external media, and other means often results in authenticity issues which put decision making and regulatory compliance at risk.
- *Data asset availability* includes data and the systems used to process the information, and the security controls used to protect data are all available and functioning correctly when the information is needed. Availability also includes limiting access to data to authorized people in authorized roles doing authorized activities.
- *Data asset non-repudiation* means one party of a transaction cannot deny having received a transaction nor can the other party deny having sent a transaction.

The protection activities detailed above apply to data assets held in storage, processing, or in transit, and covers both malicious and accidental threats. Information assurance has a broad scope and includes risk management, data governance, privacy laws, data protection laws, compliance, business continuity and disaster recovery

Risk management is a central part of any organization's information assurance strategy. Risk management can be defined as the combination of the probability of an event and its consequences. It is the process whereby the organization methodically addresses the risks attached to the storing and use of data with the goal of avoiding the bad outcomes should data be lost, exposed or misused in any way. Risk management identifies, assesses and prioritizes risks, develops an economical application of resources to minimize, monitor, and control the probability and/or impact of bad outcomes. An integral component of the risk management equation is the balance of resources allocated to avoiding the undesirable outcome, with ensuring that avoidance activities do not become too burdensome.

Enterprises must develop risk portfolios with a whole-life risk approach for data assets. These risk portfolios may be developed in a standard framework over time. An example of this framework is a firewall. A firewall controls access across an IT network boundary and its policies must reflect the needs of the business. A risk assessment of a network connection might determine that an assured product is needed. However, using the information assurance model a risk manager may additionally ask the following questions.

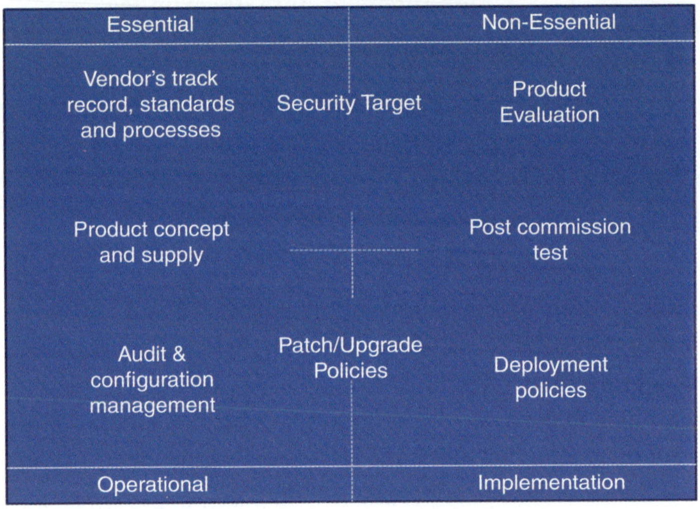

Figure 27.2 Risk management by category.

1. How much trust will be needed in the supply chain, both now and later?
2. How will upgrades or patching be performed?
3. Does the hardware or software need to be evaluated to mitigate the risks, or would regular penetration testing be more appropriate?
4. Will the firewall be integrated into the business so its security functions are not degraded?
5. Can the firewall be configured to support the access policies that the business needs?
6. How will illicit access attempts be identified and what will be done if they occur?
7. Such questions can be asked at any point in the solution lifecycle, but are most powerful when asked continuously.

27.1.7 Conclusion: importance of data governance

In a knowledge economy, information is one of the most valued enterprise assets. In a transformation economy, proper governance of data assets becomes a critical success factor. Data assets must be integrated across the enterprise to reap the financial and regulatory rewards available. Data and information will outlast applications requiring the management of data and knowledge assets through time as applications, and even business processes change.

Data governance is essential to ensuring that data are available, accurate, appropriately shared, and protected and used to further enterprise revenue generation, decision making and regulatory compliance.

27.2 Components of data governance

27.2.1 Data governance program

For data governance programs to achieve cost savings, data reuse and analytics across the enterprise, a single version of the truth must exist. Enterprise data governance programs drive a single version of the truth through the foundational activities of data integration, data quality, data models and metadata as detailed in Figure 27.3.

- Data integration refers to an enterprise's data inventory and information assets as well as the tools, strategies and philosophies by which fragmented data assets are aligned to support business goals.
- Data quality refers to the tools and processes which result in the creation of correct, *complete* and *valid* data that are required to support the enterprise mission.
- Data model is an abstraction which documents and organizes business data showing what data are needed and created by business processes and functions. The data model provides clarity for increased understanding and ease of communications between business and technical staff.
- Metadata is data about data. It is descriptive information about a particular set of data, object, or resource, including how it is formatted, and when and by whom it was collected.

If we think of data as the "nouns" of an enterprise, it is not a surprise to learn the same "noun" may be used across multiple organizational functions. For example, customer data may be created by the sales team, modified by the customer service team and removed by the order management team all acting upon the same data in alignment with their respective organizational functions.

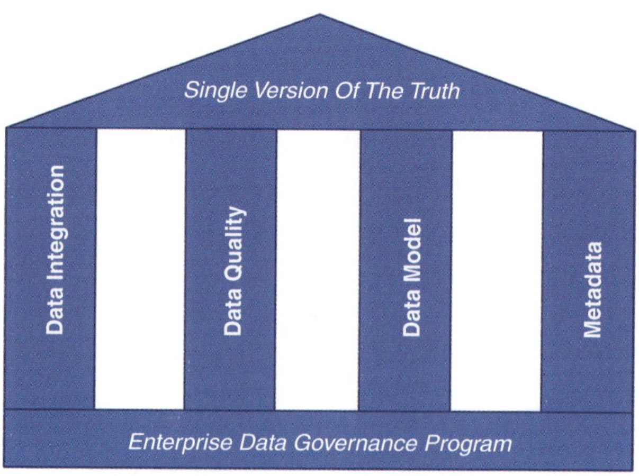

Figure 27.3 Pillars of enterprise data governance.

Data governance programs must ensure the integrity of data across the enterprise and therefore require more than just a few members of the IT staff to be successful. Data governance should be viewed as an enterprise effort which touches all functions and organizations, demanding commitment from business management, senior-level executive sponsorship, business communities and IT to achieve the single version of the truth which is a concept describing the ideal of having either a single centralized database, or at least a distributed synchronized database, which stores all of an organization's data in a consistent and non-redundant form.

It is therefore important that data governance activities be managed as a program rather than a series of disconnected projects. Program management differs from fundamentally from project management in that it manages multiple related tasks, each of which makes its own contribution to the overall data governance enterprise goal.

27.2.2 Funding model

Typically, organizations choose one of four models for creating a formal and acknowledged data governance program:

1. Fund a capital project to build the program
2. Fund as a line item on many projects
3. Fund by a stakeholder group (i.e., risk, finance, marketing, etc.)
4. Fund as part of enterprise data management or data architecture.

1. *Project to build a data governance program*: A formal project is created to design and implement the data governance program, which often includes a prototype effort to which data governance is applied, such as a JAD session of data stakeholders or data stewards to address a legacy data issue. Note: many "boilerplate" project plans are designed for the development of a piece of software, not for the development of a program. They can include steps and deliverables that do not make sense for a program development project. Also, a certain amount of flexibility needs to be built into program development project plans, or project managers can find themselves having to process change requests when efforts to define scope and focus require multiple iterations, or meetings to discuss stakeholder needs are rescheduled.
2. *Line item in other projects*: Projects that require data governance to succeed may include activities to initiate and fund the design and deployment of a formal data governance program. Examples would be SOA or master data management initiatives or data warehouse projects.
3. *Special funding from a stakeholder group*: A department or program with a major stake in the proper governance of data may initiate and fund the design and

deployment of a formal data governance program. Examples would be SOA or master data management programs, data warehouse initiatives, and compliance initiatives such as Basel II or Sarbanes–Oxley, or privacy/access control programs.

4. *Data governance included in IT/data management/data architecture efforts*: Data governance included in another technology or data-related program. If data governance is included in enterprise data management efforts, for example, the program may be given a bucket of money to implement governance without being held to the type of spending oversight generally required of projects.

27.2.3 Stakeholder identification

Before any data governance rule, a statement that defines or constrains data behaviors is created or any data-related decision is made, a prior decision must be addressed: who gets to make the decision, and when, and using what *process*? It is the responsibility of the data governance *program* to facilitate, document and catalog the collection of decision rights that are the "metadata" for data-related decisions.

Decision rights for compliance-based programs may be leveraged from statutes, regulations or processes. The decision about whether to comply with a law, regulation, or standard may be made at the organization's executive level. How to comply may require discussion of options by multiple stakeholders, however.

For other types of programs, decision rights may require ongoing negotiations. For example, who should decide the length of a data field in a new system logically falls to the data architect. However, the decision may have farther reaching impacts and would therefore also require input from other stakeholders which bring additional requirements and constraints to the decision. Data governance establishes decision rights which provide structure and processes to enable data decision making by all concerned stakeholders.

27.2.4 Roles and staffing

Who is responsible for data? The default answer tends to be IT. In reality, the IT organization does not typically have the business knowledge to explain data context and business usage. Typically IT is responsible for systems and infrastructure "the pipes and the plumbing." It does not have responsibility for the data content which runs through those pipes and is created by the business. Data is a business responsibility, since they know the data best. IT enables the collection, usage and storage of data rather than the definition and quality of data across the enterprise.

Many organizations do not have a dedicated group of people responsible and accountable for managing and improving data such as customer data, product data etc. Enterprises may lack structures for making decisions about data, for example how to prioritize data problems, what data development project to fund and resource, what standard processes and tools to use to profile and cleanse the data, etc.

Governance involves the delegation of responsibility and accountability. Governance is only effective when the decision rights have been granted by higher authority and are enforced by those who are responsible and accountable. Governance succeeds when the chain of responsibility and accountability is unbroken, and the expectations are documented and published. It is a simple concept but often overlooked when creating a data governance program. That is why many governance programs fail. Either they have not obtained the authority to govern from the *senior* leadership team, or the senior leadership fails to provide the appropriate backing. When the chain of responsibility and accountability has been interrupted then effectiveness of controls and standards has been undermined. This can be avoided by ensuring the leadership team stays engaged *and* supportive, that expectations are documented, and that compliance is monitored. These need to be non-negotiable elements of a data governance program.

Data governance programs are all unique and should be developed to support and enhance specific enterprise data assets. Data governance programs typically have the following roles with responsibilities.

Chief Data Officer: a new occupation in the career field of information systems management, a Chief Data Officer (CDO) is at the corporate level of an organization and responsible for enterprise-wide data distribution, processing and data mining information. The role of the Chief Data Officer also extends into the field of business intelligence for an organization. An individual in the position of Chief Data Officer typically reports to a Chief Information Officer (CIO) or a Chief Technology Officer (CTO). The CDO is responsible for creating strategic policies pertaining to data over a vast array of network systems for a corporation. Policies include strategies addressing security, acquisition of new systems, corporate data management policy and revenue opportunities. The CDO explains to executive management, employees and customers the value of data and the role of data as an asset and revenue generator to ensure profitable returns to an organization. The CDO also creates an annual corporate budget for data activities, equipment and processing. The CDO is responsible for system administrators, network administrators and technicians and programmers, which are in different geographical locations. The CDO constantly corresponds with managers and supervisors and may conduct monthly "how goes it" meetings to discuss items ranging from upgrading legacy data systems to implementing new systems. A CDO also conducts enquiries and research into other data companies willing to merge to create a stronger information-processing platform, which translates into more revenue. These strategies are discussed with executive and corporate management.

Executive Sponsor: sets the initial goals and direction for the data governance program, reviews and approves corporate data governance policies, procedures and standards, influences, reviews and approves corporate data governance strategy and objects, promotes and communicates the data governance program across the enterprise to ensure acceptance and success, clarifies business strategies to the corporate data governance board, influences data sponsors and users to authorize funding for corporate data governance work, i.e., budget and resource allocations.

Executive Steering Committee: approves information policy and data architecture principles enterprise-wide, provides the vision, strategic direction and objectives for the enterprise data governance program, estimates high level funding needs, requests budget from the executive sponsor and approves data quality initiatives, provides the vision, strategic direction and objectives for data initiatives across the organization (including the data governance and stewardship program), approves prioritized data initiatives, allocates corporate funding for approved data initiatives, acts as a champion for approved data initiatives, drives data initiative awareness and required change within the organization, provides final resolution for data related issues and conflicts and monitors data initiative progress and budget, and redirects if necessary.

Data Governance Board: develops and monitors an overall strategic plan for enterprise-wide data governance, sponsors, approves, prioritizes and champions data governance initiatives for all LOBs and functions, ensures scheduling and resource allocations are in place to support the data governance program, communicates with business segments to ensure expectations for data quality initiatives are in-line with enterprise reporting and ensures resolution of escalated issues.

Chair: leads and guides the corporate data governance board to achieve their objectives, directs the data governance team to estimate budget and resources, directs the data governance team to develop and implement strategy and objectives, builds consensus and resolves conflicts with data governance users, and monitors the progress of the data governance work

Business Sponsors: review and approve corporate data governance policies, procedures and standards and sponsors fixing and sustaining corporate data quality initiatives

User Representatives: prioritization of corporate data quality, data requirements – intent use of the data, reviews and approves corporate data governance policies, procedures and standards and helps sponsor fixing and sustaining corporate data quality initiatives.

Enterprise Data Architect: coordination for all enterprise data related approved initiatives, focuses on planning for infrastructure efficiencies and linkage, cleansing and usage of data, and ensures implementation of remediation and the priority of issues, ensures compliance/execution of the data governance program policies, processes and procedures across data steward, and reconciles and recreates data definition design and maintenance.

Data Ownership: holds overall responsibility for data, assigns ownership of key data elements to data stewards, accountable for business specific KPIs, assigns data element quality within business areas, implements data quality monitoring and controls, provides data quality updates to management operating committees and provides data quality feedback to business data owners.

Data Stewards: manage data quality within businesses for data owners, responsible for data element accuracy and controls, resolve data quality issues, responsible for data quality monitoring and control for all lifecycle stages: collection, keying, editing, viewing, calculation, and reporting, reports business specific KPIs and provide data quality monitoring reports to the data owner.

Data Quality Leader: leads the data quality program, the DQL has significant data management expertise and is deeply involved in all aspects of the program while also participating in the data governance board and executive steering committee, focal point for coordinating data owners and data stewards, owner of scorecard process and execution, provides scorecard feedback to all involved parties, ensures execution of policies and strategies of the data governance board and steering committee and reviews and prioritizes projects, determines funding needs and requests funding approval from the data quality steering committee.

Data Governance Technical Stewards: The technical steward is a member of the IT department and focuses on the technical metadata which needs to be captured by the data quality initiatives. Technical stewards work closely with business stewards to create metadata definitions and measurements.

27.2.5 Implementation: responsibilities and accountabilities

Data governance programs require defined data responsibilities and accountabilities to ensure successful implementation and sustainment. Data governance is one of many priorities organizations juggle each day to ensure customers are satisfied and regulations are met. Data governance responsibilities and accountabilities should be clear, measured and managed to ensure the on-going success and profitability of an organization is achieved.

To help drive data governance success, measurements and metrics may be put in place which define and structure data quality expectations across the enterprise. As David Packard once said "That which gets measured gets done." When data are assigned to data owners with specific quality requirements and supporting measurements, they will be integrated into the enterprise pathos and data governance will be successful.

Defining data ownership and quality requirements, with supporting measurements and metrics, will enable organizations to achieve the benefits of a data governance program.

27.2.6 Charter and working groups

The data governance charter is the statement of intent for the program. It should begin with the data vision and articulate the desired end state. A charter is created at the beginning of the effort, directly following the creation and approval of the program's vision, taking a strategic view of the objectives and anticipated results. A charter is sponsored by the executive sponsor and approved by the key project stakeholders, ensuring it is available for reference throughout the data governance program effort. The charter captures the goals, objectives, purpose, supporting organizations, duties and responsibilities of the data governance program. It is a high-level document and details the goals of the program and what results are expected over time.

A data governance charter should state clearly the need to raise awareness of data quality, its real importance to the organization and some goals that will be realized by the establishment of a permanent data governance effort. The charter is a first step, and serves as the foundation, for all data governance program activities. See Sections 27.5 and 27.6 for additional details.

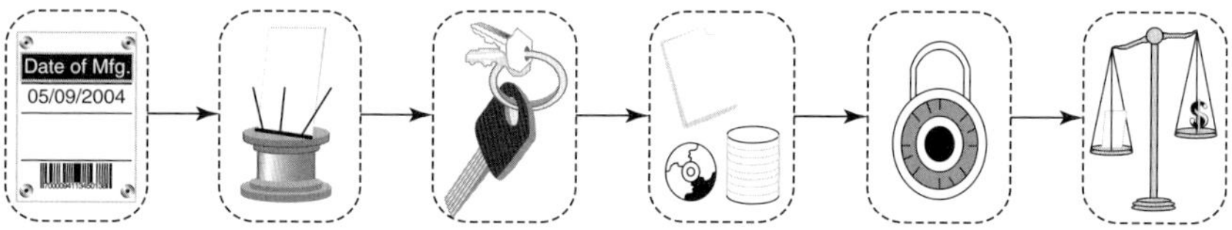

27.2.7 Data governance models

It is important to define the organizational approach and model which will be used for the data governance program execution. Things to consider in selecting an approach and model include what types of decisions the governance bodies will be called upon to make as well as the policies and standards which will be established. Data governance programs should consider the direction decisions will flow across the enterprise: top-down, bottom-up, center-out or silo-in (Data Governance Institute: Models). The data governance program may employ all four of these models at different times. Understanding them will help design structures with necessary linkages.

Top-down governance and decision flows

Top-down data governance and decision models are based upon command and control authority patterns. When decisions are made they are not typically open to debate and compliance is expected. Participants are assumed to support and implement the decisions. For the data governance framework to be successfully implemented using the top-down approach, a clear linkage between executive-level data governance committees and other elements of the program must be established. It is important that the data governance executive sponsor, his/her delegates and data governance board members have the time to establish that link.

All data governance implementation models are influenced by the alignment of executive sponsor "talk" and "walk." An official policy is important and is driven to success when leadership continually re-enforces the importance of adherence to the policy by middle management and line staff. How leadership speaks and behaves is the key to success for enterprise data governance.

Bottom-up data governance and decision flows

In contrast, the bottom-up governance model has some data-related decisions made by individuals as a part of their everyday work, and the results bubble up through an organization. Data naming standards are an example of bottom-up data governance model. Sometimes naming standards originate with a single, dedicated team. Eventually they become broader standards, and then perhaps later enterprise standards.

For the bottom-up data governance model to be successful, individuals with stewardship responsibilities must know who their point of contact is within the stewardship hierarchy and/or data governance office. Suggestions and issues and requirements must be able to "bubble up" to higher levels.

Center-out data governance and decision flows

An enterprise data governance office (or CIO or other leader) may ask experts to specify data models, to interpret compliance requirements, to design controls, or to set protocols. Center-out decisions are made by one or more centralized resources that consider options and then decide what is best for the enterprise. Leadership may ultimately issue a mandate (make a top-down decision), but before that happens it is the centralized group that considers options and their impact on stakeholders.

For center-out data governance and decision flows to be successful, multiple paths of persuasion must be set up to convince stakeholders to abide by a center-out decision. Considerations include:

- Asking leadership to set a strong "tone from the top" regarding compliance,
- Employing multiple communication channels to send clear messages about the importance of the decision and the need for compliance,
- Educating stakeholders about why the decision was made, and what alternatives were considered.

A clear exception process must be set up with issue escalation paths and issue resolution processes.

Silo-in data governance and decision flows

The silo-in decision flow brings together representatives from multiple groups to collectively agree on a course of action. Data governance and stewardship councils are examples of this model. With objectives of balancing the needs of individual silos against the needs of the entire organization, such groups meet to issue policy, set standards for the enterprise, and decide how to resolve data-related issues. In making silo-in decisions, stewardship councils are expected to provide multiple perspectives. Its members are expected to consider impacts to stakeholder groups across the enterprise. Sometimes this model is followed even though a central group of experts has agreed upon a recommendation for a course of action since, in a federated organization, members will often accept a tough decision made by a group better than an edict from a centralized source.

Silo-in decision flow models can be very effective at making "decisions that stick". They eliminate the argument that stakeholders were left out of a decision-making process and therefore should not be forced to abide by the decision. Participants become invested in the process and are more likely to promote compliance.

For this model to work, the group must be granted authority by leadership to make decisions on behalf of the enterprise. Representatives must be authorized to act on behalf of their groups, and they must be respected by those they represent so their decisions will be accepted.

Hybrid data governance models

Some data governance model decisions need to be made at multiple levels. One group may craft a recommendation, and then a higher-level group will make the final decision. When politics or budget/resource impacts are involved, a hybrid decision flow model may be called for with a center-out or silo-in recommendation followed by a top-down decision.

27.2.8 Data governance stakeholder matrix and communications plan

The data governance Responsibility, Accountability, Consulting, Information (RACI) matrix (a common responsibility matrix, see Figure 27.4), shows the chain of responsibility and accountability and serves to explain the importance of documenting the intention of controls and linking them to higher authority.

Successful data governance efforts demand and are highly dependent upon effective communications. Data cross organizational boundaries, may have different meanings across silos and are shifting the view from data as a by-product of processes to data as an enterprise asset. Therefore communication must be defined and executable. Communication plans should be divided into two categories: program communications and constituent communications. Program communications include the

Data Governance Intent Communications

Data Governance Authority

	People	Legislature	Regulators	Board of Directors	Senior Executives	Directors	Program Managers	Operations Managers	Emplo-yees
Governance Intent									
Constitution	RA	CI							
Laws	CI	RA	CI						
Rules & Regulations		CI	RA	CI					
Policy			CI	RA	CI				
Program Charters				CI	RA	CI			
Controls, Standards & Projects					CI	RA	CI		
Business Processes						CI	RA	CI	
Business Process Steps							CI	RA	CI
Business Process-Sub Steps								CI	RA

Figure 27.4 Data governance RACI matrix.

interaction that is required in order to deliver the program with quality. Included in this category are the program plan itself, the memos and status reports that the team members distribute to one another, team meetings, and the minutes that are issued from them. Constituent communications are the executive briefings, auditorium presentations, newsletters, "lunch and learn" sessions, posters, brochures, focus groups, and feedback mechanisms like email and voicemail suggestion boxes. Let us explore all these elements in a bit more detail.

Program communications

The most important component of the communications plan is the program work breakdown plan itself. It enables each data governance participant to understand in detail the goals and tasks of the program. By committing to paper the deliverables, the responsible parties, the scheduled delivery dates, and the dependencies, the program is broken down into its component parts, making it much easier to comprehend and discuss at the level of detail necessary.

Memos and status reports are important program communication channels, as they give team members the chance to raise issues, "red flags," unexpected results, and impacts on the program that were not foreseeable before they occurred.

Program meetings are also important focus points of communications. Meetings that are well run, with agendas, time contracts, full team participation, and an opportunity for honest and open exchange, can impart the overall vision to the team, build commitment, and ensure success. Meeting leaders should use the agenda to facilitate the discussion, to encourage everyone to participate, to get status reports from all so that everyone builds an overall vision of the program, and to expose and resolve open issues. Formal meetings should be captured in minutes, especially action items and deliverables. The meeting is the cornerstone for project team building and for issue resolution.

Constituent communications

Constituent communications are specifically designed to inform the client and all the engagement stakeholders. Some of the elements that a constituent communication program should contain are as follows.

- *Executive briefings*: Executive sponsors of the data governance program should be briefed periodically on the status of the project. This communication should be kept at a high level, typically in the form of a short presentation that focuses on the strategic, not the tactical, issues.
- *Assembly presentations*: Data governance is a company-wide initiative and should be presented to the employees to gain exposure and support for the program. This type of communication is typically focused on two aspects: features and benefits, otherwise known as "what's in it for me?" along with any requirements or support the program will need from them.
- *Newsletters*: Newsletter formats allow for periodic, informal communications that can keep the organization in the loop on the status and schedule of the data governance program. This is also a good forum for executives of the company to sell the program through things like open memos or "From the President's Desk" inserts.
- *Lunch-and-learn sessions*: Some organizations find it useful to set up a mini trade show environment, where representatives from the data governance program team meet with employees and answer questions, address concerns, and sell the program in an informal, unscripted environment.
- *Focus groups*: Sometimes it makes sense to deliver focused data governance presentations to specific groups or teams within the organization to address their specific concerns and questions. For instance, the marketing group might be interested in the impact the data governance program on their customer reporting and competitive analysis efforts.
- *Program marketing*: Successful data governance programs understand and embrace the need for clear and frequent communications. Clearly stated, data governance programs need to be marketed to drive awareness, initiate "buzz" and gain support. Some data governance programs have designed logos, developed "brand names" and distributed posters and brochures to advertise the features and benefits of the program.

Other communication conduits include customer surveys, electronic suggestion boxes, and intranet sites. The key point is that data governance programs succeed not only due to business and technical expertise but also due to their keen ability to share knowledge, to converse, to create a shared vision, to build team spirit and consensus – in short, to communicate.

27.2.9 Master data management

Master data may be described as data which are shared across organizational and functional boundaries and are also critical to the operations of the enterprise. In practice master data are a small portion of most enterprises data libraries. Most data do not fit into the master data category and therefore are not subject to the high data and access requirements inherent in data entities deemed master data. See Section 27.7 for additional details.

Master data can be identified and described by the way that they interacts with other data. Often master data have the same relationship data warehouse facts and dimensions share, verb-noun. For example, in a transaction system master data (nouns) are often the key data requirements: a *customer* buys a *product*, a *vendor* sells a *part*, a *partner* delivers a crate of materials to a *location,* an *employee* is related to their manager, and a *product* may be a part of multiple hierarchies describing their placement within a *store*. Transactions capture the verbs, such as sale, delivery, purchase, email, and revocation; and master data are the nouns.

Master data tend to be less dynamic than transactional data. For example, some might consider "contract" a master data element. Others might consider it a transaction. The deciding factor is the lifespan of a contract. An enterprise selling computer products would consider contracts master data. Each contract is different from the others and typically has a lifetime of greater than a year. The agency enterprise manages all contracts over time, as elements of the contract are renegotiated or products added, updated or removed. The less complex a data entity or element, the less likely it is to be master data. The more valuable data are to the enterprise, the more likely they are to be considered master data. Ultimately, when deciding on what entity types should be treated as master data, it is better to categorize them in terms of their behavior and attributes within the context of the business needs than to rely on simple lists of entity types.

A primary characteristic of master data is reuse. For example, in a simple world, the CRM system would manage everything about a customer and never need to share any information about the customer with other systems. However, in today's complex environments, customer information needs to be shared across multiple applications. That is where the trouble begins. Because – for a number of reasons – access to master customer data is not always available, people start storing master data in various locations, such as spreadsheets and application private stores. There are still reasons, such as data quality degradation and decay, to manage master data that are not reused across the enterprise. However, if a primary entity is reused in multiple systems, it is a good candidate to be tagged master data.

The challenges that enterprises experience around consistent reporting, regulatory compliance, risk assessment and management has prompted a great deal of interest in Master Data Management (MDM). It is both an analytical and technical framework which identifies and technically manages enterprise critical data. It does so by creating one critical data environment which provides a single point of truth and management. When implemented, MDM streamlines data sharing among personnel and enterprise functions and departments. MDM can facilitate computing in multiple system architectures, platforms and applications.

The benefits of the MDM paradigm increase as the number and diversity of organizational departments, worker roles and computing applications expand. For this reason, MDM is more likely to be more critical to large or complex enterprises than to small, medium-sized or simple ones. When companies merge, the implementation of MDM can minimize confusion and optimize the efficiency of the new, larger organization.

For MDM to function at its best, personnel, functions and departments should be taught how data are to be accessed and used. MDM typically has two different solution approaches, a single master hub, or distributed master data which is unified through syncing processes. Frequent and coordinated updates to master data are also essential.

What do Master Data Management (MDM) and data governance have in common? Data Governance is an over-arching framework for the ownership, quality and usage of data. Master data management provides an effective means to identify data which are shared across organizations and therefore are required to present a "single version of the truth." The costs of poor MDM are hidden in inefficient business processes and excessive data management activities and, more importantly, in the inability to use information for strategic business advantage. These all translate into lost opportunities to decrease risk and increase profitability.

MDM is a technology-enabled business discipline in which business and IT organizations work together to ensure the uniformity, accuracy, stewardship, semantic consistency and accountability of the organization's official, shared master data assets. Today, most organizations juggle multiple sets of business and data applications across corporate, regional and local systems. At the same time, customers are demanding faster and more complex responses from organizations, leading to an inconsistency that hinders the organization's ability to measure and move within the market. With MDM, CIOs can create a unified view of existing data, leading to greater enterprise agility, simplified integration and, ultimately, improved profitability.

Consider a recent report from The Data Warehousing Institute that found 83% of organizations suffer from bad data for reasons that have nothing to do with technology. Among the causes of poor quality data were inaccurate reporting, internal disagreements over which data are appropriate, and incorrect definitions rendering the data unusable. Organizations must understand that improving their data – and building the foundation for MDM – requires them to address internal disagreements and broken processes.

Staff must agree on exactly what constitutes a "customer" or a "partner," and how to resolve any disagreements across business units. Departments and divisions need to agree on hierarchies of customers and products and how to resolve duplicate records across sources. Rather than a technology-focused effort, the project becomes one of political strategy and consensus building. For any CIO who is contemplating an MDM strategy, it is important to start small and refine your approach over time.

The enterprise benefits of implementing a master data management solution are far reaching. Each source, risk analysis, compliance, reporting and analytical system will be assured high quality and accessible data supplied by one version of the truth, master data.

27.2.10 Metadata management

Metadata is data describing or specifying other data, or defining and organizing characteristics that need to be known about data, including (but not limited to) information about physical data representations, technical and business processes, rules and constraints, and data structures. Sound metadata management practices ensure that information needed to support management and business intelligence decision making is accurate, accessible, *complete*, consistent, timely, valid and relevant and assists in maintaining high integrity over production data. Metadata policy, standards and best practices should be established. The business owner should maintain business metadata, including business rules, organizational metadata and transformational metadata. Business metadata should include: who owns the data (data owner), who maintains the data (data steward), what data means (entity, attribute definitions and sample data), where the data comes from (how it is collected), where is the data stored (application system) when data is collected (transactions, daily feeds, monthly feeds) and how it is used (management information, marketing analytics). See Section 27.8 for additional details.

The IT owner should maintain application metadata, including technical and physical metadata. Application data should include who maintains the application

(IT owner), who maintains the database (production Database Administrator (DBA), and who is responsible for change control.

The IT owner is also responsible for what the data look like (logical/physical/data models), what the data means (table/column definitions and sample data), what the data profile looks like (for each table, column, the number of rows, min/max values, distinct values) and what interfaces feed into and feed out of the application.

Metadata should be adequately controlled and documented in an approved and published metadata repository and metadata back-up/restore procedures should be established. Metadata users should be assigned appropriate access rights and privileges. Metadata should be periodically reviewed by business owners for accuracy, timeliness, relevancy, consistency and validity. Processes should be established to move data model metadata to the corporate metadata repository.

Metadata should also be stored in an approved enterprise metadata repository that complies with an approved information technology standards list. Information assurance should prevent unauthorized access to the metadata repository. Data quality objectives should be included in metadata repositories. Accountabilities for each key role involved in metadata management should be defined and communicated to all stakeholders. Project management plans and project quality management plans should include provision for metadata quality assurance reviews and approvals. Processes should be defined to specify how metadata is maintained, for example additions, deletions and modifications. Human resources policies should specify that business owners are responsible for maintaining business metadata, including business rules, organizational metadata and transformational metadata, and roles, responsibilities and position descriptions should be clearly defined. The business intelligence framework should include provision for metadata management standards and best practices. The change management methodology and framework should ensure that changes to metadata follow a documented and approved change management process. The release management process should include provision for migrating metadata to production (Bowman).

27.2.11 Securing high value data

High value data come in a variety of shapes, sizes and forms in industry and are often considered intellectual property like designs or research information. It can be data that requires extra attention and control because of its high value to the enterprise and sensitivity. It can be data which is regulated by law or statute, or it might be information that is critical to the operations of the enterprise. All high value data can benefit from *applying* common and integrated data governance.

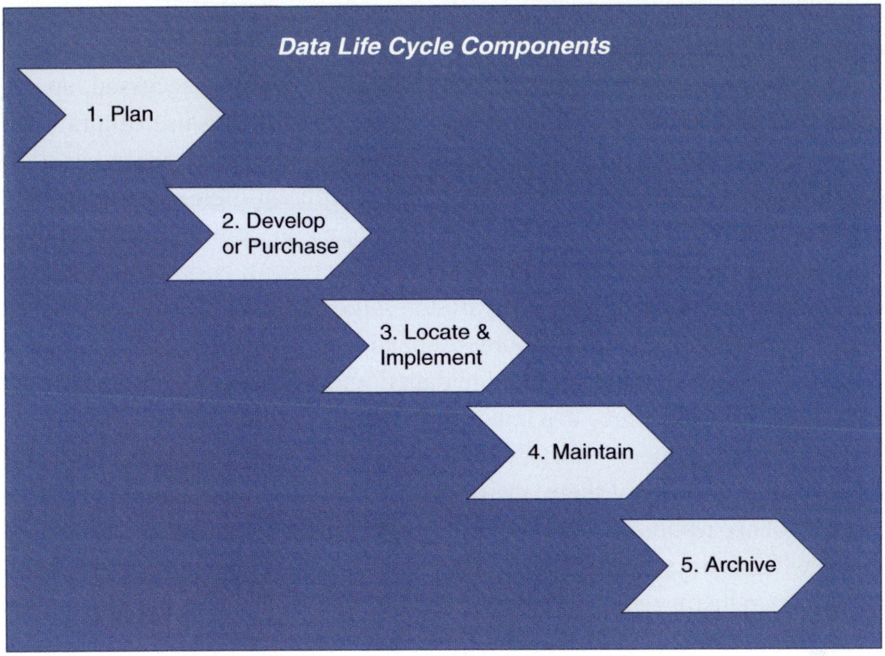

Figure 27.5 Lifecycle of high value data.

The lifecycle of high value data domains is consistent and uses common decision making and controls around the data. Looking at the lifecycle of data in Figure 27.5, we can identify some key steps in the lifecycle.

The opportunity to align and integrate efforts associated with high value data will render exponential benefits when compared to the more costly individual investments associated with individual data quality efforts.

When planning data governance for compliance oriented data, quite often it requires the same subject matter experts (SMEs) in laws, regulations, and policy. Having these SMEs look at planning holistically for high value data can help eliminate redundant decision making efforts and help identify complimentary controls that can help address needs across multiple domains. In addition, these high value data domains often require enterprise-wide interventions in order to create change (or in this case, drive new information governance). Enterprise initiatives are often difficult to move forward due their breadth and stakeholders that span multiple areas. By bringing high value data initiatives together during planning, the aggregate effort will gather larger sponsorship than having each initiative present an independent investment case.

Governance associated with the creation and acquisition of high value data may be critically important to the overall lifecycle of high value data and can help minimize risks. As an example, both privacy and security are interwoven. If an organization chooses to acquire and make use of personal data then there will be a resulting set of risks and costs associated with securing that data.

Most high value data requires limiting how it is used and accessed, specifying who can use it, determining where it can reside, and making use of audit trails to substantiate who did what, to what data. The decision making processes and how you limit access are often common across these different high value data domains. In addition, the types of technology solutions one can deploy to enable the right access controls, such as *identity* and access management, encryption, or data loss prevention are very similar and often are the same solution.

High value data disposition has interdependence across decisions, controls and approaches for each of these high value data domains. As an example, the disposition of records and adherence to a retention schedule can have direct impact, good or bad, on risk management efforts. Often, companies end up storing unnecessary volumes of data because of their lack of governance/compliance to record retention procedures. This results in exponential growth in data storage costs and greater exposure when processing legal requests.

The common themes and benefits of high value data can be summarized as follows.

- High value data often has common decision makers across the enterprise.
- Bringing high value data initiatives together drives success in mobilizing sponsorship of investments versus having each high value data area make its case on its own.
- High value data lifecycle decisions may have a significant impact on risk management.
- Technologies which enable limited access, appropriate use, and security are often the same across the high value data domains.

27.2.12 Data quality

Data quality requires the implementation of measures that a SMART (specific, measurable, actionable, relevant and timely) dashboard enables. Each data lifecycle stage should have the following required quality dimensions.

1. Creation and update:
   (a) *Accuracy*: measures the extent to which data are free of error and the degree to which they represent the corresponding real world object
   (b) *Validity*: measures the extent to which data conform to a set of domain values or rules
   (c) *Completeness*: measures the extent to which data values for an entry process have been collected and entered
   (d) *Consistency*: measures the extent to which data held in multiple database repositories conform to the same syntactic and semantic rules

   (e) *Timeliness*: measures the extent to which data are available to other processes in an acceptable timeframe following data entry or update.
2. Data integration:
   (a) *Completeness*: measures the extent to which all required data sources contain required values
   (b) *Consistency*: measures how data held in multiple database repositories conform to the same syntactic and semantic rules
   (c) *Accessibility*: measures which data sources required are available and accessible
   (d) *Timeliness*: measures the extent to which all data sources and feed are available to the data integration services in the required time frame.
3. Data presentation and data models:
   (a) *Relevance*: measures the extent to which the information is applicable and useful for the purpose intended
   (b) *Transactability*: measures the extent to which data conveys a clear, intended meaning
   (c) *Consistency*: measures how well data are presented in a consistent, unambiguous manner across every medium.
4. Data dictionaries and metadata repositories:
   (a) *Completeness*: measures modeled data definitions and relationships
   (b) *Consistency*: measures how data definitions and relationships are compatible across multiple models and dictionaries.

27.2.13 Data quality processes

Quality data is the key to effective decision-making and planning. The purpose of data governance is to have an integrated, single source of data truth to be used across the enterprise in operations, reporting and analysis. It is of critical importance that data quality focuses on two objectives: high quality and constant execution. Data quality is an on-going activity effort. Data quality is never complete.

Effective data quality depends upon the transformation and monitoring of data on an on-going basis. Data quality processes

- Provide end-to-end data quality with measurements
- Enable data quality and data profiling as an integral part component of data integration processes
- Store metadata with high quality data in a repository
- Create rules to apply to each data lifecycle and processing step.

Ensuring data quality involves the following phases:

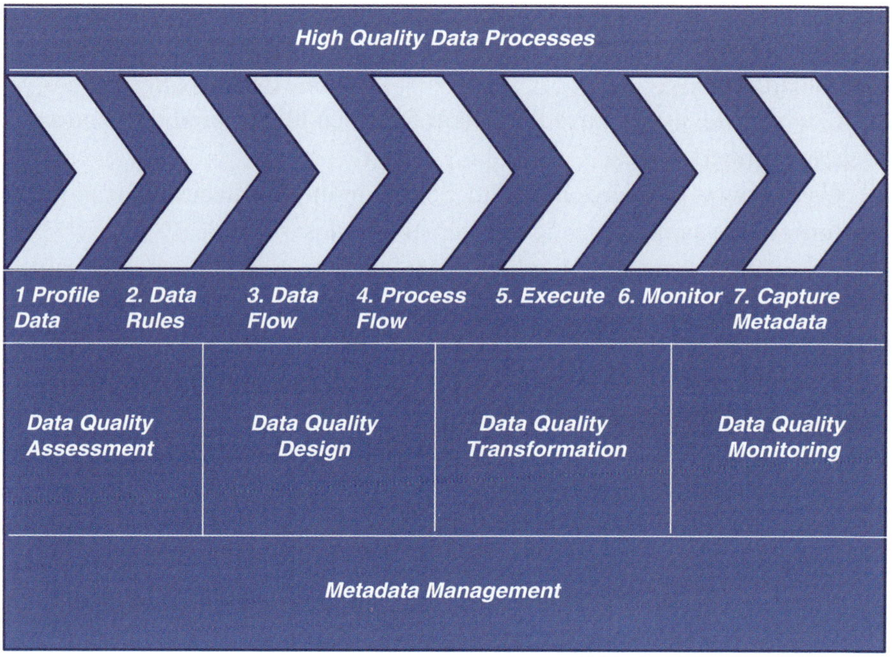

Figure 27.6 High quality data processes.

- Quality assessment
- Quality design
- Quality transformation
- Quality monitoring.

The first process in data quality establishes data sources. While data may be resident within many different transactional, reporting and analytical systems, typically the system which creates the data is nominated as the data's source system.

The second data quality process is profiling, which uncovers data anomalies, inconsistencies, and redundancies by analyzing the content, structure and relationships within the data. The data profiling analysis and data discovery techniques form the basis for data monitoring going forward.

Data profiling drives improved information quality and provides better decisions. It is a robust data analysis method to discover and measure defects in data. There are three types of data profiling.

Attribute analysis

Attribute analysis seeks to discover both general and detailed information about the structure and content of data stored within a given column or attribute. Attribute analysis looks for information about patterns, domains, data types, and unique values.

Pattern analysis attempts to discover patterns and common types of records by analyzing the string of data stored in the attribute. It identifies the percentages of data that comply with a defined regular expression format pattern found in the attribute. Using pattern results, data rules and constraints can be developed to cleanse data problems. Some commonly identified patterns include dates, email addresses, phone numbers, and social security numbers.

Domain analysis identifies a domain or set of commonly used values within the attribute by capturing the most frequently occurring values. For example, the status column in the customers table is profiled and the results reveal that 90% of the values are among the following: MARRIED, SINGLE, DIVORCED. Further data analysis reveals another 10% contains misspelled versions of these words with few exceptions.

Data type analysis enables discovery of information about data types found in the attributes. This type of analysis reveals metrics such as minimum and maximum character length values as well as scale and precision ranges.

Unique key analysis provides information to assist in determining whether or not an attribute is a unique key. It does this by looking at the percentages of distinct values that occur in the attribute. Perhaps attributes with a minimum of 70% distinct values should be flagged for unique key analysis. For example, using unique key analysis one could discover that 95% of the values in the EMP_ID column are unique. Further analysis of the other 5% reveals that most of these values are either duplicates or nulls. Derive and build a data rule that requires that all entries into the EMP_ID column be unique and not null.

Functional dependency

Functional dependency analysis reveals information about column relationships. This enables one to search for things such as one attribute determining another attribute within an object.

Referential analysis

Referential analysis attempts to detect aspects of data objects that refer to other objects. The purpose behind this type of analysis is to provide insight into how object profiling is related or connected to other objects. Because two objects are being compared in this type of analysis, one is often referred to as the parent object and the other as the child object. Some of the common things detected include orphans, childless objects, redundant objects, and joins. Orphans are values that are found in the child object, but not found in the parent object. Childless objects are values that are found in the parent object, but not found in the child object. Redundant attributes are values that exist in both the parent and child objects.

Data profiling

Data profiling is the analytical process by which data contents are examined and statistical information about the data is collected. The purpose is to discover the structure, content, and quality of the data – and to do this whenever data are being converted, migrated, warehoused or mined. Note that data profiling is defined as a process, and not a tool. The data profiling process breaks down into the following steps.

1. Identify the business requirements for data quality and business data candidates for data profiling.
2. Identify required source systems, we well as files, tables, and data elements to be included.
3. Transfer the data into an environment where large-scale data examination may be performed without impacting the transaction system or data warehouse response times.
4. Examine the contents of the database to see what is there.
5. Construct a prototype set of data quality rules to use for data comparison. These rules may be developed using profiling tools or they could be hypotheses on the part of business domain experts.
6. Compare data quality rules against the data to determine how well the proposed data quality rules match the data. Check the results with business domain experts well versed in the data.
7. Present findings to the business domain communities.
8. Gather requirements on how to use the results to improve the data quality processes.
9. If there is a large gap between the stated data quality rules and the data itself this may requirement refinement of the data quality rules by the business domain experts.

There are significant benefits to performing data profiling. By verifying the availability of information and validating the quality of that information, data usage risks are lowered across the enterprise.

Having a robust, repeatable data quality framework supported by tools and data governance processes will ensure the following benefits:

- Quality data for compliance and audit requirements
- On-going and measureable data quality measurements
- Single version of the truth to support business decision making.

There are three main types of data profiling activities: discovery of "new" data quality rules, assertion testing of stated data quality rules, and visual inspection. Data profiling discovery is the ability to inspect huge volumes of data in a database and files and look for potential data quality rules and relationships in the data. Data quality rules and relationships may well have been lost as well as represent information data analysts are not aware of. The effort requires considerable commitment and resources as well as tools, multiple processors and threads. Most large-scale installations replicate the data into another environment to avoid impacting the data warehouse or transaction environments, but this actually makes it more difficult to do data profiling because of all the overhead required for replication. Thus, a prime consideration when evaluating a profiling tool is whether it (and the supporting repository) will scale as data storage increases in size.

This discovery process uses sophisticated algorithms to detect the data quality rules and relationships, and presents the results along with violations. Results can include patterns, actual ranges of values, lists of valid values, number of nulls, incidence of repeated values, records that have a relationship, and more. The tool might detect that a field contains only four valid values. It would present the list of valid values, the distribution of those values, and information on the number of times a different value was found as well as a sample of the data records where a different value was found.

Once the tool has reported on what the data quality rules are, it is up to the data quality analyst to review those data quality rules with knowledgeable business people to discover and verify true data quality rules, and anomalies. The tool may also identify a data quality rule incorrectly. For example a misidentified narrow range of numbers versus a valid set of values for an attribute.

Data profiling assertion testing

With assertion testing, data quality rules are stated prior to examining and profiling the data. The data quality rules typically come from three sources: business analysts, documentation, and program logic. Data quality rules collected from business analysts typically are identified via complaints about perceived data quality issues, or as issues rose during a project. The complaints tend to be straightforward to resolve. A data quality analyst questions a business analyst during an interview to ascertain what the data quality rule is, as well as how important the business analyst considers the perceived violation of the data quality rule. This data quality rule should then be validated by the data steward for that element, and if the data steward agrees, the data quality rule would then be documented for use in the assertion testing process. The process for collecting the data quality rule during a project is similar, except that the issues are raised during project meetings and evaluated by the project manager with the data steward, to determine if the data quality rule is worth investigating.

Typically very little is known about a system and the data generated by the system by the people who use the system. In that case, the data quality rules may be determined by checking existing documentation. This documentation may have been supplied by the original vendor, or be part of the documentation for projects that customized the application. As with data quality rules gathered from analysts, the data quality rules have to be evaluated against business requirements to determine if the data quality rule is worth looking into.

Finally, the actual program code may be checked to see what data quality rules are present. This effort may be daunting, and requires both a programmer and a business analyst to evaluate the results. Program code has an advantage over both talking to business analysts and reading documentation as it is clear when describing what data quality rules the system is enforcing. Further, it is the running code which controls the actual data content that is produced, and is thus the final authority.

Once the data quality analyst has gathered the data quality rules from various sources and evaluated the rules, they are "programmed" into an assertion testing data profiling tool. The last step is to run the tool, and evaluate the results. The tool output indicates how well the data quality rule matched the data as well as rendering a sample of data quality rule violations.

Data profiling visual inspection

The last type of data profiling involves visual inspection of data. This technique finds data inaccuracies that are not easily formulated as boundaries, limits, or data quality rules. Inspection helps to find problems with business-related quantities, such as:

- Frequency distribution of values
- Sums and totals of values
- Comparing counts or totals from one data source to another, especially multiple systems that source the same data (or data that the business THINKS should be same)
- Values with counts less than some interesting threshold
- "Interesting" text strings, such as "Don't know," "Not Provided," "Company A," or "999–99–9999."

27.2.14 The impacts on metadata

Besides improved quality, one of the key deliverables from a data profiling effort is metadata: information about the data and the data quality rules that apply to the data.

There are two kinds of metadata related to data quality: data quality rules metadata and data quality results metadata. The data quality rule metadata records the data quality rules themselves, which may come from data stewards, documentation, or program code, and the data element upon which the quality rules are attached. Data quality rules may be complex and require a database for capture. Going through the process of identifying and capturing data quality rules enable robust implementation of datas quality across current and future activities.

Data profiling results in a determination that:

- The interpretation of the data given by the metadata is correct and the data are incorrect
- The data are correct and the metadata (data quality rules) are incorrect
- Both the interpretation and the rules are incorrect.

With so much information coming out of data profiling, it is imperative to have a "system of record" to record the data quality rules and results metadata. This metadata repository provides a single version of the truth where analysts may check current data quality rules, currently known data quality issues, and the results of data profiling including annotations from prior analysis.

Understanding column property analysis

Column property analysis analyzes individual columns (fields) to understand all the values that appear in the column. It looks at the values stored in a column independent of other columns. This type of profiling is both the easiest to do and the easiest for the business to understand. The data quality rules tend to be the most straightforward and understandable. The output provides data on the values, frequency, uniqueness, patterns, data type, size and length, ranges, and maximum and minimum values. The results of column property analysis enable analysts to:

- Discover the metadata and content quality problems
- Validate that data conform to expectations
- Compare actual data to target requirements, and
- Analyze invalid, default, and missing values.

Understanding structural analysis

Structural analysis focuses on the quality of data structures, how various types of data (such as patient and prescription, or customer and order, or borrower and account) are tied together through relationships implemented in the data capture. It looks at defining identifiers (such as customer ID) to evaluate whether these attributes are in fact unique. It also ensures that data existence is supported by data relationships. For example, this type of analysis will detect whether an account exists without a customer. It could also detect incidences where the customer should not exist because there is no account (or prescription, or order, etc.).

Structural analysis requires analysis to evaluate the results. Unlike column properties analysis, structural analysis requires that a business come to an agreement on the definition and relationships between data structures that are shared across the enterprise.

Understanding data relationship analysis

Data relationship analysis tests whether data relationships are correct. The first test analyzes the relationships within a single record or business object to ensure acceptable combinations of values. For example, a customer record might have a field for the customer's name, as well as a type code which can take the values business or personal. The data quality rule might state that if the customer type was "personal," the customer's name was mandatory. But if the customer type was "business," then the customer's name was optional.

The second type of data relationship testing requires that data values conform to the relationship over multiple tables or business objects. This finds prohibited conditions, such as a customer cannot be both a retail customer and a wholesale customer. It also finds missing or erroneous records, such as a home equity loan must have a matching collateral record. Having a tool which enables cross-comparisons of domains will help to uncover data quality issues.

Understanding domain analysis

Domain analysis looks at values to try and discover aggregations, counts, or frequencies that appear to be unreasonable. Domain analysis enables the business analyst to spot things like:

- Significant changes in volumes (25% fewer loan applications than last month)
- Significant changes in value (loan portfolio value jumped 40% since last quarter)
- "Odd" relationships (loan-to-phone number ratio is 1.02 to 1, but several phone numbers show 25 to 1 ratios).

Using the data profiling results to advantage

Data profiling leads to a concise, documented, and validated set of data quality rules and requirements. These data quality rules can be used to detect errors in the data before they are loaded into a storage environment or merged into a consolidated system. Data quality rules may be applied when the data are moved from the source to the destination and corrective action taken during this load process to call attention to the errors or prevent bad data from being transferred. To handle the detection of data quality errors during loading the following is required.

- A list of data quality rules to detect errors for. This is usually a subset of the data quality rules, as not all data quality rules are worth spending processing time to detect and handle.
- The error-handling specifications which determine what is to be done when bad data are detected. Some errors are so severe that it is worth stopping the load to figure out what is going wrong. Others merely warrant writing the errors to a special holding area where they can be analyzed later.

- A development process that links the profiling data quality rules and results to the ETL mapping specifications with appropriate tool support. The detection and handling of data quality is linked to the ETL jobs, but sits externally so that changing the enforced data quality rules does not require rebuilding the entire ETL job stream.

Data profiling is crucial to proactive data quality. Many companies are trapped in a cycle that generates poor quality data that then require clean-up and results in poor business decisions. To break the cycle, identify core business requirements, the data that are critical to support those requirements, the data quality rules for those data, and how well data quality rules match the data. The results of this data profiling can then be used to correct the data (after first determining the cause and instituting processes to keep the data clean). A proactive approach to data quality can only be achieved through the process of data profiling. Profiling data is not difficult in principle, but requires a significant commitment from the business and support from IT.

Measures, KPIs and reports

To ensure quality, data owners responsible for data quality, and data stewards accountable for data quality, should implement a data quality monitoring environment. Data quality monitoring, the process of examining data over time and developing alerts when data violate any business rules, is key to the success of any data governance program.

Data should be monitored, measured, and reported across the four lifecycle stages: creation and update, data integration, data presentation and data models, dictionaries and metadata repositories. Each data stage should provide information on the data quality dimensions detailed above in the data quality section.

Data quality is an important asset that comes with a price tag. It takes carefully managed business and technical processes to drive high data quality. To ensure the enterprise benefits of high data quality are achieved requires management attention and tools.

Key performance indicators are one management tool used to maintain high data quality. To develop useful KPIs, begin with the business impact of data quality and set priorities accordingly. Focus on high risk data including potential enterprise show-stoppers and costly data first.

Begin with business object level (data entities or tables in databases) such as customers, accounts, investments, etc. A supporting KPI could be Customer Data Quality Indicator = number of customer records with no missing or incorrect data/ the total number of customer records

Follow the business object level monitoring with attribute level KPIs which might include scoring at the most detailed data level and include accuracy, completeness, consistency, correctness, integrity, timeliness and uniqueness.

The MIT Total Data Quality Program (`http://web.mit.edu/tdqm/`) offers in-depth data quality reference materials and is a joint effort with membership including MIT Information Quality Program, CITM at the University of California at Berkeley, government agencies such as the US Navy, and industry partners. The TDQM program has evolved into four key activity areas:

- MIT Information Quality Program
- The International Conference on Information Quality
- The ACM Journal of Data and Information Quality
- Information Quality Certificate Program.

27.2.15 Summary

With global competition, demands for lowering operating costs, complying with vast regulatory statutes and the relentless need to drive revenue sets the stage for data governance.

Data governance is not a product to be purchased like a database, but rather a discipline comprising multiple component activities all designed to identify, control and use data assets to drive revenue.

Organizations that successfully implement and sustain data governance will have their data assets inventoried, controlled and available to drive revenue and reduce costs. Data governance will prove to be a business differentiator by identifying, protecting and making available the all data assets across the enterprise.

27.3 Appendix: Sample IAP processes and activities (Stevens, 2005)

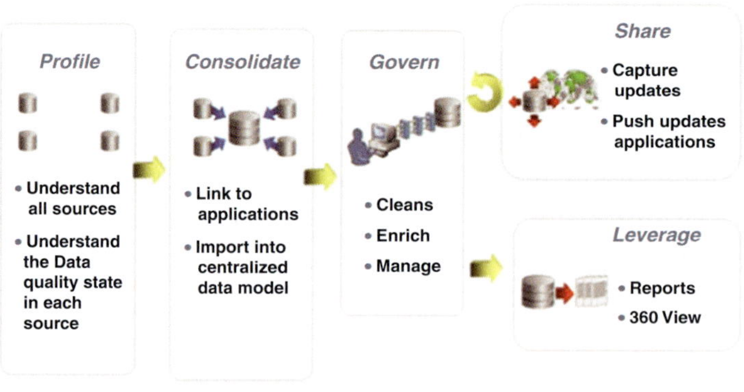

1. Capture background information

Purpose

The purpose of this step is to collect information about who is completing the information asset profile and when the profile is being completed.

Concepts

Information assets are likely to evolve over time, requiring an information asset profile to be updated or recreated. Although the information collected here is not directly used in developing information asset profiles, it may be useful for tracking purposes as the profile is used for risk assessment and other purposes, particularly as an information asset changes.

It may be necessary to investigate or know the history of an information asset profile as the asset matures. By specifying when the asset was profiled and by whom, a higher continuity is ensured. For example, enterprise managers may want to track all instances of an IAP for a specific asset to assess significant changes in ownership, custodianship, or value over the life of the asset within the enterprise.

Important tips

Documenting the IAP creation date and naming those who performed the IAP is not a trivial activity. The historical value of an asset may surpass the actual value of an asset as a commodity over time. For example, the ability to track an asset's value in importance from the asset's inception to maturity to decline may help in devising protection mechanisms or serve as a predictor for similar information assets in the future. It is important to capture the date of the IAP creation, the version information, and the names of personnel who create the IAP. Some important items are as follows.

- Recording not only the names of personnel but the roles (e.g., Director of Financial Services and asset X's owner) they currently occupy in the organization and their contact information. Whenever possible, the owner of the asset and other relevant stakeholders should be included in this process. This helps to ensure the acceptance of the output.
- Creating a version number for the IAP that follows a standard, enterprise-wide convention. The main purpose of the version number is to capture the order, change history, and version control information.
- Cross-referencing the version or date information by making reference to the IAP that has been updated or superseded by the current version.

Performing the activity

The following steps are required to complete this activity.

1. Record the names, relevant titles/roles/positions, and contact information for each of the individuals who are creating this IAP.
2. Record the date on which the profile is created.
3. Record the version of the IAP.

2. Define the information asset

Purpose

The purpose of this step is to characterize an information asset. Before any type of analysis activity (e.g., risk assessment) can be performed on an information asset, the organization must understand and agree upon what an information asset contains.

Concepts

The level of detail that is captured should assist in

- defining the content of an asset and its boundaries
- determining ownership of the asset
- determining the value (monetary or otherwise) of the asset
- determining the security requirements of the asset.

The depth to which an asset must be defined depends largely on the organization and how the information asset profile will be used. Using good judgment and being consistent in defining assets can reduce the issues related to the many combinations and permutations of combined (complex) information assets.

Some information assets are highly tangible. In other cases, an information asset may not be as visible. This may be the case for an information asset such as a "vendor database" or "customer database" that exists only in electronic form. In these cases, it takes more examination of what is contained in the vendor database to truly define the asset.

The definition of the information asset should strive to satisfy these requirements at a minimum:

- consistency (the definition does not change over short periods of time or in different settings)
- clarity (the definition lacks ambiguity and vagueness and is not subject to interpretation)
- universal understandability (the definition transcends different lexicons and technologies)

- acceptance (the definition is acceptable to all who have a need to know)
- physicality (the definition is clear as to how the asset is physically instantiated – i.e., electronically or on paper, microfiche, etc.).

Considering these qualities will provide an adequate definition that can be used throughout the organization.

Important tips

For an initial definition of an information asset, it is generally acceptable to use the highest level of description that accurately represents the boundaries of an asset and its contents. For example, in describing the vendor database, it may be perfectly acceptable to state "all of the tables, fields, and data elements in the vendor database that support our integrated financial system." However, this will not be the case in some instances. If an asset is described by its component parts there is more clarity, consistency, etc.

When developing an information asset description, it may be advantageous to initially provide more data than less and to shape and sharpen the definition as more information profiling activities are performed. However, too much detail can overwhelm those who rely on the definition.

Additional tips

- Be aware that the process of defining the information asset may result in the definition of more than one asset. Delving into the detail of an information asset may result in a better delineation of the asset into two distinct logical assets, even though the asset may be physically stored, transported, or processed on a regular basis as a single asset.
- Stay away from technical conventions when describing information assets, and only use them when they enhance (or are essential to) the definition. For example, naming all of the fields or tables in the vendor database to define the "vendor" asset may be overkill. It may provide comfort that the boundaries have been tightly defined but may provide no advantage in using the asset for risk assessment later on.
- Be sure to define both electronic and paper assets. Sometimes an information asset exists in either or both forms. This should be captured in the description because it will eventually affect risk mitigation planning to secure the assets.
- Involve the owner of the asset and other relevant stakeholders in the definition process. This will ensure the accuracy and consistency of the definition and the acceptance of stakeholders. In some cases the owner will not be able to be determined until after the asset is defined. In these cases, the definition of the information asset should be reviewed with the owner after the owner is identified in Activity 3 to ensure agreement.

In some cases it may make sense to include in the description specific information that is excluded from the asset. It is possible that an asset is easier to define by explaining what is not included as well as what is included. The purpose is to ensure that the reader understands the contents and the boundaries of the information asset. In addition to including information about content that is excluded from an asset, it is often useful to include notes explaining why decisions were made about bounding the asset. The purpose of these notes is to provide insight to others looking at the asset profile.

Performing the activity

The following steps are required to complete this activity.

1. Examine the information asset and develop a description.
2. Record the description on an information asset profile worksheet.
3. Resolve any issues regarding the description detail before you proceed.
4. Record an appropriate information asset name on an information asset profile worksheet.

3. Identify the asset owner

Purpose

The purpose of this activity is to identify and document the owner of the information asset. Ownership is important because the owner should work with the individual or group performing the IAP in the remainder of the activities.

Concepts

Ownership of information assets is often confounding for an organization. Identifying ownership is one of the most important activities in effective security and risk management of information assets that an organization can perform. Many organizations have never taken an accurate and complete inventory of their information assets. The failure to identify asset owners is one of the primary reasons why information security management is often ineffective in organizations.

The owner of an information asset should be an organizational stakeholder (or organizational unit) that is responsible and accountable for

- describing the information asset
- defining the security requirements of the information asset
- communicating the security requirements of the information asset to all custodians and users
- ensuring that the security requirements are met (via regular monitoring)
- designing an appropriate protection strategy to protect the information asset

- determining risks to the information asset
- developing strategies to mitigate risks to the information asset.

The owner of an information asset may have an organizational or a legal responsibility for ensuring the asset's viability and survivability.

Important tips

In the process of defining an asset, it may be realized that the asset has more than one owner. Often, this is an indication that the asset being defined is, in actuality, more than one asset. If this is the case, an information asset profile should be created for each asset and ownership should be documented accordingly.

Remember that in many organizations owners are often simply assumed to be those who manage the asset's containers, and thus ownership is "assigned" to those who take custody of the asset from a technical standpoint (i.e., system administrators, database administrators, and IT staff in general).

Ownership can also be confused with data creation or origination. The creators of data or information are not necessarily the owners of information assets. Ownership is often assigned in organizations without regard to who (or what system, etc.) created the information or where it originated (internally, with an outside vendor, etc.). These issues are often irrelevant as long as ownership is established and the stakeholder who has accepted ownership also has the responsibility and authority to perform ownership duties.

In capturing ownership information, the focus should be on the role or position within the organization that has ownership of an information asset and not a specific person. In many organizations the people in specific positions change much more frequently than the positions themselves.

Performing the activity

The following steps are required to complete this activity.

1. Record the owner of the information asset being profiled. The owner should be a role/position within the organization.
2. For each owner, record contact information for the person currently acting in the role as owner.

4. *Identify containers*

Purpose

The purpose of this step is to capture a list of all of the containers on which an information asset is stored, transported, or processed and the associated list of the managers of those containers.

Concepts

In an information security risk assessment, the identification of key containers is essential to identifying the risks to the information asset itself. An information asset is protected through controls implemented at the container level. The level of protection provided by the controls is directly related to how well they implement the security requirements of the information assets. Any risks to the containers on which the information asset lives are inherited by the information assets.

This activity efficiently defines the boundaries of the environment that must be examined for risk. It also describes the custodial relationships that must be understood for successfully communicating security requirements and for designing effective security controls. This is especially important in resolving some of the previously described dilemmas of data.

The containers that are captured are broken down into four categories:

- systems and applications
- hardware
- people
- other containers.

In addition to capturing the list of containers, the managers of those containers also need to be recorded. The manager of the container takes custodianship of an information asset and may be required to implement the security requirements of the information asset. It may be necessary to talk with the container managers during the risk assessment process to gather additional information.

Important tips

As with the asset definition, the level of detail necessary to be captured in this activity is highly dependent on the organization and on how the asset profiles will be used. In a small organization with only a few systems, it may be practical to consider each container individually. In organizations where there are large classes of similar systems managed by an IT department or departments, it may make sense to simply collect the class of container to be considered. Some basic questions to consider when enumerating containers are the following.

- What information systems or applications use or process this information asset?
- On what hardware platforms might the information asset be found?
- What people have access to the information asset?
- Are any automated processes reliant on the information asset?
- What media types are used to store the information asset?
- Is the information asset often printed, who would print it, and where are printed copies stored?

- Does the information asset ever enter the possession of a customer or partner?
- Are backups or offsite storage of the information asset contracted to a third-party organization?
- Are there any internal or external spaces where the information asset might be stored in physical form (paper, tape, CD-ROM, etc.)?

The information captured on the container manager should focus on the role or position within the organization that has ownership of the information asset and not a specific person, unless a specific person acts as a container for the information asset.

Performing the activity

The following steps are required to complete this activity.

1. For each category of container, record the name of all the containers on which the information asset is stored, transported, or processed.
2. For each container, record the owner of that container. The owner should usually be a recorded as a role/position within the organization and not a specific person.

5. *Identify security requirements*

Purpose

The purpose of this step is to capture the specific security requirements of the information asset.

Concepts

The security requirements of an information asset are generally defined across the dimensions of confidentiality, integrity, and availability. These dimensions are referred to as security objectives by the Federal Information Systems Management Act of 2002 and are defined as follows.

- *Confidentiality* is "preserving authorized restrictions on information access and disclosure, including means for protecting personal privacy and proprietary information . . ." [44 U.S.C., Sec. 3542].
- *Integrity* is "guarding against improper information modification or destruction, and includes ensuring information non-repudiation and authenticity preserving authorized restrictions on information access and disclosure, including means for protecting personal privacy and proprietary information . . ." [44 U.S.C., Sec. 3542].
- *Availability* is "ensuring timely and reliable access to and use of information . . ." [44 U.S.C., Sec. 3542].

It is very important that the security requirements of an information asset be accurately specified. If an owner cannot detail the security requirements for an asset, then the owner cannot expect that anyone to whom he or she grants custodial control of the asset will appropriately protect it.

Important tips

- Recognize that the information may be subject to external agreements or licensing terms that dictate additional security requirements. Information assets often include data from other information assets or from external parties. The use of these data is often covered by acceptable use policies and licensing agreements that describe who may use the data, when the data can be used, or what can be done with the data
- Recognize that the information may be subject to laws or regulations that also dictate additional security requirements. Any additional security requirements specified by a law or regulation should be included.
- As with most other steps, the use of roles is preferred to the use of individuals when describing requirements.
- Requirements should be stated as explicitly as possible.
- This is another step where examination of the asset may result in the determination that it is actually two separate assets.
- The information asset owner is responsible for specifying the security requirements for an information asset. In many organizations, the owner will delegate the responsibility for this to the team developing information asset profiles.
- The stakeholders of an asset are an excellent source of security requirements and are also excellent for checking draft security requirements.

Performing the activity

Record the confidentiality, integrity, and availability requirements for the information asset.

6. Determine the information asset valuation

Purpose

Before the risks to an information asset can be assessed, the tangible and intangible value of the asset must be known.

Concepts

The owner of the information asset and its stakeholders should determine the value of the information asset to the enterprise or business unit. The contribution of the asset to the owner's goal achievement (or the potential to impede goal achievement) should be reflected in the valuation. Determining the value is an attempt to capture how important this information is to the organization, mainly the value derived from its use but also considering the impact of its loss or unavailability.

Valuing information assets has proven to be very hard for many organizations. Information assets are not often carried on the books as capital investments, so determining a monetary equivalent is not always straightforward. Often the value of an information asset is found in the process it supports and not in the information itself. One way to consider the value of an asset is to look at the potential impact on the organization if something were to happen to it. Every organization will need to determine for itself the appropriate type of valuation.

In the federal space, a significant amount of guidance has been issued to help federal government agencies determine a valuation for their information assets. FIPS Publication 199 [NIST 04a] and the NIST Special Publication 800–60 volumes [NIST 04b] provide explicit guidance. An asset's value is determined by looking at the potential impact on the organization if the security of the asset were to be compromised. Information is first classified by type (public relations information, for example). Then for each type of information the potential impact is rated on a simple high, medium, or low value for each security objective (confidentiality, integrity, and availability).

The table, taken from the FIPS Publication 199 guidance, provides a simple example of how the potential impact is rated across each of the security objectives.

Potential impact definitions for security objectives potential impact

Security objective	Low	Moderate	High
Confidentiality	The unauthorized disclosure of information could be expected to have a limited adverse effect on organizational operations, organizational assets or individuals.	The unauthorized disclosure of information could be expected to have a serious adverse effect on organizational operations, organizational assets or individuals.	The unauthorized disclosure of information could be expected to have a severe or catastrophic adverse effect on organizational operations, organizational assets or individuals.

Security objective	Low	Moderate	High
Integrity	The unauthorized modification or destruction of information could be expected to have a limited adverse effect on organizational operations, organizational assets or individuals.	The unauthorized modification or destruction of information could be expected to have a serious adverse effect on organizational operations, organizational assets or individuals.	The unauthorized modification or destruction of information could be expected to have a severer or catastrophic adverse effect on organizational operations, organizational assets or individuals.
Availability	The disruption of access to the use of information or an information system could be expected to have a limited adverse effect on organizational operations, organizational assets or individuals.	The disruption of access to the use of information or an information system could be expected to have a serious adverse effect on organizational operations, organizational assets or individuals.	The disruption of access to the use of information or an information system could be expected to have a severe or catastrophic adverse effect on organizational operations, organizational assets or individuals.

Important tips

- Always consider the costs to the organization if the security requirements of an asset are violated.
- Talk to the owners and stakeholders of an information asset and determine what it produces or what processes rely on it. The more important the output or process, the more significant the information asset.
- Consider external impacts, such as legislation, regulation, and reputation, when developing your valuation.
- The ultimate goal of a valuation is to be able to do a cost benefit trade-off analysis. The valuation should include information to allow decision makers to make informed choices.
- Remember that this step is not concerned with an assessment of risk to the asset. The likelihood of an impact should not be considered.
- An organization should use the method that it believes will provide the most useful valuation of an information asset.

Performing the activity

Determine the value of an information asset and then capture this information on the information asset profile worksheet.

The IAP process is designed to drive data asset inventory and establishes definitions, quality requirements, ownership, usage parameters, security, protection profiles and enterprise value. Each activity in the IAP process captures additional information about an information asset, and results in a complete data asset profile.

27.4 Sample data governance charter

27.4.1 Introduction

Enterprise customers, shareholders and regulators have high expectations for how data are held and used. It is expected that data are accurate, complete and up-to-date. While these expectations are nothing new, enterprise has recognized a need to standardize the way data are managed and governed. Enterprise data governance policy includes:

1. Definition of enterprise, functional and organizational data governance
2. Definition of roles and responsibilities
3. Implementation of divisional data governance processes and procedures
4. Development of a common understanding of data
5. Management of data governance through metrics and measurements.

27.4.2 Enterprise data governance authority

The establishment of the enterprise data governance is instantiated in this document which will serve as the charter.

The members and working groups of enterprise data governance will be instrumental in developing structures, roles and mechanisms to ensure data quality across the organization. Enterprise data governance policy will direct groups to focus their efforts on the following goals:

- Manage data model changes
- Ensure data intensive projects align with architectural data principles
- Define domain ownership
- Support data standards and definitions
- Support standard tools
- Define “golden sources” of data
- Define data quality remediation solutions
- Recommend KPIs
- Monitor compliance.

27.4.3 Enterprise data design membership

Data governance is charged with developing and maintaining data quality across the enterprise. To do so requires the identification and participation of key organizations and their senior management representatives. Voting members are listed in the table.

Organization	Role	Participant
	Chair (Interim)	Mr. A
EIM	Primary	Ms. B
	Secondary	Mr. C
	Secondary	Mr. D
Retail	Primary	Mr. E
	Secondary	Ms. F
Commercial	Primary	Mr. G
	Secondary	Ms. H
Group Services	Primary	Mr. I
	Secondary	Ms. J
Risk	Primary	Mr. K
	Secondary	Mr. L

Non-voting and observer members are listed in the table.

Organization	Role	Participant
	Primary	Mr. M
ETL	Secondary	Ms. N
	Secondary	Mr. O
	Secondary	Mr. P
PMO	Primary	Mr. Q
	Secondary	Ms. R
EDW	Primary	Mr. S
	Secondary	Ms. T
Architecture	Primary	Mr. U
	Secondary	Ms. V
Security	Primary	Mr. W
	Secondary	Mr. X

The enterprise data governance committee consists of one chairman and four voting members from the organizations listed above. The data governance charter may modify and/or amend membership through a nomination originating from an active member followed by a full membership vote. The chairman serves for one year and is responsible for calling meetings, setting meeting agendas, chairing meetings, calling for votes, preparing minutes and initiating annual membership nominations.

Enterprise data governance committee voting members may nominate and approve non-voting members. These non-voting members provide consultation and execution support to the voting membership.

Enterprise data governance committee meetings are scheduled monthly. The meeting schedule may be adjusted at the discretion of the chairman.

27.4.4 Enterprise data governance expected results

With the development and implementation of enterprise data governance the following results are expected.

- Policy and processes to support logical, semantic and physical data model management including changes and updates
- The definition of business domain data ownership with responsibility for data quality in source systems, data warehouse and data marts
- The establishment and implementation of an enterprise data dictionary to capture and implement data standards and definitions across the organization
- The definition, policies and processes to support standardization of tools
- The identification of enterprise critical data
- Identification of "golden sources" for enterprise critical data
- The development of KPIs to measure critical data quality
- The establishment of data quality mechanisms to move, measure, monitor and report data quality throughout the data lifecycle.

27.4.5 Enterprise data governance operating principles

Enterprise data governance members will agree to incorporate the following principles in performing their responsibilities.

- Integrity: participants will practice integrity with their dealings with each other; they will be truthful and forthcoming when discussing drivers, constraints, options, and impacts for data-related decisions.
- Transparency: mechanisms will exhibit transparency; it should be clear to all participants and auditors how and when data-related decisions and controls were introduced into existing processes.
- Auditability: decisions, processes, and controls will be auditable; they will be accompanied by documentation to support compliance-based and operational auditing requirements.
- Accountability: define accountabilities for cross-functional data-related decisions, processes, and controls.
- Stewardship: define accountabilities for stewardship activities that are the responsibilities of individual contributors, as well as accountabilities for stewards and data owners.

- Checks-and-balances: define accountabilities in a manner that introduces checks-and-balances between business and technology teams as well as between those who create/collect information, those who manage it, those who use it, and those who introduce standards and compliance requirements.
- Standardization: introduce and support standardization of enterprise data in alignment with enterprise policies.
- Change management: support proactive and reactive change management activities for reference data values and the structure/use of master data and metadata.

27.4.6 Enterprise data owners' and stewards' responsibilities

Enterprise has outlined data ownership and data steward roles and responsibilities and will identify and assign business domain data ownership RESPONSIBLE for critical organizational data elements. Data owners shall:

- Hold overall accountability for data quality in their division
- Assign ownership of key data elements to data stewards
- Accountable for business specific KPIs
- Assign data element quality within business areas
- Implement data quality monitoring and controls
- Provide data quality feedback to data owners.

Enterprise business domain data owners will identify and assign accountability for data quality to one or more data stewards. A data steward:

- Manages data quality within businesses for data owners
- Is responsible for data element accuracy and controls
- Resolves data quality issues
- Is responsible for data quality monitoring and control for all lifecycle stages: collection, keying, editing, viewing, calculation, and reporting
- Reports business specific KPIs
- Provides data quality monitoring report to data owners.

27.4.7 Enterprise data governance requirements and measures

To ensure data quality is defined, implemented and measured, SMART (specific, measurable, actionable, relevant and timely) KPI measurements will be developed. In addition, data should be monitored, measured, and reported across the four lifecycle stages: creation and update, data integration, data presentation and data models, dictionaries and metadata repositories. Each data stage has the following required quality dimensions.

- Creation and update
- Accuracy: measures the extent to data is free of error and the degree to which it represents its corresponding real world object
- Validity: measures the extent to which data conform to a set of domain values or rules
- Completeness: measures the extent to which data values for an entry process have been collected and entered
- Consistency: measures the extent data held in multiple database repositories conform to the same syntactic and semantic rules
- Timeliness: measures the extent to which data are available to other processes in an acceptable timeframe following data entry or update
- Data integration
- Completeness: measures the extent to which all required data sources contain required values
- Consistency: measures how data held in multiple database repositories conform to the same syntactic and semantic rules
- Accessibility: measures which data sources required are available and accessible
- Timeliness: the extent to which all data sources and feed are available to the data integration services in the required time frame
- Data presentation and data models
- Relevance: the extent to which the information is applicable and useful for the purpose intended
- Transactability: measures the extent to which data convey a clear, intended meaning
- Consistency: measures how well data are presented in a consistent, unambiguous manner across every medium
- Data dictionaries and metadata repositories
- Completeness: measures modeled data definitions and relationships
- Consistency: measures how data definitions and relationships are compatible across multiple models and dictionaries.

27.4.8 Enterprise data architecture principles

The following enterprise data architecture principles provide guidance to the data governance mission by providing high quality and integrated data across the enterprise.

- Data are treated as a core business asset
- Data owners are collectively *accountable* for the quality of all data held within enterprise

- Data stewards are collectively *responsible* for the quality of all data held within enterprise
- Data are meaningful and understood
- Data are mapped to the enterprise logical data model
- Data have a designated master source (golden source)
- Data are compliant with law and regulations
- Data are shared
- Data are fit for purpose
- Data management functions are implemented on a minimum set of technologies
- Data are accessible
- Data are secure and trustworthy.

27.4.9 Enterprise data and model standards

Common data definitions are being developed through a single enterprise data model known as DM-XXX. DM-XXX is an industry standard logical data model containing definitions and relationships for approximately 3,000 data entities. It is used as a reference model to ensure robustness and flexibility. Enterprise data model standards must conform to the DM-XXX.

Data will be mapped to the Enterprise DM-XXX ensuring standard understanding and definitions.

Data will be mapped to "golden sources" which will be subsequently mapped to the DM-XXX ensuring consistency and understanding.

The Enterprise DM-XXX will be managed by a data governance data model working group responsible for developing decision criteria and decision processes designed to guide projects in their data governance and quality implementation.

27.4.10 Enterprise data dictionary

Enterprise data governance will charter a working group focused on developing and maintaining a data dictionary for enterprise. The working group with define data standardization requirements, tools and processes enabling projects to include data dictionary activities in their development and implementation.

27.4.11 Enterprise data governance initial scope

Enterprise data governance initial data scope includes data that could have a major or significant impact upon the organization. This includes data provided by external suppliers and data held on end user developed applications and includes data used for:

- Financial decision making or validation
- Regulatory reporting
- Internal/external reporting
- Data which otherwise present a reputational risk to the organization.

Enterprise data governance has a data risk assessment approach available for use to determine whether data are critical in nature and require data quality remediation.

27.4.12 Enterprise data governance implementation timetable

Data governance implementation timetable

Tasks	Description	Dec 2011	March 2012	Dec 2012
Governance structure	Establish enterprise data governance authority to manage and disseminate data quality policies and procedures	X		
Governance roles and responsibilities	Define data ownership and data stewardship roles and responsibilities to support the ongoing management of data quality	X		
	Define data ownership and data stewardship accountability	X		
Governance mechanisms	Define decisions, decision criteria, processes and procedures to support data quality activities across enterprise	X		
Implement governance	Implement data ownership and data stewardship across enterprise		X	
Data model change management	Define decisions, decision criteria, processes and procedures for maintaining enterprise data models (logical, semantic and physical)	X		
Data standards and definitions	Pilot divisional data dictionary	X		
	Division data defined in a "data dictionary" accessible to stakeholders		X	
	Data dictionary fully implemented			X
Define data domain ownership	Define data domains and establish business unit owners	X		
Golden data Sources defined	For high quality data, identify source systems of record	X		
Data quality remediation	Identify data in need of data quality processes and measures	X		
	Identify data quality issues affecting high quality data	X		
	Define data quality measures and metrics	X		
	Establish data quality methodology	X		
	Correct data quality issues	X		

Tasks	Description	Dec 2011	March 2012	Dec 2012
Data quality KPIs	Define data quality measures and metrics	X		
	Establish on-going data quality mechanisms for monitoring and reporting of metrics and KPIs	X		
	Data quality is implemented in all change programs	X		
	Full KPI suite is implemented		X	
	Data quality risk issues recorded, addressed and monitored			X
Data quality monitoring and reporting	Key data elements quality measured and reported through KPIs		X	
Third party contractors	New contracts to include data quality standards	X		
	Existing contracts to include data quality standards upon renewal		X	
Divisional assurance reviews	Assurance teams to be established to review standards of compliance to group, divisional and functional data quality standards		X	

27.5 Sample data governance formational sessions agenda

1. Meeting goals and results
2. Chairman, voting and non-voting membership
3. Data governance responsibilities
4. Data governance expected results
5. Data governance policy objectives
6. Define business data domain duties and owners
7. Define data stewards duties and domains
8. Data architectural principles
9. Data governance implementation timetable
10. Working groups definition
11. Meeting goals and results status.

27.6 Sample data architecture principles

The following enterprise data architecture principles provide guidance to the data governance mission by providing high quality and integrated data across the enterprise.

1. Data are treated as a core business asset
2. Data owners are collectively *accountable* for the quality of all data held within enterprise

3. Data stewards are collectively *responsible* for the quality of all data held within enterprise
4. Data are meaningful and understood
5. Data are mapped to enterprise logical data model
6. Data have a designated master source (golden source)
7. Data are compliant with law and regulations
8. Data are shared
9. Data are fit for purpose
10. Data management functions are implemented on a minimum set of technologies
11. Data ares accessible
12. Data are secure and trustworthy

27.7 Sample master data management capture

Master data management solution capture includes the following.

1. Profile the master data and understand all possible sources and the current state of data quality in each source
2. Consolidate master data into a central repository using a shared logical/physical data model and link it to all participating applications
3. Govern master data, cleanse it, remove duplicates and enrich with metadata from all participating systems and manage it according to defined business rules
4. Share the master data, synchronize the central master data with enterprise business processes and applications, ensure data stays in sync across the enterprise landscape
5. Leverage a single version of the truth for all master data objects

27.8 Sample metadata capture

Metadata is data about data which allows data identification, assessment, grouping, sorting, governance, security, privacy, etc. according to defined data governance metadata criteria. If data governance is to be efficient and effective, it needs to facilitate the retrieval, access and use of data according to the needs of staff and other legitimate stakeholders. Metadata plays a crucial role in achieving enterprise data usage and allows data to be evaluated with the need to read the actual data stores.

Metadata also provides essential technical information which is necessary to ensure successful sharing of data across the enterprise. It is invaluable in supporting migration of data over time as systems need to be upgraded and user needs transform over time.

Metadata has existed and been used to organize data for centuries and with the advent of computing transforms into digital requirements. Reasons for capturing and using metadata include:

- Ensuring that adequate contextual information is recorded and linked to records
- Facilitating retrieval through description
- Controlling access by defining security or legal status for retention and use
- Reducing the risk of unauthorized access and use
- Facilitating the transfer of data across the enterprise, functions and organizations
- Capturing data responsibilities over time
- Limiting the cost of data storage
- Safeguarding vital data
- Supporting data migration plans
- Providing evidence of authenticity.

It is important to remember that metadata does not need to be part of the data itself, but rather captured and tagged to the data.

Metadata should be clearly defined, captured and used to manage and monitor data quality throughout the data lifecycle.

Automated Records Management System (ARMS) suggested metadata attributes

Identifier	A unique identifier for a record or group of records
Title	The title given to the record, folder or class
Creator	The person responsible for the content of the record
Date	Date (and time) an important event occurred to a record excluding disposal events
Quality requirements	Data domain, values, ranges, etc. to ensure high quality
Data owner	Business data owner responsible for create, read, update and delete data quality
Data steward	Business data steward accountable to the business data owners for data quality
Technical stewards	The technical steward is a member of the IT department and focuses on the technical metadata which needs to be captured by the data quality initiatives; technical stewards work closely with business stewards to create metadata definitions and measurements
Record type	The recognized form a record takes, which governs its internal structure and relates to its transactional purpose or to the action or activity it documents
Relation	Identifies instances where a record has a direct relationship with that of another record or clarifies how a record relates to groups
Aggregation	Unit of measurement used to define where in the information hierarchy any records management action is carried out
Language	Language of the intellectual content of the record or resource
Security and access	Classification restrictions and permissions placed on access
Disposal	What happens to the records at the end of their lifecycle (can also be referred to as a record's retention)
Format	The format of the record or in what medium the information is contained

Subject	Keywords or phrases describing the subject content of the record
Description	Free text description of the record
Function	Enterprise business function
Location	Physical storage location
Preservation	Information on the record's description, migration, sustainability and preservation management processes that have been employed during the life of the record and its component(s), to facilitate its survival across technical platforms

27.9 Sample change management

Most organizations have change management mechanisms for technology, software and websites. Yet, many organizations do not practice structured change management for data governance.

Why is this important? Some types of data, for example master data and reference data should be tightly controlled to ensure quality and validity. Data values appear in thousands and millions of transactions; without change management, different repositories storing master and reference data will not be in sync.

Examples of data-related change management (Data Governance Institute: Guidance by Gwen):

- Changes to allowable values for reference tables
- Changes to physical data stores that impact the ability to access or protect in-scope data
- Changes to data models
- Changes to data definitions
- Changes to data structures
- Changes to data movement
- Changes to the structure of metadata repositories
- Changes to types of metadata included in a metadata repository
- Changes to stewardship responsibilities.

Some organizations use highly structured, step-by-step change management processes. These structured processes support data owners, stewards and users in their alignment of data changes. A useful by-product of structured process change is documentation and audit trails for auditors and regulators.

Another approach to data governance change management is to set requirements for change notification. Data stewards are asked to notify the data governance board/ committees about particular types of changes. The proposed changes would be communicated to all known data stakeholders and would collect feedback about potential issues.

Data governance change management is a key process supporting the sustainable success of data quality across enterprises. With data as a shared asset, it must be managed as such and change management is the technique to do so.

Bibliography

Ayres, Ian, 2007, Super Crunchers.

Bowman, David, David Bowman's Information Management Checklist, information-management-architect, March 8, 2012.

Brooks, W. Mark, Targeting High Value Data For Information Governance, `www.wmarkbrooks.com`, February 3, 2012.

Data Governance Institute, Data Governance Definitions, `datagovernance.com/adg_data_governance_definition.html`, February 29, 2012.

Data Governance Institute, Data Governance Models, `datagovernance.com/gbg_governance_models.html`, March 7, 2012.

Data Governance Institute, Guidance by Gwen, `www.datagovernance.com/gbg_change_management.html`, February 7, 2012.

Evans-Correia, 2012, Comparing Wal-Mart's 583 TB to the Library of Congress estimate of 20TB, June 20.

Gallaugher, John, 2012, *The Data Asset: Databases, Business Intelligence, and Competitive Advantage*, gallaugher.com/chapters, p. 2, March 3.

Lewis, Michael, 2003, Money Ball.

Oracle Warehouse Builders User Guide 11g release 1, Phases in the Data Quality Lifecycle, `www.docsoracle.com`, January 24, 2012.

Prime Data Consulting, Data Governance Blog – Blogging on Business Intelligence, Data Governance and Enterprise Data Management, `primedataconsulting.com/blogPDC/wp-content/uploads2011/01/RACI-Waterfall.jpg`, January 21, 2011.

Stevens, James F., 2005, Information Asset Profiling.

Teradata Corporation, Data Governance Impacts, Teradata.com, January 14, 2012.

Index